Topics in Heterocyclic Chemistry

Volume 59

The series Topics in Heterocyclic Chemistry presents critical reviews on present and future trends in the research of heterocyclic compounds. Overall the scope is to cover topics dealing with all areas within heterocyclic chemistry, both experimental and theoretical, of interest to the general heterocyclic chemistry community. The series consists of topic related volumes edited by renowned editors with contributions of experts in the field. All chapters from Topics in Heterocyclic Chemistry are published OnlineFirst with an individual DOI. In references, Topics in Heterocyclic Chemistry is abbreviated as Top Heterocycl Chem and cited as a journal.

Michael P. Doyle • Xinfang Xu
Editors

Heterocycles from Carbenes and Nitrenes

Methods, Reactions and Synthetic Applications

With contributions by

A. C. B. Burtoloso · M. P. de Jesus · M. Dehghany ·
R. Echemendía · C. Empel · D. Gryko · W. Hu · Y. Hu ·
R. M. Koenigs · K. Lee · K. O. Marichev · J. P. Milton ·
O. Reiser · J. M. Schomaker · K. Seo · K. Strunk ·
A. Trinh · J. A. M. Vargas · X. Xu · L.-W. Ye · M. Zhang ·
D. Zhu · S. Zhu

Springer

Editors
Michael P. Doyle
Department of Chemistry
University of Texas at San Antonio
San Antonio, TX, USA

Xinfang Xu
Sun Yat-sen University
Guangzhou, China

ISSN 1861-9282 ISSN 1861-9290 (electronic)
Topics in Heterocyclic Chemistry
ISBN 978-3-031-36734-2 ISBN 978-3-031-36735-9 (eBook)
https://doi.org/10.1007/978-3-031-36735-9

This Springer imprint is published by the registered company Springer Nature Switzerland AG
The registered company address is: Gewerbestrasse 11, 6330 Cham, Switzerland

Preface

Heterocyclic compounds are key ingredients for synthetic chemistry, pharmaceuticals, agrochemicals, and material science. This volume reports recent developments of practical and selective reactions and methods involving carbenes and nitrenes to supply the increasing needs for structural diversity in heterocycles. Carbenes and nitrenes are two very reactive, but controllable, intermediates that have been applied to a variety of catalytic transformations for the straightforward construction of C–C, C–N, and C–X bonds. Their expeditious assembly of heterocycles with structural diversity has been realized through versatile metal carbene reactions, including those of ylide transformations, C–H/X–H insertion, cycloaddition, rearrangement, and cascade reactions. Analogous catalytic nitrene transformations are potent complementary protocols for the construction of N-heterocycles through direct C–N bond formation reactions. In addition, with the discovery and introduction of efficient catalysts, including those of Rh-, Cu-, and Fe-complexes, enzyme, and many others, highly stereoselective methods, especially the enantioselective versions, have been enabled in carbene and nitrene chemistry. The broad spectrum of catalytic procedures using carbenes and nitrenes in the synthesis of heterocycles that have been developed and documented by experts worldwide has made possible the construction of this book.

There are many fine and specific reviews in books and journals that describe carbene and nitrene chemistries and report the synthesis and applications of heterocyclic compounds. We were surprised to find that reviews linking these two subject areas were lacking. Thus, when we were contacted by the publisher about editing a volume in the book review series Topics in Heterocyclic Chemistry, we saw an attractive opportunity to enhance knowledge and understanding in this area. We created a list of leading experts in carbene/nitrene applications to heterocyclic syntheses, and we were gratified that each of the experts agreed to contribute to this volume.

In this special volume, we have included the multiple ways in which different carbene and nitrene transformations are used for the synthesis of heterocycles. Professor Gryko's chapter on photo-induced carbene transformations describes the efficient and sustainable synthesis of heterocycles, often without the need for a catalyst. Diverse syntheses of O- and S-heterocycles via onium and ylides are described over three chapters by Professors Koneigs, Burtoloso, and Hu with the latter chapter also reviewing cascade reactions that feature gem-difunctionalization. Although cyclopropanation, at a glance, does not directly form heterocyclic compounds, Prof. Reiser describes in his chapter that cyclopropanes formed via carbene addition can be converted into a diverse variety of heterocycles. Professor Marichev provides insights on the construction of heterocycles with different ring sizes by catalytic cycloaddition reactions with vinylcarbene species. Catalytic carbene transformations beyond using diazo compounds as carbene precursors are documented in chapters by Professors L. Ye and S. Zhu, respectively, who describe the synthesis of complex heterocycles using carbene-forming reagents, including alkynes, cyclopropenes, and eneynes, that serve as practical alternatives to diazo compounds. The rapidly emerging synthesis of heterocycles by nitrene transfer, where azides, carbamates, sulfamates and even unactivated amines are nitrene precursors, is reported by Professor Schomaker. Together, these chapters provide a comprehensive view of the diversity of methods and the variety of applications in the construction of heterocycles that are made available through carbene and nitrene chemistries.

We are sincerely grateful to the authors for their contributions to the developments described in this volume that have opened new vistas for heterocyclic syntheses. We are also grateful to the Series Editors, Bert Maes and Janine Cossy for organizing this attractive series, and the members of the Springer staff for their valuable advice and patience.

San Antonio, TX, USA Michael P. Doyle
Guangzhou, China Xinfang Xu

Contents

Photo-Induced Carbene Transformations to Heterocycles

Joseph P. Milton and Dorota Gryko

Contents

Abstract Heterocycles are biologically relevant molecules with broad applications in pharmaceutical and agrochemical research. Among the countless approaches for their synthesis, those starting from various carbene/nitrene precursors occupy an important place. Undeniably, environmentally friendly methods for the generation of these reactive intermediates that exclude common transition metals, which are well known for their low abundance, high price, and toxicity, are of great interest. In this chapter, we present photochemical transformations of carbene/nitrene precursors leading to heterocyclic scaffolds. Under the irradiation of light, diazo or azido compounds are typically employed as convenient sources of these reactive species, which subsequently enable cyclizations via Wolff rearrangements, C-H and X-H insertions, or from cycloadditions, as well as various other mechanisms, with only the loss of dinitrogen.

Keywords Carbenes · Diazo compounds · Heterocycles · Photochemistry

---

J. P. Milton and D. Gryko (✉)
Institute of Organic Chemistry Polish Academy of Sciences, Warsaw, Poland
e-mail: joseph.milton@icho.edu.pl; dorota.gryko@icho.edu.pl

1 Introduction

Photochemistry is a 'green' methodology that enables the synthesis of heterocycles in an efficient and sustainable manner, often without the need for a catalyst [1]. It provides an elegant way to access intermediates and pathways that are often not available or difficult to reach with typical ground-state reactions [2]. In photochemical reactions, absorption of a photon leads to the generation of an electronically excited state of a reagent and, in fact, contrary to thermal conditions, the reaction starts from a higher energy level [3]. This characteristic often '*facilitates a thermally unachievable, energetically uphill reactions*' [4]. For this reason, the potential of photochemistry has been widely recognised and has quickly attained the state of a mature field [5–9].

Light-induced processes have also proven to be efficient for generating carbenes/nitrenes from various precursors such as diazo compounds, diazirines [10, 11], pyridotriazoles [12, 13], oxadiazolines [14], silyl substituted ketones, and azides. Their reactivity has previously been reviewed with a focus on their use in either photochemical [15–19] or metal-mediated transformations [20–22]. In general, direct photolysis of such compounds leads to the formation of singlet carbenes and requires that the species being irradiated absorbs light of the applied energy. However, for those precursors that do not absorb the light source, a photocatalyst must be added, the role of which is to transfer either energy or donate/accept electrons (Scheme 1) [23]. In the first case, the catalyst acts as a photosensitiser that enables the generation of triplet carbenes, while a photoredox catalyst gives access to open-shell intermediates [15].

Currently, the photochemical synthesis of heterocycles via carbene intermediates mostly utilises the photolysis of the aforementioned photolabile precursors, of which diazo compounds prevail as carbene sources. Stabilised diazo reagents are categorised into three types: acceptor-acceptor (A/A, two electron-withdrawing groups), acceptor (A, one electron-withdrawing group), and acceptor-donor (A/D, one electron-withdrawing group and one electron-donating group), Fig. 1a [16]. Destabilised diazo compounds (D/D, two electron-donating groups) are, in general, highly unstable, but some diaryl diazo compounds can be stored [25]. The

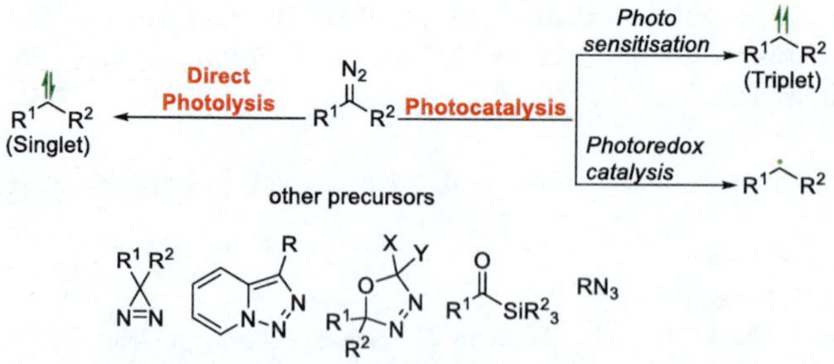

Scheme 1 Photocatalytic transformations of carbene precursors

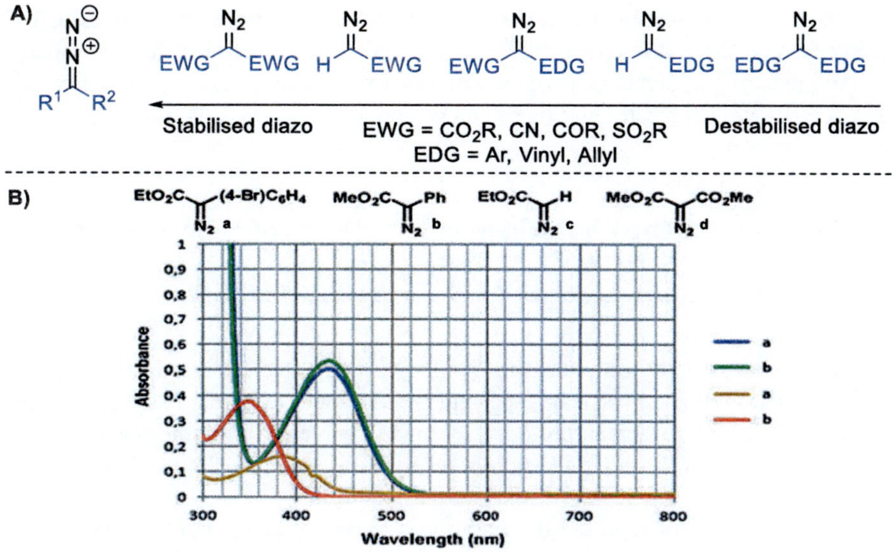

Fig. 1 (**a**) General structure of diazo compounds and their substituents, graded from 'stabilised' to 'destabilised'. (**b**) UV-vis absorbance spectra of some common diazo compounds. Courtesy of the Royal Society of Chemistry, 2018 [24]

nature of the substituents attached to the diazo group impacts their photophysical properties. Most stabilised A/A or acceptor diazo reagents do not absorb light in the visible region, while the absorption of D/A analogues is bathochromically shifted towards the blue region (Fig. 1b) [24]. On the other hand, those bearing two donor substituents, such as aryl-aryl diazoalkanes, can absorb even in the green or red region [25]. Since the seminal report by Jurberg and Davies in 2018 [24], a rise in blue light photochemistry of diazo reagents has been observed. However, even though protocols engaging triplet carbenes do exist, so far, to the best of our knowledge, they have not been utilised for the construction of heterocyclic scaffolds.

Photochemistry is regarded as an eco-friendly methodology and is of intense interest to industry. Its application to the synthesis of heterocycles, which is the core of pharmaceutical chemistry with over 85% of all biologically active compounds containing at least one heterocycle [26], is particularly relevant. Heterocycles are also widely used as agrochemicals, antioxidants, dyes, among others [27]. For decades, heterocyclic scaffolds have been constructed by catalytic reactions using transition metals. These methods are well-documented and reliable; however, they require extensive purification to ensure that the compound surpasses toxicological assessments [28], which has elevated interest in the industrial sector for photochemical methods leading to these key moieties.

Herein, we describe transformations of photochemically generated carbenes/nitrenes leading to heterocyclic scaffolds. The chapter is divided into sections based on the key carbene-mediated mechanistic step. To the best of our knowledge, it collates all examples for the formation of heterocycles. We wish to stress that it strictly covers only photochemical carbene-based transformations; there are many other reports of photochemical transformations to heterocycles, but they involve other reactive intermediates and these can be found elsewhere [29, 30].

2 Carbenes for the Photochemical Synthesis of Heterocycles

2.1 C-H Insertions Leading to Heterocyclic Scaffolds

Carbenes insert into carbon-hydrogen bonds in a concerted fashion facilitating the functionalisation of inactivated sp^3, sp^2, and sp C-H bonds (Scheme 2) [31]. Insertion is preferred for more acidic protons, so having an electron-withdrawing group adjacent to the C-H bond undergoing insertion is advantageous, although not required.

Jurberg and Davies reported the photolysis of aryl diazoacetates **1**, under blue light to generate singlet carbenes that engage in C-H, O-H and N-H insertion reactions, as well as cyclopropanation. When C-H insertion is intramolecular, it gives straightforward access to diverse heterocyclic scaffolds. For example, a photochemical intramolecular reaction of isopropyl or *tert*-butyl phenyldiazoacetate (**1a, b**) leads to the corresponding β-propiolactone **2** or γ-butyrolactone **3** in 57 and 91% yield, respectively (Scheme 3) [24]. Of note is the fact that the reaction is mostly viable for bulky alkoxy moieties, otherwise only the extrusion of dinitrogen takes place [32].

Likewise, α-diazo amides furnish lactams. Here, the amide moiety imposes close proximity of an alkyl group to the in-situ generated carbene, thus facilitating cyclisation [33, 34]. Recent ultrafast time-resolved infrared spectroscopy studies showed that these reactions happen either via a rearrangement in the excited state (RIES) or from a pathway involving a singlet carbene [35]. Along this line, the synthesis of several lactams **6–10** via photochemical C-H insertions was accomplished (Scheme 4) [36].

Scheme 2 General mechanism for a C-H insertion reaction

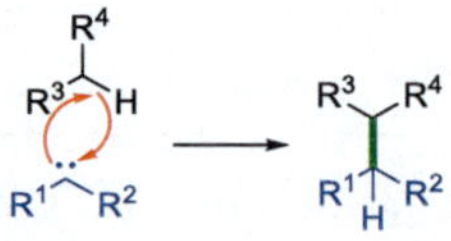

Scheme 3 Photochemical intramolecular C-H insertions affording lactones

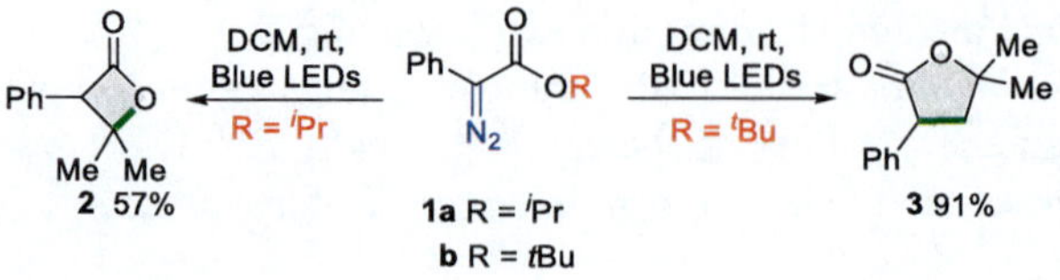

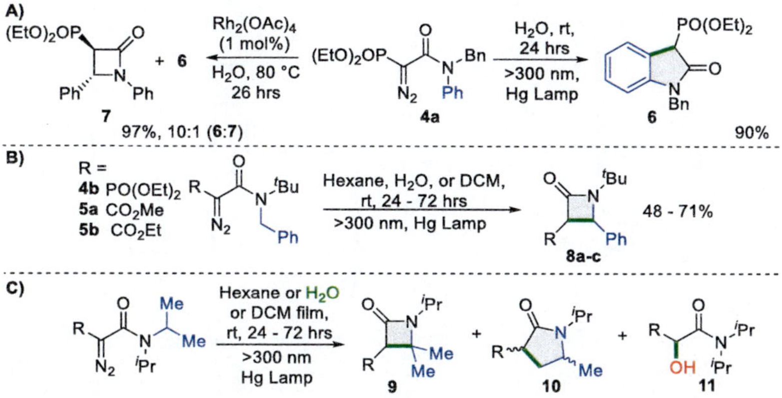

R	Solvent	9	10	11
5c CO_2Et	H_2O	58		
	Hexane	90		
	DCM film	>97		
4c $PO(OEt)_2$	H_2O	37	46	15
	Hexane	63	30	
	DCM film	39	20	

Scheme 4 Light-enabled the synthesis of lactams

While the photochemical reaction of diazo reagent **4a** yields benzo-γ-butyrolactam **6** as the sole product (90%), the reaction under dirhodium (II)-catalysis is less effective with the formation of β-lactam **7** as a pronounced side product (Scheme 4a). Diazo compounds **4b**, **5a**, and **5b** bearing a nitrogen substituted with *tert*-butyl and benzyl groups afford the corresponding β-lactams **8a–c** as the sole product (Scheme 4b). Moreover, in water, the reaction is highly *trans*-stereoselective with the *trans*-isomer being the only diastereoisomer when R = CO_2Et. The site-selectivity in these transformations is governed not only by the susceptibility of C-H bonds towards diazo insertions (Ph > Bn > ^{t}Bu) but also depends on the nature of a substituent at the position α to the diazo functionality. Consequently, for the bis(isopropyl)substituted diazo ester **5c**, only β-lactam **9** was isolated, but when R = $PO(OEt)_2$ (**4c**) a mixture of products forms including β-lactam **9** and γ-lactams **10**, as well as the respective O-H insertion product **11** when the reaction is conducted in water (Scheme 4c).

As is well represented by the studies of Lowe and Parker (Scheme 5) [37], the proximity of the reactive sites in C-H insertions is also important in determining reaction selectivity. Under UV irradiation, when the amide substituent of diazo compound **5d** is a pyrrolidine ring, the reaction favours C-H insertion into the alkyl ester substituent, analogous to Jurberg and Davies' work, forming β-propiolactone **12** (from the ethyl ester) or γ-butyrolactone **13** (from the *tert*-butyl ester). Conversely, when piperidine is the amide substituent (**5f**), C-H insertion occurs at the α-amino site forming fused β-lactam **14** in an approximate 2:1 *trans*: *cis* ratio. Corey and Felix made a similar observation for piperidinyl substituted

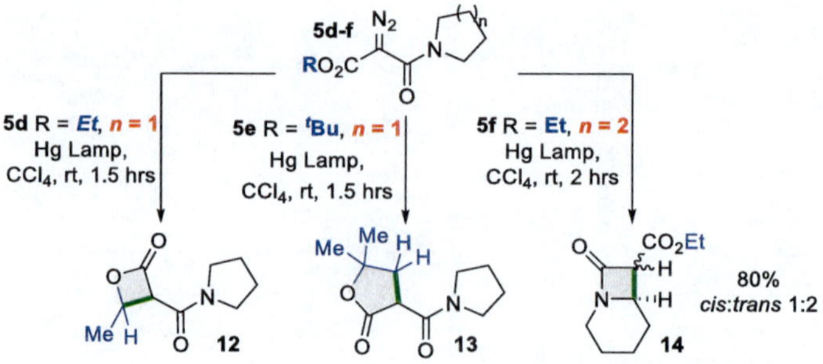

Scheme 5 The formation of 4- and 5-membered lactones and lactams via C-H insertions

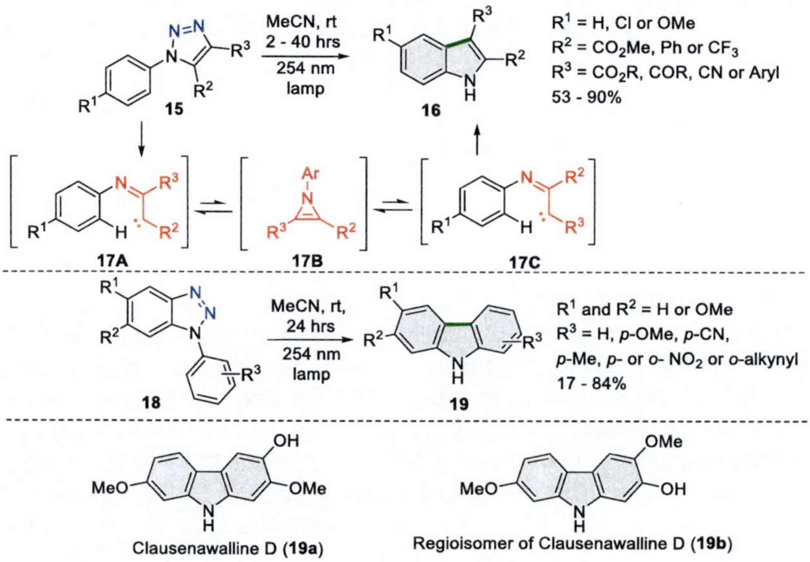

Scheme 6 Synthesis of indoles and carbazoles via irradiation of triazoles

α-diazo amide compounds, and such selectivity enabled the formation of penicillin derivatives [38].

C-H insertion reactions are also viable with respect to aromatic C-H bonds as illustrated by the synthesis of indoles and carbazoles under UV irradiation with the use of triazoles as carbene precursors (Scheme 6) [39].

Interestingly, the production of indoles **16** from the photolysis of *N*-aryl triazoles **15** is fully regioconvergent; substituents R^2 and R^3 can be organised either way, and the same product **16** forms. However, the yields are quite different due to the stability of the proposed intermediates **17A** and **17B**. The same method applied to benzotriazoles **18** gives carbazoles **19** in 22–84% yield. In this case, however, a mixture of regioisomers forms when the *N*-aryl group has two substituents. The

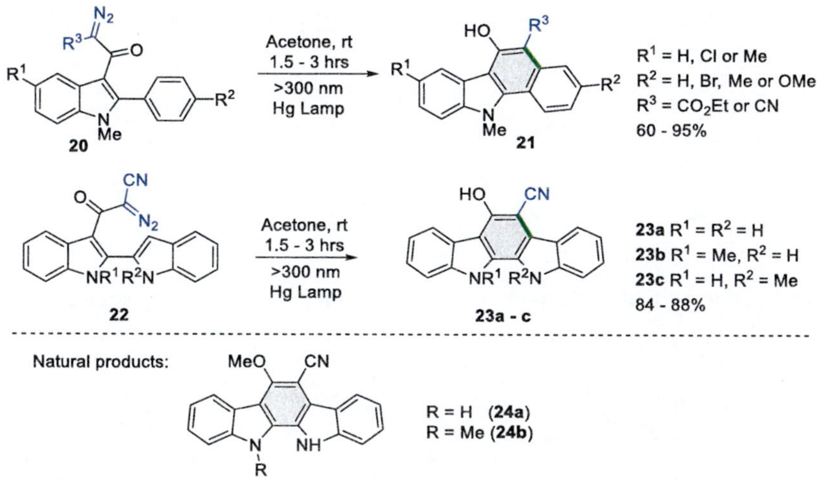

Scheme 7 Synthesis of benzocarbazoles and fused carbazoles from blue light irradiation of diazo compounds

synthetic utility of the procedure is represented by the total synthesis of Clausenawalline D **19a** and its isomer **19b**, which was isolated in six steps with an overall yield of 3% for each compound.

The photochemical methodology based on the intramolecular transformations of carbenes is also suitable for the synthesis of expanded systems, such as carbazole scaffolds (Scheme 7) [40]. With indole-based α-diazo reagent **20**, benzocarbazoles **21** form in over 60–95% yield under UV irradiation, with no major effect on the yield by changing the nature of the aryl substituents. The method also facilitates the synthesis of key intermediates **23a** and **23b** that allow the synthesis of two natural products (**24a,b**), which were isolated from the blue-green alga *Nostoc sphaericum*.

A straightforward method for carbene transformations into heterocycles would be to use α-heteroatom substituted carbene precursors. This strategy is dominated by transition metal-catalysed reactions, and they are exhaustively discussed by Cheng and Meth-Cohn in their review [41]. Until recently, photochemical transformations have not been widely utilised in this context. Along this line, under irradiation with a high-pressure mercury lamp, α-silyl diazo compounds **25a–d** afford siletanes **26** as a single *trans*-diastereoisomer from C-H insertion on alkyl groups of the silicon substituent (Scheme 8) [42]. When the silicon atom is substituted with chloride or isocyanate, only the four-membered ring is formed, contrasting with that when azido or isothiocyanate groups are present from which a mixture of 4- and 6-membered heterocycles **26** and **27** is furnished. All the siletanes can be transformed into the respective 3,4-dihydro-2H-1,2-oxasilines **27** by a thermally induced 1,3(C→O) silyl shift.

Similarly, the synthesis of oxasilolanes **28** from diazo compound **25e** can be accomplished when there is an alkoxy substituent on silicon (Scheme 9) [43]. The

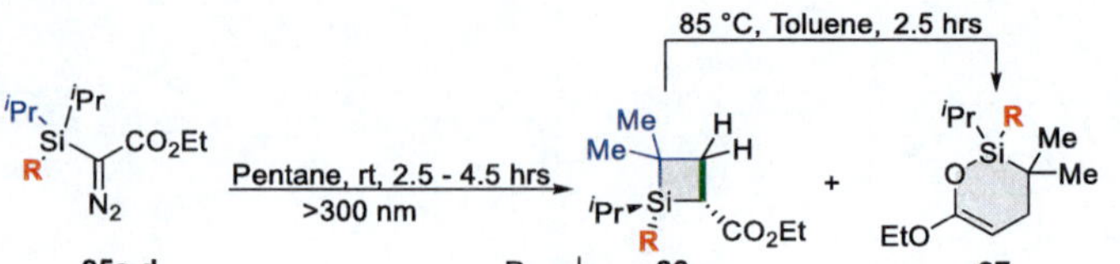

	R	26	27
a	Cl	100	-
b	N3	46	54
c	N=C=O	100	-
d	N=C=S	26	74

Scheme 8 Photochemical synthesis of four-membered silicon-containing heterocycles

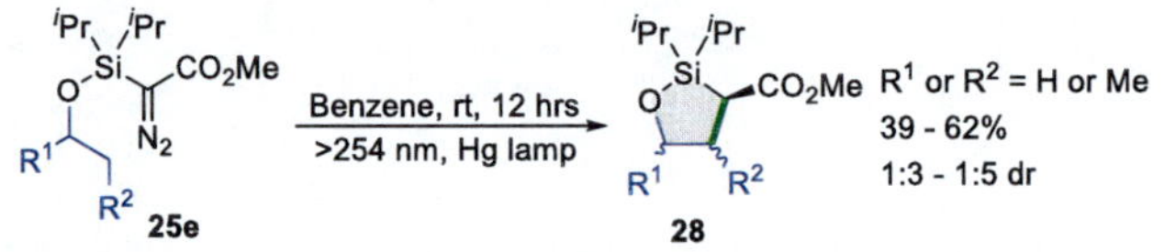

Scheme 9 Synthesis of oxasilolanes via a photochemical C-H insertion reaction

reaction, although affording product **28** in only moderate yields, is noteworthy to mention because C-H insertion typically occurs at the α-position with respect to oxygen, whereas here insertion occurs at the β-position.

2.2 *O-H Insertions in Heterocycle Synthesis*

Oxygen-hydrogen insertion represents another key reaction in the repertoire of carbene-mediated reactions. The first step involves the attack of oxygen at the carbene to form an ylide, and its negatively charged carbon centre subsequently removes the proton from oxygen to yield an ether compound (Scheme 10) [44].

This transformation is the key to a two-step, one-pot photochemical process for the synthesis of 2,3-dihydrobenzofurans **31** (Scheme 11) [45].

The use of aryl diazoacetates **1** under blue light irradiation facilitates an O-H insertion reaction on the phenol **29** which, after subsequent treatment with a base, closes the ring in a yield of up to 88% and 2.3: 1 *dr*. The presence of *tert*-butyl groups on the quinone moiety has a strong influence on the reaction, with no cyclisation possible when they are replaced with isopropyl groups.

Scheme 10 General reaction mechanism for O-H insertion

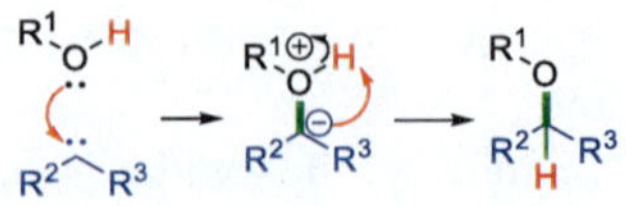

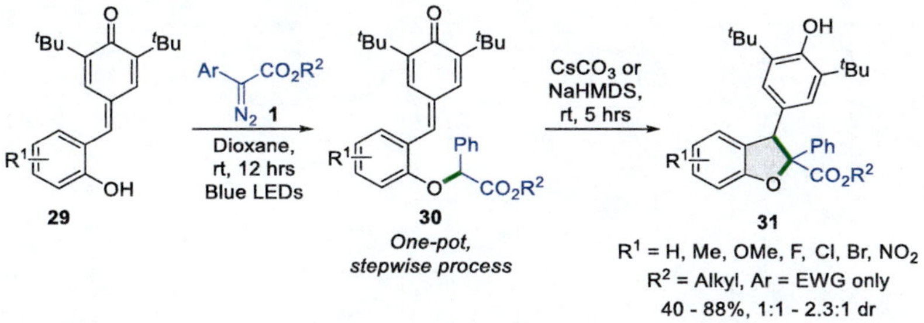

Scheme 11 Photochemical one-pot, stepwise synthesis of 2,3-dihydrobenzofuran derivatives

2.3 Photochemically Generated Ylides in Heterocycle Synthesis

Carbenes form ylides with electronegative atoms such as oxygen, sulfur, and nitrogen, providing they are unsubstituted with hydrogen (Scheme 12). Such behaviour is described by Kirmse as being analogous to the interaction between Lewis acids and Lewis bases [46].

Ylide based transformations often facilitate ring-expansion or ring-contraction processes, leading to new heterocyclic structures. Along this line, the Koenig group described the photochemical ring-expansion of oxetanes **32a** and thietanes **32b** to their respective tetrahydrofurans **33a** and tetrahydrothiophenes **33b** (Scheme 13) [25]. A wide range of substituents on the aryl ring of diazo substrate **1**, regardless of its electronic nature or position, is tolerated enabling the formation of five-membered heterocycles **33a,b** in 53–98% yield. When chiral phenyldiazoacetates (e.g. bearing

Scheme 12 Formation of ylides from carbenes

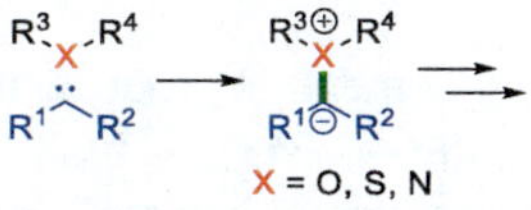

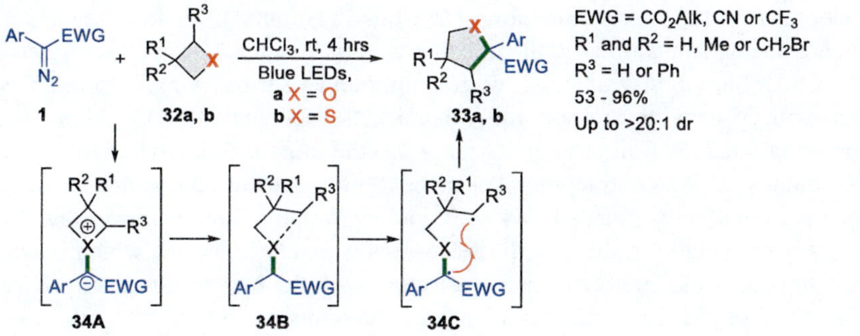

Scheme 13 Ring-expansion of oxetanes and thietanes to form the corresponding tetrahydrofuran and tetrahydrothiophene

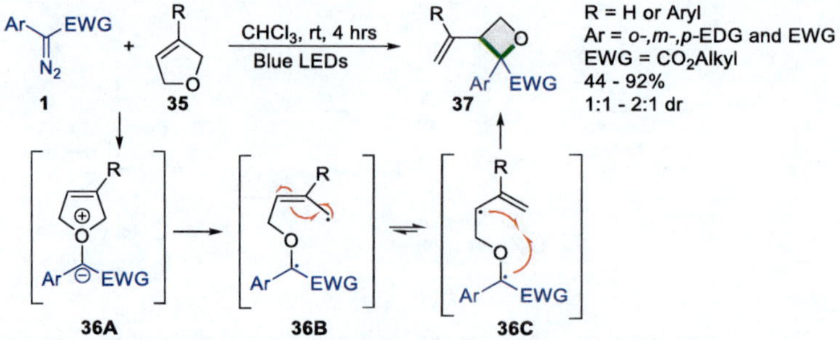

Scheme 14 Ring-contraction of 2,5-dihydrofurans to form oxetanes

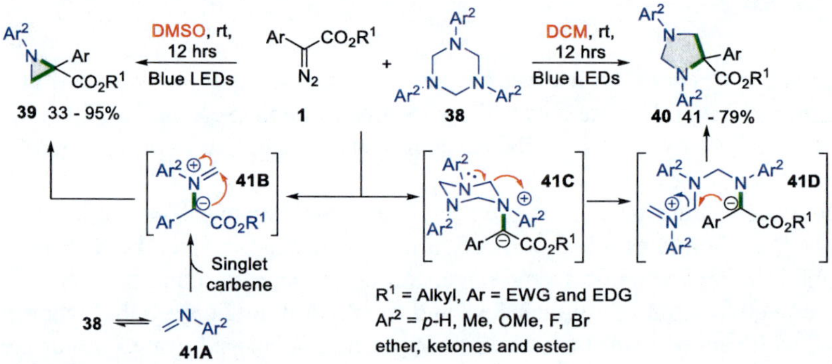

Scheme 15 Solvent-dependant reactions of diazo compounds with 1,3,5-triazinanes under blue light irradiation

a (−)-menthol group as the ester substituent) are used, the reaction leads to tetrahydrofuran **33a** in high diastereoselectivity (>20:1 *dr*). However, 3-, 5- and 6-membered heterocycles are incompatible with the ring-expansion reaction under the developed conditions.

Conversely, the ring-contraction of 2,5-dihydrofurans **35** yields oxetanes **37** for which a mechanism similar to the one depicted in Scheme 13 operates (Scheme 14) [47]. Under blue light irradiation, diazo compound **1** forms a singlet carbene, which reacts with **35** to produce oxonium ylide **36A**. Subsequently, the C–O bond breaks, generating diradical **36B**, which rearranges to the more stable allylic radical species **36C**. Finally, a ring-contraction takes place to give oxetanes **37,** which consists of a broad variety of substituents in 44–92% yield. Computational studies show that the energy barrier between the two diastereomers is nearly identical, which is why the reaction is only moderately diastereoselective with the best result of only 2:1 *dr*.

This strategy is also viable for the generation of ammonium ylides and, hence, the synthesis of *N*-heterocyclic structures. In this regard, Cheng and co-workers reported a divergent solvent-dependant synthesis of both aziridines **39** and imidazolidines **40** using the same starting materials, triazines **38** and aryl diazoacetates **1** (Scheme 15)

[48]. Under blue light irradiation, a [2+1]-cycloaddition occurs when DMSO is used as solvent, whereas a [4+1]-cycloaddition takes place in DCM. Reportedly, the solvent effect of DMSO facilitates in-situ formation of *N*-phenylmethanimine **41A**, from triazines **38**, which is not the case in DCM. Both transformations are highly tolerable with different aryl and ester substituents present in the diazo reagent **1**, as well as different aromatic groups on triazines **38**, enabling yields up to 95% for azirines **39** and up to 79% for imidazolidines **40**. The method proved to be viable for the modification of bioactive molecules.

The photochemical formation of ammonium ylides **43A** upon aniline derivative **42** is the key step in the synthesis of indolines **44** as reported by Hu and co-workers [49]. The final ring closure on the *Si*-face yields the pyrrolidine moiety in high diastereomeric ratio of >95:5 in all cases studied (Scheme 16). DFT calculations corroborate that the reaction proceeds via the formation of a singlet carbene and subsequent ylide formation. The idea of a triplet carbene pathway is excluded as it would trigger a radical mechanism; the energy barrier of such a transition state is approximately 2.5 times higher than the proposed singlet carbene route. The reaction gives indolines **44** in approximately 50% yield in most cases, although, when the aryl substituent on diazo compound **1** is highly electron-withdrawing (*p*-trifluoromethyl or *p*-nitro), the yield drops significantly to less than 30% even with the reaction time prolonged to 96 h. Interestingly, this example is rather unique as nitrogen atoms bearing hydrogen typically facilitate N-H insertion; therefore, ylide formation here is rather atypical.

However, diazo compounds are not the only source of carbenes generated in a photochemical manner. Heteroaryldiazirines **45** also give access to carbenes upon light irradiation although for their photolysis more energetic light is required as their absorption is hypsochromically shifted. Under UV light irradiation, diazirines similarly give access to carbenes; in this case, once generated they add to the imine nitrogen of substrate **46**, leading to intermediate **47A**, which triggers intramolecular Michael addition to form 2,3-dihydropyrroles **48** (Scheme 17) [50]. Concomitant expulsion of HCl then affords the respective pyrrole **49**. 2-Pyridyl and 2-thienyl

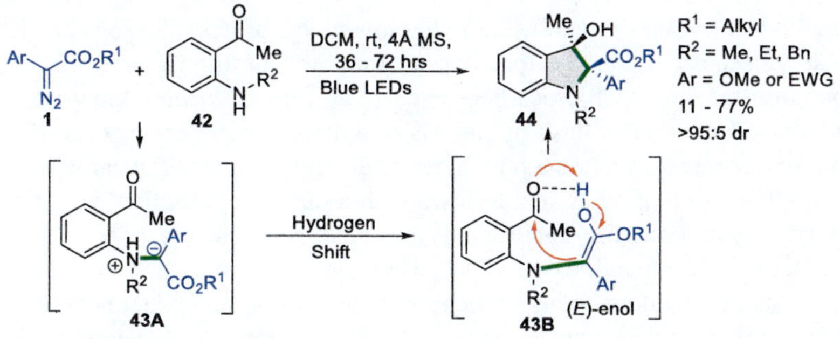

Scheme 16 Synthesis of highly substituted indolines via a formal [4+1]-annulation involving photochemically generated ylides

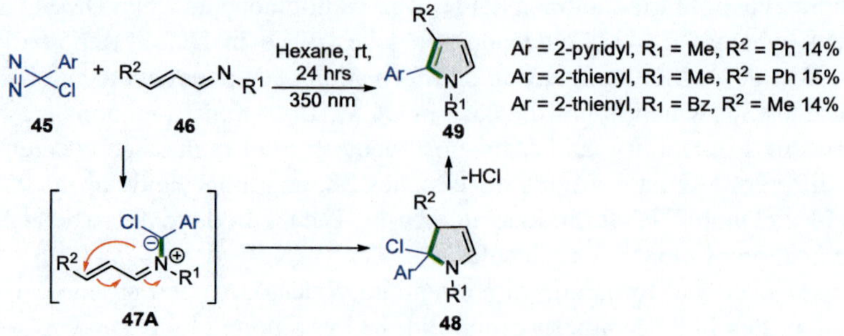

Scheme 17 Photochemical synthesis of pyrroles from diazirines and α,β-unsaturated imines

Scheme 18 Laser flash photolysis of a diazirine, pyridine and acetylene for the formation of an indolizine

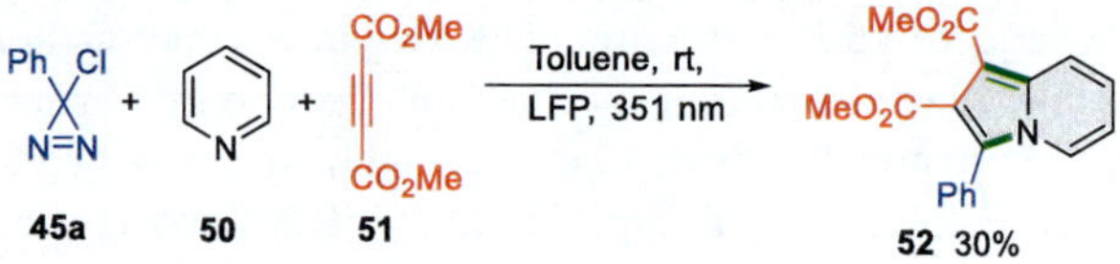

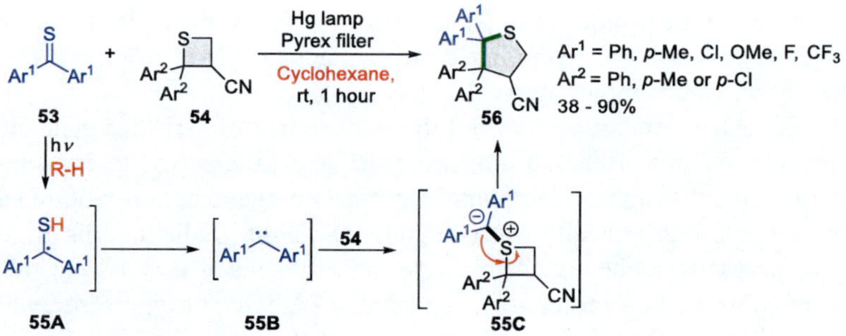

Scheme 19 Generation of carbenes from thioketones enabling ring-expansion of thietanes

pyrroles **49** were synthesised, albeit in rather low yields of less than 15% in all cases, presumably because of the high reactivity of iminium starting material **46**.

Along this line, phenylchlorocarbenes generated from diazirines **45a** gives access to indolizine **52** upon laser flash photolysis of a three-component mixture (Scheme 18) [51]. In this transformation a photochemically generated singlet carbene adds to pyridine (**50**) to form an ylide that follows with a dipolar cyclisation with dimethyl acetylenedicarboxylate (**51**) facilitating the final ring closure. The subsequent removal of HCl affords indolizine **52** in 30% yield.

Aitkene and co-workers reported the rather uncommon photochemical formation of carbene **55B** from diaryl thioketones **53** enabling the ring-expansion of thietanes **54** (Scheme 19) [52]. Substrate **53** in the triplet excited state is proposed to abstract a hydrogen from cyclohexane, which generates radical **55A**, which subsequently transforms into carbene **55B** by the concomitant loss of sulfhydryl HS·; ylide **55C** then undergoes a thia-Stevens rearrangement to form tetrahydrothiophenes **56**. The reaction is highly concentration dependant and requires highly dilute conditions (0.001 M), since at higher concentrations decomposition of thietane **54** to a 1,1-diaryl-2-cyanoethene takes place. Yields are generally good (66–90%) except when strongly electron-withdrawing trifluoromethyl groups are attached to the aryl rings of thioketone **53**.

2.4 Photochemical Wolff Rearrangements and Subsequent Transformations

The Wolff rearrangement is a classic reaction in organic chemistry introduced by Ludwig Wolff [46]. Despite being discovered over 120 years ago, the mechanism of the reaction is still heavily debated [47]. Nevertheless, it is assumed to involve nitrogen extrusion, which forms a carbene and leads to a [1,2]-rearrangement. The reaction gives access to ketenes that can be quenched by nucleophilic reagents to form carboxylic acids, amides, and other species, or can react via [2+2]-cycloaddition leading to cyclobutanones (Scheme 20).

Photochemically generated ketenes **58** are highly versatile intermediates and can be utilised in the synthesis of heterocycles of many different sizes. When nitroso compounds **59** are applied as a reaction partner, formal [2+2]-cycloadditions furnishes a four-membered oxazetidine ring (**61**) in 31–71% yield (Scheme 21) [53]. The reaction can tolerate nitroso compounds which contain strongly electron-withdrawing groups (CF_3 or CN) at the *ortho*-position. When an electron-donating group (such as methyl) is present, the resultant cycloadduct is so unstable that a retro rearrangement occurs.

Photochemical Wolff rearrangement of α-diazocarbonyl compounds followed by an intramolecular reaction with a nucleophile leads to heterocycles, as reported by Liao and co-workers (Scheme 22a) [54]. In particular, diazo ketones **62** under UV light irradiation give access to five-membered lactones **63** in 33–78% yield with high diastereoselectivity. In contrast, a rhodium catalysed reaction relies solely on an O-H insertion and gives corresponding 3-oxotetrahydrofurans **64** in excellent yields (over

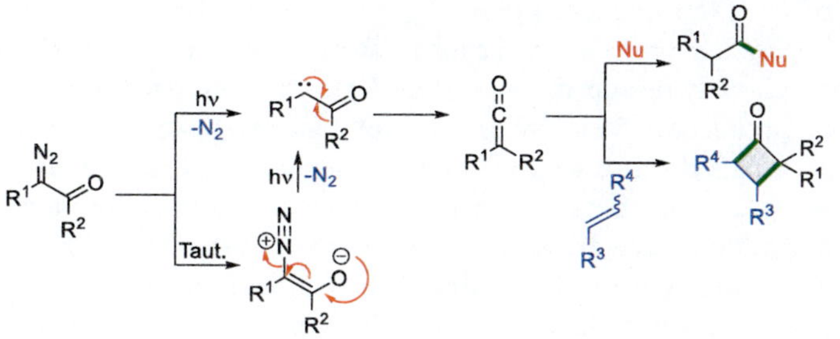

Scheme 20 Possible general mechanistic pathways for the Wolff rearrangement

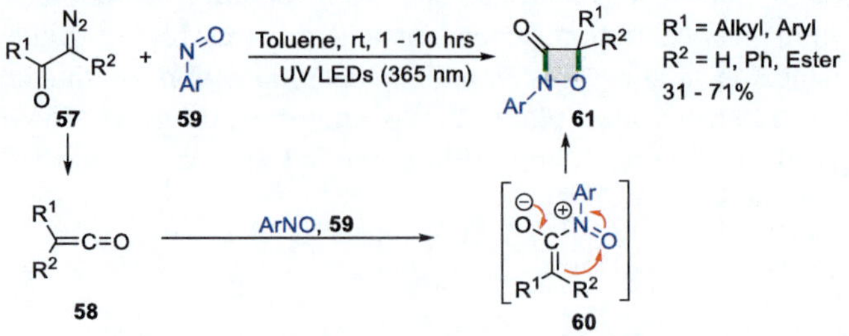

Scheme 21 Cycloaddition of nitroso aryl compounds with photochemically generated ketenes

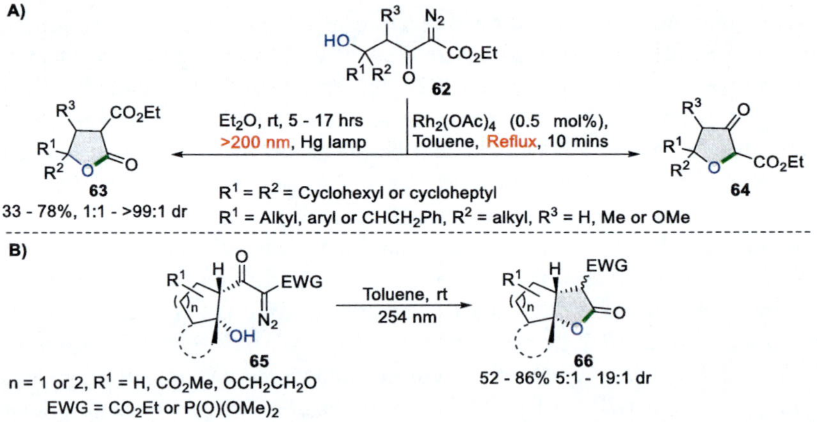

Scheme 22 Synthesis of γ-butyrolactones via photochemical Wolff rearrangements with intramolecular ring closure

90%) though with low diastereoselectivity. Along the same line, Zhang and Romo used diazo compounds **65** for the synthesis of bicyclic and tricyclic β-lactones **66** (Scheme 22b). In all cases, thermal conditions (toluene at reflux) proved less efficient in producing the desired heterocycle in comparison to the photochemical conditions. Similarly, Rh-catalysis can be utilised for such a compound through an O-H insertion to 3(*2H*)-furanones [55].

Likewise, the synthesis of γ-lactams **68** was accomplished (Scheme 23). Subjecting specifically designed diazo esters **67** to irradiation under a high-pressure mercury lamp produced 3,5-disubstitued pyrrolidones **68** in 45–84% yields whether R is an aryl or an alkyl substituent with relatively poor diastereoselectivity. Removing the ester group from lactam **68** reveals that the reaction maintains the pre-installed stereogenic centre with >95% *ee* in all cases [56]. The same group used this methodology towards the synthesis of alkaloid derivatives and even for the total synthesis of *(R)*-pyrrolam A (**71**) [57], while McDonald and co-workers prepared an enantiomerically pure *trans*-β-lactam from an α-amino acid under flow conditions [58].

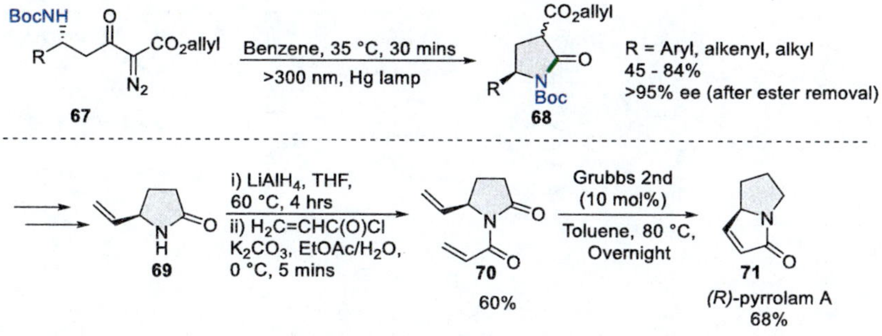

Scheme 23 Synthesis of lactams involving a photochemical Wolff rearrangement

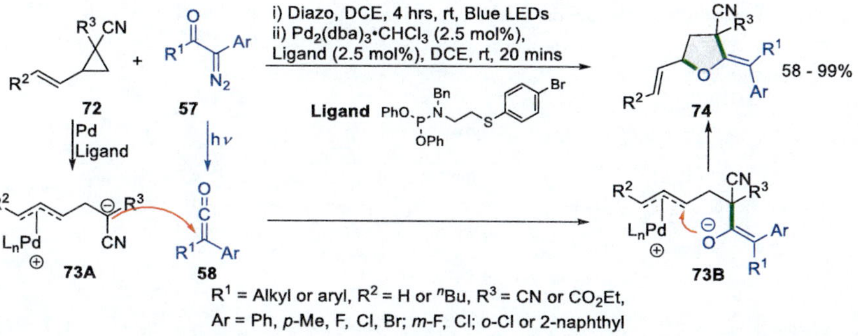

Scheme 24 Synthesis of tetrahydrofurans via a palladium-catalysed insertion into cyclopropanes followed by a chain reaction with photochemically generated ketenes

The formation of highly reactive ketenes **58** obtained from aryl diazo compounds **57** under photochemical conditions does not require any catalyst, thus predisposing them to be used in conjunction with traditional metal catalysis. The Xiao group merged a palladium-catalysed ring-opening of cyclopropanes **72** with the photochemical generation of ketenes **58** for the synthesis of tetrahydrofurans **74** (Scheme 24) [59]. The reaction is highly tolerant to a wide range of substituents on the aromatic ring of aryl diazo ketones **57** while the R group can contain double and triple bonds or an ether moiety with only a small effect on product yield. The use of a chiral ligand enabled the formation of the product also, but only with 50% *ee*.

A similar approach enabled the enantioselective synthesis of seven-membered lactones **77** with a chiral quaternary stereocenter (Scheme 25) [60]. The palladium-catalysed ring-opening of vinylethylene carbonates **75** with expulsion of carbon dioxide affords 1,5-dipolar intermediate **76A** that subsequently reacts with ketene **58** generated in the Wolff rearrangement. A number of cyclic carbonates **77** react with 2-diazo-1-phenylpropan-1-ones **57** with most proceeding in over 90% yield and *ee*, even on a gram-scale in a flow photoreactor. The reaction is highly tolerable to a range of functional groups including alkenyl, alkynyl and ethereal groups present in diazo compounds **57**. One example is given for a cyclic diazo compound that

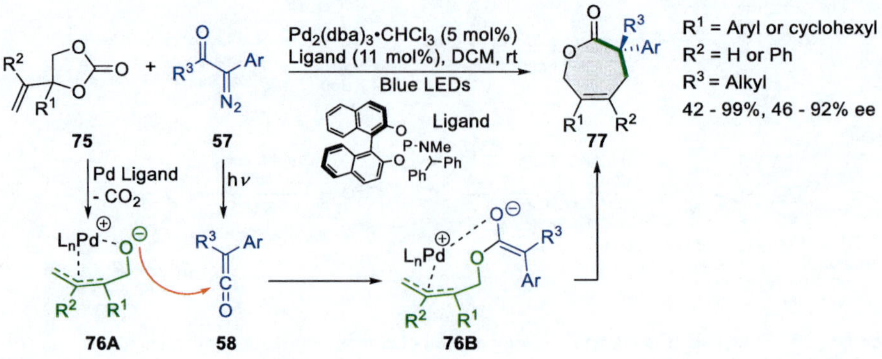

Scheme 25 Wolff rearrangement of diazo compounds with simultaneous ring-opening of vinylethylene carbonates from seven-membered lactones

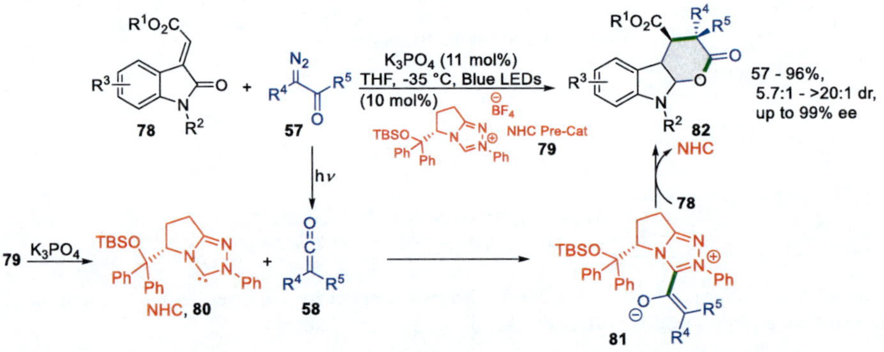

Scheme 26 Synthesis of fused tetrahydro[2,3-b] indoles via direct photolysis of diazo compound merged with NHC-catalysis

afforded a spiro-lactam in 83% yield and 76% *ee*. A similar reaction mechanism operates for the enantioselective synthesis of six-membered lactones, also by Xiao [61].

Direct photolysis of α-oxo-diazo compounds **57** can also be merged with *N*-heterocyclic carbene (NHC) catalysis (Scheme 26) [62]. In this context, tetrahydro [2,3-b]indoles **82** were synthesised in high yields and *ee* (57–96%, >20:1 *dr* and 99% *ee*) with good reaction scalability. However, the nature of the *N*-protecting group on substrates **78** is crucial for the reaction outcome, as introducing groups other than benzyl at the nitrogen atom diminished both *ee* and *dr*, with no product formation for unprotected or *N*-tosyl-indolinone being observed.

When, however, both diazo- and oxo-functionalities are present in the cyclic system, such as in diazo compound **83**, it is predisposed to undergo ring-contraction yielding new heterocyclic platforms (Scheme 27a). In this regard, Rodina and co-workers have extensively investigated the photochemical ring-contraction of tetrasubstituted 3-oxo-4-diazotetrahydrofurans **87** [63, 64]. Early data showed that

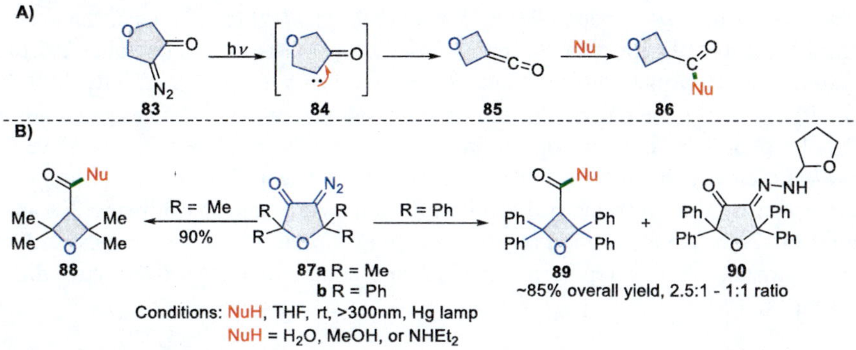

Scheme 27 Investigation into the Wolff rearrangement of tetrasubstituted 3-oxo-4-diazotetrahydrofurans

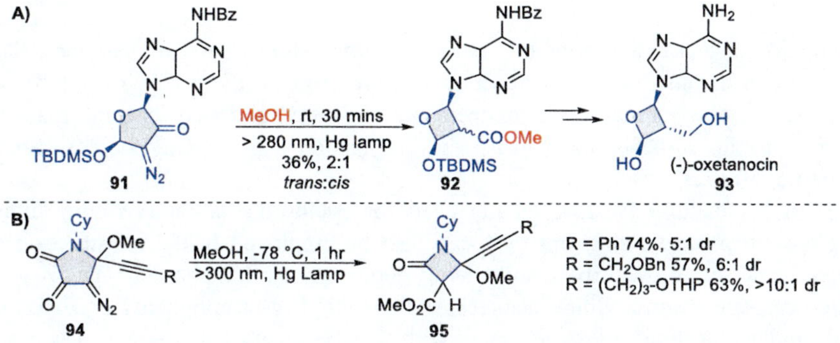

Scheme 28 Ring-contraction via a photochemical Wolff reaction for the synthesis of β-lactams

short-wave irradiation (>210 nm) of 2,2,5,5-tetraalkyltetrahydrofurans **87a** furnishes oxetanes **88** in near quantitative yield but gives lower yields in the case of tetraphenyl derivative **87b**. When the latter is exposed to high-wave irradiation (>300 nm), surprisingly it simultaneously undergoes C-H insertion into the α-position of THF with full retention of the nitrogen atoms to give side-product **90**. This is not, however, the case for the tetramethyl analogue **87a** (Scheme 27b). The ratio of products depends on the nucleophile used with a 1:1 mixture being formed when using water, whereas a 2.5:1 ratio was observed for diethylamine in favour of the ring-contracted product **89** [65]. The position of the phenyl groups does not have an impact on the reaction course, and such C-H insertion occurs independent of the aryl group position [66].

Norbeck and Kramer utilised a ring-contraction of 3-oxotetrahydrofuran **91** as the key step towards the total synthesis of (−)-oxetanocin (**93**) (Scheme 28a) [67]. The ring-forming step affords the respective oxetane **92** in 36% yield in an approximate 2:1 mixture, with the *trans,trans*-oxetane as the major product, enabling the formation of desired product **93** in 5% overall yield after 12 steps. This is a significant improvement over the previous total synthesis of (−)-oxetanocin from Niitsuma and co-workers, which took 19 steps with an overall yield of only 0.008% [68, 69]. Similarly, the photoinduced ring-contraction works for *N*-heterocycles. For example, photolysis of 4-diazopyrrolidine-2,3-diones **94** furnishes β-lactams **95** in 57–74% yield in up to >10:1 *dr* (Scheme 28b). The utility of this strategy is represented by the synthesis of bicyclic β-lactams and generated a penicillin analogue in 72% isolated yield with exclusive formation of the *trans*-product [70]. Lowe and Yeung reported a similar ring-contraction of 3-diazopyrrolidine-2,4-diones [71].

2.5 *Light-Induced [2+1]-Cycloadditions Leading to Heterocycle Rings*

Cycloadditions are debatably the most fundamental reactions of carbenes. The stereochemical outcome of the reaction is dependent on the nature of a carbene; reactions with singlet carbenes proceed in a concerted and stereospecific manner, whereas triplet carbenes react in a stepwise fashion and are only stereoselective (Scheme 29) [72].

Such cycloadditions are typically used for cyclopropanation involving either singlet or triplet carbenes, but they can also be employed in the construction of heterocyclic scaffolds. Jurberg and Davies demonstrated that photochemically generated carbenes from aryldiazoacetates, such as ethyl (4-bromophenyl)diazoacetate (**96**), under blue light irradiation react with a large array of solvents as acceptors, among them, acetone. A [2+1]-cycloaddition occurs in such a case to form epoxide **97** in 33% yield (Scheme 30) [24].

Photochemically generated carbenes were employed for the functionalisation of tethered *N*-Boc indole **98**; a three-step telescoped process formed either

Scheme 29 General mechanism for the [2+1]-cycloaddition leading to cyclopropanes

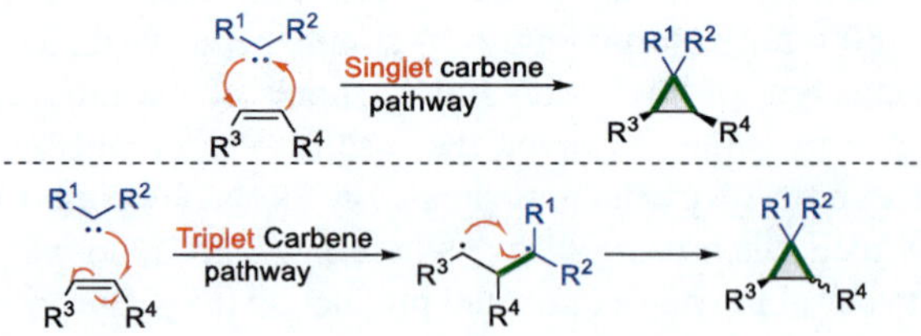

Scheme 30 [2+1]-cycloaddition of a diazo compound with acetone resulting in an epoxide

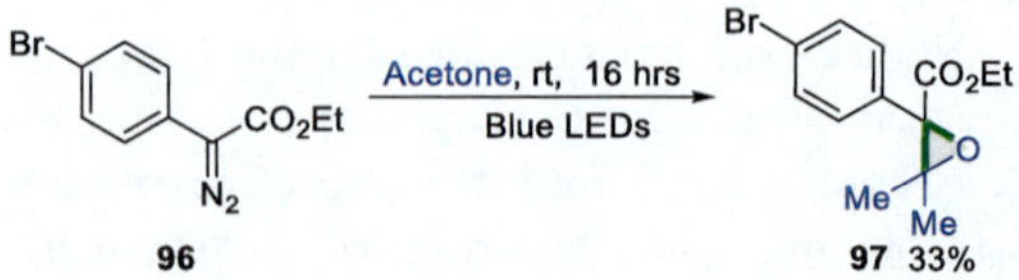

96 **97** 33%

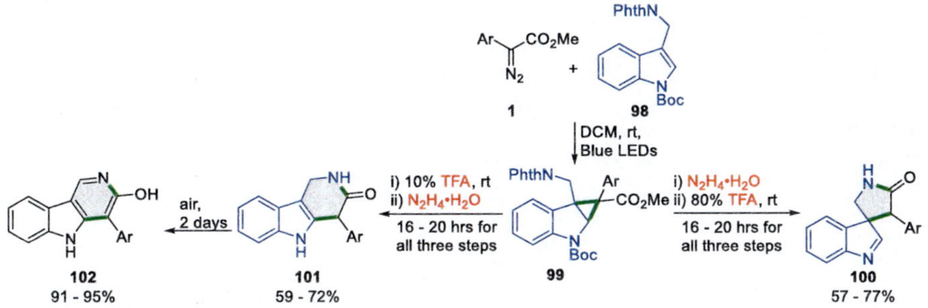

Scheme 31 Synthesis of γ-carbolinones and spiro[pyrrolidinone-3,3′]indoles via the cyclopropanation and selective deprotection of indoles

Scheme 32 A photochemical [2+1]-cycloaddition leading to a silicon-containing 'housane'

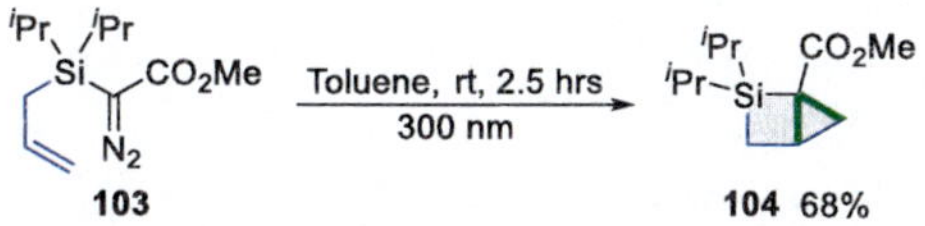

γ-carbolinones **101** or spiro[pyrrolidinone-3,3′]indoles **100**, depending on the order of removal of the protecting groups' steps (Scheme 31) [73]. In the first step, a light-induced reaction of aryl diazoacetates **1** with tethered *N*-Boc indole **98** occurs, forming the expected [2+1]-cyclopropane derivatives **99**. From this, the cycloadducts are subjected to phthalimide deprotection, followed by *N*-Boc removal which gives spiro pyrrolidones **100**. When the deprotection step follows treatment of compound **99** with TFA, γ-carbolinones **101** are formed instead. The carbolinones can be subsequently oxidised to fused pyridines **102** in over 90% yield by simple exposure to air. Of note is the fact that from the same starting materials, three distinctive heterocyclic scaffolds can be formed depending on the conditions used.

The photochemical intramolecular [2+1]-cycloaddition of specifically designed α-silyldiazo compound **103** bearing a silicon-bound allyl substituent yields 2-silabicyclo[2.1.0]pentane **104** (a silicon-containing 'housane') in 68% yield as reported by Maas and co-workers (Scheme 32) [74]. The same reaction under rhodium catalysis or under thermal conditions leads to a complex mixture of compounds, with no desired product being identified. Compound **104** is the first example of a silicon-containing housane derivative and is reported to be thermally stable, but in methanol can be opened to a non-fused cyclopropane.

The formation of 1,2-oxysila-heterocycles can be achieved by intramolecular cyclopropanation reactions of siloxy-substituted diazo compounds with tethered olefins **105**. Such substrates, when exposed to UV light, enable the formation of bicyclic oxasilolanes ($n = 1$) and an oxasilinane ($n = 2$) in 32–55% yield (Scheme 33a) [43]. Similarly, when (vinyloxy)silyldiazoacetates **107** are exposed to UV light, 2,5-dihydro-1,2-oxasiloles **108** are formed, presumably from a rearrangement of the initially formed unstable bicycloxasiletane (Scheme 33b) [75]. Such compounds are

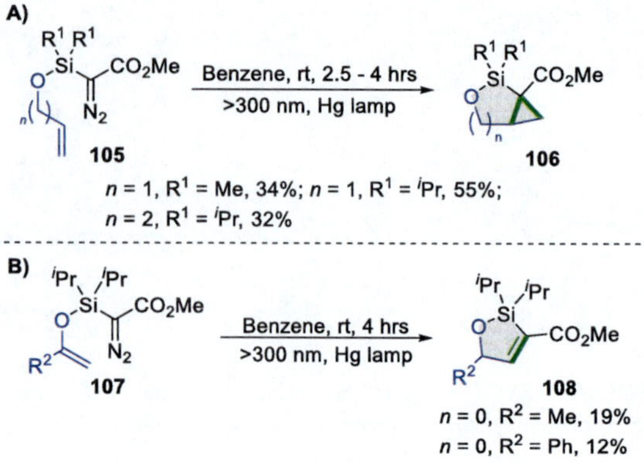

Scheme 33 Synthesis of 1,2-oxasilolanes and 1,2-oxasilinanes from intramolecular cyclopropanation or C-H insertion

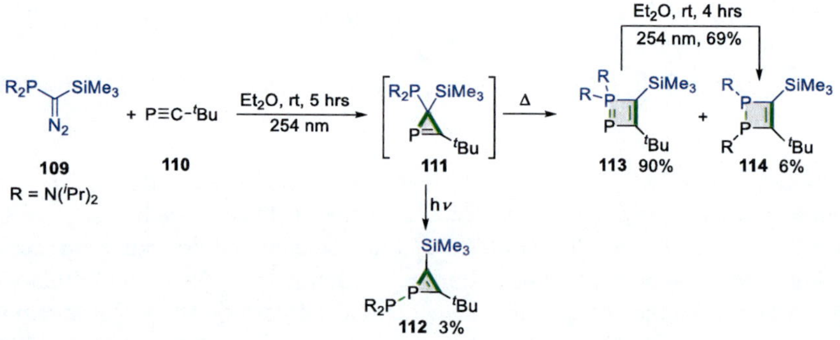

Scheme 34 Photochemical intramolecular cycloaddition of silyl diazo compounds with a phosphane

hydrolytically labile and, hence, could only be isolated in low yields, which were less than 20% for both substrates reported.

The synthesis of various phosphorus heterocycles from phosphinocarbenes, in particular those of diphosphetes **113** and dihydrophosphetes **114**, was accomplished by Sanchez and co-workers (Scheme 34) [76]. Under UV irradiation of compound **109** a singlet phosphinocarbene is produced, which undergoes a [2+1]-cycloaddition with (2,2-dimethylpropylidyne)phosphane (**110**) to furnish 2*H*-phosphirene **111** (detected by ^{31}P NMR). This intermediate **111** can undergo either a photochemical rearrangement to 1*H*-phosphirene **112** or a thermal rearrangement to a diphosphete and dihydrophosphete. Diphosphete **113**, known to act as a ligand for tungsten complexes [77], can be further converted to **114** as the exclusive product under photochemical conditions.

2.6 Carbenes Generated Via Photochemical Brook Rearrangements

The Brook rearrangement is a transformation characteristic for silyl ketones giving access to heteroatom substituted carbenes having nucleophilic character (Scheme 35). This approach has been broadly utilised for the synthesis of heterocycles including those that contain a silicon atom [41]. The formation of silyloxy carbenes from Brook rearrangements is a pleasing alternative to diazo compounds as they maintain full atom economy. These rearrangements can occur under thermal conditions that are frequently realised under microwave conditions [78].

Initial work on the photochemical Brook rearrangement from Adrian Brook demonstrated the generation of cyclic silyloxy carbenes, such as **116A** from 1,1-diphenylsilacyclohexanones **115**, which can be trapped with various reagents, including olefins, alcohols, and aldehydes (Scheme 36) [79, 80].

Predictably, [2+1]-cycloaddition of the carbene with the olefin leads to the respective 1,2-oxasila- spirocyclic scaffold **117**, while the reaction with acetaldehyde affords epoxide **118** as an inseparable mixture of diastereoisomers. Other aldehydes proved ineffective in giving the desired bicyclic heterocycle. O-H Insertion with methanol leads to 7-methoxy-1,2-oxasilepane (**119**) in 36% yield. Svarovsky and co-workers used such a transformation for the synthesis of potential silicon-containing pH-sensitive prodrugs from sugars that are active against tumours [81]. Under UV light irradiation, O-H insertion into various carbohydrates enabled the formation of the seven-membered ring in yields of over 90%.

Scheme 35 General reaction mechanism for a photochemical Brook rearrangement

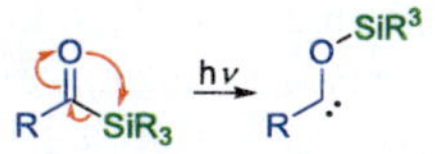

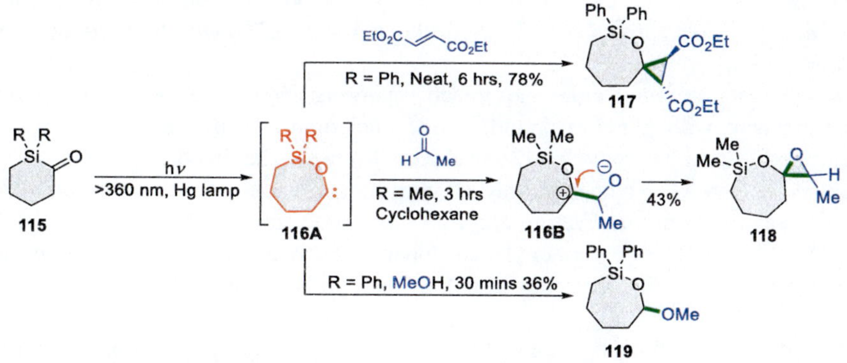

Scheme 36 Reactions of photochemically generated silyloxy carbenes

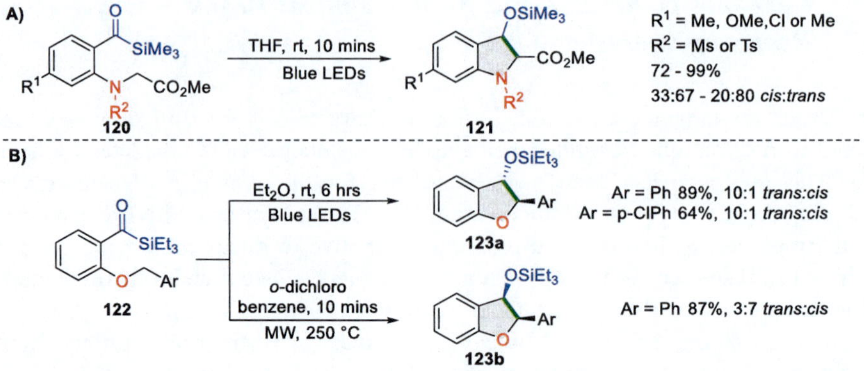

Scheme 37 Formation of five-membered heterocycles from the C-H insertion siloxy carbenes

Scheme 38 Intramolecular [2+1]-cycloaddition of siloxy carbenes to afford cyclopropyl-fused O- and N-heterocyclic scaffolds

Becker et al. showed that the use of blue light is sufficient for the generation of silyloxy species from specifically designed *ortho*-aminated substrates **120** (Scheme 37a) [82]. The Brook rearrangement followed by C-H insertion enables the synthesis of indolines **121** in high yields (99% yield in most cases). Optimisation studies revealed that diastereoselectivity is solvent-dependent with THF proving to be the solvent of choice, giving *dr* ranging from 33:67 to 20:80. Priebbenow and co-workers found that a similar C-H insertion reaction generates 2,3-dihydrobenzofurans **123a** with high *trans*-selectivity (Scheme 37b) [83]. This complements Shen and Dong's work, where the same substrate under microwave conditions at 250°C produced 2,3-dihydrobenzofurans **123b** with high *cis*-selectivity instead [78].

A similar intramolecular approach merging blue light-induced Brook rearrangement with [2+1]-cycloaddition of nucleophilic siloxy carbenes leads to cyclopropyl-fused heterocycles **125** (Scheme 38) [83]. The *exo*-diastereomer forms exclusively from (*E*)-alkenes in only 10 min; longer reaction times are required for alkenes bearing poorly activating alkyl groups (for example, when R^1 or $R^2 = Me$ and H or $R^1 = R^2 =$ cyclohexyl). In general, the reaction is high yielding and tolerates many functional groups enabling the synthesis of fused heterocycles in 69–97% yield.

For propargyl derivatives **126** reactions of siloxy carbenes result in the formation of cyclopropene **127B**, and the instability of the three-membered ring triggers a retro Brook rearrangement, which enables the synthesis of chroman-4-ones **128** (Scheme

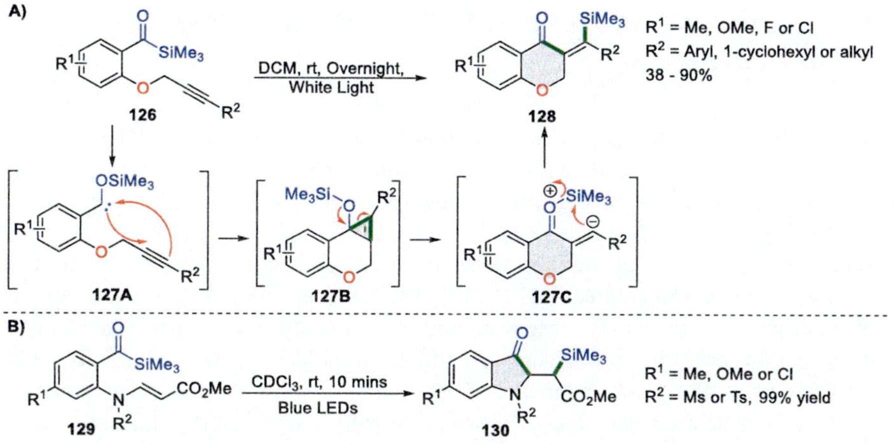

Scheme 39 Synthesis of chromon-4-ones and indolin-3-ones via cycloaddition and retro-Brook rearrangement

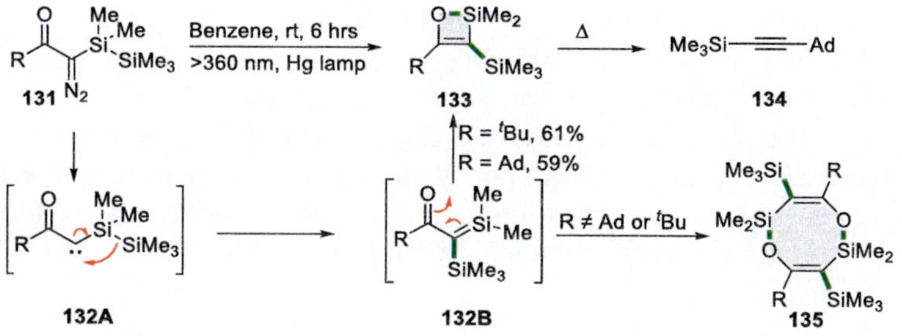

Scheme 40 Formation of 1,2-silaoxetenes and 1,5,2,6-dioxadisilocines from the silyl migration into carbenes

39a) [84]. The reaction furnishes **128** in high yields, up to 92%, with the *Z*-isomer as the only diastereoisomer in the majority of cases. The reaction is viable under solvent-free conditions although the *E*/*Z* ratio is eroded. Becker showed that this type of reactivity is also compatible with acrylates **129**, allowing the formation of 3-oxo-indolines **130** in >99% yield (Scheme 39b) [82].

While not a Brook rearrangement, the synthesis of silaoxetenes **133** can be achieved through direct photolysis of α-silyl-α'-carbonyl compounds **131** (Scheme 40) [85]. The fate of silene **132B** depends on the bulkiness of the ketone substituent. Adamantyl or *tert*-butyl substituted silanes undergo a [2+2] intramolecular cycloaddition to form 1,2-silaoxetenes **133**. Sterically less hindered substrates (R = ^{i}Pr, Me or aryl) tend to dimerize to an 8-membered ring **135**, the formation of which is unclear but believed to either arise from cyclodimerisation of 1,2-silaoxetene **133** or the silene **132B**. Sekiguchi and co-workers reported that the silaoxetene is converted to 2-adamantan-1-yltrimethylsilylacetylene (**134**) under thermal conditions [86].

2.7 Nitrenes as Intermediates in Photochemical Synthesis of Heterocycles

Nitrenes, the nitrogen analogues of carbenes, are reactive intermediates easily accessed from vinyl or aryl azides. Once formed, they subsequently react in an intramolecular manner to form azirines, which are highly strained molecules and are prone to ring-opening (Scheme 41).

Hassner and Fowler reported the photochemical synthesis of azirines **137** directly from vinyl azides **136** (Scheme 42a) [87]. While rather common, the azirines are only intermediates in various reactions and are typically not isolated. Sometimes, however, they can be isolated in high yields (over 80% yield), although when $R = R^2 =$ ethyl the yield dropped to 55%. Isomura showed that under reductive conditions, *2H*-azirines produce aziridines **138** [88]. Unexpectedly, the cyclisation of azidocyclooctenes **139** furnish aziridines **140a,b** in quantitative yields, which turned out to be highly stable even with two double bonds being present in the ring (Scheme 42b). These bicyclic heterocycles **140** readily hydrolyze under acidic conditions and treating the subsequent product with a base facilitates dimerization to yield a new heterocyclic scaffold, pyrazine **141** [87].

Under UV irradiation with a high-power mercury lamp, the direct conversion of benzene rings to pyridine rings via carbon-nitrogen exchange was initially described in 1972. However, because two azepines and two pyridines were produced as an inseparable mixture, the method was not synthetically useful [89]. It is now known, as shown by Patel and Burns, that the transformation of azidobenzenes **142** under less energetic blue light leads selectively to *3H*-azepines **144** (Scheme 43) [90]. The

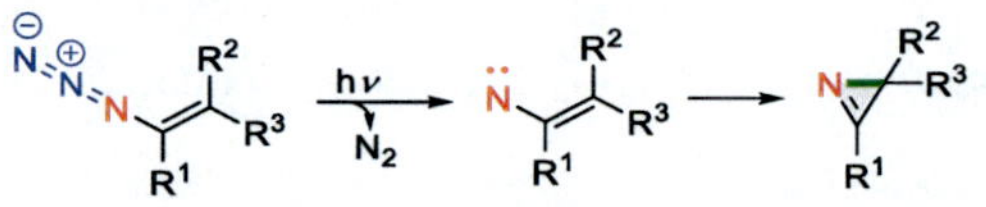

Scheme 41 General mechanism for the photochemical formation of nitrenes

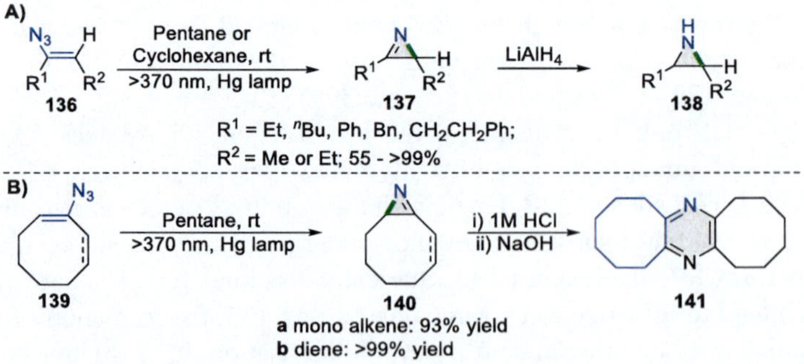

Scheme 42 Synthesis of isolatable azirines under photochemical conditions; 9-aza-bicyclo[6.1.0] nonenes were reported to be highly stable

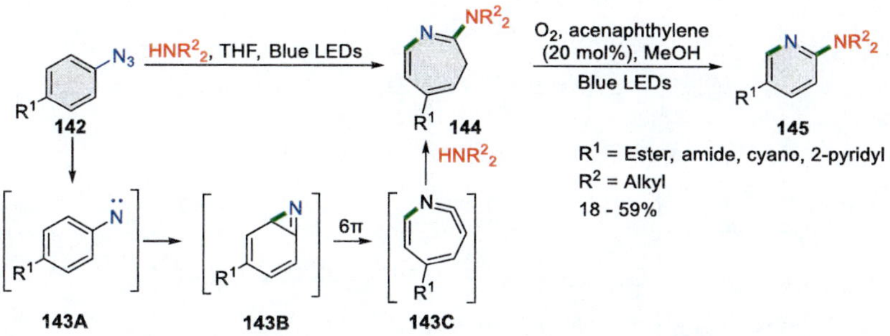

Scheme 43 Functional group interconversion of aryl azides to pyridines via an isolatable 3H-azepine

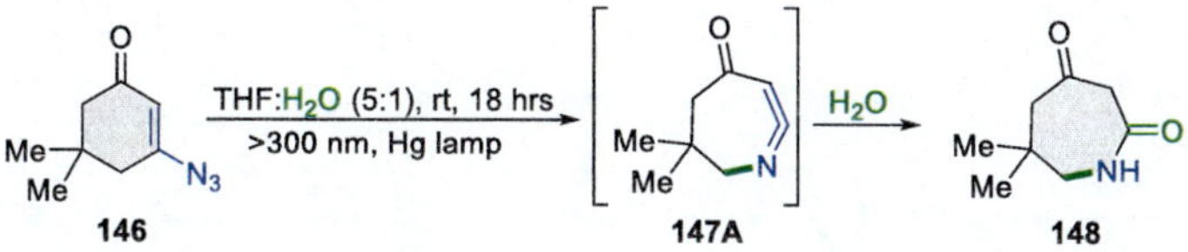

Scheme 44 Synthesis of 6,6-dimethylazepane-2,4-dione from photolysis of a vinyl azide

mechanism suggests the generation of nitrene **143A**, which subsequently produces azirine **143B**. 6π-Electrocyclization followed by attack of the amine generates azepine heterocycle **144,** which can further react with singlet oxygen to brandish a new scaffold, pyridine **145**. Both steps can be realised in a one-pot manner in yields of up to 59%.

Along this line, irradiation of cyclic vinyl azide **146** results in a ring-expansion to an azepane-2,4-ketoamide **148** (Scheme 44) [91]. Interestingly, the mechanism of the reaction is suggested not to proceed via an azirine but instead via strained intermediate **147A**, that presumably results from a Curtius-type rearrangement. Subsequent hydrolysis leads to the formation of 6,6-dimethylazepane-2,4-dione (**148**) as the only product, although no comment on the yield was made.

2.8 *Miscellaneous*

Among the many heterocyclic scaffolds that have been synthesised based on photochemical transformations, thiazolines **152** are also part of this collection. This moiety forms regioselectively under blue light irradiation of α-diazo-1,3-diketones **150** with β-ketothioamides **149** in good yields (Scheme 45) [92]. The reaction is scalable and highly tolerant with a variety of electron-donating or electron-withdrawing substituents on the aryl rings of both substrates having no detrimental

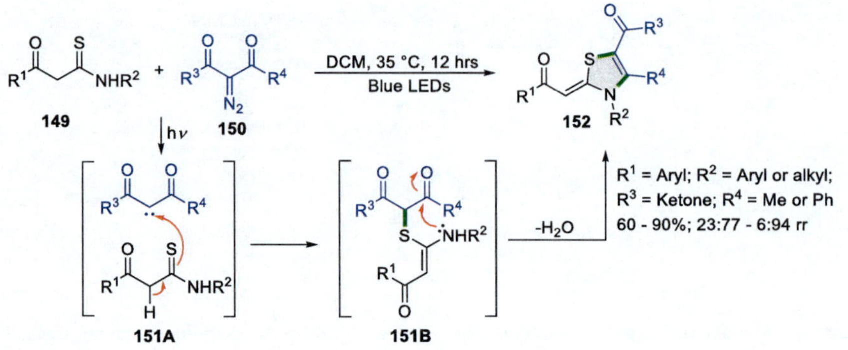

Scheme 45 Photochemical synthesis of thiazolines from diazo 1,3-dicarbonyl diazo compounds and β-ketothioamides

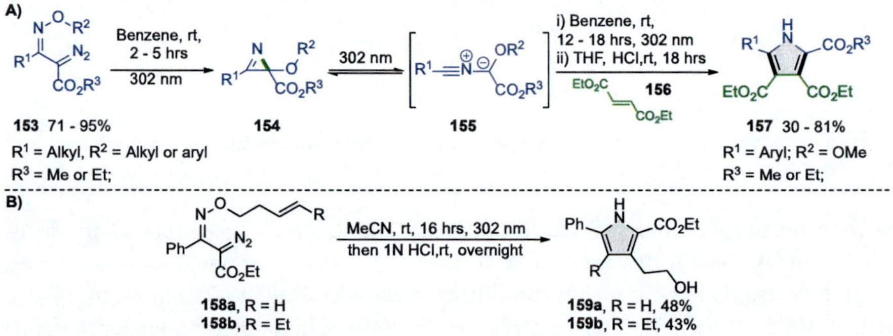

Scheme 46 Synthesis of 2H-azirines from intramolecular N-O insertion of carbenes

effect on product yields. For diaryl substituted diazo ketones **149**, regioselectivity in the formation of compound **152** erodes but is still relatively high (up to 94:6 *rr*).

Another five-membered valuable heterocycle, namely pyrrole **157**, is accessible from α-diazo-β-oxime esters **153**. Photochemically generated carbenes intramolecularly insert into the N-O bond to form 2*H*-azirines **154** in 71–95% yield (Scheme 46a) [93]. Prolonged irradiation with UV light induces rearrangement of the azirine scaffold to nitrile ylide **155** that subsequently undergoes cyclisation with diethyl fumarate (**156**) to yield a five-membered ring; treatment with hydrochloric acid forms the pyrrole **157** in 40–81% yields in a one-pot manner. The reaction is equally viable with pre-formed substrates, such as with **158a,b** that allowed an intramolecular tandem cyclization to pyrroles **159a,b** in 48 and 43% yield, respectively (Scheme 46b).

Thamattoor and co-workers explored the potential reactivity of putative alkenyl carbenes **161** originating from the light-induced extrusion of phenanthracene from **160** (Scheme 47) [94]. Upon irradiation, singlet carbene **161** is formed, which is corroborated with computational studies and subsequently undergoes a Fritsch–

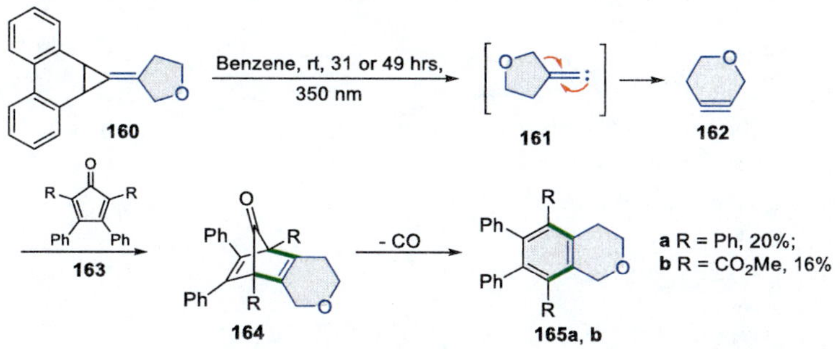

Scheme 47 Synthesis of isochromanes from trapping of in-situ generated cycloalkynes

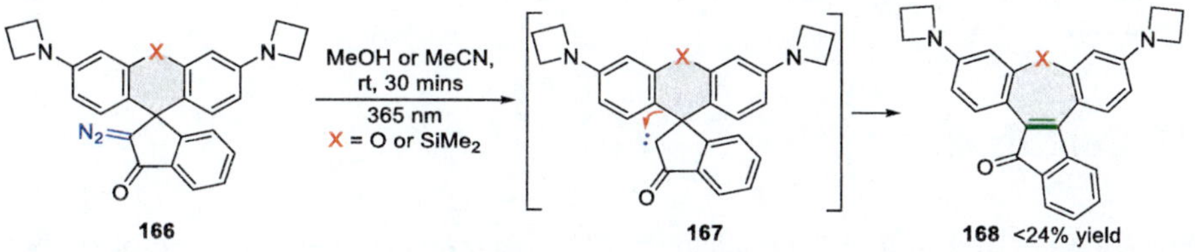

Scheme 48 Ring-expansion of diazoindanones facilitating a carbon-carbon bond migration

Buttenberg–Wiechell rearrangement to strained alkyne **162**. Such strained compounds are known to undergo Diels-Alder reactions that form isochromanes **165** in 16–20% yield. The Garg group has reported similar reactivity for the nitrogen analogue of intermediate **162** that is formed in-situ from a fluoride-induced 1,2-elimination reaction [95].

Even though complex molecules are typically regarded as unselective in photochemical activation, Grimm et al. utilised light for the modification of 'Janelia Fluor' dyes **166** [96]. In this case, the formation of carbene **167** induces an unintended unexpected ring-expansion of oxygen and silicon-containing heterocycles (Scheme 48). The generated singlet carbene facilitates a carbon-carbon bond migration transforming dibenzo[*a*,*e*]pyran **166** to the dibenzo[*b*,*f*]oxepine **168**, and a similar transformation took place for the silicon-analogue. Similar reactions were also reported by the groups of Belov [97] and Halabi [98].

The photolysis of aryldiazoacetates **1** can lead to diverse products, and they have been explored extensively. Jin and co-workers employed them in a blue light-induced three-component reaction leading to dihydroisoxaziles **173** (Scheme 49) [99]. Once the carbene forms, it abstracts hydrogen from amine **169**, leading to ion pairs **171B** and finally enamine **172** that undergoes a [3+2]-cycloaddition that furnishes 4,5-dihydroisooxazoles **173**. The role of aryldiazo compound **1** is rather unique in this case as it acts as both an electrophile and a base and is not present in the final product. The *N*-hydroximoyl chloride could be replaced with other reagents to enable the synthesis of additional heterocycles such as triazoles, furans and tetrahydroquinolines.

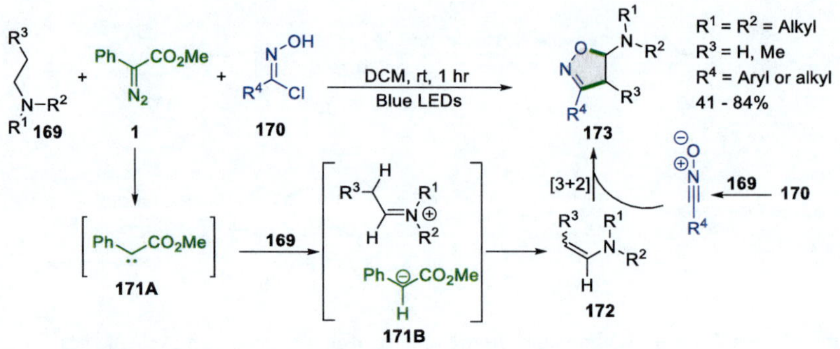

Scheme 49 Formation of 4,5-dihydroisoxazoles via [3+2]-cycloaddition of photochemically generated vinylamines

3 Conclusion

For a long period of time photochemistry has been regarded as just a curiosity, but recently it has been recognised as a valuable tool in organic chemistry. New reaction pathways and modes of reactivity have been opened giving access to structures that were inaccessible by other means. The photochemistry of carbenes/nitrenes is not an exception.

This review has covered the synthesis of heterocycles under photochemical conditions but is limited to those involving carbenes/nitrenes as the key reactive intermediates. These intermediates can be photogenerated from various precursors, among which diazo compounds and azides are dominant. Under light irradiation, dinitrogen extrusion typically occurs generating carbenes/nitrenes that subsequently form other reactive intermediates such as ylides, ketenes or diradicals giving access to a wide variety of oxygen-, nitrogen-, silicon- and phosphorus-containing heterocycles. In general, highly energetic UV light-induced reactions dominate this field, but it limits the library of potential reactions. However, since the seminal work by Davies and Jurberg, the use of visible light for activation has become the tool of choice for aryl diazo compounds, providing reactants absorbed in this region. Photocatalytic transformations are far less developed, and we look forward to seeing what developments will come in the future.

References

1. Verschueren RH, De Borggraeve WM (2019) Electrochemistry and photoredox catalysis: a comparative evaluation in organic synthesis. Molecules 24:2122
2. Crisenza GEM, Mazzarella D, Melchiorre P (2020) Synthetic methods driven by the photoactivity of electron donor-acceptor complexes. J Am Chem Soc 142:5461–5476

3. Protti S, Ravelli D, Fagnoni M, Albini A (2013) Smooth photogeneration of α,n-didehydrotoluenes (DHTs). Pure Appl Chem 85:1479–1486
4. Noyori R (2010) Insight: green chemistry: the key to our future. Tetrahedron 66:1028
5. Candish L, Collins KD, Cook GC, Douglas JJ, Gómez-Suárez A, Jolit A, Keess S (2022) Photocatalysis in the life science industry. Chem Rev 122:2907–2980
6. Allen AR, Noten EA, Stephenson CRJ (2022) Aryl transfer strategies mediated by photoinduced electron transfer. Chem Rev 122:2695–2751
7. Juliá F, Constantin T, Leonori D (2022) Applications of halogen-atom transfer (XAT) for the generation of carbon radicals in synthetic photochemistry and photocatalysis. Chem Rev 122: 2292–2352
8. Capaldo L, Ravelli D, Fagnoni M (2022) Direct photocatalyzed hydrogen atom transfer (HAT) for aliphatic C–H bonds elaboration. Chem Rev 122:1875–1924
9. Pitre SP, Overman LE (2022) Strategic use of visible-light photoredox catalysis in natural product synthesis. Chem Rev 122:1717–1751
10. Moss RA (2006) Diazirines: carbene precursors par excellence. Acc Chem Res 39:267–272
11. Ollevier T, Carreras V (2022) Emerging applications of aryl trifluoromethyl diazoalkanes and diazirines in synthetic transformations. ACS Org Inorg Au 2:83–98
12. Zhang Z, Yadagiri D, Gevorgyan V (2019) Light-induced metal-free transformations of unactivated pyridotriazoles. Chem Sci 10:8399–8404
13. Akter M, Rupa K, Anbarasan P (2022) 1,2,3-triazole and its analogues: new surrogates for diazo compounds. Chem Rev 122:13108–13205
14. Orłowska K, Santiago JV, Krajewski P, Deperansińska I, Zawada K, Chaładaj W, Gryko D (2023) UV light is no longer required for the photoactivation activation of 1,3,4-oxadiazolines. ACS Catal 13(3):1964–1973. https://doi.org/10.1021/acscatal.2c05319
15. Durka J, Turkowska J, Gryko D (2021) Lightening diazo compounds? ACS Sustain Chem Eng 9:8895–8918
16. Ciszewski ŁW, Rybicka-Jasińska K, Gryko D (2019) Recent developments in photochemical reactions of diazo compounds. Org Biomol Chem 17:432–448
17. Galkina OS, Rodina LL (2016) Photochemical transformations of diazocarbonyl compounds: expected and novel reactions. Russ Chem Rev 85:537
18. Hua T-B, Yang Q-Q, Zou Y-Q (2019) Recent advances in enantioselective photochemical reactions of stabilized diazo compounds. Molecules 24
19. Candeias RN, Afonso AMC (2009) Developments in the photochemistry of diazo compounds. Curr Org Chem 13:763–787
20. Bergstrom BD, Nickerson LA, Shaw JT, Souza LW (2021) Transition metal catalyzed insertion reactions with donor/donor carbenes. Angew Chem Int Ed 60:6864–6878
21. Xiang Y, Wang C, Ding Q, Peng Y (2019) Diazo compounds: versatile synthons for the synthesis of nitrogen heterocycles *via* transition metal-catalyzed cascade C–H activation/carbene insertion/annulation reactions. Adv Synth Catal 361:919–944
22. Li YP, Li ZQ, Zhu SF (2018) Recent advances in transition-metal-catalyzed asymmetric reactions of diazo compounds with electron-rich (hetero-) arenes. Tetrahedron Lett 59:2307–2316
23. de Frémont P, Marion N, Nolan SP (2009) Carbenes: synthesis, properties, and organometallic chemistry. Coord Chem Rev 253:862–892
24. Jurberg ID, Davies HML (2018) Blue light-promoted photolysis of aryldiazoacetates. Chem Sci 9:5112–5118
25. Jana S, Yang Z, Pei C, Xu X, Koenigs RM (2019) Photochemical ring expansion reactions: synthesis of tetrahydrofuran derivatives and mechanism studies. Chem Sci 10:10129–10134
26. Jampilek J (2019) Heterocycles in medicinal chemistry. Molecules 24:3839
27. Ashitha KT, Krishna MSA, Basavaraja D, Somappa SB (2022) Recent advances in the transition metal-free synthesis of heterocycles from α,β-unsaturated ketones. Org Chem Front 9:5306–5358

28. Kiciński W, Dyjak S (2020) Transition metal impurities in carbon-based materials: pitfalls, artifacts and deleterious effects. Carbon 168:748–845
29. Lefebvre C, Fortier L, Hoffmann N (2020) Photochemical rearrangements in heterocyclic chemistry. Eur J Org Chem 2020:1393–1404
30. Latrache M, Hoffmann N (2021) Photochemical radical cyclization reactions with imines, hydrazones, oximes and related compounds. Chem Soc Rev 50:7418–7435
31. Davies HML, Beckwith REJ (2003) Catalytic enantioselective C-H activation by means of metal-carbenoid-induced C-H insertion. Chem Rev 103:2861–2903
32. Kirmse V, Dietrich H, Bücking HW (1967) Intramolekulare reaktiønen von alkoxycarbonylcarbenen. Tetrahedron Lett 8:1833–1835
33. Rando RR (1970) Conformational and solvent effects on carbene reactions. J Am Chem Soc 92: 6706–6707
34. Rando RR (1972) Conformational and medium effects on intramolecular carbene reactions. J Am Chem Soc 94:1629–1631
35. Zhang Y, Burdziński G, Kubicki J, Platz MS (2009) Ultrafast time-resolved infrared spectroscopy study of the photochemistry of *N,N*-diethyldiazoacetamide: rearrangement in the excited state. J Am Chem Soc 131:9646–9647
36. Candeias NR, Gois PMP, Veiros LF, Afonso CAM (2008) C-H carbene insertion of α-diazo acetamides by photolysis in non-conventional media. J Org Chem 73:5926–5932
37. Lowe G, Parker J (1971) Photochemical conversion of α-diazo-amides and -esters into β-lactams and β- and γ-lactones. J Chem Soc D 197:577–578
38. Corey EJ, Felix AM (1965) A new synthetic approach to the penicillins. J Am Chem Soc 87: 2518–2519
39. Alimi I, Remy R, Bochet CG (2017) Photochemical C–H activation: generation of indole and carbazole libraries, and first total synthesis of clausenawalline D. Eur J Org Chem 2017:3197–3210
40. Yang J, Zhang Q, Zhang W, Yu W (2014) Synthesis of benzo[*a*]carbazoles and indolo[2,3-a] carbazoles via photoinduced carbene-mediated C-H insertion reaction. RSC Adv 4:13704–13707
41. Cheng Y, Meth-Cohn O (2004) Heterocycles derived from heteroatom-substituted carbenes. Chem Rev 104:2507–2530
42. Maas G, Bender S (2000) Synthesis and ring enlargement of 2-ethoxycarbonyl-1-silacyclobutanes. Chem Commun:437–438
43. Maas G, Krebs F, Werle T, Gettwert V, Striegler R (1999) Silicon-oxygen heterocycles from thermal, photochemical, and transition-metal-catalyzed decomposition of α-(alkoxysilyl and alkenyloxysilyl)-α-diazoacetates. Eur J Org Chem:1939–1946
44. Clayden J, Greeves N, Warren S (2012) Organic chemistry.2nd edn. Oxford University Press
45. Zhou S, Cai B, Hu C, Cheng X, Li L, Xuan J (2021) Visible light and base promoted O-H insertion/cyclization of Para-quinone methides with aryl diazoacetates: an approach to 2,3-dihydrobenzofuran derivatives. Chin Chem Lett 32:2577–2581
46. Kirmse W (1971) Chapter 1 – Introduction. In: Carbene chemistry, 2nd edn. Academic Press
47. Qi D, Bai J, Zhang H, Li B, Song Z, Ma N, Guo L, Song L, Xia W (2022) Photoinduced synthesis of functionalized oxetanes via diradical-mediated ring contraction. Green Chem 24: 5046–5051
48. Cheng X, Cai BG, Mao H, Lu J, Li L, Wang K, Xuan J (2021) Divergent synthesis of aziridine and imidazolidine frameworks under blue LED irradiation. Org Lett 23:4109–4114
49. Chen J, Liu S, Lv X, Hong K, Lei J, Xu X, Hu W (2020) Blue light-promoted formal [4+1]-annulation of diazoacetates with *o*-aminoacetophenones: synthesis of polysubstituted indolines and computational study. J Org Chem 85:13920–13928
50. Romashin YN, Liu MTH, Ma W, Moss RA (1999) New synthesis of 2,2′-heteroarylpyrroles from heteroarylchlorocarbenes. Tetrahedron Lett 40:7163–7165

51. Jackson JE, Soundararajan N, Platz MS, Liu MTH (1988) Pyridine ylide formation by capture of phenylchlorocarbene and tert-butylchlorocarbene. Reaction rates of an alkylchlorocarbene by laser flash photolysis. J Am Chem Soc 110:5595–5596
52. Kassir AF, Guillot R, Scherrmann MC, Boddaert T, Aitken DJ (2020) Formation of tetrahydrothiophenes *via* a thia-Paternò-Büchi-initiated domino photochemical reaction. Org Lett 22:8522–8527
53. Roy S, Kumar G, Chatterjee I (2021) Photoinduced diverse reactivity of diazo compounds with nitrosoarenes. Org Lett 23:6709–6713
54. Liao M, Dong S, Deng G, Wang J (2006) Synthesis of oxygen-containing heterocyclic compounds based on the intramolecular O-H insertion and Wolff rearrangement of α-diazocarbonyl compounds. Tetrahedron Lett 47:4537–4540
55. Zhang W, Romo D (2007) Transformation of fused bicyclic and tricyclic β-lactones to fused γ-lactones and 3(2H)-furanones via ring expansions and O-H insertions. J Org Chem 72:8939–8942
56. Dong C, Mo F, Wang J (2008) Highly diastereoselective addition of the lithium enolate of α-diazoacetoacetate to *N*-sulfinyl imines: enantioselective synthesis of 2-oxo and 3-oxo pyrrolidines. J Org Chem 73:1971–1974
57. Mo F, Li F, Qiu D, Wang J (2010) Enantioselective synthesis of condensed and transannular ring skeletons containing pyrrolidine moiety. Tetrahedron 66:1274–1279
58. Vaske YSM, Mahoney ME, Konopelski JP, Rogow DL, McDonald WJ (2010) Enantiomerically pure *trans*-β-lactams from α-amino acids *via* compact fluorescent light (CFL) continuous-flow photolysis. J Am Chem Soc 132:11379–11385
59. Liu J, Li MM, Le QB, Lu LQ, Xiao WJ (2019) A photoinduced Wolff rearrangement/Pd-catalyzed [3+2] cycloaddition sequence: an unexpected route to tetrahydrofurans. Chem Commun 55:2031–2034
60. Wei Y, Liu S, Li M-M, Li Y, Lan Y, Lu L-Q, Xiao W-J (2019) Enantioselective trapping of Pd-containing 1,5-dipoles by photogenerated ketenes: access to 7-membered lactones bearing chiral quaternary stereocenters. J Am Chem Soc 141:133–137
61. Li MM, Le QB, Xiao YQ, Xiao WJ, Lu LQ (2021) Enantioselective trapping of palladium-stabilized oxo-1,4-dipoles with photochemically generated ketenes. Sci Bull 66:1719–1722
62. Wang C, Wang Z, Yang J, Shi SH, Hui XP (2020) Sequential visible-light and *N*-heterocyclic carbene catalysis: stereoselective synthesis of tetrahydropyrano[2,3-*b*]indoles. Org Lett 22: 4440–4443
63. Korobitsina IK, Rodina LL (1965) Zh Org Khim 1:932
64. Korobitsina IK, Rodina LL, Shuvalova VG (1968) Zh Org Khim 4:2016
65. Rodina LL, Mishchenko VL, Malashikhin SA, Platz M, Nikolaev VA (2003) New photochemical reaction in the series of diazofuranones. Russ J Org Chem 39:1530–1532
66. Rodina LL, Malashikhin SA, Galkina OS, Nikolaev VA (2009) Photochemical reactions of regioisomeric 2,2-dimethyl-5,5-diphenyl- and 5,5-dimethyl-2,2-diphenyl-substituted diazo ketones of a tetrahydrofuran series. Helv Chim Acta 92:1990–1998
67. Norbeck DW, Kramer JB (1988) Synthesis of (−)-oxetanocin. J Am Chem Soc 110:7217–7218
68. Niitsuma S, Ichikawa Y, Kato K, Takita T (1987) Studies on the total synthesis of oxetanocin; I. The first synthesis of a nucleoside having oxetanosyl-*N*-glycoside. Tetrahedron Lett 28:3967–3970
69. Niitsuma S, Kato K, Takita T (1987) Studies on the total synthesis of oxetanocin; II. Total synthesis of oxetanocin. Tetrahedron Lett 28:4713–4714
70. Moore HW, Arnold MJ (1983) Photolysis of 4-diazopyrrolidine-2,3-diones. A new synthetic route to mono- and bicyclic β-lactams. J Org Chem 48:3365–3367
71. Lowe G, Yeung HW (1973) Synthesis of a β-lactam related to the cephalosporins. J Chem Soc Perkin Trans 1:2907–2910
72. Thu Le NN, Klinkerch EJ, Thamattoor DM (2019) Stereochemistry of dichlorocarbene addition to alkenes: a collaborative, discovery-based experiment for the organic chemistry laboratory. J Chem Educ 96:1727–1730

73. Guha S, Gadde S, Kumar N, Black DS, Sen S (2021) Orthogonal syntheses of γ-carbolinone and spiro[pyrrolidinone-3,3′]indole derivatives in one pot through reaction telescoping. J Org Chem 86:5234–5244
74. Maas G, Daucher B, Maier A, Gettwert V (2004) Synthesis and ring opening reactions of a 2-silabicyclo[2.1.0]pentane. Chem Commun 4:238–239
75. Daucher B, Gettwert V, Striegler R, Maas G (2004) Intramolecular carbene and carbenoid reactions of α-(vinyloxy)silyl-α-diazoacetates. Zeitschrift fur Naturforsch Sect B J Chem Sci 59:1444–1450
76. Sanchez M, Réau R, Marsden CJ, Regitz M, Bertrand G (1999) From 2-phosphino-2*H*-phosphirene to 1-phosphino-1*H*-phosphirene, $1\lambda^5,2\lambda^3$-diphosphete, and 1,2-dihydro-$1\lambda^3,2\lambda^3$-diphosphete: an experimental and theoretical study. Chem – A Eur J 5:274–279
77. Armbrust R, Bergsträsser U, Regitz M, Sanchez M, Réau R, Bertrand G (1995) Synthesis of a stable $1\lambda^5,2\lambda^3$-diphosphete: the first codimer of a λ^3-phosphaalkyne with a λ^5-phosphaalkyne. J Am Chem Soc 117:10785–10786
78. Shen Z, Dong VM (2009) Benzofurans prepared by C-H bond functionalization with acylsilanes. Angew Chem Int Ed 48:784–786
79. Brook AG, Kucera HW, Pearce R (1971) Photolysis of 1,1-diphenylsilacyclohexanone in diethyl fumarate. The trapping of a siloxycarbene by an electron-deficient olefin. Can J Chem 49:1618–1621
80. Brook AG, Pearce R, Pierce JB (1971) Nucleophilic attack of siloxycarbenes on carbonyl groups. The formation of oxiranes. Can J Chem 49:1622
81. Svarovsky SA, Taraban MB, Barchi JJ (2004) Facile photochemical synthesis of mixed siloxyacetal glycosides as potential pH-sensitized prodrugs for selective treatment of solid tumors. Org Biomol Chem 2:3155–3161
82. Becker P, Pirwerdjan R, Bolm C (2015) Acylsilanes in iridium-catalyzed directed amidation reactions and formation of heterocycles *via* siloxycarbenes. Angew Chem Int Ed 127:15713–15716
83. Bunyamin A, Hua C, Polyzos A, Priebbenow DL (2022) Intramolecular photochemical [2 + 1]-cycloadditions of nucleophilic siloxy carbenes. Chem Sci 13:3273–3280
84. Zhang HJ, Becker P, Huang H, Pirwerdjan R, Pan FF, Bolm C (2012) Photochemically induced silylacylations of alkynes with acylsilanes. Adv Synth Catal 354:2157–2161
85. Schneider K, Daucher B, Fronda A, Maas G (1990) Acylsilenes (3-Oxo-1-sila-1-propenes) from acyl(disilanyl)carbenes. Chem Ber 123:589–594
86. Akira S, Ando W (1984) 1,2-silaoxetene. J Am Chem Soc 106:1486–1487
87. Hassner A, Fowler FW (1968) Synthesis and reactions of 1-azirines. J Am Chem Soc 90:2869–2875
88. Isomura K, Okada M, Taniguchi H (1969) Azirine formation by decomposition of terminal vinyl azides. Tetrahedron Lett 10:4073–4076
89. Sundberg RJ, Suter SR, Brenner M (1972) Photolysis of *ortho*-substituted aryl azides in diethylamine. Formation and autoxidation of 2-diethylamino-1*H*-azepine intermediates. J Am Chem Soc 94:513–520
90. Patel SC, Burns NZ (2022) Conversion of aryl azides to aminopyridines. J Am Chem Soc 144: 17797–17802
91. Sato S (1968) Azirines. IV. The photolysis of β-azidovinyl ketones. Bull Chem Soc Jpn 41: 2524–2525
92. Ansari MA, Yadav D, Singh MS (2020) Visible-light-driven photocatalyst- and additive-free cross-coupling of β-ketothioamides with α-diazo 1,3-diketones: access to highly functionalized thiazolines. Chem Eur J 26:8083–8089
93. Qi X, Xu X, Park CM (2012) Facile synthesis of 2-alkyl/aryloxy-2H-azirines and their application in the synthesis of pyrroles. Chem Commun 48:3996–3998
94. Fan R, Wen Y, Thamattoor DM (2017) Photochemical generation and trapping of 3-oxacyclohexyne. Org Biomol Chem 15:8270–8275

95. McMahon TC, Medina JM, Yang YF, Simmons BJ, Houk KN, Garg NK (2015) Generation and regioselective trapping of a 3,4-piperidyne for the synthesis of functionalized heterocycles. J Am Chem Soc 137:4082–4085
96. Grimm JB, English BP, Choi H, Muthusamy AK, Mehl BP, Dong P, Brown TA, Lippincott-Schwartz J, Liu Z, Lionnet T, Lavis LD (2016) Bright photoactivatable fluorophores for single-molecule imaging. Nat Methods 13:985–988
97. Belov VN, Wurm CA, Boyarskiy VP, Jakobs S, Hell SW (2010) Rhodamines NN: a novel class of caged fluorescent dyes. Angew Chem Int Ed 49:3520–3523
98. Halabi EA, Thiel Z, Trapp N, Pinotsi D, Rivera-Fuentes P (2017) A photoactivatable probe for super-resolution imaging of enzymatic activity in live cells. J Am Chem Soc 139:13200–13207
99. Luo K, Zhao Y, Tang Z, Li W, Lin J, Jin Y (2022) Visible-light-induced dual $C(sp^3)$-H bond functionalization of tertiary amine *via* hydrogen transfer to carbene and subsequent cycloaddition. Org Lett 24:6335–6340

Heterocycles from Onium Ylides

Claire Empel and Rene M. Koenigs

Contents

Abstract Ylide intermediates are important in a variety of rearrangement reactions. In this chapter, we discuss recent advances in how ylide intermediates, derived from the reactions of diazoalkanes or triazoles as carbene precursors, react in the formation of heterocycles. We discuss the reaction of ammonium, oxonium, sulfonium, and selenonium ylides, which display a divergent reactivity and are key intermediates in the de novo formation of new heterocycles but also in the ring-expansion reaction of heterocycles.

Keywords Carbene · Copper · Diazo compounds · Rearrangement · Rhodium · Triazole · Ylides

C. Empel and R. M. Koenigs (✉)
Institute of Organic Chemistry, RWTH Aachen University, Aachen, Germany
e-mail: claire.empel@rwth-aachen.de; rene.koenigs@rwth-aachen.de

1 Introduction

Onium ylides are species bearing a positive and a negative charge on adjacent atoms, where one of the two atoms is a heteroatom. Depending on the nature of this heteroatom, the ylide is described as an oxonium, ammonium, or sulfonium ylide, etc. Such ylides are highly valuable reaction intermediates and contribute to a vast array of important organic transformation – the most important one being the Wittig olefination that proceeds via a phosphonium ylide (Fig. 1a).

Ylides can be accessed through a classic nucleophilic substitution reaction using alkyl halides and an appropriate heteroatom nucleophile to form an onium salt, which necessitates a subsequent deprotonation reaction to form the pivotal ylide reagent (Fig. 1b). As a consequence, functional group tolerance under basic conditions is limited, which may impact applications in organic synthesis. A more recent and alternative strategy circumvents such multi-step procedures and gives direct access to the ylide reagent via reaction of the heteroatom nucleophile with an electrophilic carbene intermediate. The latter can be accessed from a variety of different precursors, such as diazoalkanes, hydrazones, or triazoles in the presence of suitable transition metal catalysts or under photochemical conditions (Fig. 1c).

The reaction of onium ylides is strongly dependent on the nature of the positively charged heteroatom forming the onium ylide and on the substitution pattern of the negatively charged carbon atom. While phosphonium ylides are well-known for their reactivity in Wittig olefination (Fig. 2a), the corresponding ylides based on oxygen, nitrogen, or sulfur undergo distinct transformations and preferentially undergo a pericyclic reaction, or more specifically a sigmatropic rearrangement

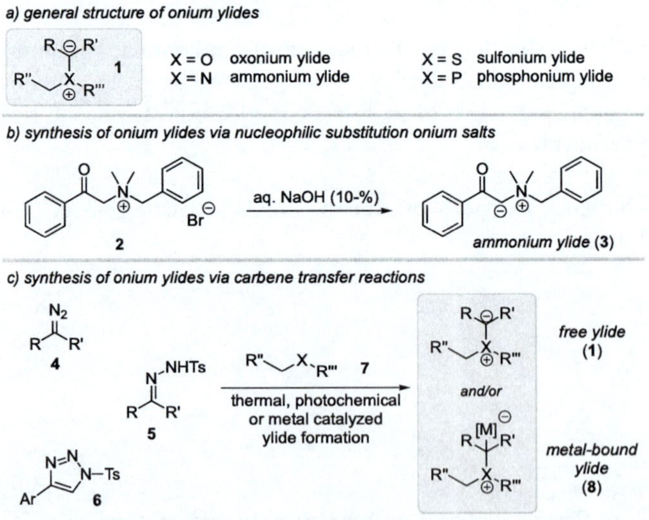

Fig. 1 Onium ylides; (**a**) types of onium ylides; (**b**) synthesis of ylides via nucleophilic substitution; (**c**) synthesis of ylides via carbene transfer reactions

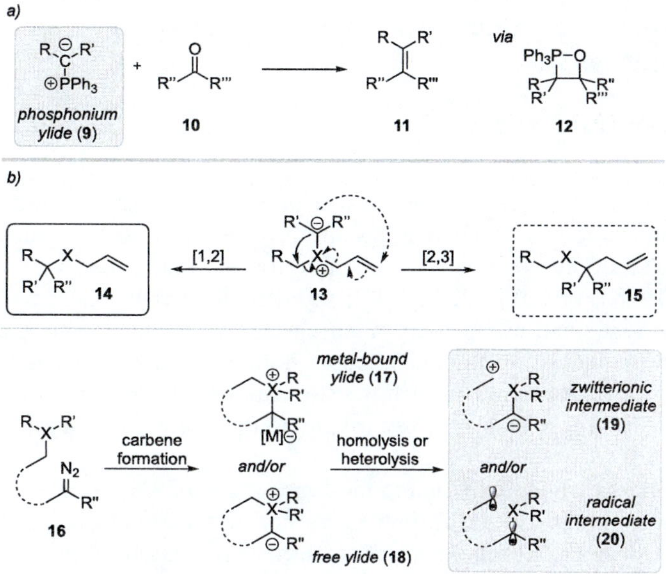

Fig. 2 Reactions of onium ylides; (**a**) phosphonium ylide and Wittig reaction; (**b**) rearrangement reactions of onium ylides via carbene transfer reactions

[1]. In this case, a strong dependency on the carbon substituents is observed and either [1,2]- or [2,3]-sigmatropic rearrangement reactions may readily occur, and the reaction mechanism may involve charged (2 electrons) or radical (1 electron) pathways (Fig. 2b). When free carbenes are employed, e.g. by photolysis of a diazoalkane or diazirine, such ylides are accessed directly and the sigmatropic rearrangement only depends on the nature of substituents. However, the intricacies of such carbene-initiated rearrangement reactions are further complicated when metal catalysts are employed. In this case, an intermediate electrophilic metal-carbene complex first undergoes addition with the heteroatom nucleophile to form either a metal-bound ylide or a free ylide intermediate [2, 3]. Latest studies employing computational methods show that there is a delicate interplay between the nucleophilic properties of the ylide and the electrophilic nature of the metal catalyst (Fig. 2b) [3, 4].

In the past years, significant advances were achieved on applications of sigmatropic rearrangement reactions via onium ylide intermediates. For a concise overview on these advances, the reader is referred to recent review articles covering this area [1, 5–7]. This chapter will emphasize applications of sigmatropic rearrangement reactions via carbene transfer reactions for the construction of heterocycles.

2 Heterocycle Synthesis via Onium Ylides from Diazoalkanes

2.1 Ammonium Ylides

The reaction of ammonium ylides is one of the best-studied examples for the synthesis of heterocycles and requires the presence of a tertiary amine for the formation of the ylide intermediate. Building on the initial discovery by Stevens et al. from 1928 (Fig. 3a) [8], a key [1,2]-sigmatropic rearrangement can be used to rapidly construct small nitrogen-containing heterocycles via carbene transfer reactions to basic amines. In a subsequent report, Bamford and Stevens later showed the applicability of diazoalkanes to such transformations [9]. This approach is of high interest as *N*-heterocycles play a key role in a large variety of natural products or drugs (Fig. 3b).

An important early contribution to the construction of *N*-heterocycles was made by West and Naidu in 1993 [10]. In the presence of a rhodium catalyst, the authors showed that diazoketones undergo intramolecular formation of cyclic ammonium ylide **28**. The latter undergoes a subsequent [1,2]-sigmatropic rearrangement reaction to form saturated *N*-heterocycles. The length of the tether between the diazo functional group and the nucleophilic amine directly impacted the ring size of the newly formed *N*-heterocycles, and either pyrrolidines or piperidines were formed (Fig. 4a). The limitation of this work lies within the migratory group, which is

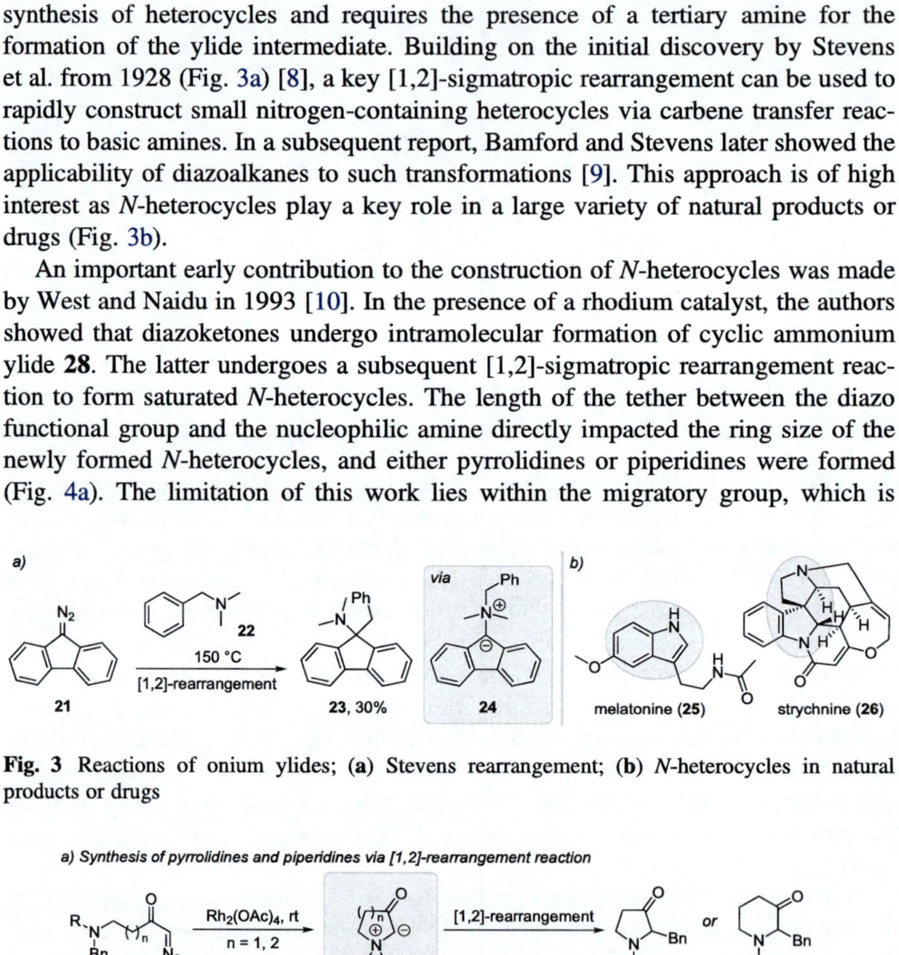

Fig. 3 Reactions of onium ylides; (**a**) Stevens rearrangement; (**b**) *N*-heterocycles in natural products or drugs

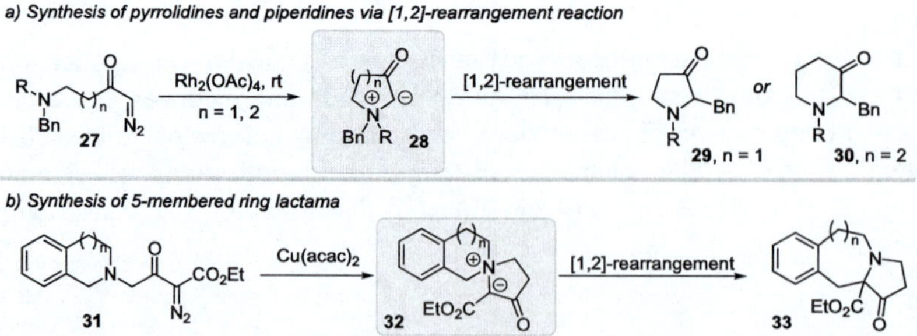

Fig. 4 Reactions of ammonium ylides; (**a**) Synthesis of pyrrolidines and piperidines via rearrangement of benzyl-substituted ammonium ylides; (**b**) Synthesis of 5-membered ring lactams

limited to benzylic substituents. Further important advances in this research area were reported by Padwa and co-workers, who could show the selective formation of an ammonium ylide via rhodium-catalyzed reaction of diazomalonate derivative **31**, followed by the sigmatropic rearrangement to give 5-membered ring lactam **33** (Fig. 4b) [11–13].

McMills and co-workers demonstrated an expansion to allyl amines (aka 2-vinyl pyrrolidines), which form a spirocyclic ammonium ylide intermediate. The small steric demand of the vinyl group, however, results in two diastereomorphic transition states that lead to two diastereoisomers of the key spirocyclic ammonium ylide intermediate (**35** vs. **36**). These differ in the spatial orientations of the vinyl group, which then leads to either [1,2]- or [2,3]-sigmatropic rearrangement reaction and the formation a mixture of two products (**37**, **38**), of which azacyclooctene **38** is formed via a [2,3]-sigmatropic rearrangement as the major product (Fig. 5a) [14].

This concept was further elaborated in a series of subsequent reports by the West and Saba groups [15, 16]. Of particular importance is the migratory aptitude and steric demand of the substituents of the ammonium ylide intermediate to achieve a chemoselective ylide formation and rearrangement reaction. For this purpose, ester and silicon-based groups were investigated, which gave a significant improvement in chemoselectivity (Fig. 5b). In both cases, the [1,2]-sigmatropic rearrangement was observed as the predominating reaction pathway. The ester or silicon group influenced the formation of the ammonium ylide in such a way, that only a single spirocyclic ylide can be formed. This is considered to be a result of the orientation of the metal carbenoid and caused by the pyramidal geometry of the nitrogen nucleophile, which outmaneuver the effect of steric repulsion. This adapted strategy could then be used in the synthesis of a large variety of different bicyclic, saturated *N*-heterocycles such as quinolizidines, indolizidines, and pyrrolizidine. Moreover, several naturally occurring alkaloids, such as (−)-lupinine, (±)-turneforcidine, or (±)-platynecine could be accessed via this method (Fig. 5c).

While early examples most commonly discuss the reaction of acceptor diazoalkanes, the reactivity of donor/acceptor diazoalkanes only recently attracted interest. In 2001, Padwa and co-workers reported the reaction of heterocycle-substituted donor/acceptor diazoalkane **58**. In the presence of a rhodium catalyst, intermediate spirocyclic ammonium ylide **59** is formed that undergoes further rearrangement to isoindolobenzazepines (Fig. 6a) [17]. More recently, Lacour and co-workers reported a detailed study on the rhodium-catalyzed intermolecular reaction of simple and readily available donor/acceptor diazoalkanes with Tröger's base (**61**) [16]. In this case, an intriguing ring expansion via a [1,2]-sigmatropic rearrangement occurred in high yield and excellent diastereoselectivity. An important observation on substituent effects of donor/acceptor diazoalkanes was made by the authors when using enantiomerically pure Tröger's base. With electron-withdrawing substituents on the aryl ring of the aryldiazoacetate, a reduced enantiomeric excess was observed in the reaction product, while a high enantiomeric excess was observed for electron-donating substituents. This observation may be reasoned by stabilization of radical or zwitterionic intermediates (**64** vs. **65**) with electron-withdrawing substituents, which in turn may lead to isomerization. The

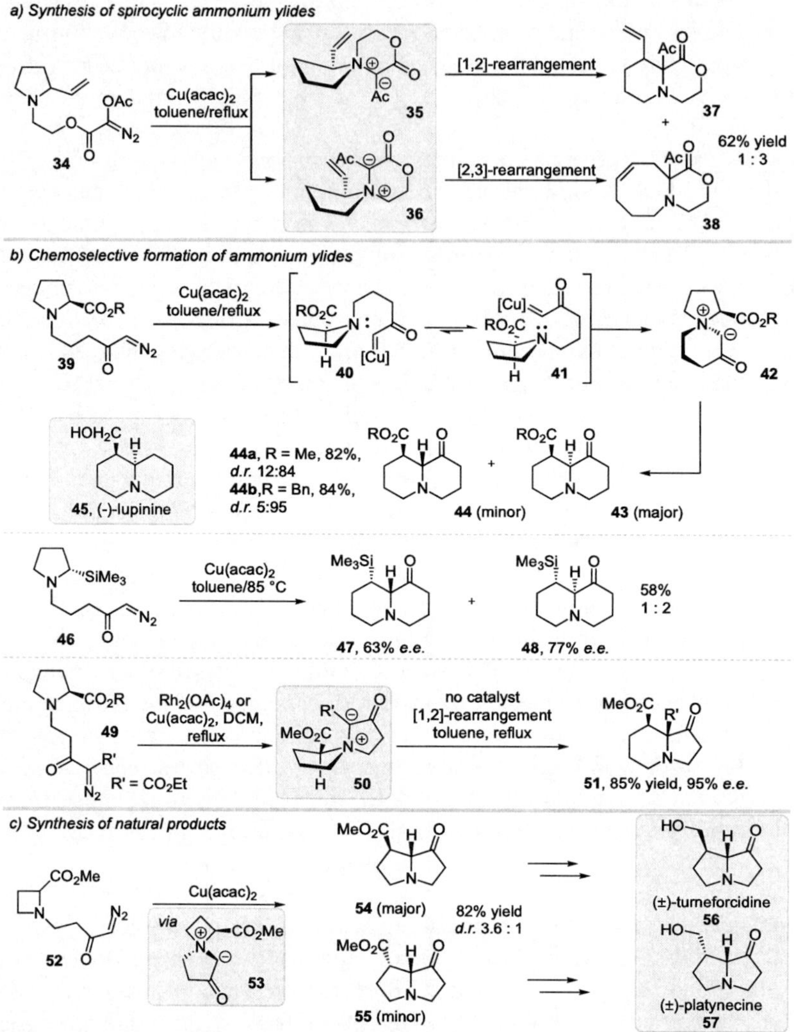

Fig. 5 Reactions of spirocyclic ammonium ylides; (**a**) Approaches toward spirocyclic ammonium ylides and limitations; (**b**) Chemoselective formation of ammonium ylides and heterocycle synthesis; (**c**) Natural products obtained by this approach

involvement of either an ionic or radical reaction pathway is an intricate detail for the mechanism of [1,2]-sigmatropic rearrangement reactions that requires further studies, e.g. by computational studies (Fig. 6b).

The formation of aziridinium ylides opens up further opportunities in the formation of heterocycles. Here, the ring strain of the three-membered aziridine ring allows for an intriguing rearrangement reaction, where medium-sized heterocycles are formed. Specifically, Schomaker and co-workers demonstrated the

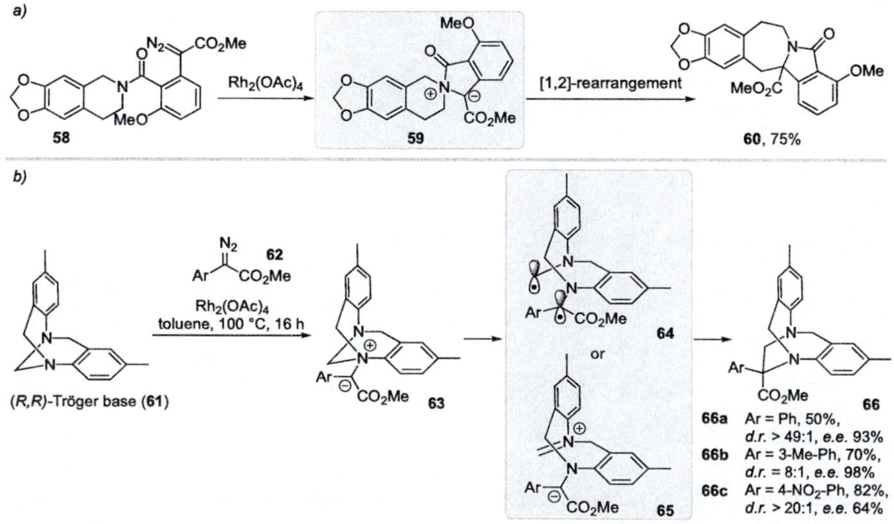

Fig. 6 Reaction of donor/acceptor diazoalkanes in ammonium ylide formation and ring-expansion reactions; (**a**) Synthesis of isoindolobenzazepines; (**b**) Ring-expansion reactions of Tröger's base

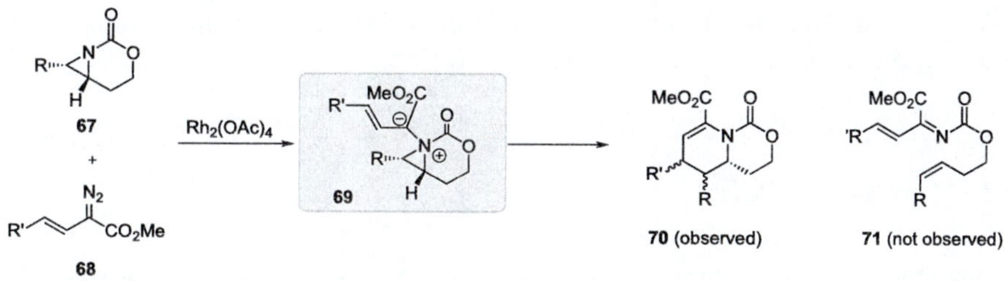

Fig. 7 Reaction of vinyl-substituted diazoester in the rhodium-catalyzed formation of aziridinium ylides and subsequent rearrangement reaction

rhodium-catalyzed reaction of vinyl diazoester **68** with aziridine **67** to give an aziridinium ylide. The latter can react in a sigmatropic rearrangement reaction with concomitant ring opening of the aziridine ring. The nature of the ylide intermediate here dictates the ring size and formation of 6-membered ring heterocycles via an intermolecular [3 + 3] ring-expansion reaction. Such important aziridinium ylides were more extensively studied by Schomaker and co-workers in a variety of further transformations and represent a fascinating starting point for a wide range of ylide-mediated transformations (Fig. 7) [18, 19].

In a recent study, Shaw and co-workers reported a significant advancement into the direction of donor/donor diazoalkanes. Such donor/donor diazoalkanes are more demanding in synthesis and often lack stability during their synthesis, work-up, or storage. Such diazoalkanes are therefore often accessed in situ, e.g. by oxidation of a hydrazone precursor. In this context, Shaw and co-workers described the synthesis

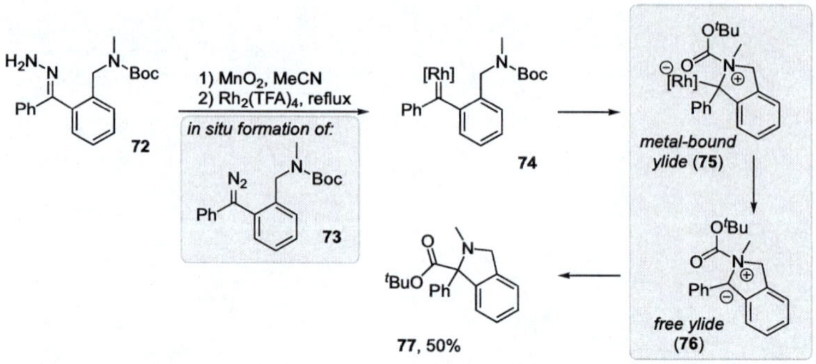

Fig. 8 Reaction of an in situ generated donor/donor diazoalkane in an intramolecular ammonium ylide-rearrangement reaction

of isoindoline via an intramolecular carbene transfer reaction and rearrangement of an intermediate ammonium ylide. In situ oxidation of hydrazone **72** with MnO_2 provided the desired donor/donor diazoalkane **73**, which furnished the key ammonium ylide in the presence of a rhodium catalyst. Final [1,2]-sigmatropic rearrangement gave the isoindoline products in high yield. DFT studies on the reaction point at the formation of free ylide intermediate **76**, which further reacts in the [1,2]-sigmatropic rearrangement (Fig. 8) [20].

2.2 *Oxonium Ylides*

In this chapter section, we focus on the reaction of oxonium ylides, which are key intermediates for the synthesis of oxygen-containing heterocycles. The latter are of high relevance in natural products and drug discovery. Similar to ammonium ylides, ring formation reactions can be achieved using acyclic substrates bearing both carbene precursor and an ether nucleophile in one carbon chain. As an alternative, ring-expansion reactions can be achieved in both intra- and intermolecular fashion, if cyclic ethers are employed. The reaction of oxonium ylides received significant attention in the past years, and the focus of this section will be on the latest advances. For a concise overview on the development of oxonium ylide-mediated reactions, the reader is referred to recent review articles [7].

Two early examples of such reactions are briefly highlighted here and foreshadow future developments. Hillman reported one of the first examples of an intermolecular ring-expansion reaction via thermal decomposition of ethyl diazoacetate (**78**) in the presence of dioxolane **80** at 150 °C. Under these conditions, a carbene is formed initially, which undergoes oxonium ylide formation with the dioxolane followed by ring expansion to give 1,4-dioxane product **81** (Fig. 9a) [21]. The intramolecular formation of oxygen heterocycles was unveiled by West and co-workers only in the 1990s and is a much more demanding reaction due to competing side reactions. In a

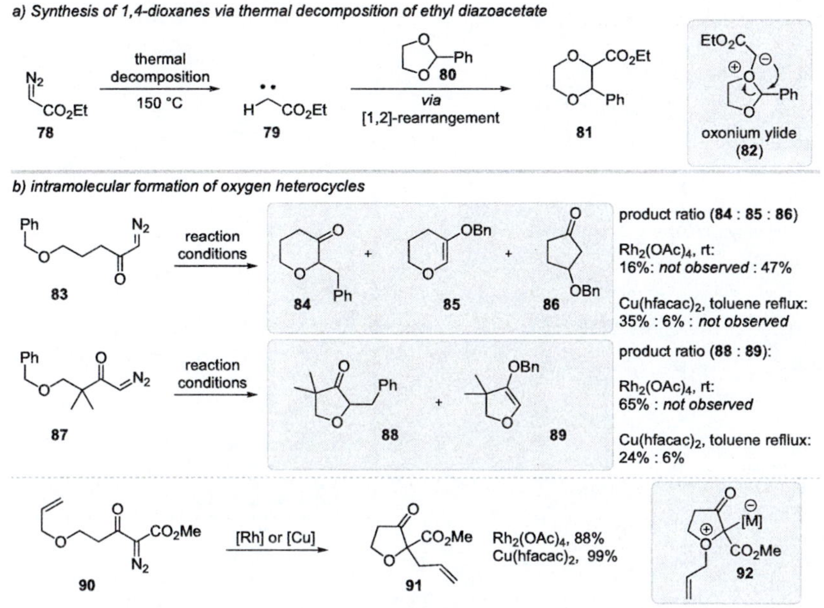

Fig. 9 Early examples on the synthesis of oxygen heterocycles via ylide intermediates; (**a**) Intermolecular ring expansion via thermal decomposition of ethyl diazoacetate; (**b**) Intramolecular cyclization reactions

first report, West and co-workers could show the ring formation reaction using ether-tethered diazoketones. This reaction was strongly dependent on the catalyst and on substituents along the tether, which hampered further development of this transformation. In contrast, derivatives of acceptor/acceptor diazoalkanes proved much more versatile and gave the desired rearrangement product in good yield and selectivity (Fig. 9b) [22, 23].

More recently, Doyle and co-workers reported the rhodium-catalyzed reaction of elaborate tetrahydropyranone-derived diazo compounds. These are derivates of diazo acetoacetates and as such are representatives of acceptor/acceptor diazoalkanes. In the presence of a rhodium catalyst, an intramolecular oxonium ylide formation occurs, which then undergoes [1,2]-sigmatropic rearrangement to give oxabicyclo[4.2.1]nonanes (Fig. 10a) [24]. While this first study focused on the influence of aryl-substituted pyranone derivatives **97**, a second study describes the reaction of related vinyl-substituted pyranones. In this case however, both [1,2]- and [2,3]-sigmatropic rearrangement occurs and leads to an unselective reaction process (Fig. 10b).

Recent progress on onium ylide-mediated heterocycle formation ring reactions mainly focused on ring-expansion reactions. In this context, the ring expansion of small 4-membered ring heterocycles received significant extension. Initially discovered by Noyori in 1966 [25] this reaction received some attention in the meantime (Fig. 11a) [26–28].

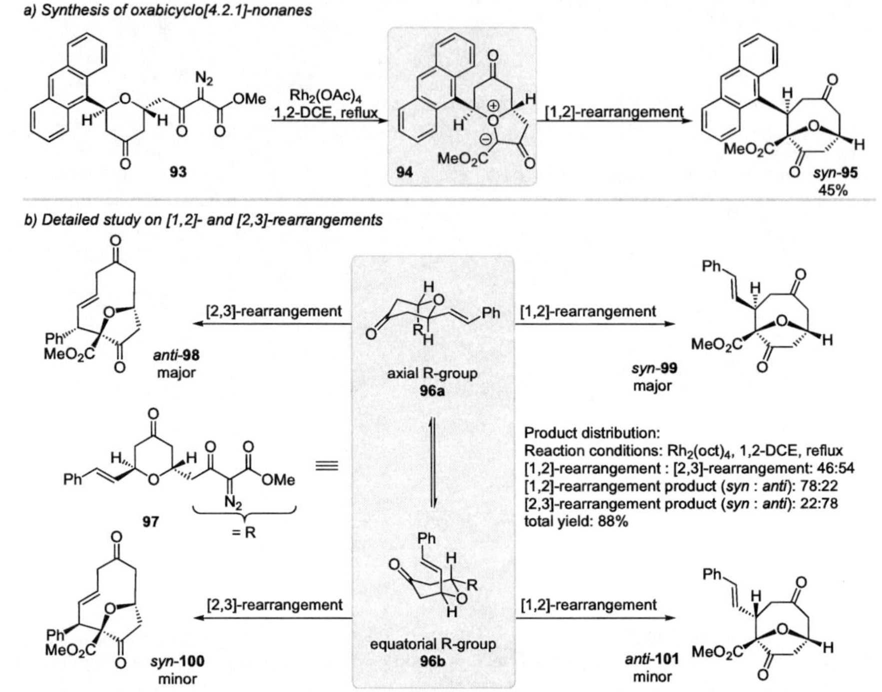

Fig. 10 Synthesis of oxabicyclo[4.2.1]-nonanes; **(a)** Studies on the influence of the aryl-substituted pyranone derivative; **(b)** Reaction of vinyl-substituted pyranone derivatives

A concise study on the general reactivity of oxonium ylides in ring-expansion reactions was reported by Njardarson and co-workers in 2011. The authors studied the copper-catalyzed reaction of ethyl diazoacetate **(78)** with oxygen heterocycles bearing a pendant vinyl substituent. This design now allows comparative studies on two competing pathways. While a [1,2]-sigmatropic rearrangement reaction leads to ring expansion of the oxygen heterocycle, a [2,3]-sigmatropic rearrangement involves the participation of the pendant vinyl group and leads to a ring expansion by three carbon atoms. The oxetane heterocycle underwent preferential ring expansion to the tetrahydrofuran, while the larger furan ring underwent preferential [2,3]-sigmatropic rearrangement. In this context, the authors also studied symmetric ethers bearing two substituted allyl groups. In this case, a [2,3]-sigmatropic rearrangement followed by ring-closing metathesis could be employed to furnish 2,3-disubstituted pyrans in a highly diastereoselective fashion (Fig. 11b) [29].

However, a concise study on this reaction was published only recently by Koenigs, Xu, and co-workers. In this work, the authors employed the recently emerging strategy of catalyst-free, photochemical carbene transfer reactions [30–32] on the ring-expansion reaction of 4-membered ring heterocycles [33]. While oxetanes underwent ring expansion to tetrahydrofurans, the related thietane gave a thiolane product. In both cases, the ring expansion occurred with high

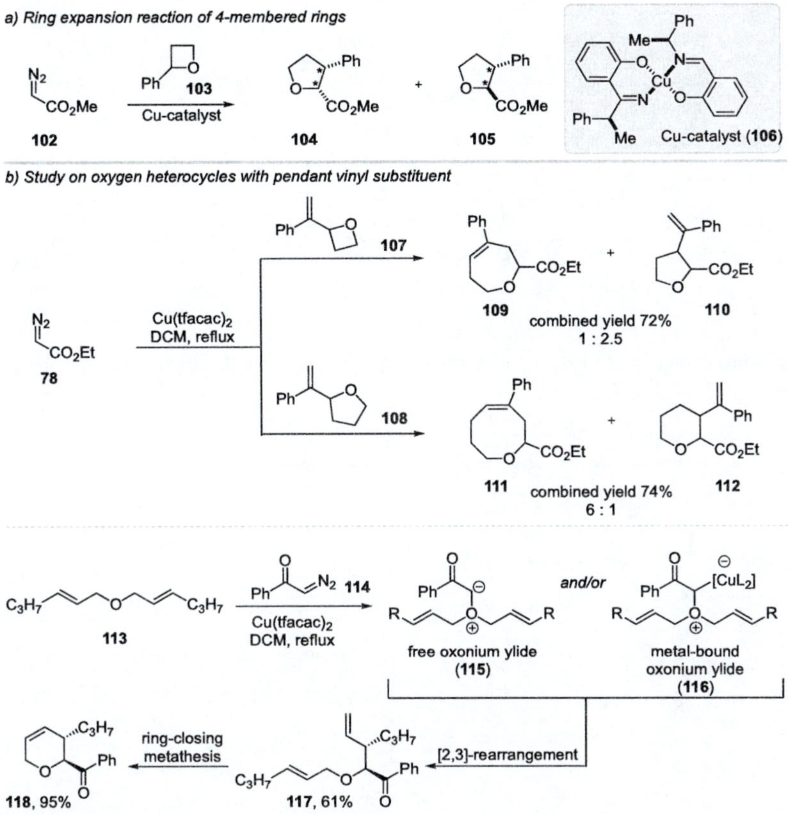

Fig. 11 Ring expansion of oxetanes; (**a**) Early studies by Noyori and co-workers; (**b**) Studies on ring-expansion reaction of heterocycles with pendant vinyl substituent

regioselectivity and only one ring expansion product was observed. However, when diastereoselective ring-expansion reactions were studied, a highly diastereoselective reaction was only obtained in the case of oxetanes, while thietanes gave low diastereoselectivity. Computational studies suggest that in both cases, a radical ring-expansion reaction is at play and that the observed differences in diastereoselectivity can be rationalized by longer bond lengths (1.86 Å and 1.76 Å for sulfur vs. 1.35 Å and 1.45 Å for oxygen) in the case of thietane, which leads to little face differentiation and, consequently, low diastereoselectivity in the ring closing step (Fig. 12).

Previously described reactions build on the ring expansion of 4-membered to 5-membered ring heterocycles. An important discovery on the ring expansion of THF was made in 2014 by Kitamura and co-workers when studying the reactions of diazo naphthalene derivatives [34]. These diazo compounds undergo a palladium-catalyzed reaction with one or two molecules of THF via an intermediate oxonium ylide, which then reacts in a rearrangement reaction to form either 8-membered or

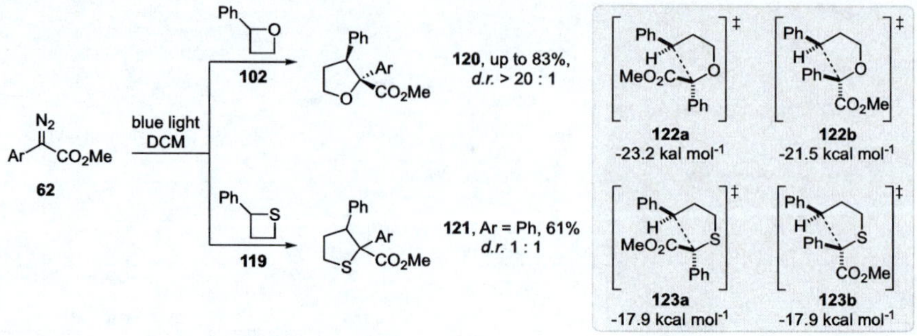

Fig. 12 Ring expansion of oxetane and thietane heterocycles via photochemical carbene transfer reactions

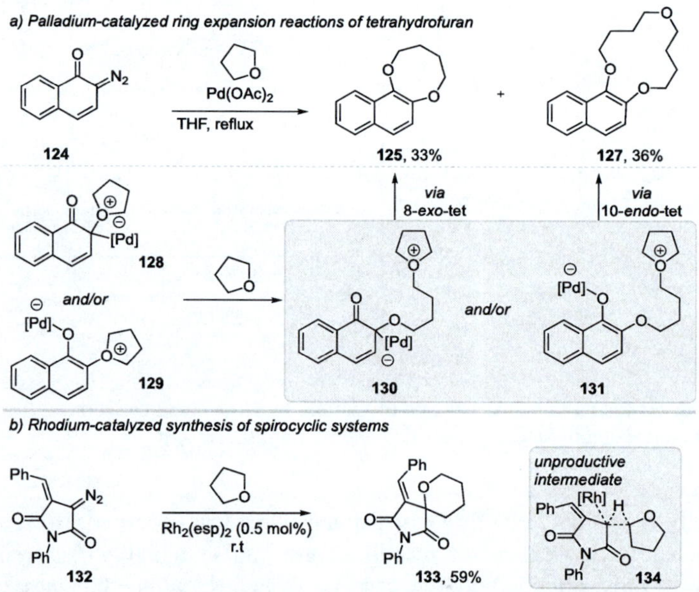

Fig. 13 Ring expansion of tetrahydrofuran; (**a**) palladium-catalyzed reaction with diazo naphthalenes; (**b**) rhodium-catalyzed reaction with cyclic diazo carbonyl compounds

13-membered oxygen-containing macrocycles, respectively (Fig. 13a). Tetrahydropyran reacts in an analogous fashion to deliver the ring enlarged products. This reactivity of THF was further studied by Dar'in, Krasavin, and co-workers using cyclic diazo carbonyl compounds [35]. In this case, a rhodium catalyst was employed and, instead of macrocycle formation, spirocycle **133** was formed (Fig. 13b).

2.3 Sulfonium Ylides

The reaction of sulfur-containing heterocycles in onium ylide-mediated reactions closely resembles the reaction of oxygen-containing heterocycles. As detailed in the previous section, one of the key differences lies within the bond lengths of C-O and C-S bond with the latter being around 0.4–0.5 Å longer as shown in a recent study by Koenigs, Xu, and co-workers [33]. An important initial application of sulfonium ylides in ring-expansion reactions was reported already in 1984 by Doyle and co-workers. 1,3-Dithianes underwent a regioselective ring-expansion reaction with ethyl diazoacetate (**78**) in the presence of rhodium acetate as the catalyst (Fig. 14) [36]. Building on this work, many groups reported on their studies on related reactions in the years after on which a comprehensive account may be found in recent review articles [5, 7].

In the context of recent work, such sulfonium ylide-mediated rearrangement reactions found important applications in total synthesis (Fig. 15). Zakarin and co-workers studied the reaction of thietane heterocycles with vinyl-substituted diazo compounds as a strategic entry for the synthesis of symmetric Nuphar thioalkaloids, yet low enantioinduction and diastereoselectivity outline clear limitations of processes based on sulfonium ylides [37]. Furnier, Zakarian, and co-workers further investigated such ring-expansion reaction in the total synthesis of unsymmetric Nuphar thioalkaloids [38]. Their strategy involves two advanced

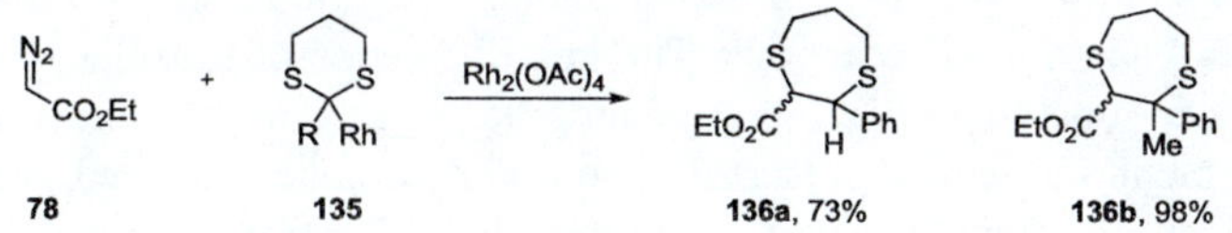

Fig. 14 Ring expansion of 1,3-dithianes

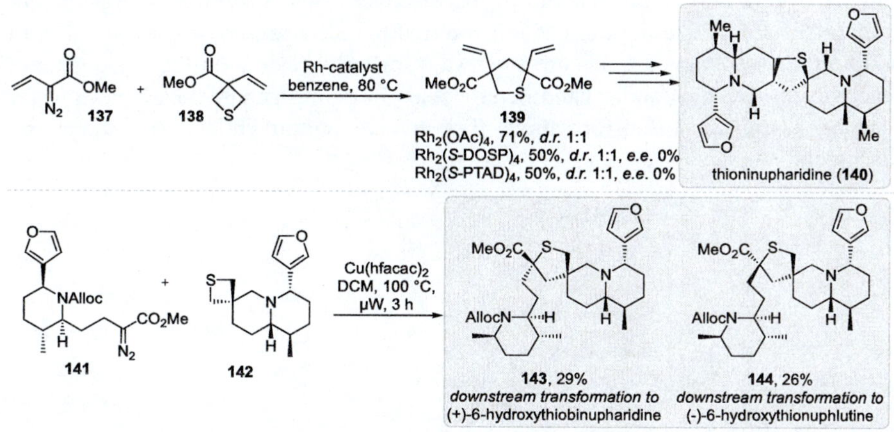

Fig. 15 Applications of sulfonium ylides in the total synthesis of Nuphar thioalkaloids

building blocks, one bearing a diazo functional group (**141**), the other bearing a thietane heterocycle (**142**). These two fragments are coupled in a copper-catalyzed ring-expansion reaction of the thietane heterocycle via a sulfonium ylide intermediate. This example showcases the virtue of sulfonium ylides in synthesis. The high nucleophilicity of the sulfur atom can be strategically employed in the presence of multiple other functional groups to selectively form a sulfonium ylide with a carbene intermediate.

3 Heterocycle Synthesis via Onium Ylides from Triazoles

Diazo compounds are most commonly employed in carbene transfer reactions, and the past decades witnessed significant strategic advances in accessing carbenes from diazo compounds and controlling their reactivity in carbene transfer reactions. Despite these advances, diazo compounds are often associated with safety risks in organic synthesis that led to research on substituting these reagents in organic synthesis. However, it is worth highlighting that a recent study on the safety of diazoalkanes showed only very few selected examples of diazoalkanes that pose serious risk hazards [39].

In this context, the triazole motif (**145**) has emerged in recent years as a powerful surrogate to diazoalkanes. Triazoles undergo Dimroth rearrangement at elevated reaction temperatures to expose a transient diazo functional group (**146**) that can be used in carbene transfer reactions. The key difference of triazoles to commonly employed ester-substituted diazo compounds lies in the substitution pattern of the carbene fragment. In the case of triazoles, an imino-carbene is formed, which can be used as a linchpin reagent to enable denitrogenative 1,3-difunctionalization reactions of the parent triazole heterocycle. The involvement of an imine functional group renders this approach particularly attractive in heterocycle synthesis (Fig. 16). For a concise overview on steps leading to the discovery of the reaction of triazoles in carbene transfer reactions, the reader is referred to recent review articles [40, 41]. In addition, these review articles are suggested for an overview on the more general reactivity of triazoles and related heterocycles in denitrogenative carbene transfer reactions; here, only transformations that involve onium ylide intermediates are discussed.

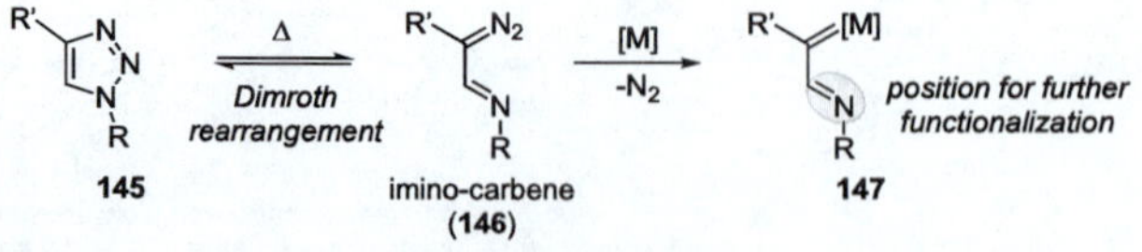

Fig. 16 Triazoles in thermal Dimroth rearrangement reactions, followed by the formation of a metal-carbene complex

3.1 Nitrogen-Based Ylides

In their seminal publication, Fokin and Gevorgyan reported in 2008 an initial application of triazoles in denitrogenative carbene transfer reactions [42]. At elevated reaction temperatures and in the presence of a rhodium catalyst, a rapid reaction with organic nitriles was observed in chloroform solvent to give a variety of imidazole heterocycles with a high tolerance toward functional groups. This reaction proceeds via initial thermal Dimroth rearrangement to expose an imino-substituted diazo compound followed by formation of rhodium-carbene complex **151**. Nucleophilic addition of the nitrile furnishes a nitrilium ylide, which upon cyclization and release of the rhodium catalyst delivers imidazole heterocycle **150** (Fig. 17). Further research into the reaction of triazoles with nitriles has involved the use of cyanogen bromide [43], *N*-cyano sulfoximines [44], or perfluoroalkyl-substituted triazoles [45] to deliver key building blocks for coupling chemistry or drug research. Moreover, Lewis acid catalysts such as $BF_3 \cdot OEt_2$ proved compatible to conduct such heterocycle synthesis via carbene transfer reactions and ylide formation from triazole precursors [46].

In further applications, this approach was extended from nitriles as nitrogen-based nucleophiles toward a whole plethora of different *N*-nucleophiles. Imines can be considered close analogs of nitriles. In 2013, Fokin and Zibinsky showed that this concept could be further extended to aldimines [47]. In this case, an onium ylide is formed through the reaction of an intermediate rhodium-carbene complex with the imine. Subsequent cyclization followed by elimination of sulfinic acid under basic conditions furnished the imidazole heterocycle.

An intriguing development in ylide-mediated transformations was made when employing azirines as substrates. These strained 3-membered azacycles can undergo a related ylide formation with metal-bound imino-carbenes. In this case, however, an ylide bearing a pendant strained ring system is formed and the latter can readily ring open to a plethora of downstream transformations. In this context, Lee and co-workers reported on the reaction of triazoles with such azirines, where the reaction product is pyrazine heterocycle **156** [48]. Tang and Shi further investigated this transformation and showed that this reaction could be further employed to manipulate downstream reaction pathways, and that either pyrazines **157** or pyrroles

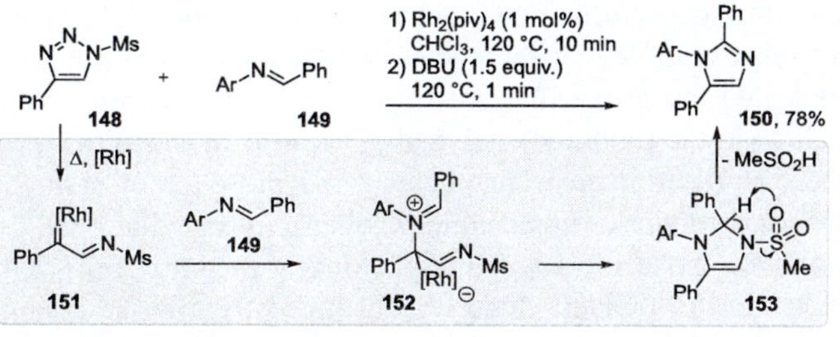

Fig. 17 Triazoles in the synthesis of imidazole heterocycles

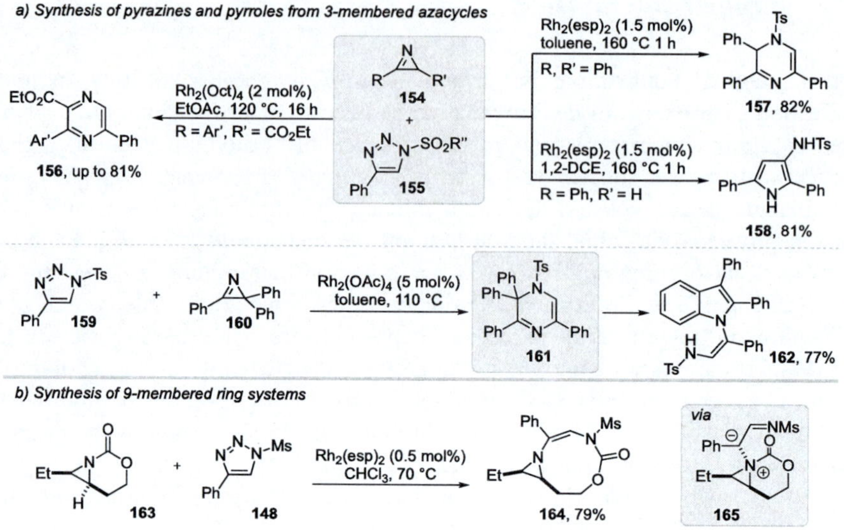

Fig. 18 Triazoles in the reaction with 3-membered azacycles; (**a**) Synthesis of pyrazine, pyrrole, or indole heterocycles; (**b**) Ring-expansion reaction

158 can be obtained as reaction products [49, 50]. Novikov and co-workers showed that the pyrazine heterocycle can further undergo ring-opening and another cyclization step to form indole heterocycle **162** (Fig. 18a) [51].

The aziridine is a heterocycle closely related to azirines. As outlined in the previous section, Schomaker and co-workers recently described the formation of aziridinium ylides via reaction of strained aziridine heterocycles with a carbene fragment [52]. In the context of reactions with triazole heterocycles, a distinctive reaction outcome is observed. The authors showed that bicyclic aziridine **163** reacts under rhodium-catalyzed conditions at elevated temperature with triazoles to form imino-substituted aziridinium ylide **165**. The latter then reacts in a migration reaction of the *N*-protecting group to the nitrogen atom of the imino-carbene fragment. This migration leads to formation of a new 9-membered ring. This reaction pathway resembles the 1,3-difunctionalization reaction of triazoles with a range of different amphiphilic reaction partners (Fig. 18b) [41].

In the context of ylide formation reactions with heterocycles, important contributions on the reaction of *N*,*O*-heterocycles were made recently. In 2015, Tang and co-workers revealed that isoxazoles are suitable reagents to form a nitrogen-based ylide intermediate in the rhodium-catalyzed reaction with triazoles. As in the previous cases, thermal Dimroth rearrangement in the presence of a rhodium catalyst initially furnishes a rhodium-bound imino-carbene intermediate, which undergoes ylide formation with isoxazole **167**. Ring opening of the latter followed by cyclization and proton transfer delivers densely substituted pyrroles in a single synthetic step (Fig. 19a) [53]. The reaction of related benzisoxazole **172**, however, gives rise to an imidazole reaction product, or a quinazoline, depending on the reaction

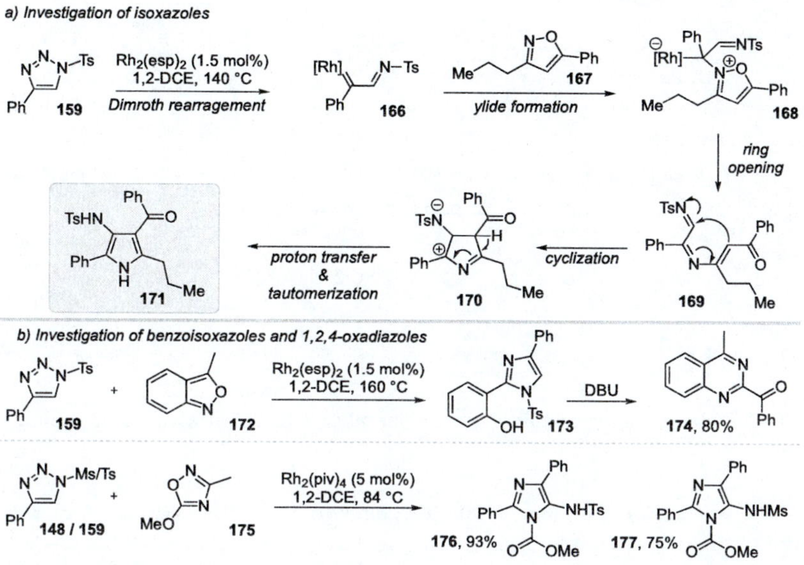

Fig. 19 Triazoles in the reaction with *N,O*-heterocycles; (**a**) Reaction with isoxazoles; (**b**) Reaction with benzisoxazoles

conditions [54]. Finally, Novikov recently disclosed that 1,2,4-oxadiazole **175** reacts in a similar fashion via formation of a nitrogen-based ylide intermediate. In this case, the ring opening of the oxadiazole heterocycle followed by cyclization furnishes an imidazole heterocycle (Fig. 19b) [55].

An intriguing feature of ylide-mediated reactions can be achieved when using pyridine and its derivatives as ylide-forming reaction partners. Yoo and co-workers reported on the reaction of pyridine with triazoles in the presence of a rhodium catalyst to furnish valuable azomethine ylide intermediate **180**. This azomethine ylide is an air-stable ylide and can further react in cycloaddition reactions, e.g. with alkynes, to give 1,4-diazepines via a [5 + 2]-cycloaddition reaction [56]. This intriguing chemistry was further developed in the reaction with a diverse set of reaction partners, such as enol diazoacetates [57], diazoesters [58], and alkynes [59, 60] to furnish a diverse set of heterocycles (Fig. 20).

Simpler aliphatic amines were studied for the formation of aromatic *N*-heterocycles in investigations of carbene transfer reactions with triazoles. These require a more subtle control on downstream reaction steps once an ylide is formed. In this context, Bao and co-workers studied 1,3,5-triazinanes **190** under rhodium-catalyzed reaction conditions [61]. In this case, following the formation of a rhodium-bound imino carbene the 1,3,5-triazinane undergoes an addition reaction to form ammonium ylide **193**. This ylide ring opens to form a heminal followed by formation of a nine-membered ring with the rhodium catalyst still bound. In a series of steps, the final reaction product, octahydro-1*H*-purine derivative **191** is formed (Fig. 21a). For

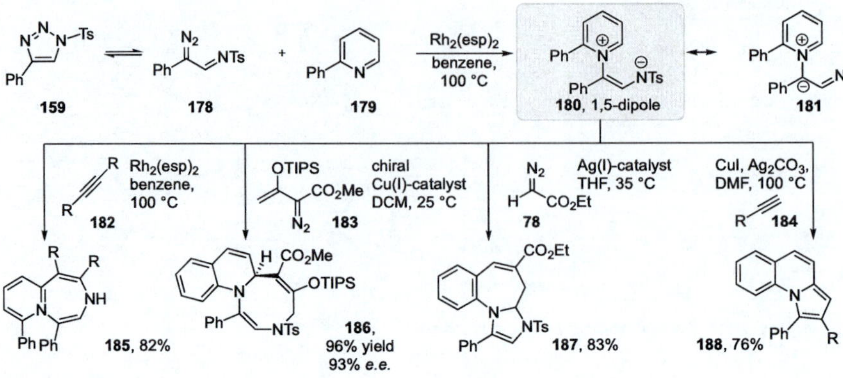

Fig. 20 Triazoles in the synthesis of azomethine ylide intermediates and their application in synthesis

a detailed discussion of the reaction mechanism and DFT studies, the reader is referred to the original literature.

The reaction of such ammonium ylides was similarly studied by Lacour and co-workers using Tröger's base (**61**). In this case, a related reaction occurs, where – upon formation of the ammonium ylide – a C-N bond is cleaved. However, in this case, a subsequent *aza*-Mannich – Friedel–Crafts – Grob fragmentation type cascade engages to provide the final reaction product [62]. More recently, Lacour and co-workers demonstrated the application of ammonium ylide-mediated transformations in the synthesis of chiral hemicyanine fluorophores [63]. Furthermore, the same authors subjected the imidazoline framework to the rhodium-catalyzed reaction with triazoles. As in the previous cases, a complex cascade reaction was observed that showcases the intriguing nature of imino-carbene intermediates in ylide formation and rearrangement reactions. Here, pyrazino-indolines or triazocines were obtained as the reaction products, depending on the reaction conditions (Fig. 21b) [64].

Another important feature of ylide intermediates lies within their ability to undergo [2,3]-sigmatropic rearrangement. This rearrangement can be selectively addressed in the presence of allyl substituents at the ylidic heteroatom and is one of the best-studied reactions of ylide intermediates via carbene transfer reactions [2]. In the context of ammonium ylides and heterocycle formation, an important recent contribution from Xu, Shen, and co-workers describes the intramolecular reaction of triazoles with *N*-allyl *N*-aryl amines. The aryl substituent prevents undesired side reactions and therefore, upon rhodium-catalyzed ylide formation, a [2,3]-sigmatropic rearrangement readily occurs to form a densely substituted pyrrolidine intermediate, which can either undergo cyclization or hydrolysis (Fig. 22a) [65]. More lately, a related reaction using benzothiazinones and triazoles as substrates gave interesting benzannelated 9-membered ring heterocycles in intermolecular reactions (Fig. 22b) [66].

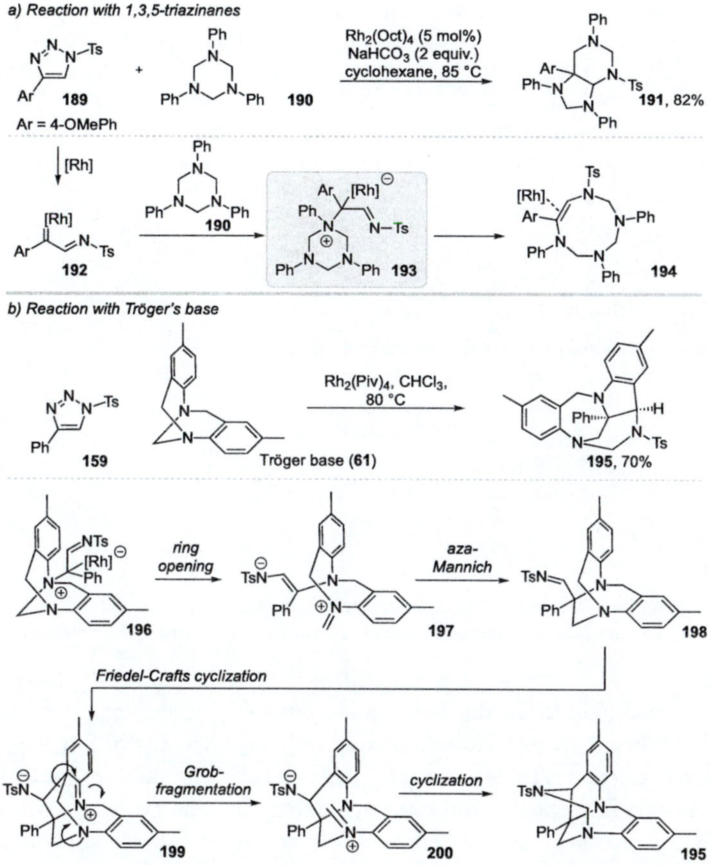

Fig. 21 Ring-expansion reactions of *N*-heterocycles; (**a**) Synthesis of octahydro-1*H*-purine derivatives; (**b**) Triazoles in the reaction with Tröger's base

3.2 Oxonium Ylides

In this part, we focus on the reaction of oxonium ylides that are obtained through carbene transfer reactions with triazoles. In a similar fashion, as in the previous subchapter, oxonium ylides derived from triazoles take advantage of the linchpin reactivity of imino-carbenes and are, therefore, excellent starting points for the synthesis of a diverse array of oxygen-containing heterocycles. The reader is also referred to recent review articles covering the whole breadth of reactions involving oxygen-nucleophiles in the reaction with triazoles [40, 41].

In their report on the reaction of triazoles with imines, Fokin and Zibinsky also described their reaction with related aldehydes [47]. The reaction proceeds in a similar fashion, via formation of a rhodium imino-carbene complex that undergoes addition with aldehydes to form an oxonium ylide. Cyclization and elimination of

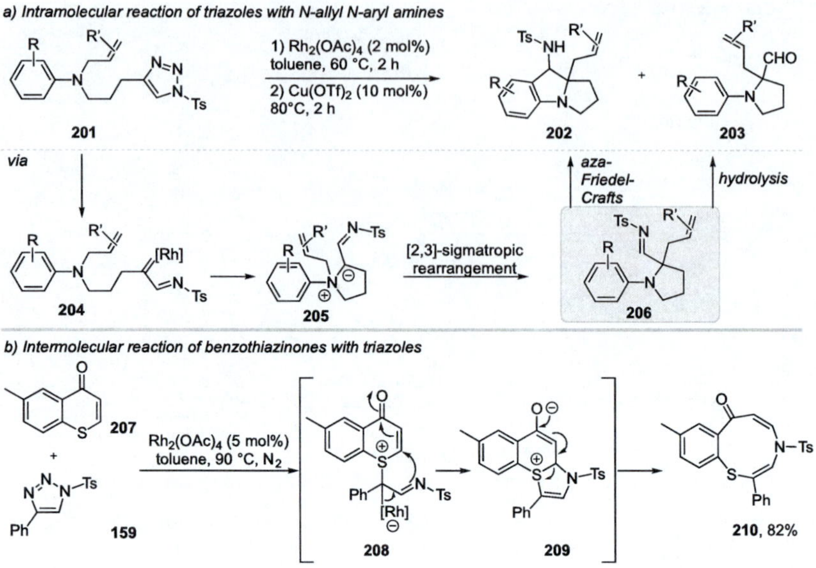

Fig. 22 Reactions with allyl-substituted ylidic heteroatoms; (**a**) intramolecular reaction of triazoles with *N*-allyl *N*-aryl amines; (**b**) Intermolecular reaction of triazoles with benzothiazinones

the rhodium complex forms the 4-oxazoline product. An aromatic divergent reactivity was observed by Miura, Murakami, and co-workers in the reaction of α,-β-unsaturated aldehydes [67]. In this case, the oxazoline product is formed initially; however, further ring opening followed by ionic cyclization gives a pyrrole reaction product. This reaction concept was further followed up in the reaction with aromatic aldehydes [68], ketoesters [69–72], unsaturated ketones [69], or in a variety of intramolecular reactions [73–75]. In all cases, the formation of an initial oxonium is key to drive the downstream reactivity and the transformation toward heterocyclic systems; depending on the substitution pattern, furans, pyrroles, or related heterocycles are obtained. Here, a noteworthy example by Gong, Yang, and co-workers describes an intramolecular cascade reaction of a triazole bearing a pendant ketone to initially give an ylide intermediate. Instead of rearrangement, this ylide was shown to react in an intramolecular dipolar cycloaddition reaction with a tethered olefin to give polycyclic oxygen heterocycles (Fig. 23) [75].

Similar to carbonyl groups, oxygen-containing heterocycles can react in ylide formation with triazoles in the presence of appropriate catalysts. In this context, Wang, Chen, and co-workers described the reaction of styrene oxides **216** with triazoles in the presence of a chiral rhodium catalyst. Upon formation of a rhodium imino-carbene complex, the epoxide undergoes nucleophilic addition to form strained oxonium ylide **218**. Intramolecular addition results in ring opening of the three-membered ring and formation of 3,4-dihydro-2*H*-1,4-oxazines (Fig. 24a) [76]. Further studies by Lee expanded the applicability of this approach to the reaction of glycidols and the formation of 6- or 7-membered ring heterocycles,

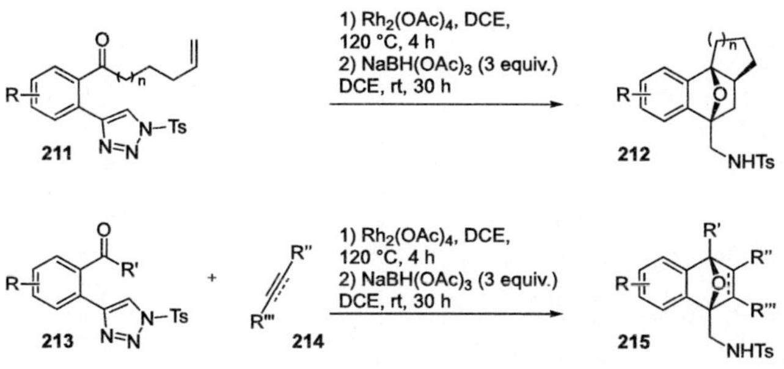

Fig. 23 Reaction of triazoles bearing a pendant carbonyl group in the formation of oxygen heterocycles

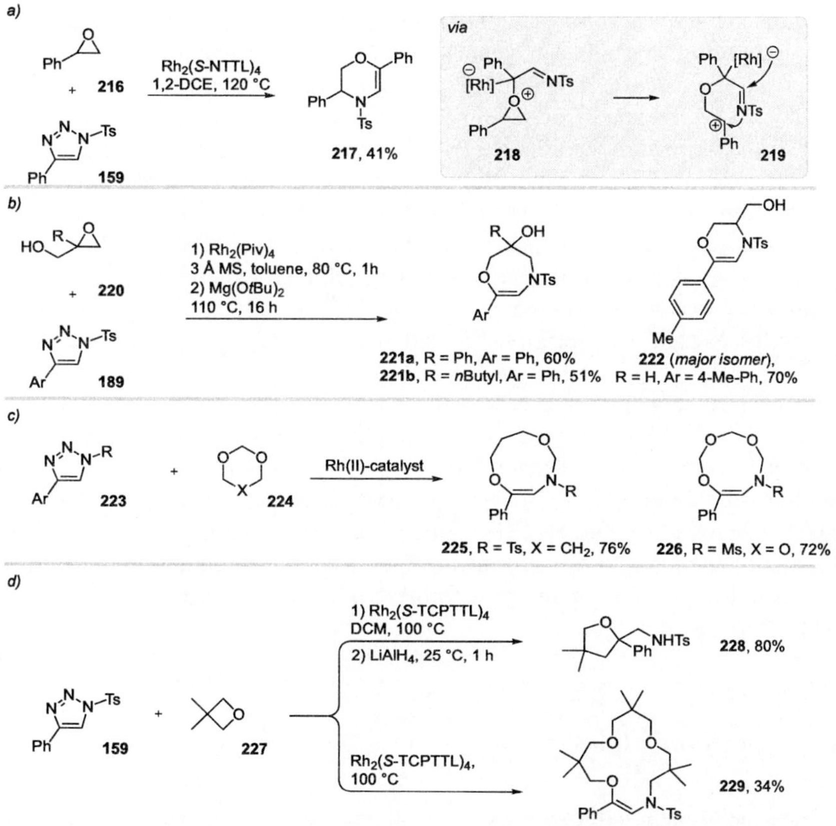

Fig. 24 Applications of triazoles in ring-expansion reactions of oxygen heterocycles; (**a**) Synthesis of 3,4-dihydro-2*H*-1,3-oxazines; (**b**) Synthesis of 6- and 7-membered ring heterocycles; (**c**) Synthesis of medium-sized oxygen heterocycles; (**d**) Synthesis of tetrahydrofurans and cyclic polyethers

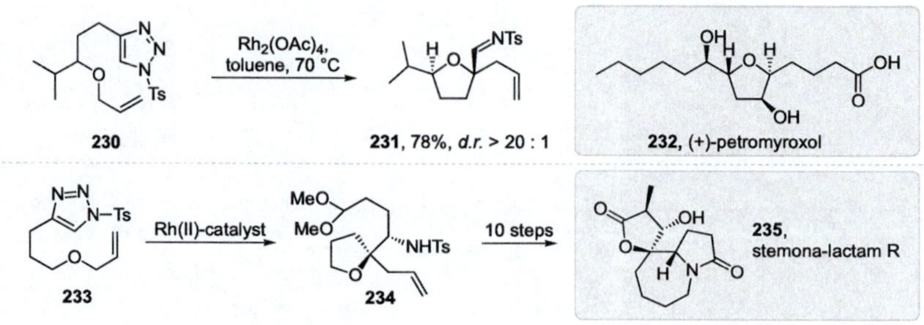

Fig. 25 Triazoles in intramolecular reactions for the synthesis of furans with application in total synthesis

depending on the substitution pattern of the glycidol (Fig. 24b) [77]. The Lacour and Beller groups further studied oxygen heterocycles in the reaction of acetals with triazoles. In this case, the acetal group underwent ring expansion to give medium-sized oxygen heterocycles (Fig. 24c) [78, 79]. In the reaction of acetals, the formation of an oxonium ylide is followed by a facile ring opening of the acetal group to expose an oxonium ion, which readily undergoes cyclization with the nucleophilic nitrogen atom of the imino-carbene fragment. In addition to acetals, Lacour and co-workers reported that oxetane heterocycles can be used in related ylide formation reactions. Here, depending on the reaction conditions, either a [1,2]-sigmatropic rearrangement occurs to form tetrahydrofuran **228**, or multiple units of the oxetane heterocycles engage in a selective formation of cyclic polyether framework **229** (Fig. 24d) [80].

In a manner similar to that described in the subchapter on ammonium ylides, oxonium ylides can be used in [2,3]-sigmatropic rearrangement reactions to furnish heterocycles. Again, the presence of an allyl substituent represents one of the main limitations of this approach. Boyer described the rhodium-catalyzed reaction of triazoles tethered to allyl ethers [81]. Upon formation of an oxonium ylide intermediate, a [2,3]-sigmatropic rearrangement gives access to furan heterocycles, which can be used in applications in the total synthesis of (+)-Petromyroxol [82] or stemona lactams (Fig. 25) [83].

3.3 *Sulfonium Ylides*

Similar to the formation of oxygen ylides with triazoles, the related sulfonium ylides can be obtained in thermal reactions of triazoles with suitable sulfur-based nucleophiles [41]. Here, one important example by Miura and Murakami for the synthesis of heterocycles via ylide intermediates is highlighted: in the rhodium-catalyzed reaction of triazoles with thionoesters important thiazole heterocycles can be obtained in high yield and proceed via the typical formation of a rhodium-carbene complex, followed by formation of a sulfonium ylide, which then cyclizes to give the

Fig. 26 Rhodium-catalyzed reaction of triazoles with thionoesters; (**a**) Synthesis of thiazoles; (**b**) Expansion toward ring-expansion reaction

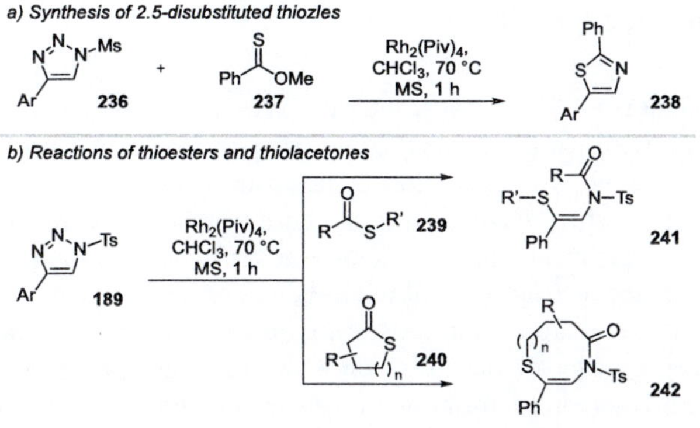

Fig. 27 Triazoles in the reaction with selenium compounds for the synthesis of dihydropyrroles

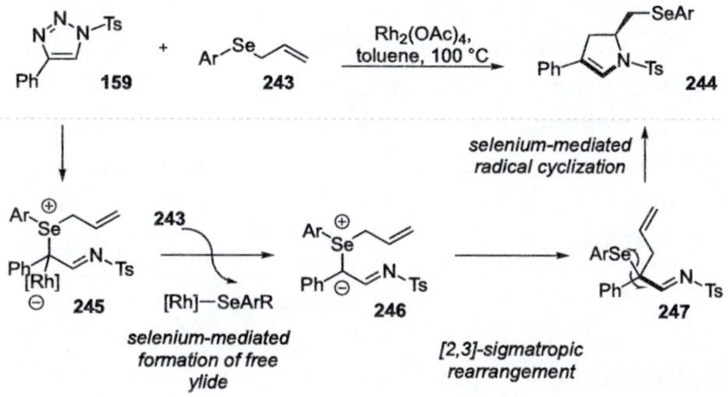

thiazole product (Fig. 26a). Importantly, the reaction with thioesters, despite formation of a sulfonium ylide, does not result in the formation of the thiazole; instead, a 1,3-difunctionalization of the imino-carbene occurs, which displays an important reaction pathway for this class of carbene compounds (Fig. 26b) [84, 85].

3.4 Selenonium Ylides

The reaction of selenonium ylides is a vastly underexplored area of research, and until today only very few studies report on carbene transfer reactions for selenonium ylide chemistry. A concise study on sigmatropic rearrangement reactions of these ylides was reported by Koenigs and co-workers [86]. In the context of heterocycle synthesis, a recent study by the Koenigs group has described the reaction of allyl selenides **243** and triazoles **159**. In the presence of a rhodium catalyst, selenonium ylide **245** is formed initially, followed by the expected [2,3]-sigmatropic rearrangement. However, in this case, a downstream transformation occurs and homolytic cleavage of the C-Se bond exposes a selenium-centered radical, that initiates a radical cyclization reaction to give a dihydropyrrole reaction product (Fig. 27) [4].

4 Conclusion

Onium ylides are key intermediates to access a broad variety of rearrangement reactions. In the context of heterocycles synthesis, ylides are of high relevance for the de novo construction of heterocycles or in ring-expansion reactions of small, strained heterocycles. Significant advances via carbene transfer reactions of diazo compounds or triazoles have been achieved over the past years. Of key relevance is here the intermediacy of metal-bound or free ylide intermediates. Initial results using combined approaches of experimental and theoretical studies suggest a strong dependency of the occurrence of either intermediate depending on the nature of the heteroatom forming the ylide and the metal catalyst being employed. More in-depth studies toward a generalization are needed in this regard. At present, the vast majority of reactions concern the use of ammonium, oxonium, or sulfonium ylides, while selenonium or other heavy atom analogs are much less investigated. It can be expected that research focusing on such ylide intermediates will come into the focus of chemists in the upcoming years. Lastly, precursors that allow the facile synthesis of ylide intermediates beyond classic diazo-based or -related carbene precursors are in high demand, to reduce risks associated to the chemistry of diazo compounds.

References

1. Jana S, Guo Y, Koenigs RM (2021) Recent perspectives on rearrangement reactions of ylides *via* carbene transfer reactions. Chem A Eur J 27:1270–1281
2. Hock KJ, Koenigs RM (2017) Enantioselective [2,3]-sigmatropic rearrangements: metal-bound or free ylides as reaction intermediates? Angew Chem Int Ed 56:13566–13568
3. Laconsay CJ, Tantillo DJ (2021) Metal bound or free ylides as reaction intermediates in metal-catalyzed [2,3]-sigmatropic rearrangements? It depends. ACS Catal 11:829–839
4. Li F, Pei C, Koenigs RM (2021) Rhodium-catalyzed cascade reactions of triazoles with organoselenium compounds – a combined experimental and mechanistic study. Chem Sci 12: 6362–6369
5. Zhang Y, Wang J (2010) Catalytic [2,3]-sigmatropic rearrangement of sulfur ylide derived from metal carbene. Coord Chem Rev 254:941–953
6. Lee H, Kim KT, Kim M, Kim C (2022) Recent advances in catalytic [3,3]-sigmatropic rearrangements. Catalysts 12:227
7. Empel C, Jana S, Koenigs RM (2021) Advances in [1,2]-sigmatropic rearrangements of onium ylides *via* carbene transfer reactions. Synthesis 53:4567–4587
8. Stevens TS, Creighton EM, Gordon AB, MacNicol M (1928) Degradation of quaternary ammonium salts. Part I. J Chem Soc:3193–3197
9. Bamford WR, Stevens TS (1952) A new reaction of aliphatic diazo-compounds. J Chem Soc:4675–4678
10. West FG, Naidu BN (1993) New route to substituted piperidines *via* the Stevens [1,2]-shift of ammonium ylides. J Am Chem Soc 115:1177–1178
11. Beall LS, Padwa A (1998) An approach to the cephalotaxine ring skeleton using an ammonium ylide/Stevens [1, 2]-rearrangement. Tetrahedron Lett 39:4159–4162

12. Padwa A, Snyder JP, Curtis EA, Sheehan SM, Worsencroft KJ, Kappe CO (2000) Rhodium(II)-catalyzed equilibration of push-pull carbonyl and ammonium ylides. A computationally based understanding of the reaction pathway. J Am Chem Soc 122:8155–8167
13. Kappe CO (1997) Ammonium ylide versus carbonyl ylide formation in the rhodium(ll)-catalyzed decomposition of diazoacetylureas. Generation and X-ray structure of a stable five-membered N-acyl ammonium ylide. Tetrahedron Lett 38:3323–3326
14. Wright DL, Weekly RM, Groff R, McMills MC (1996) A metallocarbenoid approach to the formation of spirocyclic ammonium ylides leading to the preparation of medium-sized azacane rings. Tetrahedron Lett 37:2165–2168
15. Muroni D, Saba A, Culeddu N (2004) A concise highly enantioselective cascade synthesis of indolizidine alkaloids with a quaternary stereocenter. Tetrahedron Asymmetry 15:2609–2614
16. Vanecko JA, West FG (2005) Ring expansion of azetidinium ylides: rapid access to the pyrrolizidine alkaloids turneforcidine and platynecine. Org Lett 7:2949–2952
17. Padwa A, Beall LS, Eidell CK, Worsencroft KJ (2001) An approach toward isoindolobenzazepines using the ammonium ylide/stevens [1,2]-rearrangement sequence. J Org Chem 66:2414–2421
18. Eshon J, Nicastri KA, Schmid SC, Raskopf WT, Guzei IA, Fernández I, Schomaker JM (2020) Intermolecular [3+3] ring expansion of aziridines to dehydropiperi-dines through the intermediacy of aziridinium ylides. Nat Commun 11:1273
19. Dequina HJ, Schomaker JM (2020) Aziridinium ylides: underused intermediates for complex amine synthesis. Trends Chem 2:874–887
20. Nickerson LA, Bergstrom BD, Gao M, Shiue YS, Laconsay CJ, Culberson MR, Knauss WA, Fettinger JC, Tantillo DJ, Shaw JT (2020) Enantioselective synthesis of isochromans and tetrahydroisoquinolines by C–H insertion of donor/donor carbenes. Chem Sci 11:494–498
21. Gutsche CD, Hillman M (1954) Reactions of ethyl diazoacetate with aromatic compounds containing hetero atoms attached to the benzyl carbon. J Am Chem Soc 76:2236–2240
22. West FG, Naidu BN, Tester RW (1994) Profound catalyst effects in the generation and reactivity of carbenoid-derived cyclic ylides. J Org Chem 59:6892–6894
23. Murphy GK, West FG (2006) [1,2]- or [2,3]-rearrangement of onium ylides of allyl and benzyl ethers and sulfides via in situ-generated iodonium ylides. Org Lett 8:4359–4361
24. Jaber DM, Burgin RN, Hepler M, Zavalij P, Doyle MP (2011) Control of selectivity in the generation and reactions of oxonium ylides. Chem Commun 47:7623–7625
25. Nozaki I, Moriuti S, Takaya H, Noyori R (1966) Asymmetric induction in carbenoid reaction by means of a dissymmetric copper chelate. Tetrahedron Lett 43:5239–5244
26. Ito K, Yoshitake M, Katsuki T (1996) Enantiospecific ring expansion of oxetanes: Stereoselective synthesis of tetrahydrofurans. Heterocycles 42:305–317
27. Friedrich K, Jansen U, Kirmse W (1985) Oxygen ylides −1. Reactions of carbenes with oxetane. Tetrahedron Lett 26:193–196
28. Lo MMC, Fu GC (2001) Applications of planar-chiral heterocycles in enantioselective catalysis: Cu(I)/bisazaferrocene-catalyzed asymmetric ring expansion of oxetanes to tetrahydrofurans. Tetrahedron 57:2621–2634
29. Mack DJ, Batory LA, Njardarson JT (2012) Intermolecular oxonium ylide mediated synthesis of medium-sized oxacycles. Org Lett 14:378–781
30. Yang Z, Stivanin ML, Jurberg ID, Koenigs RM (2020) Visible light-promoted reactions with diazo compounds: a mild and practical strategy towards free carbene intermediates. Chem Soc Rev 49:6833–6847
31. Ciszewski ŁW, Rybicka-Jasińska K, Gryko D (2019) Recent developments in photochemical reactions of diazo compounds. Org Biomol Chem 17:432–448
32. Empel C, Pei C, Koenigs RM (2022) Unlocking novel reaction pathways of diazoalkanes with visible light. Chem Commun 58:2788–2798
33. Jana S, Yang Z, Pei C, Xu X, Koenigs RM (2019) Photochemical ring expansion reactions: synthesis of tetrahydrofuran derivatives and mechanism studies. Chem Sci 10:10129–10134

34. Kitamura M, Kisanuki M, Kanemura K, Okauchi T (2014) Pd(OAc)$_2$-catalyzed macrocyclization of 1,2-diazonaphthoquinones with cyclic ethers. Org Lett 16:1554–1557
35. Dar'in D, Kantin G, Bakulina O, Inyutina A, Chupakhin E, Krasavin M (2020) Spirocyclizations involving oxonium ylides derived from cyclic α-diazocarbonyl compounds: an entry into 6-oxa-2-azaspiro[4.5]decane scaffold. J Org Chem 85:15586–15599
36. Doyle MP, Griffin JH, Chinn MS, Van Leusen D (1984) Rearrangements of ylides generated from reactions of diazo compounds with allyl acetals and thioketals by catalytic methods. Heteroatom acceleration of the [2,3]-sigmatropic rearrangement. J Org Chem 49:1917–1925
37. Lu P, Herrmann AT, Zakarian A (2015) Toward the synthesis of *nuphar* sesquiterpene thioalkaloids: Stereodivergent rhodium-catalyzed synthesis of the thiolane subunit. J Org Chem 80:7581–7589
38. Lacharity JJ, Fournier J, Lu P, Mailyan AK, Herrmann AT, Zakarian A (2017) Total synthesis of unsymmetrically oxidized Nuphar thioalkaloids *via* copper-catalyzed thiolane assembly. J Am Chem Soc 139:13272–13275
39. Green SP, Wheelhouse KM, Payne AD, Hallett JP, Miller PW, Bull JA (2020) Thermal stability and explosive hazard assessment of diazo compounds and diazo transfer reagents. Org Process Res Dev 24:67–84
40. Davies HML, Alford JS (2014) Reactions of metallocarbenes derived from N-sulfonyl-1,2,3-triazoles. Chem Soc Rev 43:5151–5162
41. Akter M, Rupa K, Anbarasan P (2022) 1,2,3-Triazole and its analogues: new surrogates for diazo compounds. Chem Rev 122:13108–13205
42. Horneff T, Chuprakov S, Chernyak N, Gevorgyan V, Fokin VV (2008) Rhodium-catalyzed transannulation of 1,2,3-triazoles with nitriles. J Am Chem Soc 130:14972–14974
43. Lee E, Ryu T, Shin E, Son JY, Choi W, Lee PH (2015) Synthesis of 2-bromoimidazoles from alkynes, *N*-sulfonylazides, and bromocyanides. Org Lett 17:2470–2473
44. Kim S, Kim JE, Lee J, Lee PH (2015) *N*-imidazolylation of sulfoximines from *N*-cyano sulfoximines, 1-alkynes, and *N* -sulfonyl azides. Adv Synth Catal 357:3707–3717
45. Motornov V, Markos A, Beier P (2018) A rhodium-catalyzed transannulation of *N*-(per)-fluoroalkyl-1,2,3-triazoles under microwave conditions – a general route to *N*-(per)-fluoroalkyl-substituted five-membered heterocycles. Chem Commun 54:3258–3261
46. Yang D, Shan L, Xu ZF, Li CY (2018) Metal-free synthesis of imidazole by $BF_3 \cdot Et_2O$ promoted denitrogenative transannulation of *N*-sulfonyl-1,2,3-triazole. Org Biomol Chem 16: 1461–1464
47. Zibinsky M, Fokin VV (2013) Sulfonyl-1,2,3-Triazoles: convenient synthones for heterocyclic compounds. Angew Chem Int Ed 52:1507–1510
48. Ryu T, Baek Y, Lee PH (2015) Synthesis of pyrazines from rhodium-catalyzed reaction of 2*H*-azirines with *N*-sulfonyl 1,2,3-triazoles. J Org Chem 80:2376–2383
49. Wang Y, Lei X, Tang Y (2015) Rh(II)-catalyzed cycloadditions of 1-tosyl 1,2,3-triazoles with 2*H*-azirines: switchable reactivity of Rh-azavinylcarbene as [2C]- or aza-[3C]-synthon. Chem Commun 51:4507–4510
50. Zhao YZ, Yang HB, Tang XY, Shi M (2015) Rh^{II} -catalyzed [3+2] cycloaddition of 2*H*-azirines with *N*-sulfonyl-1,2,3-triazoles. Chem A Eur J 21:3562–3566
51. Khaidarov AR, Rostovskii NV, Zolotarev AA, Khlebnikov AF, Novikov MS (2019) Synthesis of 1-(2-aminovinyl)indoles and 1,3′-biindoles by reaction of 2,2-diaryl-substituted 2*H*-azirines with α-imino Rh(II) carbenoids. J Org Chem 84:3743–3753
52. Dequina HJ, Eshon J, Schmid SC, Raskopf WT, Sanders KM, Fernández I, Schomaker JM (2022) Re-evaluation of product outcomes in the Rh-catalyzed ring expansion of aziridines with *N*-sulfonyl-1,2,3-triazoles. J Org Chem 87:10902–10907
53. Lei X, Li L, He YP, Tang Y (2015) Rhodium(II)-catalyzed formal [3 + 2] cycloaddition of *N*-sulfonyl-1,2,3-triazoles with isoxazoles: entry to polysubstituted 3-aminopyrroles. Org Lett 17: 5224–5227
54. Lei X, Gao M, Tang Y (2016) Rh(II)-catalyzed transannulation of *N*-sulfonyl-1,2,3-triazoles with 2,1-benzisoxazoles or 1,2-benzisoxazoles. Org Lett 18:4990–4993

55. Strelnikova JO, Rostovskii NV, Starova GL, Khlebnikov AF, Novikov MS (2018) Rh(II)-catalyzed transannulation of 1,2,4-oxadiazole derivatives with 1-sulfonyl-1,2,3-triazoles: Regioselective synthesis of 5-sulfonamidoimidazoles. J Org Chem 83:11232–11244
56. Lee DJ, Han HS, Shin J, Yoo EJ (2014) Multicomponent [5 + 2] cycloaddition reaction for the synthesis of 1,4-diazepines: isolation and reactivity of azomethine ylides. J Am Chem Soc 136: 11606–11609
57. Lee JY, Varshnaya RK, Yoo EJ (2022) Synthesis of chiral diazocine derivatives *via* a copper-catalyzed dearomative [5+3] cycloaddition. Org Lett 24:3731–3735
58. Kim J, Yoo EJ (2021) Catalytic ring expansion of activated heteroarenes enabled by regioselective dearomatization. Org Lett 23:4256–4260
59. Lee JY, Samala S, Kim J, Yoo EJ (2021) Contractions of 1,4-diazepines to pyrroles triggered by valence tautomerization: a one-pot approach and mechanism. Org Lett 23:9006–9011
60. De N, Ko D, Yeol BS, Oh C, Kim J, Baik MH, Yoo EJ (2020) Cu(I)-catalyzed enantioselective [5 + 1] cycloaddition of N-aromatic compounds and alkynes via chelating-assisted 1,2-dearomative addition. ACS Catal 10:10905–10913
61. Ge J, Wu X, Bao X (2019) Rhodium(II)-catalyzed annulation of *N*-sulfonyl-1,2,3-triazoles with 1,3,5-triazinanes to produce octahydro-1*H*-purine derivatives: a combined experimental and computational study. Chem Commun 55:6090–6093
62. Bosmani A, Guarnieri-Ibáñez A, Goudedranche S, Besnard C, Lacour J (2018) Polycyclic indoline-benzodiazepines through electrophilic additions of α-imino carbenes to Tröger bases. Angew Chem Int Ed 57:7151–7155
63. Saleh N, Bosmani A, Besnard C, Bürgi T, Jacquemin D, Lacour J (2020) Access to chiral rigid hemicyanine fluorophores from tröger bases and α-imino carbenes. Org Lett 22:7599–7603
64. Guarnieri-Ibáñez A, de Aguirre A, Besnard C, Poblador-Bahamonde AI, Lacour J (2021) Regiodivergent synthesis of pyrazino-indolines *vs.* triazocines *via* α-imino carbenes addition to imidazolidines. Chem Sci 12:1479–1485
65. Xu HD, Jia ZH, Xu K, Zhou H, Shen MH (2015) One-pot protocol to functionalized benzopyrrolizidine catalyzed successively by $Rh_2(OAc)_4$ and $Cu(OTf)_2$: a transition metal–Lewis acid catalysis relay. Org Lett 17:66–69
66. Jablasone ST, Ye Z, Duan S, Xu ZF, Li CY (2021) Synthesis of benzothiazonine by rhodium-catalyzed denitrogenative transannulation of 1-sulfonyl-1,2,3-triazole and thiochromone. Org Biomol Chem 19:5758–5761
67. Miura T, Tanaka T, Hiraga K, Stewart SG, Murakami M (2013) Stereoselective synthesis of 2,3-dihydropyrroles from terminal alkynes, azides, and α,β-unsaturated aldehydes *via N*-sulfonyl-1,2,3-triazoles. J Am Chem Soc 135:13652–13655
68. Li J, Zhu SR, Xu Y, Lu XC, Wang ZB, Liu L, Xu DF (2020) Synthesis of 2,5-diaryloxazoles through rhodium-catalyzed annulation of triazoles and aldehydes. RSC Adv 10:24795–24799
69. Ma X, Liu L, Wang J, Xi X, Xie X, Wang H (2018) Rhodium-catalyzed annulation of α-imino carbenes with α,β-unsaturated ketones: construction of multisubstituted 2,3-dihydropyrrole/pyrrole rings. J Org Chem 83:14518–14526
70. Li Y, Zhang R, Ali A, Zhang J, Bi X, Fu J (2017) Synthesis of 6-substituted piperidin-3-ones via Rh(II)-catalyzed transannulation of *N*-sulfonyl-1,2,3-triazoles with electron-rich aromatic nucleophiles. Org Lett 19:3087–3090
71. Cheng W, Tang Y, Xu ZF, Li CY (2016) Synthesis of multifunctionalized 2-carbonylpyrrole by rhodium-catalyzed transannulation of 1-sulfonyl-1,2,3-triazole with β-diketone. Org Lett 18: 6168–6171
72. Mi P, Yuan H, Wang H, Liao P, Zhang J, Bi X (2017) Divergent reactions between α-imino rhodium carbenoids and 1,3-diketones: substrate-controlled O-H versus C-H insertion. Eur J Org Chem:1289–1293
73. Shen H, Fu J, Gong J, Yang Z (2014) Tunable and chemoselective syntheses of dihydroisobenzofurans and indanones via rhodium-catalyzed tandem reactions of 2-triazole-benzaldehydes and 2-triazole-alkylaryl ketones. Org Lett 16:5588–5591

74. Zhang YS, Tang XY, Shi M (2014) Unprecedented synthesis of aza-bridged benzodioxepine derivatives through a tandem Rh(II)-catalyzed 1,3-rearrangement/[3+2] cycloaddition of carbonyltriazoles. Chem Commun 50:15971–15974
75. Yuan H, Gong J, Yang Z (2016) Stereoselective synthesis of oxabicyclo[2.2.1]heptenes via a tandem dirhodium(II)-catalyzed triazole denitrogenation and [3 + 2] cycloaddition. Org Lett 18: 5500–5503
76. Ma X, Pan S, Wang H, Chen W (2014) Rhodium-catalyzed transannulation of *N*-sulfonyl-1,2,3-triazoles and epoxides: Regioselective synthesis of substituted 3,4-dihydro-2 *H*-1,4-oxazines. Org Lett 16:4554–4557
77. Ko YO, Jeon HJ, Jung DJ, Kim UB, Lee SG (2016) Rh(II)/Mg($O^tBu)_2$-catalyzed tandem one-pot synthesis of 1,4-oxazepines and 1,4-oxazines from *N* -sulfonyl-1,2,3-triazoles and glycidols. Org Lett 18:6432–6435
78. Medina F, Besnard C, Lacour J (2014) One-step synthesis of nitrogen-containing medium-sized rings via α-imino diazo intermediates. Org Lett 16:3232–3235
79. Pospech J, Ferraccioli R, Neumann H, Beller M (2015) Rhodium(II)-catalyzed annulation of azavinyl carbenes through ring-expansion of 1,3,5-trioxane: rapid access to nine-membered 1,3,5,7-trioxazonines. Chem Asian J 10:2624–2630
80. Guarnieri-Ibáñez A, Medina F, Besnard C, Kidd SL, Spring DR, Lacour J (2017) Diversity-oriented synthesis of heterocycles and macrocycles by controlled reactions of oxetanes with α-iminocarbenes. Chem Sci 8:5713–5720
81. Boyer A (2014) Rhodium(II)-catalyzed stereocontrolled synthesis of 2-tetrasubstituted saturated heterocycles from 1-sulfonyl-1,2,3-triazoles. Org Lett 16:5878–5881
82. Boyer A (2015) Enantioselective synthesis of (+)-petromyroxol, enabled by rhodium-catalyzed denitrogenation and rearrangement of a 1-sulfonyl-1,2,3-triazole. J Org Chem 80:4771–4775
83. Fu J, Shen H, Chang Y, Li C, Gong J, Yang Z (2014) Concise stereoselective synthesis of oxaspirocycles with 1-tosyl-1,2,3-triazoles: application to the total syntheses of (±)-tuberostemospiroline and (±)-stemona-lactam R. Chem A Eur J 20:12881–12888
84. Miura T, Funakoshi Y, Fujimoto Y, Nakahashi J, Murakami M (2015) Facile synthesis of 2,5-disubstituted thiazoles from terminal alkynes, sulfonyl azides, and thionoesters. Org Lett 17:2454–2457
85. Miura T, Fujimoto Y, Funakoshi Y, Murakami M (2015) A reaction of triazoles with thioesters to produce β-sulfanyl enamides by insertion of an enamine moiety into the sulfur–carbonyl bond. Angew Chem Int Ed 54:9967–9970
86. Jana S, Koenigs RM (2019) Rhodium-catalyzed carbene transfer reactions for sigmatropic rearrangement reactions of selenium ylides. Org Lett 21:3653–3657

Heterocycles from Sulfur Ylides

Antonio C. B. Burtoloso, Jorge A. M. Vargas, Matheus P. de Jesus, and Radell Echemendía

Contents

Abstract Sulfur ylides (SY) are a special class of zwitterionic compounds that possess the ability to mediate different transformations. Represented by sulfonium and sulfoxonium ylides, these compounds have been employed for many years in traditional epoxidation, cyclopropanation, and aziridination reactions, as well as in some sigmatropic rearrangements. The use of sulfur ylides to generate metal-carbenes dates back more than 50 years, but only recently has this chemistry become more mature, especially with the use of α-carbonyl sulfoxonium ylides. This chapter will disclose examples of heterocycle syntheses, containing mainly N, O, and S, where the transformation is mediated by metal-carbenes derived from sulfur ylides (i.e., a carbene formed from the ylidic bond). The chapter will be divided into two main categories that lead to these heterocycles: (1) direct metal-carbene formation from SY in the first step, and (2) metal-catalyzed C–H activation of a substrate in the first step, followed by metal-carbene formation with the SY.

Keywords C–H activation · Cyclization · Heterocycles · Metal-carbenes · Sulfur ylides

A. C. B. Burtoloso (✉)
São Carlos Institute of Chemistry, University of São Paulo, São Carlos, SP, Brazil
e-mail: antonio@iqsc.usp.br

1 Introduction

The importance of sulfur ylides (SY) as versatile platforms in synthesis is documented by several reviews that include reactivity patterns, preparation, and transformations [1–5]. The reader is invited to consult them for a deep understanding of this class of compounds.

In 1930, Ingold and Jessop [6] described the first example of a stable sulfur ylide (a sulfonium ylide) that could be isolated. However, it was only in the 60s that these compounds, both sulfonium and sulfoxonium, started to be explored in more detail with the efforts from A. Johnson and Lacount [7], Corey and Chaykovsky [8–10], Franzen [11, 12], Trost [13], and C. Johnson [14–16]. Sulfonium and sulfoxonium ylides differ from each other in reactivity/stability, the latter being more stable and less nucleophilic due to the presence of the electronegative oxygen atom attached to the sulfur. Another important difference is the type of leaving group on the SY. While reactions with sulfonium ylides liberate a sulfide as the leaving group, reactions with sulfoxonium ylides furnish a sulfoxide.

As with any ylide or 1,1′-dipolar species, SY can effect two functionalizations on the same carbon center (reaction with an electrophile furnishes an electrophilic intermediate with a positively charged sulfur that as a leaving group can be displaced by a nucleophile; Scheme 1). Although the use of this type of ambiphilic reactivity to perform geminal difunctionalizations still dominates, this intrinsic reactivity has also been successfully applied in the generation of metal-carbenes. The generation of free carbenes photochemically is also possible, but less common, and this topic will not be covered in the chapter.

As basically all reactions with SY start with the ylide acting as a C-nucleophile, information on the nucleophilicity and basicity of these compounds is extremely important and useful. Figure 1 illustrates data of the gas phase proton affinity (charts A and B), p*K*a of protonated SY (chart C), and HOMO energies (chart D) of some SY published in the literature [17, 18], highlighting the higher reactivity of sulfonium over sulfoxonium ylides. The substituent attached to the anionic carbon as well

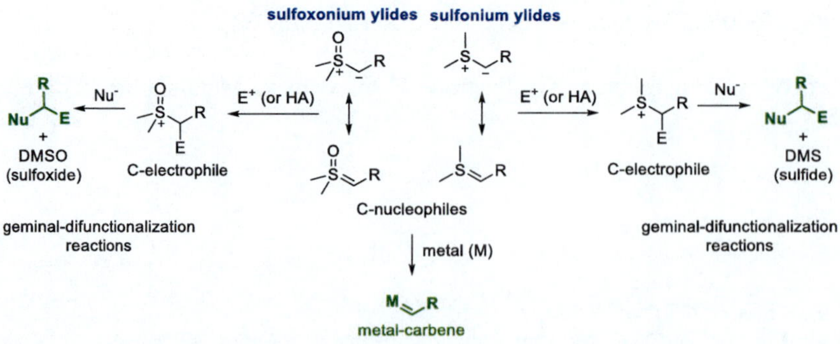

Scheme 1 Schematic representation of the ambiphilic nature of sulfur ylides and their main type of products

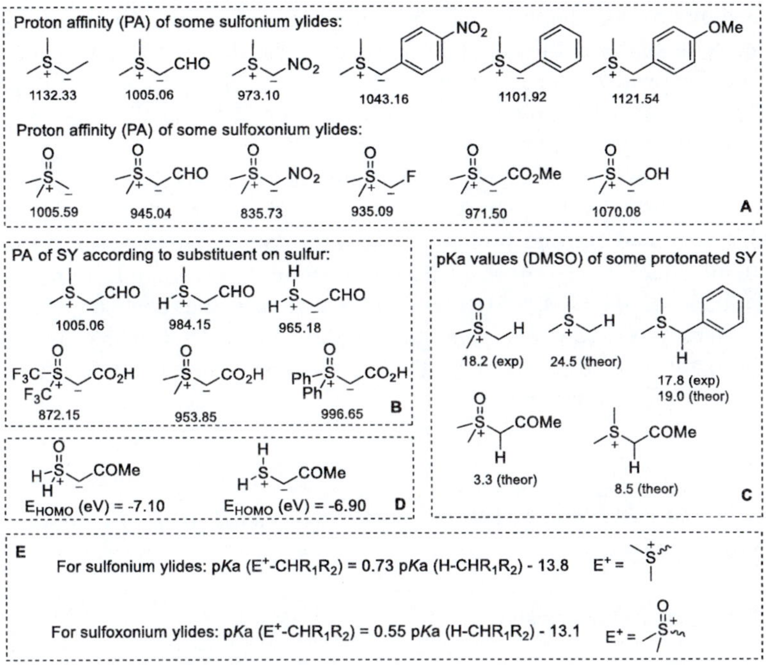

Fig. 1 Charts **A** and **B**: Gas phase Proton affinities of several sulfonium and sulfoxonium ylides calculated by DFT methods (SOGGA11-X) [17]. Chart **C**: Experimental and theoretical p*Ka* values in DMSO for some protonated SY [18]. Chart **D**: HOMO (highest occupied molecular orbital) energies of a carbonyl-sulfoxonium and sulfonium ylide [17]. Chart **E**: developed equations to predict the p*Ka* of any protonated SY [18]

as to the sulfur atom also alters the reactivity patterns of these SY. The choice of both substituents on sulfur and carbon adjusts SY reactivity to reach a desired specific transformation. Interestingly, and based on these theoretical studies, two equations were also developed to predict the p*Ka* of any protonated SY [18] (chart E).

2 Heterocycle Synthesis from Sulfur Ylide Mediated by Carbenes

Initial evidence that both sulfoxonium and sulfonium ylides could promote the formation of carbenes was described by Corey in 1964 [9] and Trost in 1966 [13]. Based on the observed products, Corey claimed that free carbenes could be involved when α-keto sulfoxonium ylides were exposed to UV light. Trost employed light (or copper sulfate and heating) in the presence of a α-keto sulfonium ylide and observed the formation of a cyclopropane as the main product. This gives strong evidence of the intermediacy of carbenes. According to the authors, a dimeric conjugated olefin was initially formed from the coupling of the formed keto carbene with a molecule of the sulfonium ylide. This olefin was then captured by the carbene to furnish the observed cyclopropane. The use of sulfonium ylides to prepare cyclopropanes was later explored in detail by Cohen [19] and Cimetiére and Julia [20], employing copper acetonates. However, it was only after the seminal contributions from Baldwin in 1993 [21] and Mangion in the 2000s [22, 23], with the use of α-keto and α-ester sulfoxonium ylides in combination with Rh, Ir, or Au, that the use of SY as metal-carbene promoters became more popular. Since α-carbonyl sulfoxonium ylides have proven to be the best substrates for insertion reactions into polar bonds (N-H, O-H, S-H), these ylides have been the substrates of choice to prepare heterocycles. Moreover, the good stability and the fact that an odorless sulfoxide is released after the formation of the metal-carbene make the use of sulfoxonium ylides more attractive.

2.1 *Category 1: Direct Metal-Carbene Formation from Sulfur Ylides*

2.1.1 Nitrogen Heterocycles

The first nitrogen heterocycle that was synthesized via a sulfur ylide, where the key transformation involved a metal-carbene, was described by Baldwin in 1993 [21]. The transformation involved the synthesis of a pyrrolidinone and a piperidinone in a one-carbon homologation-cyclization. Keto sulfoxonium ylide **1**, containing a terminal protected-amino group, was efficiently prepared by the ring opening of a chiral azetidine-2-one in the presence of dimethyl sulfoxonium methylide in 97% yield (Scheme 2). Reactions of the synthesized keto sulfoxonium ylide with a series of copper and rhodium catalysts were carried out, revealing that the use of rhodium trifluoroacetate provided the best results for the desired cyclization. Yields of the 4-pyrrolidinone **2** could be improved to 77% by the slow addition of the ylide to a suspension of the catalyst in refluxing 1,2-dichloroethane (Scheme 2). Application of the same sequence from 2-pyrrolidinone **3** provided 5-piperidinone **4** in 48% overall yield. Although diazo compounds were already known to perform this type of transformation, this was the first example of an N-H insertion reaction where a sulfoxonium ylide was employed. The mechanism is believed to follow a similar path to the ones already established for X-H (polar) insertion with diazo compounds: metal-carbene formation, cyclization by the formation of a ylide (ammonium ylide in this case), and hydrogen transfer [24].

Nearly two decades later in 2009, Mangion at Merck disclosed a series of studies on the use of ester- and keto-sulfoxonium ylides in insertion reactions into polar bonds (both inter- and intra-molecular). The use of ruthenium and iridium [22] or gold salts [23] in noncoordinating solvents such as CH_2Cl_2 proved to be optimal for insertion reactions with amines, alcohols, and thiols (insertion into C–H bonds and cyclopropanation reactions failed and dimers, formed from the coupling of two ylide

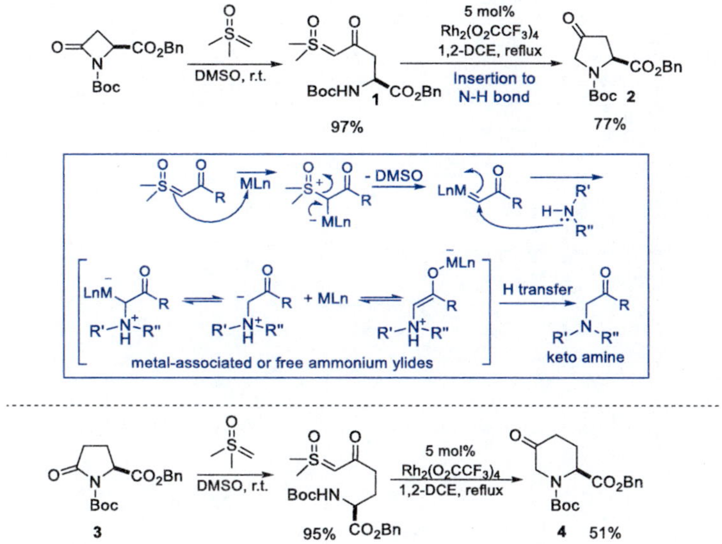

Scheme 2 5- and 6-membered nitrogen heterocycles prepared from keto sulfoxonium ylides, mediated by a rhodium carbene

molecules, were the only products observed). The use of dirhodium compounds, efficient catalysts to generate carbenes from diazo compounds, provided low yields of the insertion products (control experiments revealed that DMSO, the by-product of the ylide decomposition, was deactivating the rhodium catalysts). Although the use of gold catalysts such as AuCl(SMe_2) required milder reaction conditions and gave better yields in intermolecular reactions when compared to those with [Ir(cod)Cl]$_2$, the gold catalysts were not suitable for the intramolecular reactions. Considering that, iridium was the metal of choice to prepare several *N*-heterocycles with 5, 6, and 7-membered rings in yields varying from 67 to 90% (Scheme 3). Moreover, the application of this methodology in the total synthesis of two important heterocycles (MK-7655 and MK-7246) [25, 26] in more than 100 kg from ylides **5** and **6** (Scheme 3) represented the first practical application of N-H insertion reactions from SY. MK-7655, known as relebactam, is a β-lactamase inhibitor used in combination therapy to treat a resistant bacterial infection and MK-7246 is a potent and selective CRTH2 antagonist to be used in respiratory diseases. The exposure of keto sulfoxonium ylides **5** and **6** to a 1 mol% loading of [Ir(cod)Cl]$_2$ in toluene under heating provided the piperidinones **7** and **8** in 87% and 83% yield, respectively.

The straightforward construction of indoles and pyrroles could also be achieved with the use of sulfoxonium ylides in combination with iridium catalysis. In 2017 Vaitla and co-workers [27] demonstrated that when sulfoxonium ylides are mixed with [Ir(cod)Cl]$_2$, a Brønsted acid such as *p*-TSA and anilines, under microwave

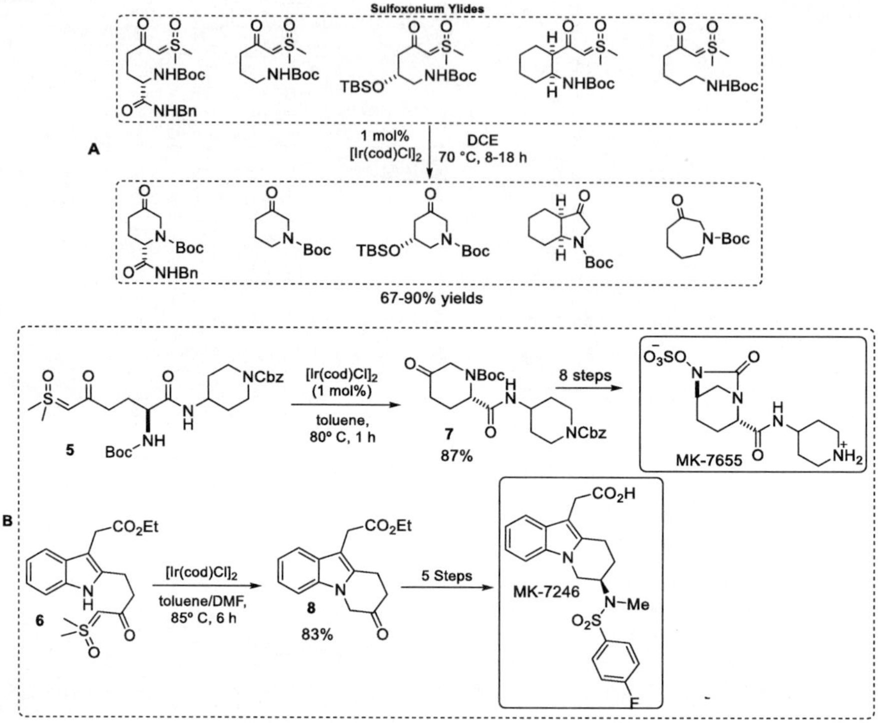

Scheme 3 Iridium-catalyzed N-H insertion from sulfoxonium ylides

irradiation at 140°C for 45 min, substituted indoles are formed in good yields. Replacing anilines with β-enamino esters, keeping the same reaction conditions, furnished trisubstituted pyrroles. The mechanism of these reactions is believed to start with the iridium carbene from the sulfoxonium ylides. In the case of the use of anilines, N-H insertion gives the classical keto amine. Imine formation with another aniline molecule, followed by a Friedel-Crafts-type alkylation, aromatization, and tautomerization led to indoles (Scheme 4a). In the case of enamines, a formal C–H insertion reaction with the iridium carbene, hydrogen transfer, and cyclization provided trisubstituted pyrroles (Scheme 4b). These reactions can also be performed under conventional heating (reflux toluene) for 16 h.

An interesting application of sulfoxonium ylides in cascade reactions for the synthesis of 2,3-diaroylquinolines was disclosed by Zou and co-workers [28]. The annulation of excess sulfoxonium ylides in the presence of anthranils, catalyzed by copper(0) and silver salts, provided a series of 2,3-diaroylquinolines in yields varying from 25 to 68% (Scheme 5). This unprecedented [4 + 1 + 1] annulation, according to the authors, starts with the oxidation of copper powder in the presence of silver triflate and air (the absence of the silver salt gave only a trace of the desired products). Formation of the copper carbene from sulfoxonium ylide **9** followed by coupling with another molecule of ylide **9** surprisingly gave 1,4-dicarbonyl ylide **10**.

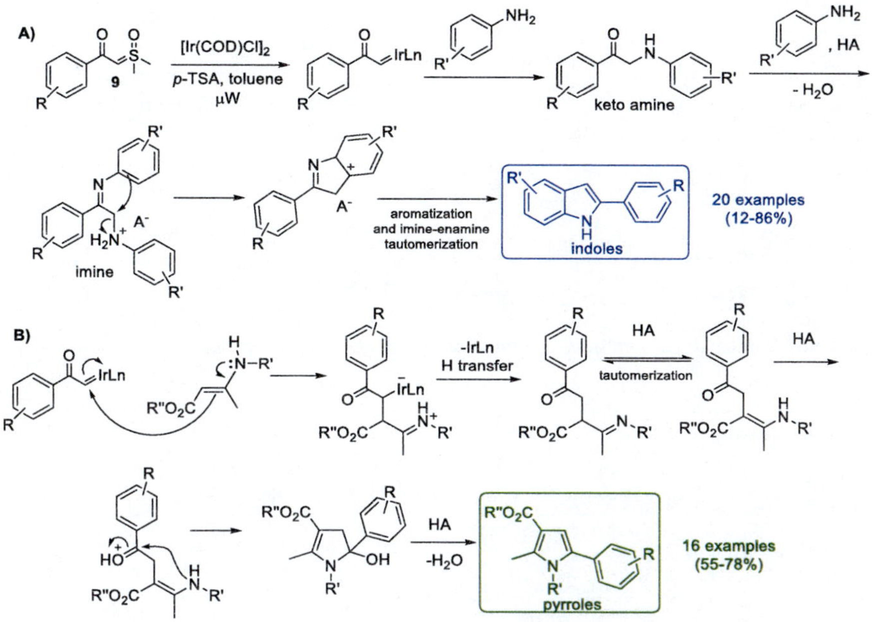

Scheme 4 Iridium and Brønsted acid catalyzed the synthesis of indoles and pyrroles

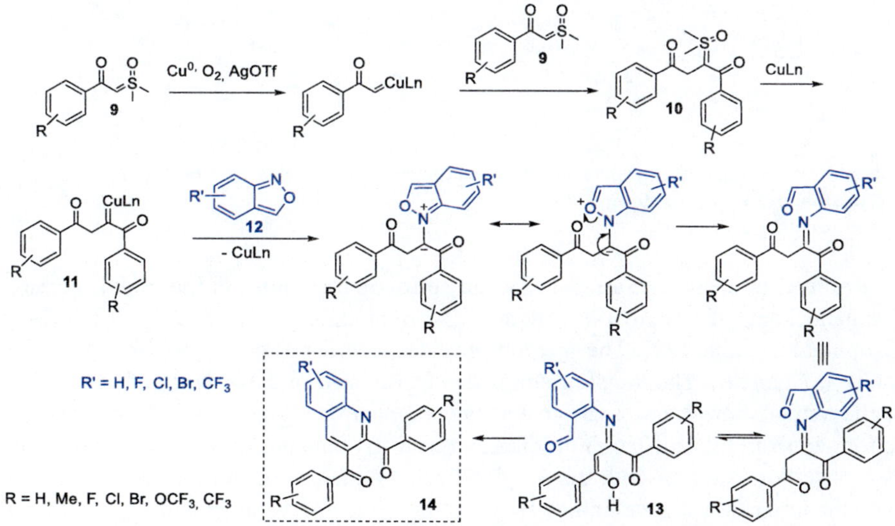

Scheme 5 Sulfoxonium ylides in cascade reactions for the synthesis of 2,3-diaroylquinolines

Formation of the new copper carbene **11** from **10**, and reaction with anthranils **12** (probably via an ammonium ylide) furnished intermediate **13** that after carbonyl addition and dehydration provided the 2,3-diaroylquinolines **14**.

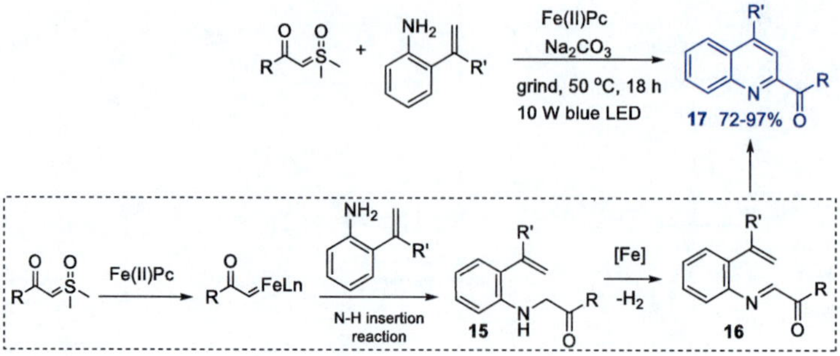

Scheme 6 Iron-catalyzed synthesis of quinolines from sulfoxonium ylides

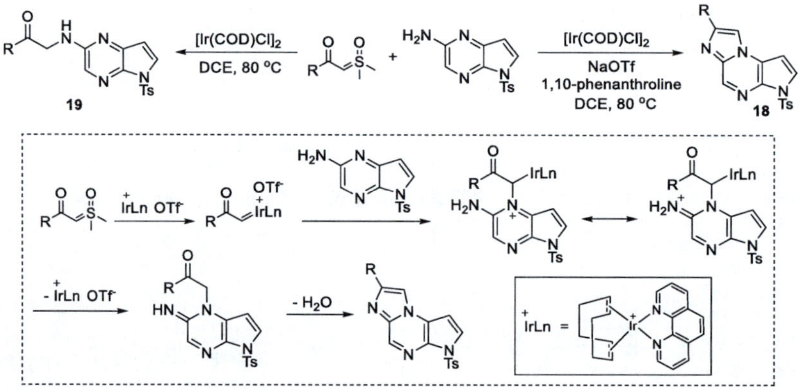

Scheme 7 Synthesis of tricyclic azoles from sulfoxonium ylides catalyzed by iridium. Proposed mechanism given for azole formation

Another annulation strategy based on sulfoxonium ylides is the photo-thermo-mechanochemical reaction with 2-vinylanilines catalyzed by iron (II) phthalocyanine [29]. The method provides a series of quinolines in very good yields (Scheme 6). The transformation also works well in the absence of light, but a higher temperature is necessary to achieve comparable high yields. An iron carbene is formed from the sulfoxonium ylide and iron (II) phthalocyanine, followed by an N-H insertion reaction with the vinylanilines. Oxidation of the formed amines **15** with the iron catalyst gave imines **16** that after an electrocyclic reaction provided the quinolines **17**.

The reaction of other aromatic amines, such as aryl amidines, with sulfoxonium ylides under catalysis by iridium provided several tricyclic azoles **18** [30] (Scheme 7). Interestingly, the reaction catalyzed with the traditional $[Ir(cod)Cl]_2$ in the absence of any additional ligand gives, almost exclusively, the expected N-H insertion product (keto amine **19**). However, the use of 1,10-phenanthroline in

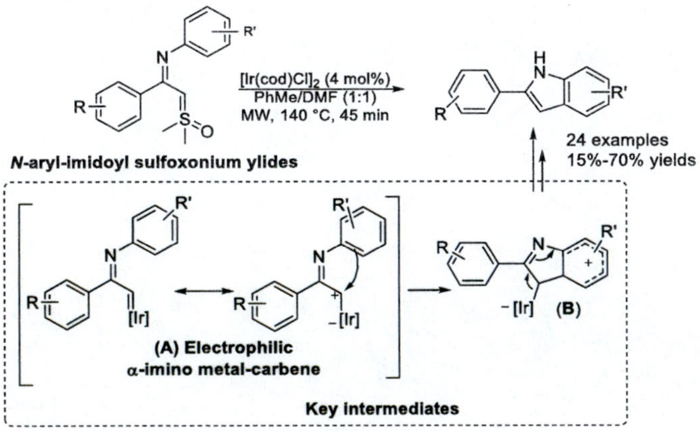

Scheme 8 Synthesis of indoles from *N*-aryl-imidoyl sulfoxonium ylides

combination with the noncoordinating triflate anion gives preference for the formation of the azole (insertion occurs with the nitrogen of the ring instead of the free amino group) via a proposed four-coordinate iridium complex.

In 2020, the Burtoloso group disclosed an interesting approach to access indoles in one step from *N*-aryl-imidoyl sulfoxonium ylides via an intramolecular C–H functionalization catalyzed by $[Ir(cod)Cl]_2$, through the formation of α-imino metal-carbene intermediates (Scheme 8) [31]. The authors employed the protocol described by Gilchrist and co-workers to prepare the *N*-aryl-imidoyl sulfoxonium ylides [32, 33], but they also developed an alternative synthetic route to produce these imino ylides from the reaction of keto sulfoxonium ylides and anilines in the presence of $TiCl_4$ as a Lewis acid. Indoles were obtained in moderate to high yields employing a catalytic amount of the Ir(I)-catalyst under microwave irradiation at 140°C for 45 min. Catalysts like $Rh_2(OAc)_4$ $AuCl(SMe)_2$, $[RuCl_2(PPh_3)_3]$ were also studied but were inefficient for this transformation. A series of control experiments suggested that after the formation of an α-imino metal-carbene species **A** (a prior metal-catalyzed C–H activation in the aromatic ring was ruled out) an intramolecular cyclization process occurred, leading to the formation of species **B**. Subsequently, regeneration of the catalyst with the aromatization of the system provided the desired product.

2.1.2 Oxygen Heterocycles

Examples in the synthesis of oxygen heterocycles, where the carbene is formed at the carbon of the ylidic bond is less common when compared to nitrogen heterocycles. For example, Skrydstrup (Scheme 9a) described an interesting method to prepare 2,4-disubstituted furans from keto sulfonium ylides and alkynes, catalyzed by gold. The reaction, however, proceeds from a mechanism where the gold carbene is formed on the terminal alkyne carbon, with subsequent formation of carbonyl

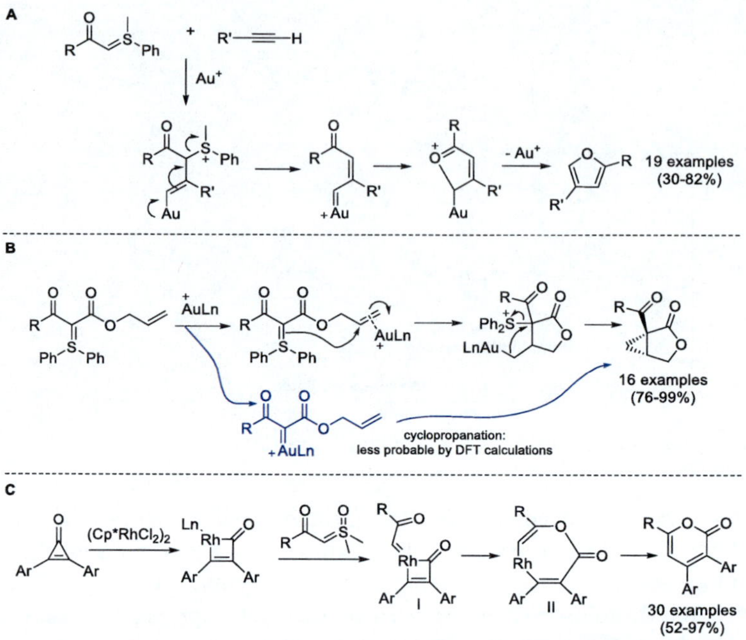

Scheme 9 Chart **A** synthesis of furans from gold carbenes that were not formed in the ylidic bond. Chart **b** cyclopropanation reactions to form bicyclic lactones, where the formation of a carbene from the ylidic bond is unlikely to occur. Chart **C** Synthesis of pyrones with the prior formation of a rhodium carbocycle from cyclopropenones, followed by rhodium carbene generation

ylides [34]. Maulide and co-workers (Scheme 9b) described a novel gold-catalyzed cyclopropanation reaction from allyloxycarbonyl diphenylsulfonium ylides [35]. Interestingly, DFT calculations demonstrated that gold carbenes (formed in the ylide carbon) may not be involved in the transformation. Instead of that, olefin activation by the metal, followed by subsequent intramolecular C-C bond formation, with the release of diphenyl sulfide, is a more reasonable explanation for this reaction. Another approach, by Jiang and Li, employs keto sulfoxonium ylides in combination with cyclopropenones to prepare trisubstituted 2-pyrones, by a Rh(III)-catalyzed [3 + 3] annulation [36]. Cyclometallation between the cyclopropenone and the rhodium complex, followed by Rh-carbene formation from the sulfoxonium ylide gives intermediate **I**. Migratory insertion of an oxygen atom into the Rh-CO bond gives intermediate **II** that, after reductive elimination, provides 2-pyrones (Scheme 9c).

Other examples, now with the intermediacy of a metal-carbene from the ylide carbon, involve the use of allyl-ester sulfonium ylides via cyclopropanation or C–H insertion reactions and were described by Müller [37] (this approach gives better results when diazo compounds are used). In this work, lactones **22** and **23** were prepared in 40% and 9% yield, from ylides **20** and **21**, respectively, using rhodium

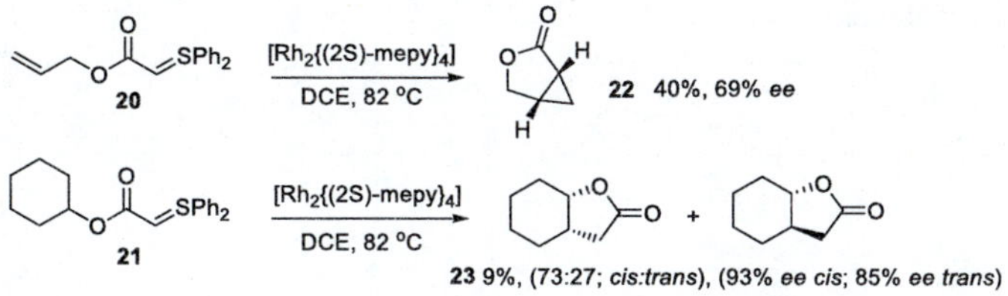

Scheme 10 Allyl-ester sulfonium ylides in asymmetric cyclopropanation or C–H insertion reactions

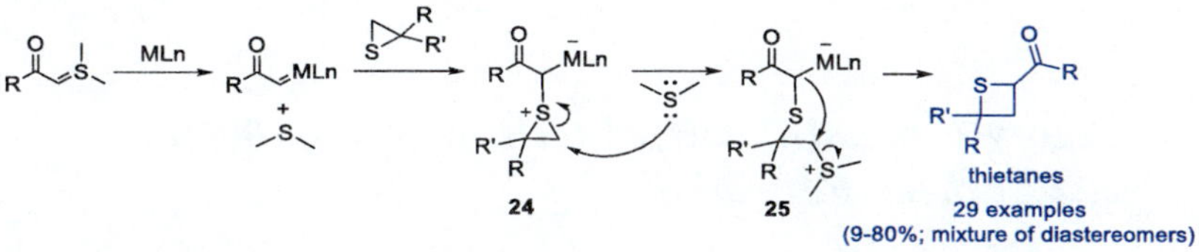

Scheme 11 Metal-catalyzed synthesis of thietanes from sulfonium ylides

catalysis. The same reactions with the use of the diazo carbonyl surrogates furnished these compounds in 75% and 11–65% yields, respectively (Scheme 10).

2.1.3 Sulfur-Containing Heterocycles

The synthesis of several substituted thietanes from sulfonium ylides and thiiranes by rhodium and iridium catalysis has been described [38]. Formation of a metal-carbene from the dimethylsulfonium ylide, followed by reaction with the thiirane, gives a new metal-associated sulfur ylide **24**. Released dimethyl sulfide, during metal-carbene formation, then attacks the three-membered ring of **24**, giving opened intermediate **25**. In an intramolecular fashion, a second nucleophilic substitution process provides the substituted thietane (Scheme 11).

The direct synthesis of 2-aminothiazoles was achieved from aryl-keto sulfoxonium ylides and thioureas, via a rhodium carbene insertion to the S-H bond, followed by an annulation reaction (Scheme 12) [39]. Electron-withdrawing and releasing groups on the aromatic ring of the sulfoxonium ylides were well tolerated. The use of alkyl and vinyl keto sulfoxonium ylides were also explored for three examples in 23–50% yields.

Scheme 12 Rh-catalyzed synthesis of 2-aminothiazoles from sulfonium ylides

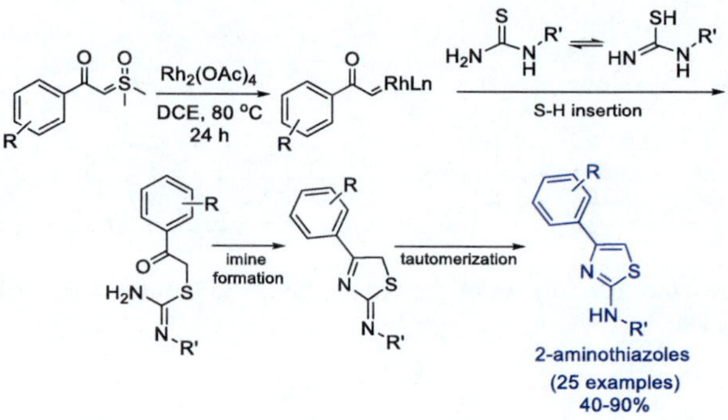

2.2 *Category 2: Metal-Catalyzed C–H Activation Before Metal-Carbene Formation*

The metal-catalyzed C–H activation/annulation with sulfoxonium ylides is a very attractive strategy to access structurally diverse heterocyclic compounds. In general, this approach starts with the selective C–H activation [40–43] of an aromatic compound (C–H bond *ortho* to a directing group) by catalytic amounts of a transition metal, typically Rh(III), Ru(II), and Ir(III). Due to the effective coordination of these directing groups with the metal, the catalyst can easily insert into the near *ortho* C–H bond with high regioselectivity to form a metallacycle [40]. Subsequent coupling with the sulfoxonium ylide, and annulation by different ways, creates 5- and 6-membered rings (Scheme 13a). Scheme 13b shows a general and detailed mechanism for these transformations. First, as mentioned above, the substrate reacts with the activated metal (usually generated via ligand exchange) to provide an electrophilic five-membered metallacycle **A** through C–H metalation. After that, attack of the nucleophilic sulfoxonium ylide to the metal center of intermediate **A** furnishes **B**, which then undergoes extrusion of DMSO to give the metal-carbene species **C**. Next, migratory insertion in the metal-carbene affords the six-membered metallacycle intermediate **D**. The subsequent protodemetallation of **D** gives the key intermediate **E** and releases the metal catalyst back to the catalytic cycle. At the end, **E** can be isolated as a final product or cyclized to provide the heterocycle (generally with formation of a 5- or 6-membered ring). As this C–H functionalization/annulation process follows very similar mechanisms (just varying the type of X-Y directing group), the reader is invited to return to Scheme 13 during the discussion of this section.

2.2.1 Nitrogen Heterocycles

Nitrogen-based heterocyclic compounds occupy an exclusive position as a valuable source of therapeutic agents in medicinal chemistry. These structural units are broadly distributed in natural products and are used as the building blocks of new drug candidates [44]. A comprehensive compilation of the structural diversity at the

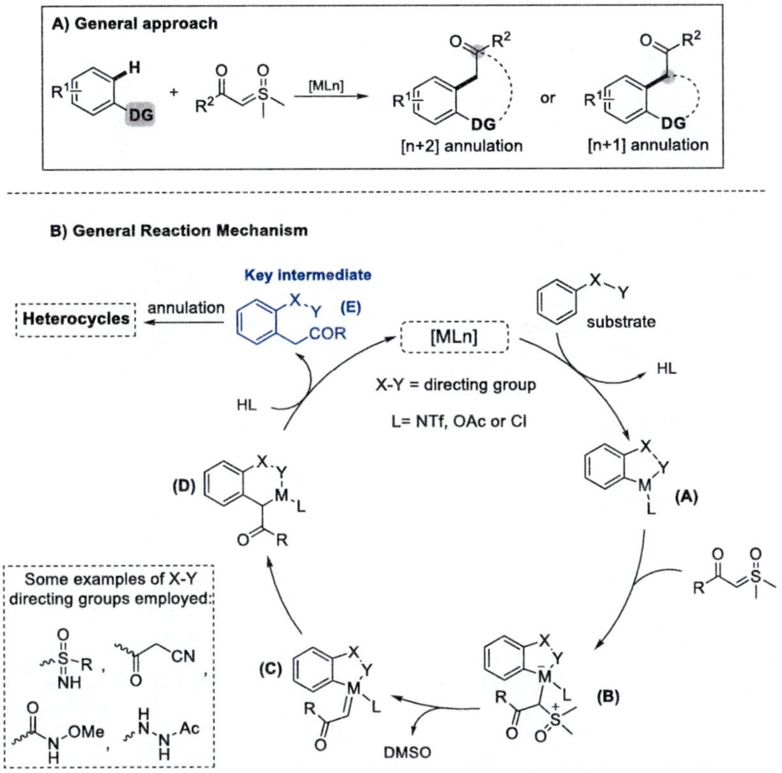

Scheme 13 The general approach and mechanism for sulfoxonium ylide-based heterocycle synthesis via C–H functionalization and annulation

FDA (Food and Drug Administration) databases reveals that around 75% of unique small-molecule drugs contain a nitrogen heterocycle [45]. Consequently, significant efforts have been devoted to developing efficient routes for synthesizing this family of compounds [46].

In 2018, Fan and co-workers reported the selective synthesis of benzo[α]carbazoles and indolo[2,1-α]-isoquinolines through Rh(III)-catalyzed cascade reactions of 2-arylindoles with sulfoxonium ylides (Scheme 14) [47]. This reaction protocol allows the modification in both 2-aryl indoles and sulfoxonium ylides to afford benzo[α]carbazole (**26**) in good yields. To expand the substrate scope of the methodology, the authors installed different groups at the C-3 position of the indole moiety. Aryl indoles bearing a cyano and a formyl group on this position of the indole were employed in this transformation and 5-acylbenzo[α]carbazoles (**27**) were obtained with moderate to good yields. Interestingly, when 2-arylindoles with a methyl group attached to the C3-position were used, indolo[2,1-α] isoquinolines (**28**) could be obtained.

The Cheng group designed a related process for the synthesis of polysubstituted carbazoles [48]. As illustrated in Scheme 15, a formal Rh(III)-catalyzed annulation of 3-(1H-indol-3-yl)-3-oxopropanenitriles **29** with sulfoxonium ylides enables the

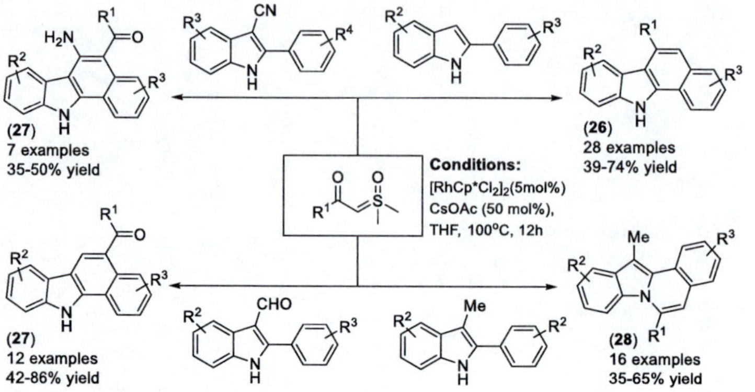

Scheme 14 Selective synthesis of dissimilarly substituted benzo[α]carbazoles and indolo[2,1-α] isoquinolines

Scheme 15 The annulation of aryl 3-oxopropanenitriles with sulfoxonium ylides toward polysubstituted carbazole

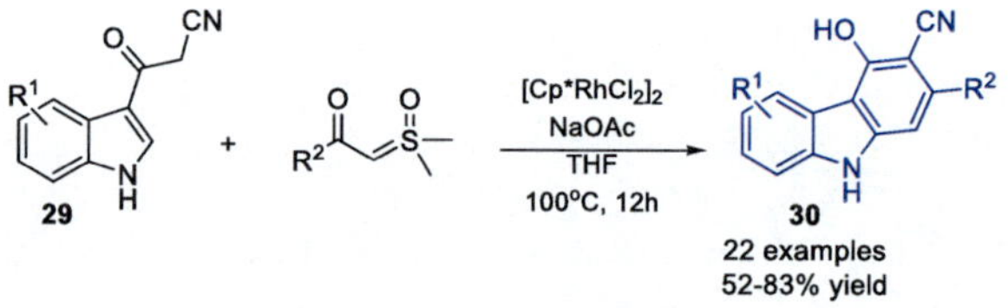

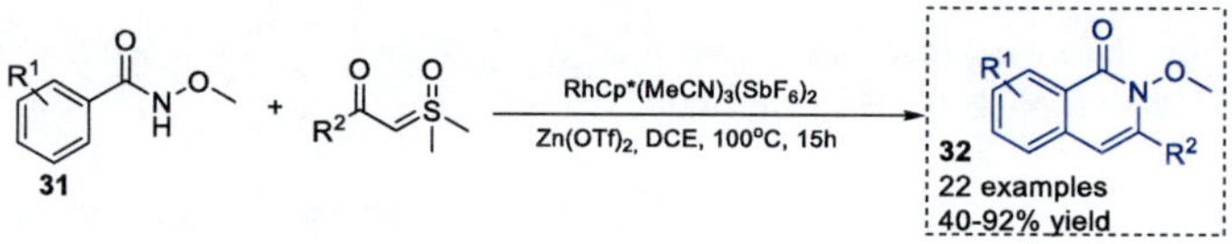

Scheme 16 Chemodivergent synthesis of isoquinolinones

synthesis of complex carbazoles. The substrate scope and functional group tolerance of the reaction are quite extensive and 22 examples of polysubstituted carbazoles **30** were prepared. A plausible reaction pathway is based on previous reports of this work [49].

The C–H functionalization approach to isoquinolinones synthesis developed by Li and co-workers is shown in Scheme 16 [50]. Here, an Rh(III) annulation between *N*-methoxybenzamides **31** and sulfoxonium ylides under a Lewis acid-controlled condition promotes a chemodivergent cyclization. A variety of isoquinolinones **32** were synthesized in moderate to excellent yields using the catalyst RhCp* $(MeCN)_3(SbF_6)_2$ and $Zn(OTf)_2$ as the Lewis acid. In contrast, the sulfoxonium ylide with the sterically hindered *tert*-butyl group failed to undergo the C–H functionalization/annulation reaction. Notably, when other aliphatic sulfoxonium

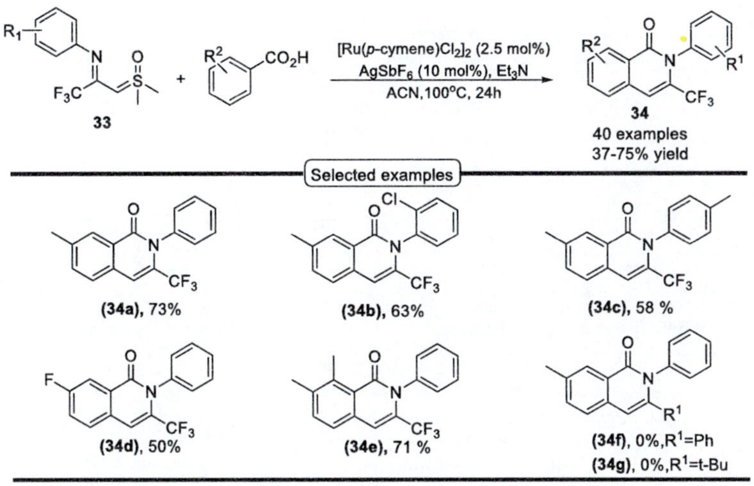

Scheme 17 Trifluoromethyl-containing isoquinolinones via ruthenium(II)-catalyzed C–H alkylation/amidation

ylides and PivOH were used under the optimal reaction conditions, isoquinolinone products were formed in high yields. To gain some insight into the reaction mechanism, the authors conducted some deuterium labeling experiments, and the results suggested that C–H bond cleavage is not involved in the turnover-limiting step.

More recently, Cheng and co-workers constructed a series of CF_3 isoquinolinones **34** through the Ru(II)-catalyzed C–H functionalization/annulation approach [51]. To this end, a range of CF_3-imidoyl sulfoxonium ylides (**33**) was selected as carbene precursors together with carboxylic acids that serve as a weakly coordinating directing group to facilitate the desired *ortho*-C–H activation by the metal (Scheme 17). The optimized reaction condition showed that the better additives were $AgSbF_6$ (10 mol%) and triethylamine (2 equiv). In the absence of a base or an additive, the yield of the desired product was significantly decreased. Aromatic sulfoxonium ylides with electron-donating and electron-withdrawing groups at the *ortho*, *meta*, and *para* positions give the desired product in moderate to good yields. Interestingly, highlighting the importance of the CF_3 group in the success of this reaction, the authors observed that phenyl and *t*Bu-containing imidoyl sulfoxonium ylides failed to give the isoquinolinone products. A deuterium incorporation experiment with CD_3OD as a co-solvent showed that H/D exchange occurred at the *ortho*-position of the carboxylic acid. Furthermore, the kinetic isotope effect (KIE = 3.0) indicated that C–H bond cleavage is involved in the rate-limiting step.

Rhodium(III)-catalyzed chemoselective [5 + 2] cycloaddition between *N*-(cyanoacetyl)indolines and sulfoxonium ylides has also been demonstrated [52]. As illustrated in Scheme 18, $Cp^*Rh(OAc)_2 \cdot H_2O$, LiOAc, and *N*-(cyanoacetyl)-indolines **35** reacted with sulfoxonium ylides to produce seven-membered ring fused tricyclic skeletons **36** in moderate to good yields. However, five- and six-membered fused rings (**37**) were also isolated as minor products in some cases. According to the

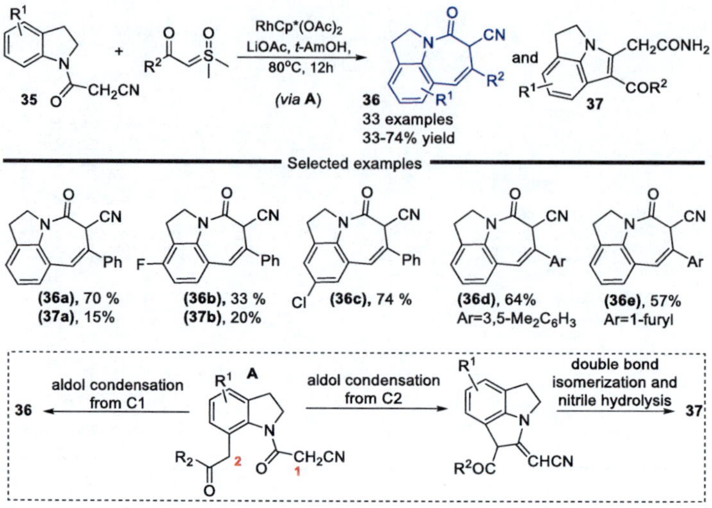

Scheme 18 Rh (III)-catalyzed synthesis of hydrogenated tricyclic indoles

authors, temperature was a determining factor for the reaction efficiency, and heating to 80°C was optimal.

Isoquinolines represent important structural motifs frequently found in natural products, functional materials, and pharmaceuticals [53, 54]. In 2018 the groups of Li and Wang independently reported the synthesis of a variety of isoquinoline derivatives [55, 56]. Li and co-workers demonstrated the use of [RhCp*$(CH_3CN)_3(SbF_6)_2$] with $Zn(OTf)_2$ as an additive for the C–H bond functionalization and annulation of *N*-aryl or -alkyl benzamidines with sulfoxonium ylides [55]. Thereafter, the Wang group described a ruthenium-catalyzed direct mono-C–H functionalization/annulation reaction of benzimidates and sulfoxonium ylides in the presence of an organic acid additive [56]. In both methodologies, the nitrogen of the imino function was used as a DG for the selective coordination and C–H activation step at the ortho-position in the aromatic ring. The desired isoquinolines (**38** and **39**) were achieved in moderate to good yields (Scheme 19). However, in the Rh-catalyzed version longer reaction times (36 h) are needed.

Three years later, Zhang and co-workers presented a rhodium(III)-catalyzed annulation of enamides and sulfoxonium ylides [57]. For this, a low catalytic amount of $[RhCp^*Cl_2]_2$ (2.5 mol%) was used, and no additive or base was necessary. In addition, the nature of the solvent was a determining factor to obtain the desired product, and hexafluoroisopropyl alcohol proved to be optimal (Scheme 20).

The development of novel and creative methodologies for the synthesis of indoles involving sulfoxonium ylides is highly appealing in the synthetic organic community. This ubiquitous structural motif is widely present in biologically active compounds and natural products [58]. The groups of Yu and Wang described almost at the same time the synthesis of 1-pyridin-indole derivatives [59, 60]. Interestingly, *N*-aryl-2-aminopyridine was used as the substrate and directing group at the same time.

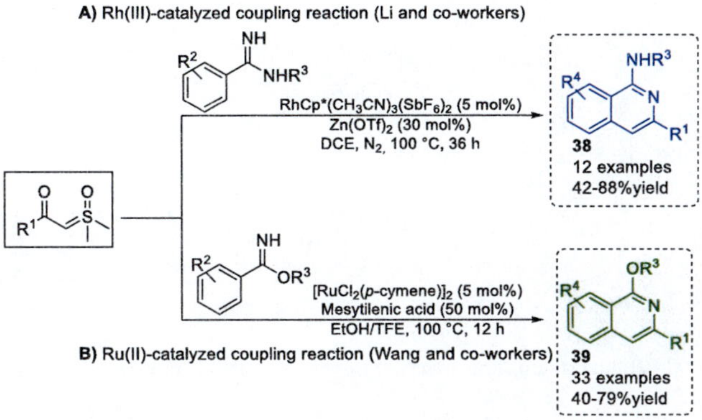

Scheme 19 Synthesis of isoquinolines catalyzed by Rhodium (III) and Ruthenium (II)

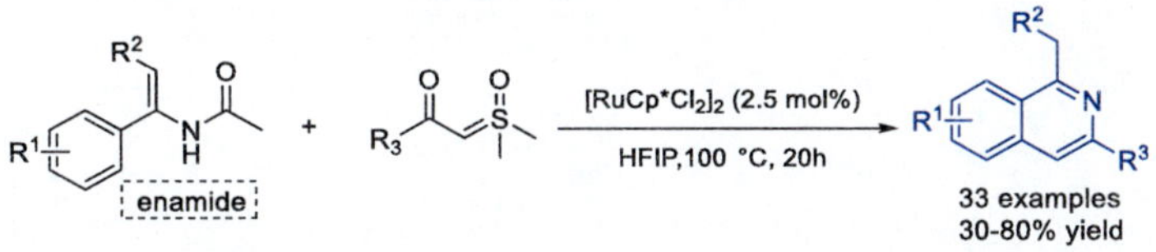

Scheme 20 Rhodium(III)-catalyzed annulation of enamides with sulfoxonium ylides

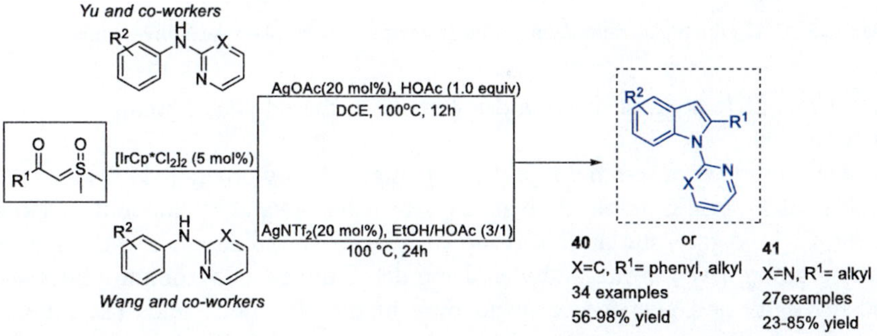

Scheme 21 Iridium (III)-catalyzed [3 + 2] annulation reaction to *N*-pyridin indole synthesis

Either protocol used $[IrCp^*Cl_2]_2$ as the metal catalyst and silver salts as an additive (Scheme 21). Yu and co-workers reported that aryl and alkyl β-ketosulfoxonium ylides participate in C–H functionalization with *N*-phenylpyridin-2-amines to afford the desired products (**40**) in good to excellent yields [59]. On the other hand, Wang and co-workers used more sterically hindered sulfoxonium ylides to access 2-alkyl indoles (**41**) via a [3 + 2] cycloaddition reaction [60]. The substrate scope with sulfoxonium ylides was limited to those that are aliphatic. When a sulfoxonium ylide with an aryl group was employed, a low yield (23%) of the indole was observed. Moreover, in both reports the kinetic isotope effect was measured, and the results

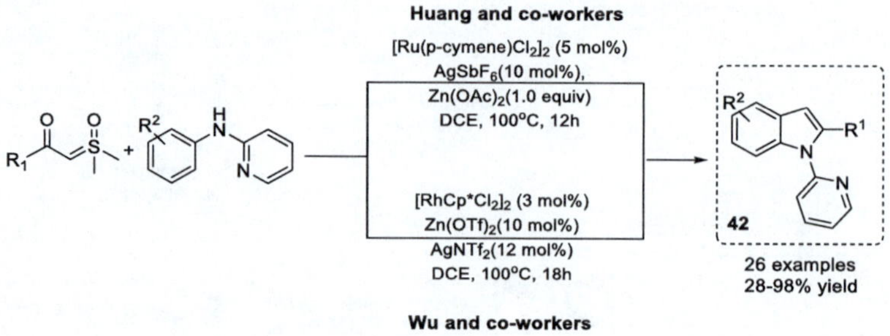

Scheme 22 Metal-catalyzed coupling reaction between *N*-aryl-2-pyridinamines and sulfoxonium ylides

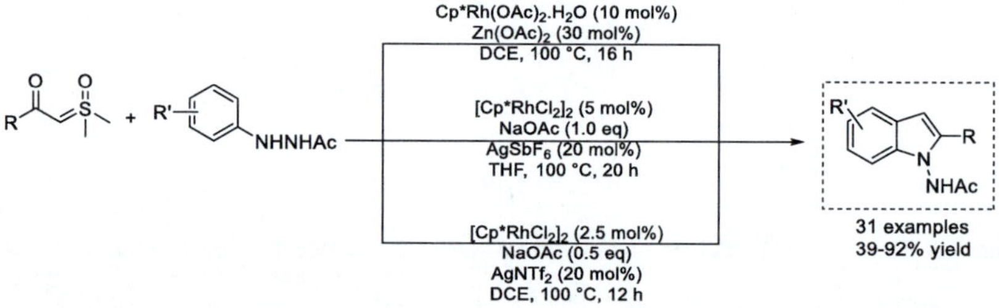

Scheme 23 C–H Functionalization/annulation reactions with *N*′-phenylacetohydrazines

indicate that C–H bond cleavage did not occur in the rate-determining step of this transformation.

In the same year, a seminal report by Huang and co-workers [61] described the synthesis of 1-pyridin-indole derivatives (**42**) using a catalytic amount of [Ru(*p*-cymene)Cl_2]$_2$. Again, the amino group of *N*-aryl-2-pyridinamines was used as a directing group. Even *ortho*-methylated and disubstituted sulfoxonium ylides show good reactivity and give the corresponding indoles in good yields (Scheme 22). Soon after, the Wu group developed a similar protocol only differing in the catalytic system and reaction time [62].

Arylhydrazines are also suitable substrates for directed C–H functionalization/annulation reactions. This feature was documented by the groups of Xie and Zhang that reported a Rh(III) catalyzed methodology for the synthesis of an extensive family of 1-aminoindoles [63, 64]. To this end, substituted *N*′--phenylacetohydrazines reacted with a wide range of keto sulfoxonium ylides (aromatic and aliphatic) with Rh-(III) catalysis at 100°C (Scheme 23). One year after these reports, Liu and co-workers described a very similar process with lower reaction times [65].

One of the main approaches to the synthesis of 3-ketoindoles (**43**) is the Friedel–Crafts acylation between indoles and oxalyl chloride. However, it suffers from poor selectivity, low yield, and multistep operations [66]. An efficient methodology for

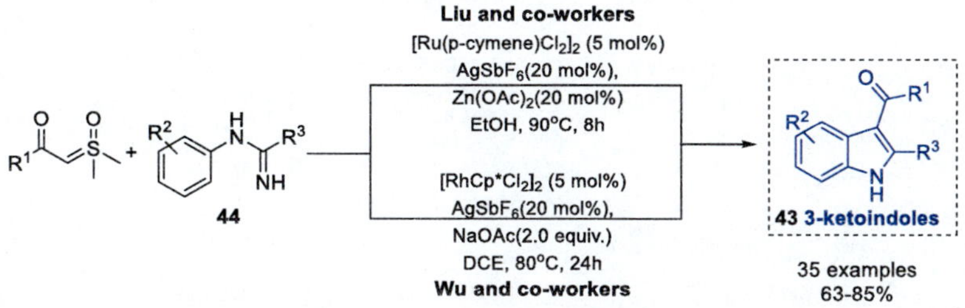

Scheme 24 Synthesis of 3-ketoindoles from aryl imidamides

the construction of these compounds has been revealed by Zhou and co-workers [67]. The reaction of several aryl imidamides (**44**) with sulfoxonium ylides in the presence of $[Ru(p\text{-cymene})Cl_2]_2/AgSbF_6$ as the catalyst was studied (Scheme 24). The reaction yield was strongly influenced by the solvent, with ethanol being the most appropriate. This reaction protocol was compatible with various substituted imidamides and sulfoxonium ylides with moderate to good yields. Surprisingly, this transformation allows selective [4 + 1]-cycloaddition to yield 3-ketoindoles avoiding the [4 + 2]-cycloaddition to isoquinoline products. Similarly, the Hai group also used N-arylamidines with catalyst $[RhCp^*Cl_2]_2/AgSbF_6$ in the presence of NaOAc as an additive [68].

Other valuable substrates for the synthesis of indoles are the *N*-nitroso anilines (**45**). An exciting example of the use of these substrates was described by Cui and co-workers [69] who reported that the *N*-nitroso group works not only as a directing group but also as an internal nitrosation agent. According to the authors, screening various protonic acids showed that strong acidic conditions were conducive in achieving the conversion to 3-nitrosoindoles **46** (Scheme 25). In addition, this protocol allows wide reaction scope with different aromatic sulfoxonium ylides in the presence of $[Cp^*RhCl_2]_2$ with HFIP as the best solvent. In contrast, the use of aliphatic or sterically hindered sulfoxonium ylides has a detrimental effect on the reaction yield.

Pawar and co-workers described an elegant approach for the synthesis of indoloindolones [70]. The authors developed a Cp*Co(III)-catalyzed C–H functionalization cascade process using anthranils (**47**), wherein both the functionalities amine and aldehyde were utilized for functionalization with aryl sulfoxonium ylides. This Cp*Co(III)-catalyzed [4 + 1] cycloaddition process was performed in DCE as solvent and heated in an oil bath to 110°C (sealed vial), achieving indoloindolone derivatives (Scheme 26a). In addition, Wu and co-workers described a similar work to form indoloindolones via rhodium-catalyzed [4 + 1] cycloaddition in good yields using silver salt and HOAc as additives (Scheme 26b) [71].

Another example to illustrate the synthesis of one nitrogen containing heterocyclic compounds was developed by the group of Lee in 2021 [72]. Rhodium(III)-catalyzed chemodivergent synthesis of azulenolactams was achieved from the reaction of *N*-methoxyazulene-1-carboxamides with sulfoxonium ylides (Scheme 27). This reaction was revealed to be quite efficient in terms of yields.

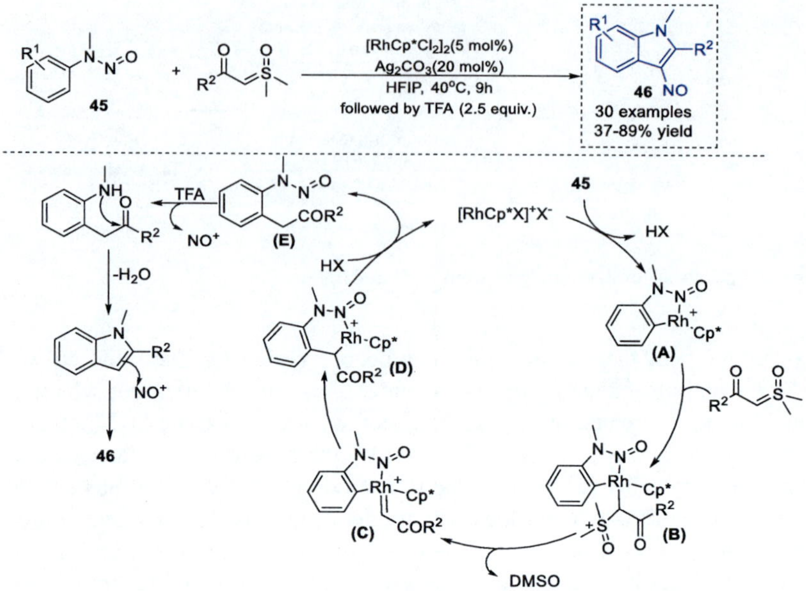

Scheme 25 Rh(III)-mediated synthesis of 3-nitrosoindoles

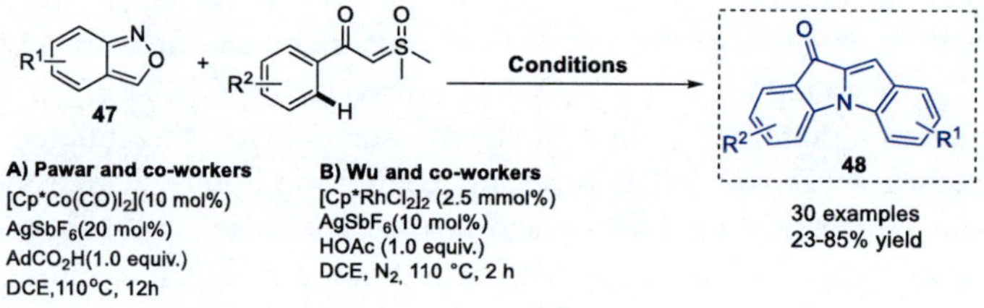

Scheme 26 Cp*Co(III)-catalyzed C–H amination/annulation for the synthesis of indoloindolones

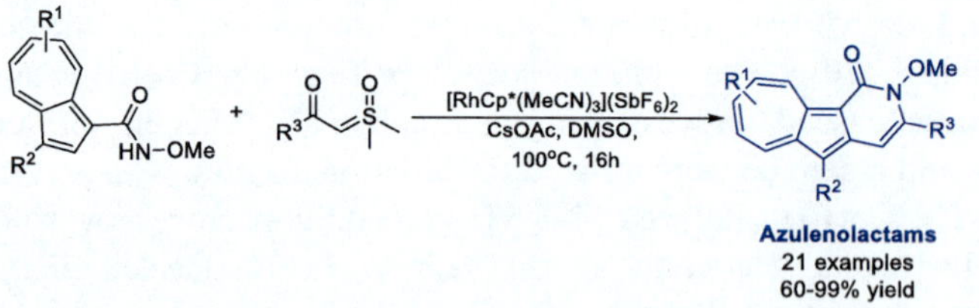

Scheme 27 Synthesis of azulenolactams from Rh(III)-catalyzed cyclization

2.2.2 Oxygen Heterocycles

Oxygen-containing heterocycles constitute an important class of organic compounds since they are present in numerous natural products, drugs, and molecules with biological activities [73]. Among the existing synthetic methodologies for the synthesis of isocoumarins, reactions based on transition-metal catalyzed C–H bond activation are the most important and effective since they enable the preparation of products in a high atom and step-economical manner. As already mentioned, sulfoxonium ylides have recently become an important tool for safely and practically accessing carbene precursors to synthesize heterocycles [74].

In 2018, Li and co-workers reported a series of chemodivergent and redox-neutral annulations between *N*-methoxy-benzamides and sulfoxonium ylides via Rh(III)-catalyzed C–H activation (Scheme 28) [50]. From the same starting materials, the authors obtained isocoumarins (**49**) and isoquinolones (**50**), another class of heterocycle that is present in numerous pharmaceuticals. The transformation showed high selectivity for isocoumarin when PivOH was employed since the acid acts to activate the amide carbonyl group toward oxygen attack.

In the same year Ackermann and co-workers also reported annulations of benzoic acids and sulfoxonium ylides providing isocoumarins with ample scope (Scheme 29a) [75]. Catalysts of Ru outperformed those of Rh and Ir, furnishing key intermediates by C–H/O–H functionalizations through weak *O*-coordination. In addition, the authors verified that doubling the number of equivalents of the sulfoxonium ylide led to doubling the C–H functionalized products.

In 2019, Wang and co-workers reported a protocol for Rh(III)-catalyzed annulation reactions between 2-aryl-oxazolines and sulfoxonium ylides [76]. After the attachment of the methylene carbonyl group from the sulfoxonium ylide, cyclization occurred from the attack of the enol form to the oxazoline imine group, followed by hydrolysis, under acid catalysis (Scheme 29b).

Similarly, in 2022 Li and co-workers developed a Ru(II)-catalyzed C–H bond functionalization of *N*-iminopyridinium ylides with sulfoxonium ylides (Scheme 29c) [77] that also furnished isocoumarins. When compared to Wang's methodology, the main differences were the absence of silver salt and the use of Ru (II) instead of Rh(III). Both methodologies showed a wide substrate scope and good tolerance of the functional group, in addition to being able to be performed on a gram scale.

Based on a different strategy, Zhou and co-workers developed a Ru(II)-catalyzed homocoupling method to access substituted isocoumarins **50** in 2019 (Scheme 30a) [78]. This efficient and step-economic method has the advantage that the sulfoxonium ylide acts both as the aromatic substrate and the acyl methylating

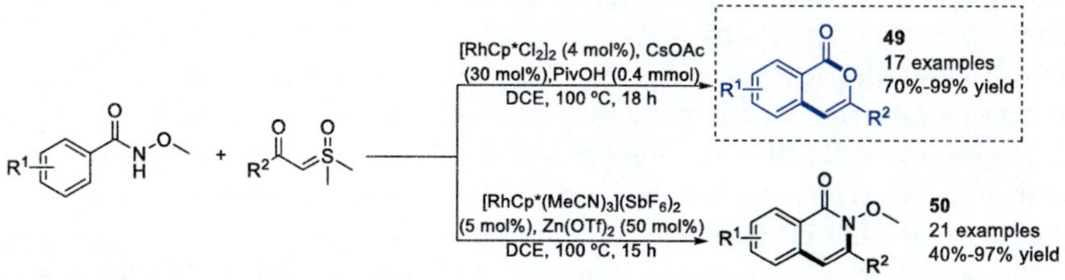

Scheme 28 Synthesis of isocoumarins and isoquinolones from *N*-methoxybenzamides and sulfoxonium ylides via C–H functionalization

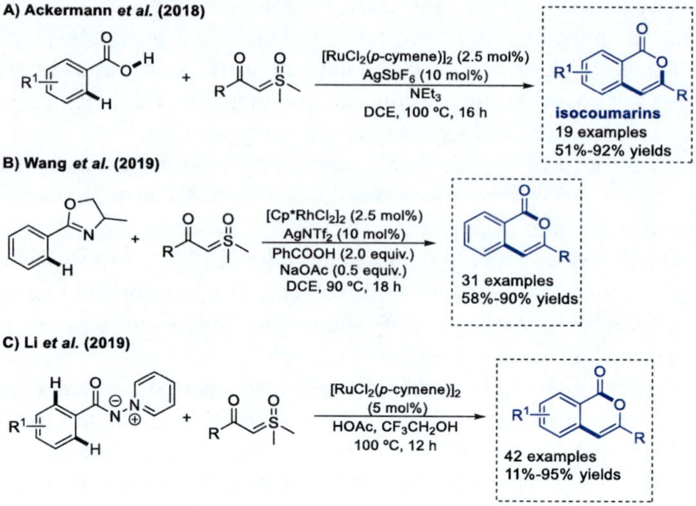

Scheme 29 Ru- and Rh-catalyzed annulation reactions employing sulfoxonium ylides

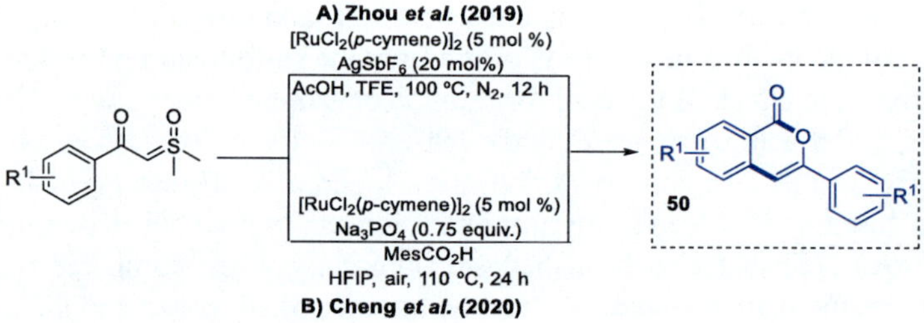

Scheme 30 Homocoupling reactions of sulfoxonium ylides toward substituted isocoumarins

reagent in the reaction. One year later, in 2020, Cheng and co-workers also reported this Ru(II)-catalyzed annulation between two molecules of sulfoxonium ylides, but in this case, without the use of silver salt (Scheme 30b) [79].

Aiming to obtain other useful heterocycles, in 2018, Li and colleagues developed a methodology for the synthesis of naphthols (**51**) and 2,3-dihydronaphtho[1,8-bc] pyrans (**52**) via multiple C–H functionalizations (Scheme 31) [80]. Naphtho[1,8-bc] pyrans were obtained by the subsequent addition of TfOH to the benzoylacetonitriles. Interestingly, replacing the cyano group with acetyl, ester, and arenesulfonyl did not allow the formation of product. The mechanism proposal was based on a series of control experiments, as well as from the studies already described in literature. After formation of intermediate **A**, a cyclization-condensation gives the 1-naphthol intermediate **51**. Finally, from **51**, the hydroxyl intramolecularly attacks the carbonyl group in R to form the cyclic hemiacetal **52** (Scheme 31).

Accessing the same scaffold, but with different substituents, Wu and co-workers reported a Ru(II)-catalyzed *peri*-selective C–H functionalization/annulation of

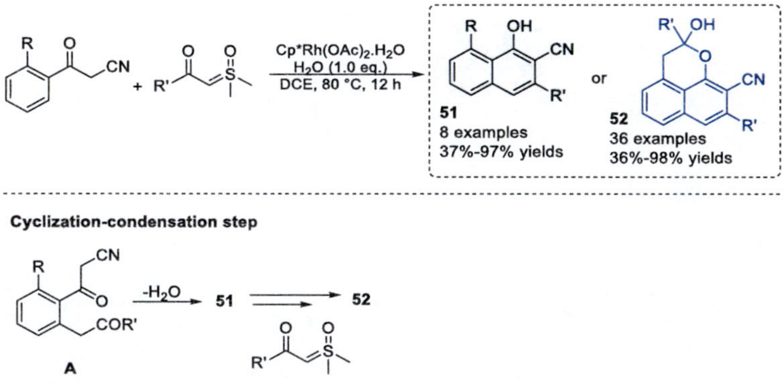

Scheme 31 Rh(III)-catalyzed cascade C−H functionalization of benzoylacetonitriles, followed by annulation, with sulfoxonium ylides

Scheme 32 Ru(II)-catalyzed C−H functionalization of naphthols, followed by annulation, with imino sulfoxonium ylides

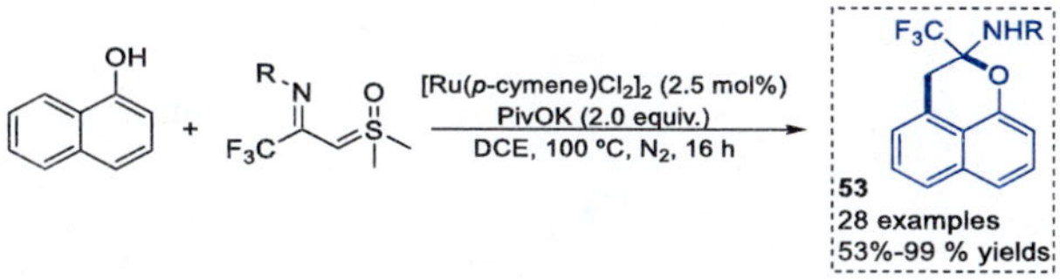

1-naphthols with sulfoxonium ylides for the synthesis of 2-(trifluoromethyl)-2,3-dihydrobenzo[*de*]chromen-2-amines **(53)** in 2022 (Scheme 32) [81]. It is worth mentioning that heterocycles with the CF_3 group represent an important class of organic compounds with wide applications in pharmaceuticals, agrochemicals, and functional material.

In 2019, Prabhu and co-workers developed a method to obtain lactone derivatives, containing naphthol moieties (Scheme 33) [82], which are present in natural products and pharmaceuticals. Employing functionalized alkynes such as **(54)** and sulfoxonium ylides, furanone-fused naphthols **(55)** were obtained by a domino Rh (III)-catalyzed C−H functionalization, regioselective annulation, and lactonization. Interestingly, the products obtained are analogues of the natural product Fimbricalyx lactone A (Scheme 33). In addition, the methodology proved to be effective in synthesizing bromo-lactones, which are used as intermediates in synthesizing photochromic dichroic materials.

Although the mechanism is similar to the general one already presented in Scheme 13, it differs in some points that deserve illustration (for example, C–H functionalization occurs in the aromatic ring of the sulfoxonium ylide, which also acts as the directing group. Moreover, it is not the ylide that inserts into the formed metallacycle after the C–H activation step, but the alkyne) (Scheme 33). The first step is the formation of catalyst **A** in the presence of $AgNTf_2$ and $ClCH_2COOH$. Then, rhodacycle **B** is formed by the coordination of the oxygen of the sulfoxonium ylide enolate to an active metal complex, followed by C–H activation by the metal. Later, alkyne insertion to the metallacycle **B** leads to the formation of seven-

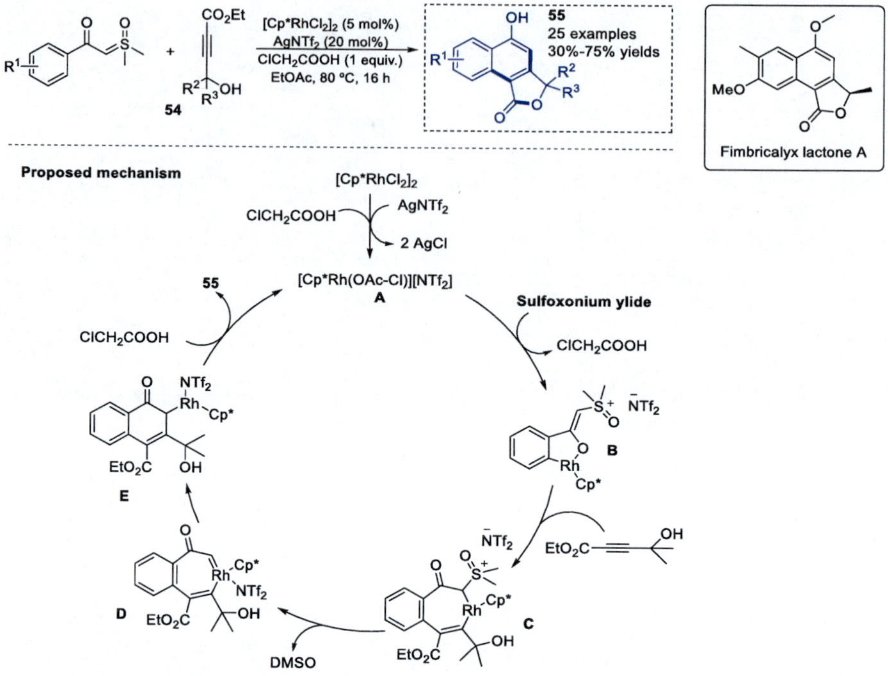

Scheme 33 Rh(III)-catalyzed synthesis of lactones, bearing a naphthol ring, from alkynes and sulfoxonium ylides

membered rhodacycle **C**, which eliminates DMSO, furnishing rhodacycle **D**. Reductive elimination leads to the intermediate **E** and finally, in the presence of chloroacetic acid, protodemetalation and regeneration of active catalyst **A** lead to the formation of the desired product. The regioselectivity observed (see **C**) is probably due to the stronger secondary interaction between the metal and the hydroxyl group of the alkyne, as well as to the steric hindrance caused by the alkyne geminal dimethyl groups. Although not proposed, a cationic Rh, with the weak-coordinating $^{-}NTf_2$ group as the counteranion (in **D** and **E**) can also be considered.

In 2019, Yao and co-workers designed rhodium(III)-catalyzed C–H functionalization/annulation of salicylaldehydes with sulfoxonium ylides (Scheme 34) [83], furnishing 2-substituted chromones in good yields and broad functional group tolerance. This transformation is interesting because there are only a few examples in the literature of synthetic methods to directly access 2-substituted chromones. Regarding the mechanism, after the formation of the key intermediate (via general mechanism), an intramolecular dehydrative condensation promoted by silver salt furnishes chromone product.

Another example of this same transformation was reported in the same year by Yu and co-workers [84]. The main differences from the previous work were the absence of silver salt, the use of CsOAc, and a mixture of H_2SO_4 and AcOH for the acid-promoted cyclization. Although cyclization reactions catalyzed by transition

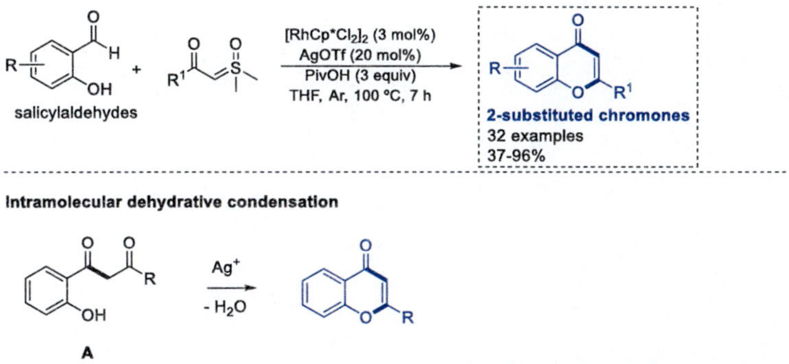

Scheme 34 Synthesis of chromones through C–H functionalization/annulation of salicylaldehydes

metals have already been widely explored, the C–H functionalization process is almost always restricted to benzenoid aromatic compounds. Although less common, compounds such as azulene, for example, are also likely to undergo C–H functionalization. However, regarding the resonance structures of azulene, the most stable one has a formal positive charge on the 7-membered ring and a formal negative charge on the 5-membered ring, giving the molecule a dipole moment of 1.08 D. This is why the 2-position of azulene (see resonance structure in Scheme 35) is less reactive than 1- and 3-position, which makes it very difficult to introduce substituents at the 2-position. In this sense, in 2020 Lee and co-workers studied the reaction between aromatic azulenes and sulfoxonium ylides (Scheme 35) [85]. They developed a regioselective and chemodivergent reaction between *N*-methoxyazulene-1-carboxamides (**56**) and sulfoxonium ylides leading to the selective formation of azulenolactones (**57**) and azulenolactams (**58**) from the same starting material. In this transformation, sulfoxonium ylides act as a precursor of a secondary carbene, providing functionalization of the less reactive 2-position of azulene. With respect to the mechanism, after the initial general steps already discussed previously in Scheme 13, the acylmethylated intermediate **A** is formed (Scheme 35). At this point, two pathways are possible. In pathway a, activation of the amide group by PivOH supports a nucleophilic attack by the enol oxygen, leading to the formation of azulenolactones with the elimination of protonated NH_2OMe. On the other hand, in pathway b the intramolecular nucleophilic addition from de amide nitrogen atom, followed by dehydration, affords azulenolactam.

In 2022 Shankaraiah and co-workers reported a Ru-catalyzed carbene annulation of sulfoxonium ylides on the olefinic bond of *cis*-stilbene acid to obtain biologically relevant α-pyrone skeletons (**59**) [86] (Scheme 36). The authors also employed a hypervalent iodonium ylide as a carbene precursor to obtain chromene-2-one derivatives, but it did not show compatibility with the Ru catalyst system. As observed from the substrate scope of sulfoxonium ylides, good yields and broad functional group tolerance were obtained. According to the general mechanism, intermediate

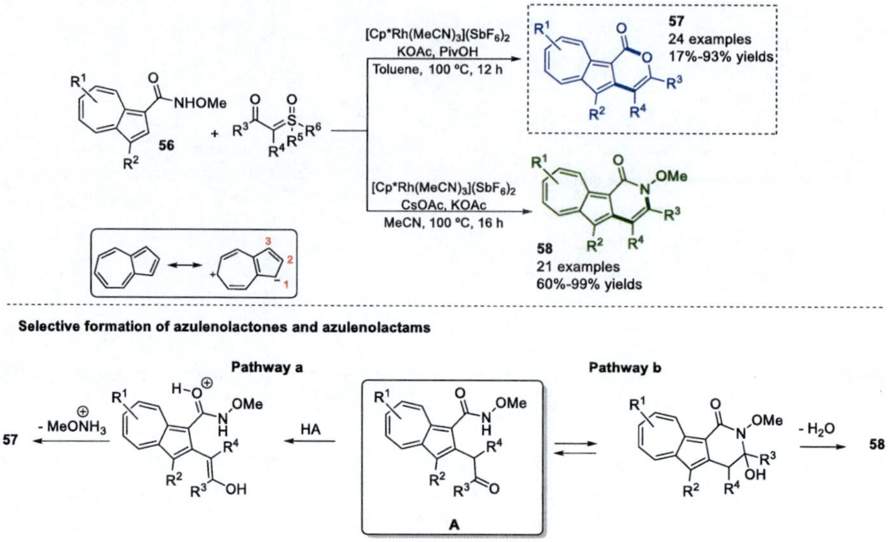

Scheme 35 Selective formation of azulenolactones and azulenolactams from sulfoxonium ylides and *N*-methoxyazulene-1-carboxamides

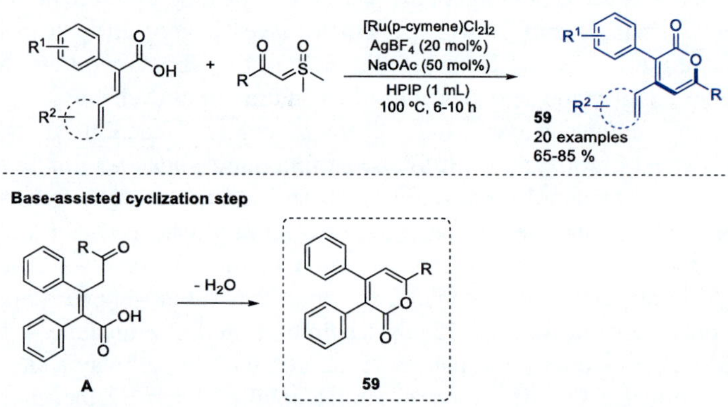

Scheme 36 Synthesis α-pyrones by transition-metal catalyzed annulations of sulfoxonium ylides and *cis*-stilbene acids

A is formed after the protodemetalation step. Next, cyclization and elimination afford product **(59)** (Scheme 36).

2.2.3 Heterocycles Containing Two or More Heteroatoms

Heterocycles containing two or more heteroatoms incorporating S, N, and/or O are an interesting class of organic compounds with fundamental significance in medicinal chemistry due to their presence in many natural and bioactive compounds [87, 88]. Different routes to obtain these compounds are being documented [89], and C(sp^2)-H functionalization with sulfur ylides represents one of the strategies to access this class of heterocycles. In this regard, Rh(III)- and Ru(II)-catalysts are the catalysts of choice in C(sp^2)-H functionalization using sulfur ylides as a substrate.

For example, six-membered ring heterocycles containing S and N can be obtained by the coupling-cyclization reaction between sulfoximines (known to be a nucleophilic directing group in C–H functionalization) and β-ketosulfoxonium ylides (bifunctional C_2 or C_1 synthons) catalyzed by Ru(II) or Rh(III) [90, 91]. In 2018, Zheng and co-workers described the synthesis of sulfoximine-containing benzothiazines using Rh(III)-catalysis via an annulative C–H functionalization reaction (Scheme 37a) [90]. In the presence of a catalytic amount of [RhCp* $(CH_3CN)_3(SbF_6)_2$] and $Zn(OTf)_2$ as an additive, several 1,2-benzothiazine products **61** were obtained in good to excellent yields. This protocol allowed modifications in both sulfoximine and β-ketosulfoxonium ylides. The scope was not limited to aryl-substituted sulfoxonium ylides; alkyl-substituted sulfoxonium ylides were also well tolerated under the optimized reaction conditions. However, longer reaction times (36 h) and higher temperatures (100°C) were required. Interestingly, the protocol allows the use of other arenes, such as substituted *N*-aryl or alkyl benzoamidines and benzophenone imines to provide isoquinolines in yields up to 88%.

In the same year an important report from Xie and co-workers disclosed that benzothiazines could be obtained using Ru(II)-catalyzed coupling-cyclization of S-aryl sulfoximines and sulfur ylides (Scheme 37b) [91]. 1,2-benzothiazines (**61**) were formed using a catalytic amount of [$RuCl_2$(*p*-cymene)]$_2$ in the presence of a silver salt additive. However, the addition of an acid is required to achieve high reaction yields. The scope of the reaction was demonstrated for a range of sulfoxonium ylides and *S*-aryl sulfoximines. The corresponding 3-aryl 1,2-benzothiazines were obtained with moderate to excellent yields up to 98%. However, the use of sterically hindered *ortho*-substituted benzoyl or α-fattyacyl sulfoxonium ylides provides the corresponding 3-aryl or 3-alkyl 1,2-benzothiazine

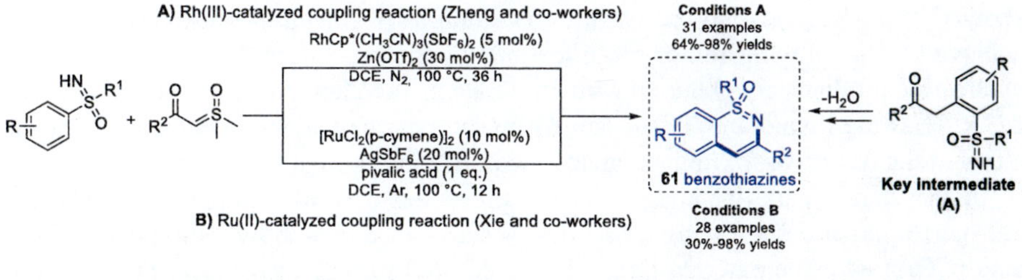

Scheme 37 Metal-catalyzed coupling reaction between sulfoximines and β-ketosulfoxonium ylides

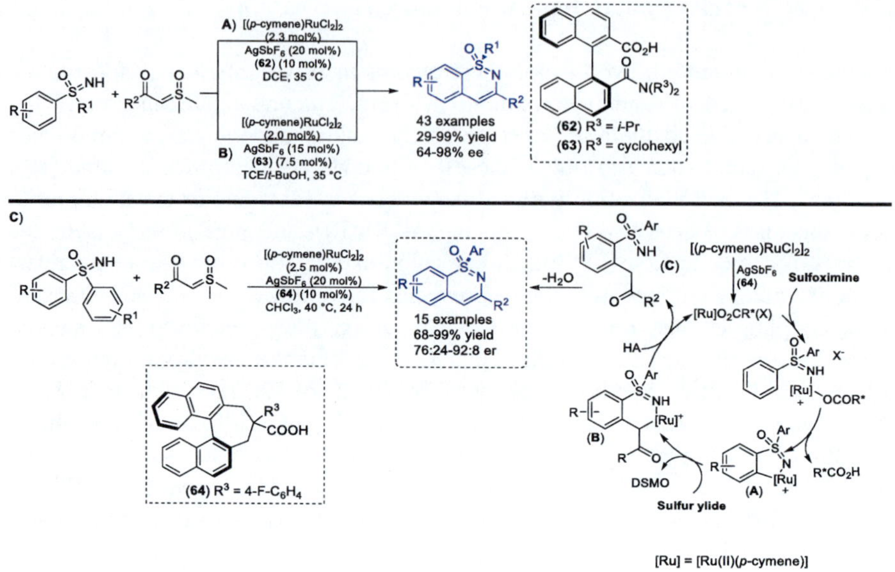

Scheme 38 Asymmetric synthesis of 1,2-benzothiazines

with poor to moderate yields. In addition, *S*-alkyl-substituted-*S*-phenylsulfoximines (alkyl = benzyl, ethyl, and cyclopropyl) were also tolerated under the optimized reaction conditions. However, the corresponding products were achieved in poor to moderate yields. Mechanistically, an intermediate (**A**) is released at the end of the catalytic cycle (see Scheme 13), which undergoes intramolecular carbonyl addition from the imine nitrogen, followed by a dehydration reaction to provide the benzothiazine product.

More recently, Zhou and co-workers reported the first asymmetric method to prepare a range of sulfoximines with excellent yields and enantioselectivities, employing a chiral carboxylic acid (CCA) (**62**) as an external ligand in the Ru(II)-catalyzed enantioselective C–H functionalization/annulation reaction between sulfoximines with sulfoxonium ylides (Scheme 38a) [92]. Under optimized conditions, 3-alkyl-1,2-benzothiazines were prepared in good yields (80–98%) and demonstrated excellent enantioselectivity (97–99% *ee*). In the same work, the authors showed that Ru(II)-catalyzed kinetic resolution of racemic sulfoximines was achieved by employing minor modifications in the procedure (Scheme 38b). The cyclization products are obtained with moderate to excellent enantioselectivity using CCA **63** as the ligand and TCE/*t*-BuOH mixture as the solvent. This protocol also allowed the use of sulfoximines having two different aromatic groups.

In the same year Matsunaga's group presented a similar protocol to achieve 1,2-benzothiazines employing a pseudo-C_2-symmetric binaphthyl chiral carboxylic acid (CCA) **64** as an external ligand (Scheme 38c) [93]. This protocol allowed the synthesis of several benzothiazines with good yields (up to 99%) and modest to good

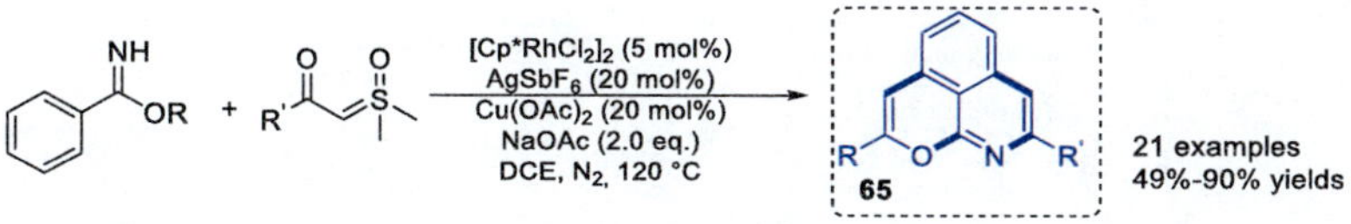

Scheme 39 Rh(III)-catalyzed coupling reaction between alkyl benzimidates and sulfur ylides

enantioselectivities (76:24 to 92:8 er). Scope modification for both sulfoximines and sulfoxonium ylides was also reported. The mechanistic catalytic cycle was also reported by the authors. This reaction was postulated to involve an initial formation of an active Ru-carboxylate catalyst from $[RuCl_2(p\text{-cymene})]_2$, $AgSbF_6$ and CCA **64**. After coordination with the sulfoximine, followed by C–H functionalization, five-membered metallacycle **A** was formed. Reaction with the sulfoxonium ylide to provide the metal-carbene and subsequent migratory insertion afforded intermediate **C**. The protonolysis reaction of **C** provides intermediate **D** that after intramolecular condensation gives the desired product.

Following the theme of accessing heterocycles containing two or more heteroatoms, the Cheng group reported a method for the synthesis of functionalized isoquinazolones **65** from ethyl benzimidates and α-aroyl sulfur ylides via rhodium-catalyzed dual *ortho*-C(sp^2)-H functionalization and annulation reaction (Scheme 39) [94]. The nature of the solvent is a determining factor to obtain the desired product, and DCE proved to be optimal. In addition, additives such as silver and copper salts and a base are required to improve the reaction efficiency. This approach allowed the modification of both ethyl benzimidates and α-aroyl sulfur ylides providing several functionalized isoquinazolones in moderate to excellent yields.

In 2018, the Ellman group reported a series of protocols giving access to several [5,6]-bicyclic heterocycles with *N*-fused bridgeheads. Initially, they reported the preparation of azolopyrimidines **66** by the Rh(III)-catalyzed direct imidoyl C(sp^2)-H activation of *N*-azolo imines and sulfoxonium ylides, followed by intramolecular condensation (Scheme 40a) [95]. The optimized reaction conditions showed that the addition of PivOH, NaOAc, and molecular sieves was required to improve the reaction yields. Different imines derived from 2-amino-imidazole and benzaldehydes were tested; electron-rich and electron-deficient benzaldimines were also tolerated affording the product in moderate to good yields. Aryl and alkyl sulfoxonium ylides were well tolerated. In addition, this reaction protocol allows the use of diazoketones and alkynes instead of sulfur ylides.

Another research by the same group highlighted an approach in which different substituted bridgehead *N*-fused [5,6]-bicyclic heterocycles **(67)** were accessed from alkenyl azoles and sulfoxonium ylides via a Rh(III)-catalyzed coupling reaction (Scheme 40b) [96]. This was readily achieved using toluene as the solvent at 120° C for 16 h. In the same way, PivOH and NaOAc proved to be important for improving the reaction yields. Solvents like dioxane, acetonitrile, and DCE were unsuitable for this reaction. This protocol allowed modification in both sulfoxonium ylides and alkenyl azoles, affording the desired product in poor to excellent yields. A

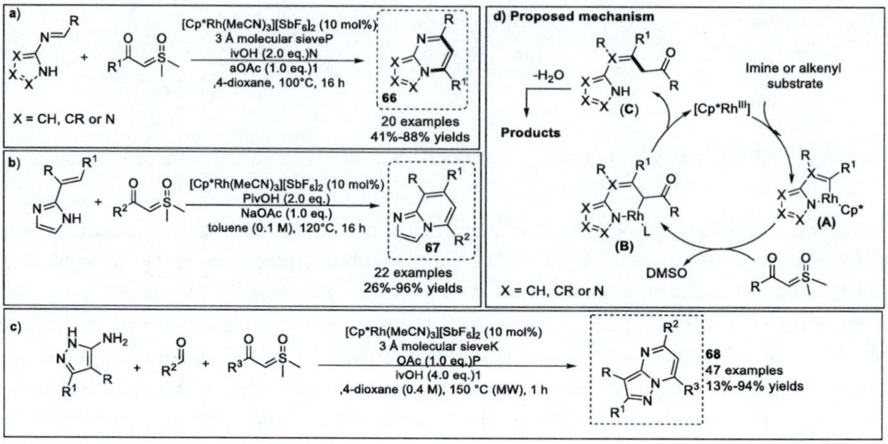

Scheme 40 Synthesis of azolopyrimidines and proposed mechanism

particularly attractive synthetic feature of this reaction is the fact that it is regioselective.

A three-component coupling reaction between aminopyrazoles, aldehydes, and sulfoxonium ylides was employed as an interesting strategy to access several pyrazolo[1,5-α]pyrimidines (**68**) in good yields via Rh(III)-catalyzed annulation reaction (Scheme 40c) [97]. The effect of catalysts and other parameters was explored in the optimization process. In a one-pot reaction, and under microwave conditions at 150 °C, the desired product could be obtained using the commercially available and air-stable cationic catalyst [Cp*Rh(MeCN)$_3$(SbF$_6$)$_2$ in the presence of KOAc, PivOH, 3 Å sieves, and dioxane as the solvent. Formyl ylides were also substrates in this reaction; however, a two-step and one-pot sequence had to be employed with them. Also, an ester group on the pyrimidine ring of the product could be installed using ethyl glyoxylate as the aldehyde in the two-stage strategy. In addition, the mechanisms for *N*-fused [5,6]-bicyclic heterocycles formation were described (Scheme 40d) using a concerted metalation/deprotonation reaction of the imine or alkenyl azole substrates with the Rh catalyst to produce intermediate (**A**). Next, metal-carbene formation from the sulfoxonium ylide, followed by migratory insertion, gave intermediate **B**. Protonolysis released ketone **C** and regenerated the catalyst. Finally, a cyclodehydration reaction of **C** furnished the product (Scheme 40d). In the case of the three-component coupling reaction, the mechanism starts with the in-situ formation of an imine from aldehyde and the aminopyrazole.

The transition-metal catalyzed C–H functionalization/annulation reaction between *N*-carbamoylindoles and sulfoxonium ylides to form heterocycles-fused indoles was reported by the Yu and Zhang groups in 2021 (Scheme 41) [98, 99]. Initially, both groups found that the use of a catalytic amount of a Rh(III)-catalyst led to a mixture of two cyclization products. After optimization studies, fused tricyclic [1,3]oxazino[3,4-*a*]indol-1-ones **69** could be accessed with high selectivity using [Rh*CpCl$_2$]$_2$ or [Cp*Rh(CH$_3$CN)$_3$](SbF$_6$)$_2$ as the catalyst in presence of additives

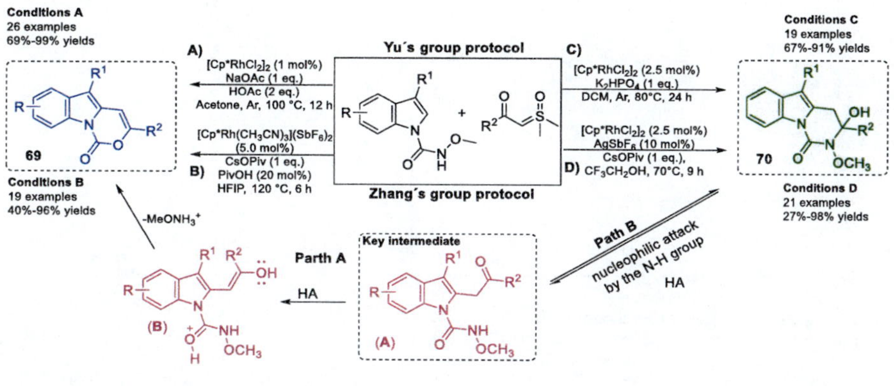

Scheme 41 Synthesis of heterocycles-fused indoles

like PivOH or HOAc and CsOPiv or NaOAc. High temperatures were required in this reaction to achieve high reaction yields 100–120 °C (Scheme 41a, b). On the other hand, dihydropyrimido[1,6-α]indol-1(2*H*)-ones (**70**) could be obtained in high selectivity and moderate to good yields when $[Cp^*RhCl_2]_2$ was employed as the catalyst in basic conditions, using K_2HPO_4 or CsOPiv (Scheme 41c, d). In both cases, the optimized reaction conditions allowed modification in sulfoxonium ylides and *N*-carbamoylindoles. Catalysts $[Cp^*IrCl_2]_2$, $MnBr(CO)_5$, and $[RuCl_2(p\text{-cymene}]_2$ were ineffective. Mechanistically, this process was postulated to involve an initial ligand exchange between the Rh(III) catalyst with NaOAc, K_2HPO_4, or CsOPiv to form the active Cp^*RhX_2 catalyst. Intermediate **A** is formed at the end of the catalytic cycle, which can undergo two different paths to form the products. The first path involves an intramolecular nucleophilic attack of the enol to the urea carboxyl group (see intermediate **B**) to yield the product **69**. The second path considers the nucleophilic attack by the N-H group on the carbonyl group to give product **70.**

The importance of *N*-oxide heterocycles has been exemplified by their application as chiral ligands and biological activities. New methods to access these molecules have been described. One example was reported by Cui and Huang in 2020, using a Rh(III)-catalyzed C–H acylmethylation/cyclization reaction of *N*-nitrosoanilines and sulfoxonium ylides to form indole *N*-oxides **71** (Scheme 42a) [100]. This reaction proceeded using a catalytic combination of $Cp^*Rh(OAc)_2{\cdot}H_2O$ and $Cu(OAc)_2$. In addition, the use of an oxidant is required to improve the reaction yields. Catalysts like $[Cp^*Co(CO)I_2]$, $[RuCl_2(p\text{-cymene})]_2$, and $[Cp^*IrCl_2]_2$ were also investigated; however, none were found to be efficient for the reaction. Different indole *N*-oxides were obtained with the optimized reaction conditions with moderate to high. After the acylmethylation reaction, $Cu(OAc)_2/Ag_2O$ are important additives to form the final product by cyclization.

In 2018, Cheng and Kim separately reported a Rh(III)-catalyzed C–H functionalization reaction between azobenzenes and sulfoxonium ylides leading to 3-alkyl-(2*H*)-indazoles (**72**) (Scheme 42b) [101, 102]. Both groups demonstrated

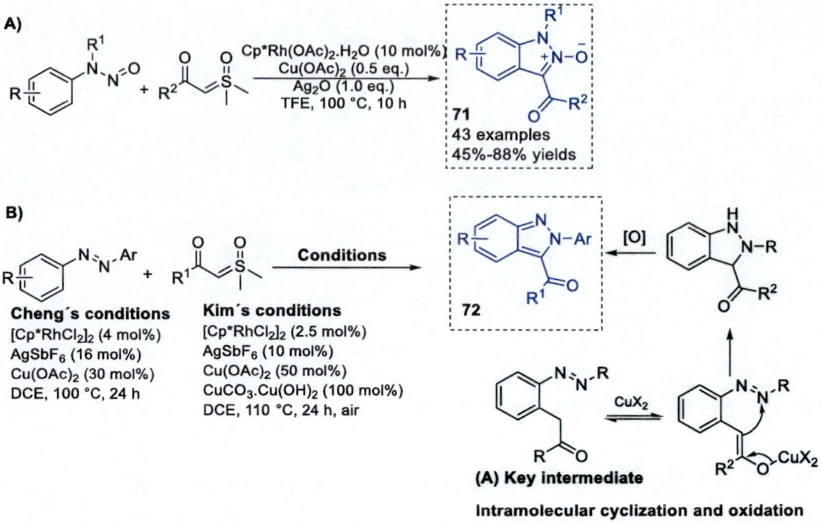

Scheme 42 Synthesis of indoles *N*-oxides and (2*H*)-indazoles via Rh(III)-catalyzed C–H functionalization/annulation approach

that the desired product could be obtained using a catalytic amount of Cp*$RhCl_2$ in presence of silver and copper salts. High temperatures and long reaction times (24 h) are required to achieve high reaction yields. Substituted azobenzenes and different alky, aryl, and heteroaryl α-carbonyl sulfoxonium ylides worked well under the optimized reaction conditions, affording the corresponding indazole in moderate to excellent yields. However, electron-deficient azobenzenes provide indazoles in poor to moderate yields. Unsymmetrical azobenzenes could also be used as the substrate, and the selectivity of the reaction is controlled by the electronic nature of its substituents. The C–H functionalization reaction can occur predominantly on the substituted electron-rich aryl ring. In the mechanism of the reaction after a Rh(III)-catalyzed [4 + 1] annulation and C–H functionalization between azobenzenes and sulfoxonium ylides, an alkylated intermediate (**A**) is released by reductive elimination. This intermediate undergoes oxidation and aromatization to provide the product. Copper additives play an important role in the second step.

2-Arylimidazoles, and more specifically 2-aryl-1*H*-benzo[d]imidazoles, have recently been employed in a novel route to access imidazo[2,1-α]isoquinolines (**73**) (Scheme 43a). In 2018, Yang and co-workers developed a rhodium(III)-catalyzed annulation between 2-arylimidazoles and α-aroyl sulfoxonium ylides to yield this class of heterocycles [103]. The nature of the solvent is a determining factor, and DCE proved to be optimal for this transformation. Ruthenium, cobalt, and iridium catalysts showed low activity in this process. This approach allowed modification in both substrates affording the desired product in moderate to excellent yields ranging between 45% and 93%. In addition, a similar protocol to access arylindazolo[2,3-α]quinoline was described by Chidrawar and co-workers (Scheme

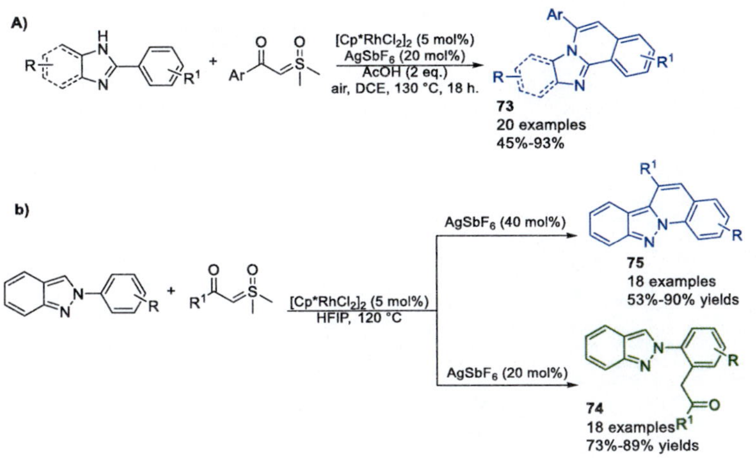

Scheme 43 Rh(III)-catalyzed synthesis of imidazo[2,1-*α*]isoquinolines

43b) [104]. In this protocol, 2-arylindazoles were coupled with β-ketosulfoxonium ylides using Rh(III) catalyst in DCM at high temperatures. However, a mixture of the *ortho*-alkylated product (**74**) and fused indazolo[2,3-α]quinolines (**75**) was obtained. A series of different catalysts, additives, and solvents were examined in the optimization process, and $[Cp^*RhCl_2]_2$ in the presence of $AgSbF_6$ showed the highest catalytic activity, and hexafluoroisopropyl alcohol (HFIP) was the optimum solvent. Interestingly, the selectivity of the reaction was determined by the amount of silver salt. Both products could be obtained in moderate to good yields under the optimized reaction conditions.

The iridium-catalyzed tandem annulation between pyrazolones and sulfoxonium ylides was employed as a strategy to form pyrazolo[1,2-α]cinnoline derivatives (**76**) (Scheme 44a) [105]. The Dong group demonstrated that the starting material can be coupled using a catalytic amount of $[Cp^*IrCl_2]_2$ in the presence of a silver salt and a Lewis acid. The screening of the reaction parameters showed that the use of TsOH was also necessary to form the desired products with excellent yields. Substituted substrates gave the corresponding products with poor to excellent yields. Alkyl, aryl, and heteroaryl sulfoxonium ylides were tolerated. The authors proposed that the ketone carbonyl intermediate, released after the catalytic cycle, underwent keto-enol tautomerism and cyclization in the presence of TsOH or the Lewis acid to yield the desired product.

More recently, Hu and co-workers reported a method that gives access to several pyrazolo[1,2-α]cinnoline derivatives (**77**) via Rh(III)-catalyzed C–H functionalization/annulation of pyrazolidinones and sulfoxonium ylide (Scheme 44b) [106]. The authors showed that the desired product could be obtained in moderate to good yield when different aryl, alkyl, or heteroaryl sulfoxonium ylides and pyrazolidinones were treated with a catalytic amount of a Rh(III) catalyst in presence of NaOAc in DCE at 60°C. However, a second cyclization product **78** was

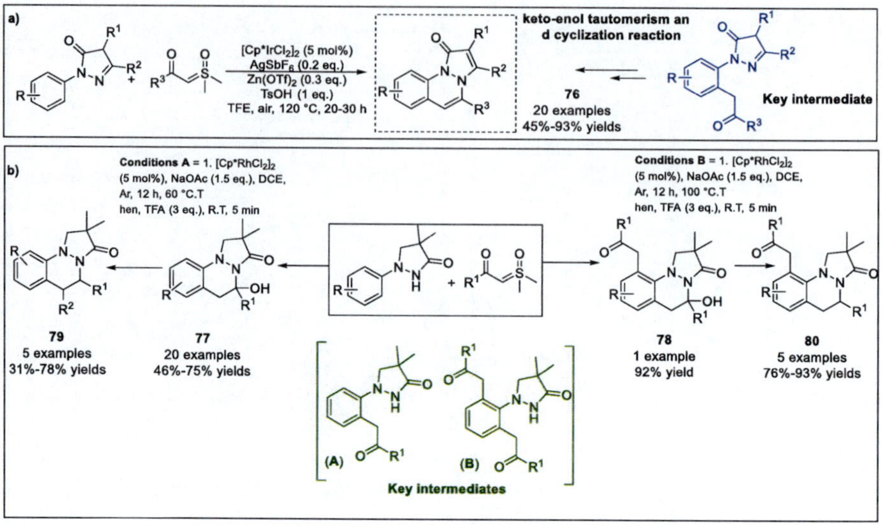

Scheme 44 Synthesis of cinnoline-fused pyrazolidinones catalyzed by iridium and rhodium

observed when the reaction was carried out at 80°C. Pyrazolo[1,2-α]cinnoline **77** could be obtained with good yields increasing the reaction temperature and the number of equivalents of the sulfoxonium ylide. The authors also reported that dehydration products (**79**) and (**80**) could be obtained in moderate to excellent yields by the addition of three equivalents of trifluoroacetic acid (TFA). Mechanistically, intermediate **A** plays an important role to form all the desired products. This intermediate is formed after the first Rh(III)-catalyzed C(sp^2)-H activation of pyrazolidinones and subsequent [4 + 2]-annulation with the sulfoxonium ylide. Subsequently, in path A, intermediate **A** undergoes an intramolecular nucleophilic attack by the N-H group to afford **77**, which after dehydration catalyzed by TFA provides **79**. In path B, intermediate **A** coordinates with an active Rh(III) catalyst species via a second C–H functionalization. Then, addition of a second sulfoxonium ylide provides intermediate **B** (by another C–H functionalization, metal-carbene formation, migratory insertion, and protonolysis). Intermediate **B** undergoes an intramolecular nucleophilic attack by the N-H group to give **78,** and further dehydration led to **80**.

In the same year, Jin and co-workers developed a rapid procedure for the synthesis of cinnoline-fused pyrazolidinones **81** by the Ru-catalyzed C–H functionalization/annulation of *N*-aryl-pyrazolidinones and sulfoxonium ylides (Scheme 45a) [107]. Using [RuCl$_2$(*p*-cymene)]$_2$ as the catalyst in DCE and Zn(OTf)$_2$ as an additive, cinnoline-fused pyrazolidinones were produced in high yield. A particularly attractive synthetic feature of this reaction is the fact that it proceeds with short reaction times. Triflate salts (Zn(OTf)$_2$) were an important additive to forming the desired product; however, other triflates like AgOTf, Cu(OTf)$_2$, In(OTf)$_3$, and Sc(OTf)$_3$ were not efficient for this reaction. From the substrate scope,

Scheme 45 Ru(II)-catalyzed C–H functionalization/annulation to form cinnoline-fused pyrazolidinones

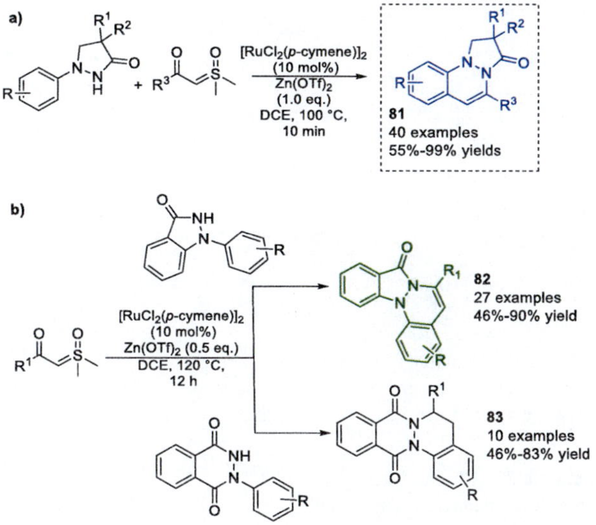

several sulfoxonium ylides could be employed as coupling partners to give the corresponding product with moderate to excellent yields. In addition, the substrate scope is wide and compatible with a range of useful functional groups. Using similar reaction conditions, Pan, Yuan, and Yu reported a Ru(II)-catalyzed C–H functionalization/intramolecular annulation of *N*-phenylindazoles or 2-arylphthalazine-1,4-diones with sulfoxonium ylides to form tetracyclic-fused cinnolines (**82** and **83**) (Scheme 45b) [108]. $[RuCl_2(p\text{-cymene})]_2$ was the optimum catalyst for this transformation. The reaction was performed in DCE and $Zn(OTf)_2$ as the solvent and additive, respectively, and the products could be obtained in moderate to good yields under the optimized reaction conditions.

Quinazolines are an interesting class of heterocycles that contain two heteroatoms in their core. They can also be accessed using transition-metal catalyzed C–H functionalization strategies. For example, Zhang and co-workers described a protocol to access C6-substituted isoquinolino[1,2-*b*]quinazolinones **84** via Rh(III)-catalyzed coupling reaction between 2-arylquinazolin-4(3H)-ones and sulfoxonium ylides (Scheme 46a) [109]. They found that $[Cp^*RhCl_2]_2/AgSbF_6$ was the best catalyst system for this transformation, and high temperatures are required for efficient reactions; at a lower temperature the product yield decreased significantly. Under optimized reaction conditions different isoquinolino[1,2-*b*]quinazolinones could be constructed with C6-regioselectivity by nitrogen-directed C−H acylmethylation in good to excellent yields. The authors reported the cytotoxic activity of the synthesized compounds.

In the same year, Wang and co-workers reported a similar strategy to access polycyclic quinazolinones (**85**) from 2-aryl quinazolin-4(3*H*)-ones and sulfoxonium ylides via rhodium-catalyzed *ortho*-C–H activation, followed by a [4 + 1]-annulation reaction (Scheme 46b) [110]. A series of different substituted sulfoxonium ylides and 2-aryl quinazolin-4(3*H*)-ones were tested, and the corresponding products were

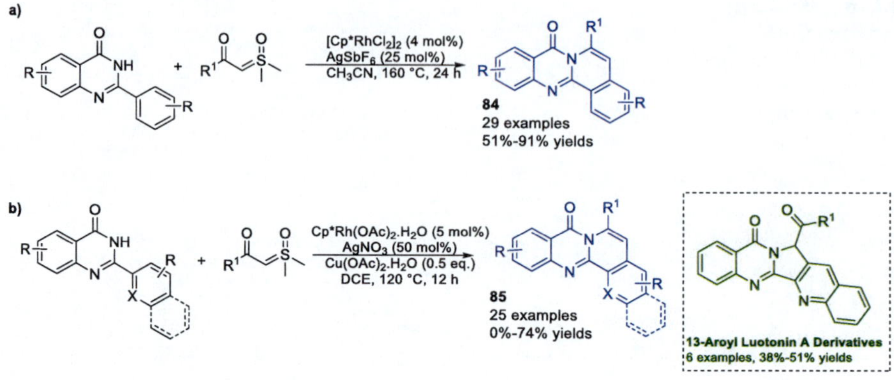

Scheme 46 Rh-catalyzed C–H functionalization strategies to form quinazolines

Scheme 47 Rh-catalyzed C–H functionalization between sulfoxonium ylides and *N*-alkyl aryl hydrazines

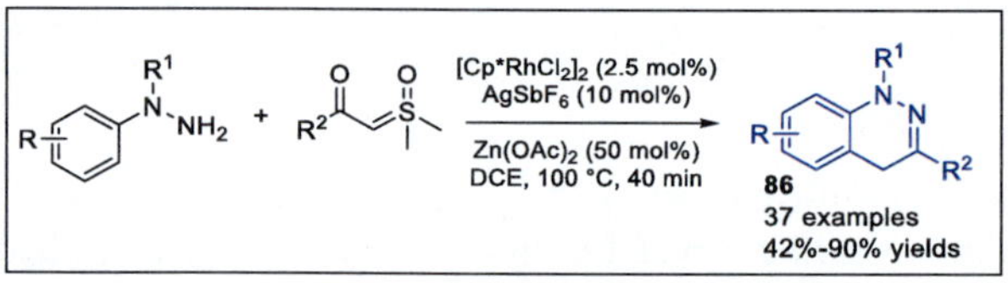

obtained in moderate to good yields. Sulfoxonium ylides with a nitro-substituent in the aryl ring did not work under the reaction conditions. In addition, the author demonstrated that this protocol allows access to 13-aroyl Luotonin A derivatives with moderate yields (up to 51%).

The Gopinath group reported a new route to building dihydrocinnolines **86** via Rh-catalyzed [4 + 2]-annulation of *N*-alkyl arylhydrazines with sulfoxonium ylides (Scheme 47) [111]. $[Cp^*RhCl_2]_2/AgSbF_6$ proved a highly effective catalytic system in combination with $Zn(OAc)_2$ for the C–H functionalization reaction. Furthermore, from the substrate scope, a variety of alkyl and α-aryl sulfoxonium ylides containing electron-donating and electron-deficient groups could be employed as coupling partners to give the corresponding dihydrocinnolines, and different *N*-alkyl arylhydrazines were tolerated. Additionally, the authors explored the photophysical properties of the synthesized dihydrocinnolines, and they found that electron-rich sulfoxonium ylides were more reactive than their electron-deficient counterparts. An intermolecular kinetic isotopic effect (KIE) showed a high k_H/k_D value indicating a primary kinetic isotopic effect and C–H activation as the rate-determining step.

In 2020, Cui and co-workers reported a Ru(II)-catalyzed annulation of 1-(2-aminophenyl)pyrroles with α-carbonyl sulfoxonium ylides to form pyrrolo [1,2-*a*]quinoxaline **87**, which are structures found in many natural and pharmaceuticals products (Scheme 48a) [112]. In the presence of a catalytic amount of $[Ru(p\text{-}cymene)Cl_2]_2$ and $AgNTf_2$, quinoxalines were obtained in moderate to high yields. Different substituted aryl and heteroaryl sulfoxonium ylides were tolerated under the optimized reaction conditions. Interestingly, α-ester sulfoxonium ylides also proved

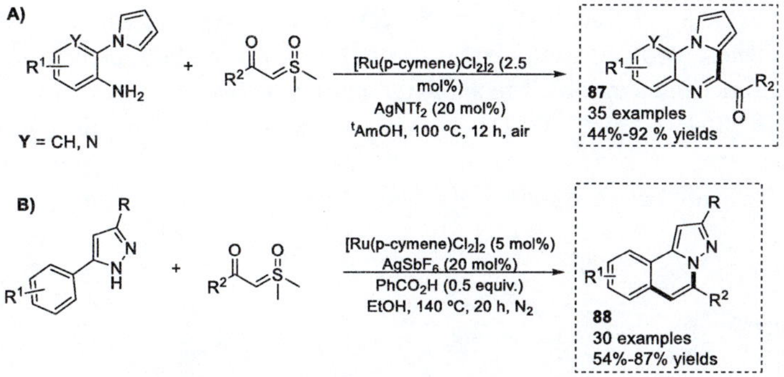

Scheme 48 Synthesis of quinoxaline derivatives

to be effective in this transformation. In the same year, Chen and co-workers explored the sulfoxonium ylides reactivity to generate metal-carbenes and then promote an annulation reaction with aryl-substituted pyrazoles to obtain pyrazolo-[1,5-*a*]isoquinolines (**88**) (Scheme 48b) [113]. Their optimum catalyst system employed the ruthenium dimer [Ru(*p*-cymene)Cl_2]$_2$ (5 mol%) and $AgSbF_6$ (20 mol%). The highest yields were obtained with EtOH as the solvent and $PhCO_2H$ as an additive.

3 Summary and Outlook

This chapter illustrates the utility of sulfur ylides as important platforms to reach different types of heterocycles, covering only the reactions mediated by metal-carbenes. Intramolecular heteroatom X-H insertion reactions, as well as metal-catalyzed C–H functionalization and subsequent metal-carbene formation from a metallacycle, constitute the main methods to reach a vast array of heterocycles from SY. In the examples explored in this chapter, the reader can see a preference in using a sulfoxonium rather than a sulfonium ylide. This is due to the higher stability and easier preparation of the former. Moreover, generation of metal-carbenes from sulfoxonium ylides liberates a sulfoxide instead of volatile and odiferous sulfides, which is the case when sulfonium ylides are employed as carbene precursors. When compared to diazo compounds, SY are one-step behind as carbene precursors. For example, the fact that the extruded sulfoxide or sulfide (during the metal-carbene formation) can deactivate many metals is one reason and limits the number of catalysts to be used. Formation of metal-carbenes is also slower with SY. On the other hand, especially for large-scale applications and in the industry, SY (mainly sulfoxonium) has advantages over diazo compounds as it does not lead to the production of gas or rapid exotherms. The fact that the majority of SY are crystalline solids is also an advantage for its choice in industry. One important feature of SY is its higher nucleophilicity and basicity when compared to structurally similar diazo compounds. These characteristics make SY more interesting in ionic reactions, especially when two functionalizations on the same carbon center are desired (geminal difunctionalization). Mechanistic studies to better understanding the role of carbene formation from SY with certain metals, as well as the discovery of new catalytic systems to generate carbenes from SY, are still desirable. Moreover, methods to prepare new and more complex sulfoxonium ylides need to be explored.

References

1. Trost M, Melvin LS (1975) In sulfur ylides: emerging synthetic intermediates. Academic Press, New York
2. Burtoloso ACB, Dias RMP, Leonarczyk IA (2013) Sulfoxonium and sulfonium ylides as diazocarbonyl equivalents in metal catalyzed insertion reactions. Eur J Org Chem 2013:5005–5016
3. Caiuby CAD, Furniel LG, Burtoloso ACB (2022) Asymmetric transformations from sulfoxonium ylides. Chem Sci 13:1192–1209
4. Bisag GD, Ruggieri S, Fochi M, Bernardi L (2020) Sulfoxonium ylides: simple compounds, chameleonic reactivity. Biomol Chem 18:8793–8809
5. Neuhaus JD, Oost R, Merad J, Maulide N (2018) Sulfur-based ylides in transition-metal-catalysed processes. Top Curr Chem 376:15
6. Ingold CK, Jessop JA (1930) XCV. Influence of poles and polar linkings on the course pursued· by elimination reactions. Part IX. Isolation of a substance believed to contain a semipolar double linking with participating carbon. J Chem Soc:713–718
7. Johnson AW, LaCount RB (1961) The chemistry of ylids. VI. Dimethylsulfonium fluorenylide-A synthesis of epoxides. J Am Chem Soc 83:417–423
8. Corey EJ, Chaykovsky M (1962) Dimethylsulfonium methylide, a reagent for selective oxirane synthesis from aldehydes and ketones. J Am Chem Soc 84:3782
9. Corey EJ, Chaykovsky M (1964) Formation and photochemical rearrangement of β’--ketosulfoxonium ylides. J Am Chem Soc 86:1640–1641
10. Corey EJ, Chaykovsky M (1965) Dimethyloxosulfonium methylide ($(CH_3)_2SOCH_2$) and dimethylsulfonium methylide ($(CH_3)_2SCH_2$). Formation and application to organic synthesis. J Am Chem Soc 87:1353–1364
11. Franzen V, Schimidt HJ, Mertz C (1961) Untersuchungen über carbene, IX. Carbene aus sulfoniumsalzen. Chem Ber 94:2942
12. Franzen V, Driesen HE (1963) Umsetzung von sulfonium-yliden mit polaren doppelbindungen. Chem Ber 96:1881–1890
13. Trost BM (1966) Decomposition of sulfur ylides. Evidence for carbene intermediates. J Am Chem Soc 88:1587–1588
14. Johnson CR, Janiga ER, Haake M (1968) Ylides from salts of sulfoximines. J Am Chem Soc 90:3890–3891
15. Johnson CR, Haake M, Schroeck CW (1970) Preparation and synthetic applications of (dimethylamino) phenyloxosulfonium methylide. J Am Chem Soc 92:6594–6598
16. Johnson CR, Rogers PE (1973) Preparation and reactions of stabilized (dialkylamino)-methyloxosulfonium methylides. Synthesis of 1,3-oxathiole 3-oxides. J Org Chem 38:1798–1803
17. Yang Y, Zheng W, Ren L, Xu X (2022) A theoretical study on the proton affinity of sulfur ylides. New J Chem 46:7910–7921
18. Fu Y, Wang HJ, Chong SS, Guo QX, Liu L (2009) An extensive ylide thermodynamic stability scale predicted by first-principle calculations. J Org Chem 74:810–819

19. Cohen T, Herman G, Chapman TM, Kuhn D (1974) A laboratory model for the biosynthesis of cyclopropane rings. Copper-catalyzed cyclopropanation of olefins by sulfur ylides. J Am Chem Soc 96:5627–5628
20. Cimetiére B, Julia M (1991) Alkene cyclopropanation with sulfonium ylids. Synlett 4:271–272
21. Baldwin EJ, Adlington RM, Godfrey CRA, Gollins DW, Vaughan JG (1993) A novel entry to carbenoid species via β-ketosulfoxonium ylides. J Chem Soc Chem Commun:1434–1435
22. Mangion IK, Nwamba IK, Shevlin M, Huffman MA (2009) Iridium-catalyzed X−H insertions of sulfoxonium ylides. Org Lett 11:3566–3569
23. Mangion IK, Weisel M (2010) Gold (I) catalysis of X-H bond insertions. Tetrahedron Lett 51: 5490–5492
24. Burtoloso ACB, Santiago JV, Bernardim B, Talero AG (2015) Advances in the enantioselective metal-catalyzed n-h and o-h insertion reactions with diazocarbonyl compounds. Curr Org Chem 12:650–659
25. Mangion IK, Ruck RT, Rivera N, Huffman MA, Shevlin M (2011) A concise synthesis of a β-lactamase inhibitor. Org Lett 13:5480–5483
26. Molinaro C, Bulger PG, Lee EE, Kosjek B, Lau S, Gauvreau D, Howard ME, Wallace DJ, O'Shea PD (2012) CRTH2 antagonist mk-7246: a synthetic evolution from discovery through development. J Org Chem 77:2299–2309
27. Vaitla J, Bayer A, Hopmann KH (2017) Synthesis of indoles and pyrroles utilizing iridium carbenes generated from sulfoxonium ylides. Angew Chem Int Ed 56:4277
28. Zhu S, Shi K, Zhu H, Jia ZK, Xia XF, Wang D, Zou L (2020) Copper-catalyzed annulation or homocoupling of sulfoxonium ylides: synthesis of 2,3-diaroylquinolines or α,α,β-tricarbonyl sulfoxonium ylides. Org Lett 22:1504–1509
29. Liu L, Lin J, Pang M, Jin H, Yu X, Wang S (2022) Photo-thermo-mechanochemical approach to synthesize quinolines via addition/cyclization of sulfoxonium ylides with 2-vinylanilines catalyzed by iron(II) phthalocyanine. Org Lett 24:1146–1151
30. Phelps AM, Chan VS, Napolitano JG, Krabbe SW, Schomaker JM, Shekhar S (2016) Ligand-controlled synthesis of azoles *via* Ir-catalyzed reactions of sulfoxonium ylides with 2-amino heterocycles. J Org Chem 81:4158–4169
31. Caiuby CAD, de Jesus MP, Burtoloso ACB (2020) α-Imino iridium carbenes from imidoyl sulfoxonium ylides: application in the one-step synthesis of indoles. J Org Chem 85:7433–7445
32. Faragher R, Gilchrist TL (1977) Imidoyl-substituted oxosulphonium ylides: preparation and reaction with nitrile oxides. J Chem Soc Perkin Trans 1:1196–1200
33. Faragher R, Gilchrist TL, Southon IW (1981) [3,2] Sigmatropic rearrangements of some allylic oxosulphonium ylides. A route to trienes and azatrienes, and their cyclic isomers. J Chem Soc Perkin Trans 1:2352–2356
34. Kramer S, Skrydstrup T (2012) Gold-catalyzed carbene transfer to alkynes: access to 2,4-disubstituted furans. Angew Chem Int Ed 51:4681–4684
35. Huang X, Klimczyk S, Veiros LF, Maulide N (2013) Stereoselective intramolecular cyclopropanation through catalytic olefin activation. Chem Sci 4:1105
36. Zhou P, Yang WT, Rahman AU, Li G, Jiang B (2020) Rh(III)-catalyzed [3+3] annulation reaction of cyclopropenones and sulfoxonium ylides toward trisubstituted 2-pyrones. J Org Chem 85:360–366
37. Müller P, Fernandez D, Nury P, Rossier JC (1999) Transition-metal-catalyzed carbenoid reactions of sulfonium ylides. Helv 82:935–945
38. Dong J, Du H, Xu J (2019) Synthesis of functionalized thietanes via electrophilic carbenoid-induced ring expansion of thiiranes with sulfonium acylmethylides as carbene precursors. J Org Chem 84:10724–10739
39. Chen Y, Lv S, Lai R, Xu Y, Huang X, Li J, Lv G, Wu Y (2021) Synthesis of 2-aminothiazoles *via* rhodium-catalyzed carbenoid insertion/annulation of sulfoxonium ylides with tioureas. Chin Chem Lett 32:2555–2558

40. Kuhl N, Hopkinson MN, Wencel-Delord J, Glorius F (2012) Beyond directing groups: transition-metal-catalyzed C–H activation of simple arenes. Angew Chem Int Ed 51:10236–10254
41. Carvalho RL, Gleiston GD, Pereira CLM, Ghosh P, Debabrata M, Silva Júnior EN (2021) A catalysis guide focusing on C–H activation processes. J Braz Chem Soc 32(5):917–952
42. Gensch T, Hopkinson MN, Glorius F, Wencel-Delord J (2016) Mild metal-catalyzed C–H activation: examples and concepts. Chem Soc Rev 45:2900-2936
43. Dalton T, Faber T, Glorius F (2021) C–H activation: toward sustainability and applications. ACS Cent Sci 7(2):245–261
44. Kerru N, Gummidi L, Maddila S, Gangu K, Jonnalagadda S (2020) A review on recent advances in nitrogen-containing molecules and their biological applications. Molecules 25: 1909
45. Vitaku E, Smith DT, Njardarson JT (2014) Analysis of the structural diversity, substitution patterns, and frequency of nitrogen heterocycles among U.S. FDA approved pharmaceuticals. J Med Chem 57:10257–10274
46. Kamtea KL, Kant R, Penoni A, Maspero A, Scapinello L (2022) *N*-heterocycles. Synthesis and biological evaluation. Springer, Singapore
47. Chen G, Zhang X, Jia R, Li B, Fan X (2018) Selective synthesis of benzo[*a*]carbazoles and indolo[2,1-*a*]-isoquinolines *via* Rh(III)-catalyzed C-H functionalizations of 2-arylindoles with sulfoxonium ylides. Adv Synth Catal 360:3781–3787
48. Xiao Y, Xiong H, Sun S, Yu J, Cheng J (2018) Rh(III)-catalyzed dual C–H functionalization of 3-(1*H*-indol-3-yl)-3-oxopropanenitriles with sulfoxonium ylides or diazo compounds toward polysubstituted carbazoles. Org Biomol Chem 16:8715–8718
49. Hu P, Zhang Y, Xu Y, Yang S, Liu B, Li X (2018) Construction of (dihydro)naphtho[1,8-bc] pyrans via rh(III)-catalyzed twofold C–H activation of benzoylacetonitriles. Org Lett 20: 2160–2163
50. Xu Y, Zheng G, Yang X, Li X (2018) Rhodium(III)-catalyzed chemodivergent annulations between *N*-methoxybenzamides and sulfoxonium ylides *via* C–H activation. Chem Commun 54:670–673
51. Wen S, Zhang Y, Tian Q, Chen Y, Cheng G (2022) Ruthenium(II)-catalyzed synthesis of CF_3-isoquinolinones *via* C–H activation/annulation of benzoic acids and CF_3-imidoyl sulfoxonium ylides. Org Chem Front 9:4388–4393
52. Hu P, Zhang Y, Liu B, Li X (2018) Facile construction of hydrogenated azepino[3,2,1-hi] indoles by Rh(III)-catalyzed C–H activation/[5 + 2] annulation of *N*-cyanoacetylindolines with sulfoxonium ylides. Org Chem Front 5:3263–3266
53. Bentley KW (1998) The isoquinoline alkaloids. Harwood Academic Publishers, Amsterdam
54. Twardy DJ, Wheeler KA, Török B, Dembinski R (2022) Recent advances in the synthesis of isoquinoline-fused benzimidazoles. Molecules 27:2062
55. Zheng G, Tian M, Yu Y, Chen X, Li X (2018) Rhodium(III)-catalyzed annulative coupling between arenes and sulfoxonium ylides *via* C–H activation. Org Chem Front 5:998–1002
56. Shi X, Wang R, Zeng X, Zhang Y, Hu H, Xie C, Wang M (2018) Ruthenium (II)-catalyzed oxidant-free coupling/cyclization of benzimidates and sulfoxonium ylides to form substituted isoquinolines. Adv Synth Catal 360:4049–4053
57. Hong C, Yu S, Liu Z, Zhang Y (2021) Rhodium(III)-catalyzed annulation of enamides with sulfoxonium ylides toward isoquinolines. RSC Adv 11:11490–11494
58. Kochanowska-Karamyan AJ, Hamann MT (2010) Marine indole alkaloids: potential new drug leads for the control of depression and anxiety. Chem Rev 110:4489–4497
59. Zhang L, Chen J, Chen J, Jin L, Zheng X, Jiang X, Yu C (2019) Synthesis of 2-substituted indoles by iridium (III)-catalyzed C-H functionalization of *N*-phenylpyridin-2-amines. Tetrahedron Lett 60:1053–1056
60. Luo Y, Guo L, Yu X, Ding H, Wang H, Wu Y (2019) Cp^*Ir^{III}-catalyzed [3+2] annulations of *N*-aryl-2-aminopyrimidines with sulfoxonium ylides to access 2-alkyl indoles through C–H bond activation. Eur J Org Chem:3203–3207

61. Cui XF, Ban ZH, Tian WF, Hu FP, Zhou XQ, Ma HJ, Zhan ZZ, Huang GS (2019) Ruthenium-catalyzed synthesis of indole derivatives from *N*-aryl-2-aminopyridines and alpha-carbonyl sulfoxonium ylides. Org Biomol Chem 17:240–243
62. Shen Z, Pi C, Cui X, Wu Y (2019) Rhodium(III)-catalyzed intermolecular cyclization of anilines with sulfoxonium ylides toward indoles. Chin Chem Lett 30:1374–1378
63. Xie W, Chen X, Shi J, Li J, Liu R (2019) Synthesis of 1-aminoindole derivatives via Rh(III)-catalyzed annulation reactions of hydrazines with sulfoxonium ylides. Org Chem Front 6: 2662–2666
64. Lv N, Chen Z, Liu Z, Zhang Y (2019) Redox-neutral rhodium(III)-catalyzed annulation of arylhydrazines with sulfoxonium ylides to synthesize 2-arylindoles. J Org Chem 84:13013–13021
65. Li H, Lu Y, Jin X, Sun S, Duan L, Liu J (2020) Rhodium(III)-catalyzed C–H annulation of 2-acetyl-1-arylhydrazines with sulfoxonium ylides: synthesis of 2-arylindoles. RSC Adv 10: 39708–39711
66. Okauchi T, Itonaga M, Minami T, Owa T, Kitoh K, Yoshino H (2000) A general method for acylation of indoles at the 3-position with acyl chlorides in the presence of dialkylaluminum chloride. Org Lett 2:1485–1487
67. Wu C, Zhou J, He G, Li H, Yang Q, Wang R, Zhou Y, Liu H (2019) Ruthenium(II)-catalyzed selective C–H bond activation of imidamides and coupling with sulfoxonium ylides: an efficient approach for the synthesis of highly functional 3-ketoindoles. Org Chem Front 6: 1183–1188
68. Lai R, Wu X, Lv S, Zhang C, He M, Chen Y, Wang Q, Hai L, Wu Y (2019) Synthesis of indoles and quinazolines via additive-controlled selective C–H activation/annulation of *N*-arylamidines and sulfoxonium ylides. Chem Commun 55:4039–4042
69. Wu Y, Pi C, Cui X, Wu Y (2020) Rh(III)-catalyzed tandem acylmethylation/nitroso migration/ cyclization of *N*-nitrosoanilines with sulfoxonium ylides in one pot: approach to 3-nitrosoindoles. Org Lett 22:361–364
70. Aher YN, Pawar AB (2021) Cp*Co(III)-catalyzed C–H amination/annulation cascade of sulfoxonium ylides with anthranils for the synthesis of indoloindolones. Chem Commun 57: 7164–7167
71. Wu X, Xiao Y, Sun S, Yu JT, Cheng J (2019) Rhodium-catalyzed reaction of sulfoxonium ylides and anthranils toward indoloindolones *via* a (4+1) annulation. Org Lett 21:6653–6657
72. Lee SC, Son JY, Kim JK, Eom H, Jang SB, Lee PH (2021) Regioselective and chemodivergent synthesis of azulenolactones and azulenolactams from rhodium(III)-catalyzed reactions of azulenecarboxamides with sulfoxonium ylides. Adv Synth Catal 363:512–524
73. Venkatachalam H, Kumar NVA (2020) The oxygen-containing fused heterocyclic compounds. IntechOpen
74. Zhang X, Li X, Li D (2022) Ru(II)-Catalyzed C–H bond activation/annulation of *N*-iminopyridinium ylides with sulfoxonium ylides. Org Biomol Chem 20:1475–1479
75. Ackermann L, Rogge T, Yang L, Liang YF (2018) Ruthenium(IV) intermediates in c-h activation/annulation by weak o-coordination. Chem A Eur J 24:16548–16552
76. Wang Q, Song H, Lyu X, Huang Y (2019) Sulfoxonium ylides as carbene precursors: rhodium (III)-catalyzed sequential C−H functionalization, selective enol oxygen-atom nucleophilic addition, and hydrolysis. Adv Synth Catal 361:5272–5276
77. Li X, Li D, Zhang X (2022) Ru(II)-catalyzed C–H bond activation/annulation of n-iminopyridinium ylides with sulfoxonium ylides. Org Biomol Chem 20:1475–1479
78. Zhou MD, Peng Z, Wang H, Wang ZH, Hao DJ, Li L (2019) Ruthenium(II)-catalyzed homocoupling of weakly coordinating sulfoxonium ylides via C−H activation/annulations: synthesis of functionalized isocoumarins. Adv Synth Catal 361:5191–5197
79. Cheng G, Lv W, Ba D, Zhao Z, Chen Y, Wen S (2020) Ruthenium(II)-catalyzed construction of isocoumarins via dual C–H/C–C activation of sulfoxonium ylides. J Org Chem 85:1216–1223

80. Li X, Liu B, Yang S, Xu Y, Zhang Y, Hu P (2018) Construction of (dihydro)naphtho[1,8-*bc*] pyrans via Rh(III)-catalyzed twofold C−H activation of benzoylacetonitriles. Org Lett 20: 2160–2163
81. Wu XF, Chen Z, Tang J, Yang Z (2022) Ruthenium-catalyzed hydroxyl-directed peri-selective C−H activation and annulation of 1-naphthols with CF_3−imidoyl sulfoxonium ylides for the synthesis of 2-(trifluoromethyl)-2,3-dihydrobenzo[*de*]chromen-2-amines. Org Lett 24:7288–7293
82. Prabhu KR, Kumar A, Hanchate V (2019) Synthesis of naphthols by Rh(III)-catalyzed domino C−H activation, annulation, and lactonization using sulfoxonium ylide as a traceless directing group. Org Lett 21:8424–8428
83. Yao H, Lin A, Chen J, Zhu X, Cai L (2019) Rh(III)-catalyzed C–H activation/annulation of salicylaldehydes with sulfoxonium ylides for the synthesis of chromones. Org Chem Front 6: 3688–3369
84. Yu C, Cheng K, Chen J, Jin L, Zhou J, Jiang X (2019) Rhodium(III)-catalyzed one-pot synthesis of flavonoids from salicylaldehydes and sulfoxonium ylides. J Chem Res 43:9–10
85. Lee PH, Jang SB, Eom H, Kim JY, Son JY, Lee SC (2021) Regioselective and chemodivergent synthesis of azulenolactones and azulenolactams from rhodium(III)-catalyzed reactions of azulenecarboxamides with sulfoxonium ylides. Adv Synth Catal 363:512–524
86. Shankaraiah N, John SE, Bora D, Dastari S, Valapil DG (2022) Synthesis of alpha-pyrones and chromen-2-ones by transition-metal catalyzed annulations of sulfoxonium and iodonium ylides with cis-stilbene acids. New J Chem 46:19722–19730
87. Soroha B, Kumar G, Kumar R, Kumari M, Kumar S (2022) A minireview of 1,2,3-triazole hybrids with *O*-heterocycles as leads in medicinal chemistry. Chem Biol Drug Des 100:843–869
88. Jampilek J (2019) Heterocycles in medicinal chemistry. Molecules 24(21):3839
89. Poonam NJ, Dhadda S, Jangid DK (2022) Recent advances in the synthesis of five- and six-membered heterocycles as a bioactive skeleton: a concise overview. ChemistrySelect 7: e202103139
90. Zheng G, Tian M, Xu Y, Chen X, Li X (2018) Rhodium(III)-catalyzed annulative coupling between arenes and sulfoxonium ylides *via* C–H activation. Org Chem Front 5:998–1002
91. Xie H, Lan J, Gui J, Chen F, Jiang H, Zeng W (2018) Ru (II)-catalyzed coupling-cyclization of sulfoxonimines with alpha-carbonyl sulfoxonium ylides as an approach to 1,2-benzothiazines. Adv Synth Catal 360:3534–3543
92. Zhou T, Qian PF, Li JY, Zhou YB, Li HC, Chen HY, Shi BF (2021) Efficient synthesis of sulfur-stereogenic sulfoximines *via* Ru(II)-catalyzed enantioselective C–H functionalization enabled by chiral carboxylic acid. J Am Chem Soc 143:6810–6818
93. Huang LT, Hirata Y, Kato Y, Lin L, Kojima M, Yoshino T, Matsunaga S (2022) Ruthenium (II)/chiral carboxylic acid catalyzed enantioselective C–H functionalization of sulfoximines. Synthesis 54:4703–4710
94. Wu X, Xiong H, Sun S, Cheng J (2018) Rhodium-catalyzed relay carbenoid functionalization of aromatic C–H bonds toward fused heteroarenes. Org Lett 20:1396–1399
95. Halskov KS, Witten MR, Hoang GL, Mercado BQM, Ellman JA (2018) Rhodium(III)-catalyzed imidoyl C–H activation for annulations to azolopyrimidines. Org Lett 20:2464–2467
96. Hoang GL, Ellman JA (2018) Rhodium(III)-catalyzed C–H functionalization of *C*-alkenyl azoles with sulfoxonium ylides for the synthesis of bridgehead *N*-fused [5,6]-bicyclic heterocycles. Tetrahedron 74:3318–3324
97. Hoang GL, Streit AD, Ellman JA (2018) Three-component coupling of aldehydes, aminopyrazoles, and sulfoxonium ylides *via* rhodium(III)-catalyzed imidoyl C–H activation: synthesis of pyrazolo[1,5-*a*]pyrimidines. J Org Chem 83:15347–15360
98. Chen J, Zhong T, Zheng X, Yin C, Zhang L, Zhou J, Jiang X, Yu C (2021) Selective synthesis of fused tricyclic [1,3]oxazino[3,4-a]indolone and dihydropyrimido [1,6-a]indolone *via* Rh (III)-catalyzed [3+3] or [4+2] C−H annulation. Adv Synth Catal 363:446–452

99. Xie H, Zhong M, Kang J, Shu B, Zhang SS (2021) A cascade Rh(III)-catalyzed C−H activation/chemodivergent annulation of *N*-carbamoylindoles with sulfoxonium ylides for the synthesis of dihydropyrimidoindolone and tricyclic [1,3]oxazino[3,4-*a*]indol-1-ones derivatives. Adv Synth Catal 363:1436–1442
100. Cui XF, Huang GS (2020) Rhodium-catalyzed tandem acylmethylation/annulation of *N*-nitrosoanilines with sulfoxonium ylides for the synthesis of substituted indazole *N*-oxides. Org Biomol Chem 18:4014–4018
101. Zhu J, Sun S, Cheng J (2018) Rh(III)-catalyzed [4 + 1]-annulation of azobenzenes with α-carbonyl sulfoxonium ylides toward 3-acyl-(2H)-indazoles. Tetrahedron Lett 59:2284–2287
102. Oh H, Han S, Pandey AK, Han SH, Mishra NK, Kim S, Chun R, Kim HS, Park J, Kim IS (2018) Synthesis of (2*H*)-indazoles through Rh(III)-catalyzed annulation reaction of azobenzenes with sulfoxonium ylides. J Org Chem 83:4070–4077
103. Yang R, Wu X, Sun S, Yu JT, Cheng J (2018) Rhodium-catalyzed annulation of 2-arylimidazoles and α-aroyl sulfoxonium ylides toward 5-arylimidazo[2,1-*a*]isoquinolines. Synthesis 50:3487–3492
104. Chidrawar A, Devulapally YC, Balasubramanian S, Basireddy VSR (2021) Rh(III)-catalyzed oxidative annulation of 2-arylindazoles with β-ketosulfoxonium ylides. ChemistrySelect 6: 13046–13050
105. Liu CF, Liu M, Dong L (2019) Iridium(III)-catalyzed tandem annulation synthesis of pyrazolo [1,2-α]cinnolines from pyrazolones and sulfoxonium ylides. J Org Chem 84:409–416
106. Hu S, Han X, Xie X, Fang F, Wang Y, Saidahmatov A, Liu H, Wang J (2021) Synthesis of pyrazolo[1,2-a]cinnolines via rhodium(III)-catalyzed [4+2] annulation reactions of pyrazolidinones with sulfoxonium ylides. Adv Synth Catal 363:3311–3317
107. Jin HS, Du YZ, Zhao QY, Zhao LM (2021) Ru(II)-Catalyzed C–H activation/annulation reactions of N-aryl-pyrazolidinones with sulfoxonium ylides: synthesis of cinnoline-fused pyrazolidinones. Org Chem Front 8:6350–6355
108. Pan C, Yuan C, Yu JT (2022) Ruthenium-catalyzed C−H functionalization/annulation of *N*-aryl indazoles/phthalazines with sulfoxonium ylides to access tetracyclic-fused cinnolines. Asian J Org Chem 11:e202200346
109. Zhang J, Wang X, Chen D, Kang Y, Ma Y, Szostak M (2020) Synthesis of C6-substituted isoquinolino[1,2-*b*]quinazolines *via* Rh(III)-catalyzed C–H annulation with sulfoxonium ylides. J Org Chem 85:3192–3201
110. Wang C, Qian PC, Chen F, Cheng J (2020) Rhodium-catalyzed [4+1] annulation of sulfoxonium ylides: sequential *ortho*-C-H functionalization/carbonyl α-amination toward polycyclic quinazolinones. Tetrahedron Lett 61:152441
111. Ganesh PSK, Muthuraja P, Gopinath P (2022) A hydrazine-directed Rh(III) catalyzed (4+2) annulation with sulfoxonium ylides: synthesis and photophysical properties of dihydrocinnolines. Chem Commun 58:4211–4214
112. Cui XF, Hu FP, Zhou XQ, Huang GS (2020) Ruthenium-catalyzed synthesis of pyrrolo[1,2-a] quinoxaline derivatives from 1-(2-aminophenyl)pyrroles and sulfoxonium ylides. Synlett 31: 1205–1210
113. Chen Z, Ding Z, Chen T, Meng L, Wang G (2020) Ruthenium-catalyzed α-carbonyl sulfoxonium ylides annulations with aryl substituted pyrazoles *via* C–H/N–H bond functionalizations. Org Biomol Chem 18:8486–8490

Heterocycles from Cascade Reactions via Carbene *gem*-Difunctionalization

Mengchu Zhang, Xinfang Xu, and Wenhao Hu

Contents

Abstract Cascade reactions via metal carbene *gem*-difunctionalization, which are initiated by metal-catalyzed carbene formation, followed by nucleophilic attack to produce ylides or zwitterionic intermediates, and terminated by electrophilic interception by different electrophiles, are useful and practical tools for the synthesis and modification of heterocycles. This method opens a new door for the rapid assembly of complex molecules imbedded with a variety of heterocyclic units in a highly regioselective manner. This chapter summarized the advances in this area contributed by our group and other groups for the efficient synthesis of multi-functional heterocycles via carbene cascade reactions in the literature through 2022; other related advances will be mentioned with a brief comment when necessary.

M. Zhang, X. Xu, and W. Hu (✉)
School of Pharmaceutical Sciences, Sun Yat-sen University, Guangzhou, China
e-mail: huwh9@mail.sysu.edu.cn

Keywords Cascade reaction · *gem*-difunctionalization · Heterocycles · Metal carbene · Ylide intermediate · Zwitterionic intermediate

1 Introduction

Heterocycles are common and important scaffolds that widely exist in natural products and biologically active molecules. Tremendous efforts have been devoted to the development of effective synthetic protocols for the construction of heterocyclic structures with broad substrate generality and a high level of structural complexity under mild conditions [1–3]. Among them, cascade reactions, which benefit from their atom- and step-economy through a one-pot-process, have become one of the most important synthetic methods for the rapid assembly of heterocycles [4–6]. These cascade reactions are environmentally-benign and cost-saving processes due to their one-pot procedure. In this regard, cascade reactions via metal carbene *gem*-difunctionalization have emerged as a robust approach for the construction of complicated molecules that are otherwise difficult to access. In these transformations, at least two C-C bonds or one C-C and one C-X bonds are formed in one-pot with concomitant generation of tertiary or/and quaternary carbon centers, such as the Kirmse-Doyle rearrangement [7, 8], multi-component reactions involving a ylide interception process [4], cycloaddition via carbonyl ylide species [9], and coupling reactions involving a migration insertion process [10] (Scheme 1).

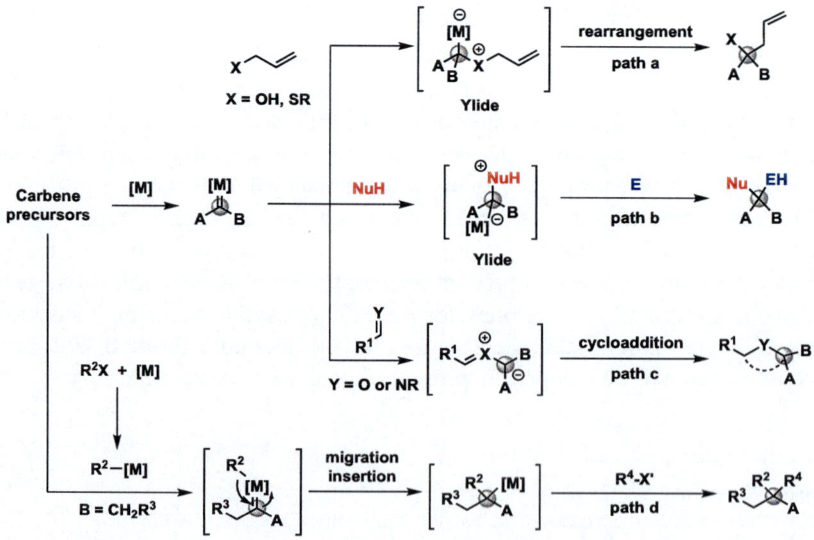

Scheme 1 Cascade reactions involving a carbene *gem*-difunctionalization process

Metal carbenes, one of the most active synthons in modern organic synthesis, are generally derived from the catalytic decomposition of diazo compounds or other carbene precursors [11–20]. Based on their pronounced utility, metal carbenes attract considerable interest in industry and academia. Recently, metal carbenes have been successfully applied to cascade reactions for the rapid construction of heterocyclic molecules via a sequential *gem*-difunctionalization process [4, 6, 10, 21, 22]. For example, metal carbenes undergo nucleophilic attack on carbonyl compounds or imines to form carbonyl or azomethine ylides, followed by 1,3-dipolar cycloadditions with different species of dipolarophiles to afford various heterocycles with structural diversity (Scheme 1, path c). Representative works in this area have been independently reported by Padwa [9, 23], Doyle [24–26], Davies [27–30], and López [31]. Moreover, Wang and coworkers established a series of transition-metal-catalyzed cross-couple reactions involving carbene migratory insertion processes (Scheme 1, path d) [10].

During the last few decades great efforts have been made in the area of carbene chemistry, and many important reviews have summarized according to the catalytic methods and different types of reactions [6, 16–18, 20]. However, reviews that focused on the assembly of polyfunctional heterocycles from carbene cascade reactions have not been reported. In this chapter, we will summarize advances in the synthesis of heterocycles via carbene cascade reactions of different carbene precursors with nucleophiles and electrophiles, which are initiated by metal-catalyzed carbene formation, followed by nucleophilic attack to produce ylides or zwitterionic intermediates, and terminated by electrophilic interception with different electrophiles (Scheme 1, path **b**). Based on the different types of electrophiles used in these cascade reactions, the electrophilic trapping process comprises Mannich-, aldol-, Michael-type addition transformations or through substitution. These diverse cascade reactions provide a fundamental platform for the synthesis of interesting heterocycles, which could be divided into three parts according to the key step for the ring construction/modification (Scheme 2): (a) heterocyclic structure formation via intramolecular nucleophilic/electrophilic addition; (b) cyclization through carbene *gem*-difunctionalization process, and (c) ring modification through carbene cascade reactions using different heterocycles as reagents. In this chapter, the pioneering and representative examples in the literature through 2022 will be the main focus, and other related advances will be mentioned with brief comments. We hope that this review will provide a general image of the application of carbene cascade reactions for the assembly of heterocycles with structural complexity and diversity that are otherwise difficult to access and inspire potential exploration of this method in synthetic and medical chemistry.

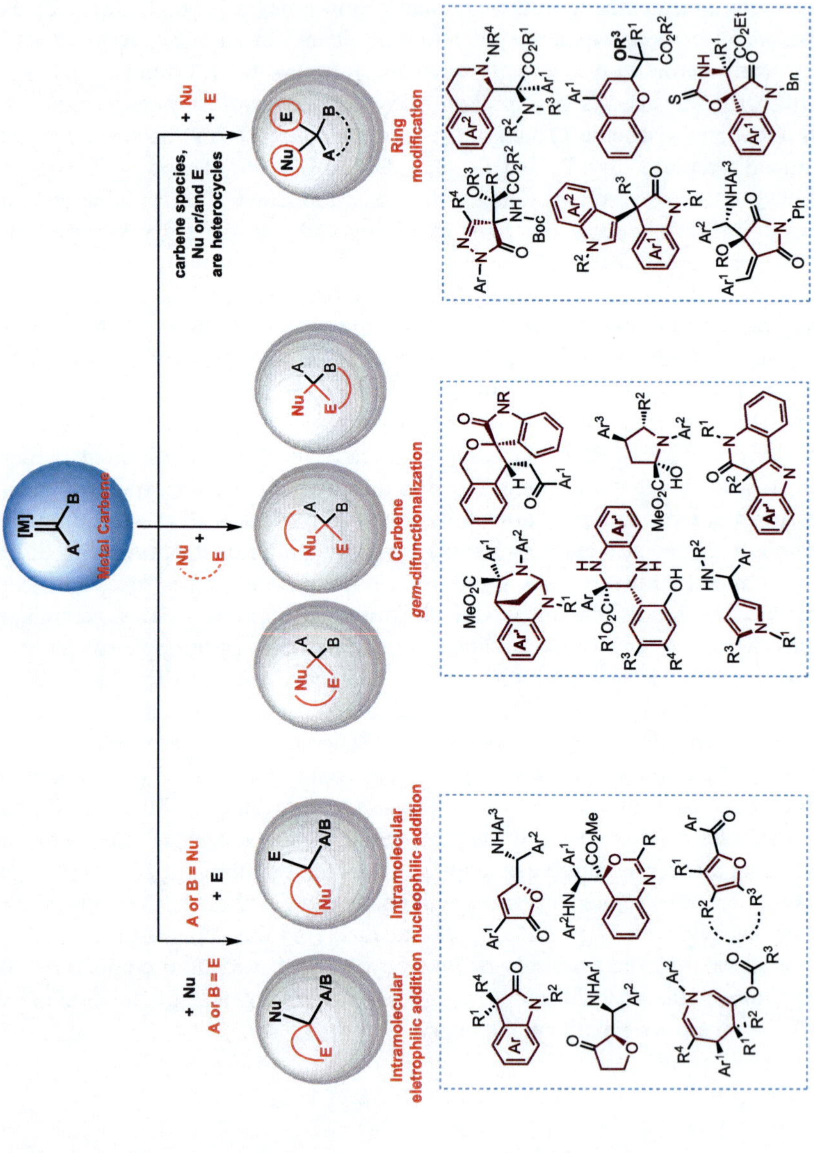

Scheme 2 General reaction patterns for the construction/modification of heterocycles via carbene cascade reactions

2 Heterocyclic Skeleton Construction via Intramolecular Cyclization

Intramolecular cyclization is an efficient way for the rapid assembly of complex heterocycles. In this part two cyclization approaches will be discussed involving carbene intermediates derived from different precursors, including diazo compounds, alkynes, and cyclopropenes (Scheme 3): metal carbene nucleophilic cyclization with tethered nucleophiles onto the corresponding carbene precursor; electrophilic cyclization of in-situ formed nucleophilic species with tethered functionality on the carbene precursor.

2.1 Heterocyclic Skeleton Construction via Nucleophilic Cyclization

Metal carbene addition with a tethered nucleophilic functionality could generate heterocyclic ylides or zwitterionic intermediates. These active intermediates could be effectively intercepted by different electrophiles to afford diverse heterocyclic skeletons (Scheme 4).

2.1.1 Diazo Compounds as Carbene Precursors

Early in 1988, Doyle's group reported an important type of diazo compounds, *N*-aryl diazoamides **1**. In the presence of rhodium(II) acetate, $Rh_2(OAc)_4$, diazoamides **1** went through an attractive intramolecular aromatic substitution to form 2(3*H*)-indolinones **2** in good to excellent yields, which might involve a zwitterionic intermediate, followed by a 1,2-proton transfer process (Scheme 5) [32].

In virtue of the highly reactive and unstable character of this in-situ formed zwitterionic intermediate, it is challenging to find suitable reactive electrophiles to

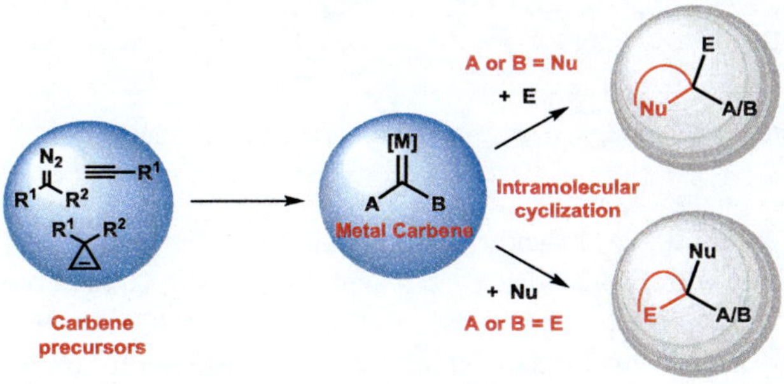

Scheme 3 General strategies for the construction of heterocyclic skeleton via metal carbene intramolecular cyclization process

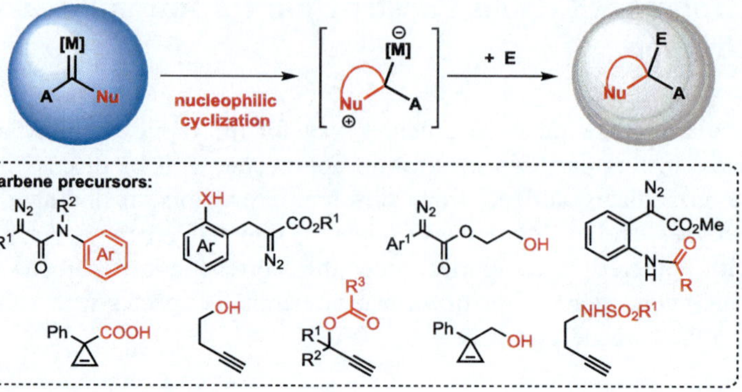

Scheme 4 Intramolecular carbene nucleophilic cyclization for construction of heterocyclic skeleton

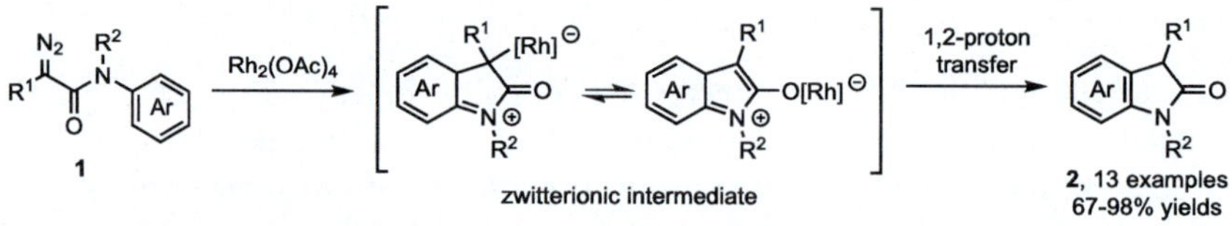

Scheme 5 Catalytic formation of zwitterionic intermediates from diazoacetamides

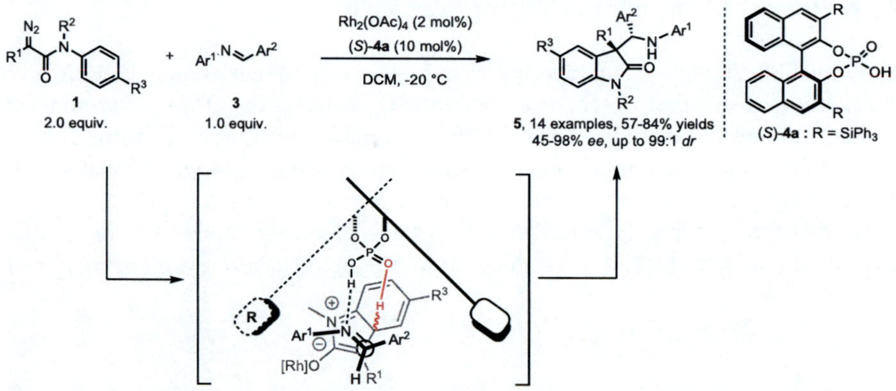

Scheme 6 Asymmetric interception of zwitterionic intermediates with imines

trap this species. In 2012, Hu and coworkers successfully intercepted this zwitterionic intermediate with Brønsted acid-activated imines [33]. Under co-catalysis with integrated rhodium acetate and chiral phosphoric acid (CPA), the zwitterionic species derived from *N*-aryl diazoamides **1** could be efficiently trapped by imines **3** through a Mannich-type addition to deliver a series of polyfunctionalized oxindole derivatives **5** in good yields with up to 98% *ee* and 99:1 *dr* (Scheme 6).

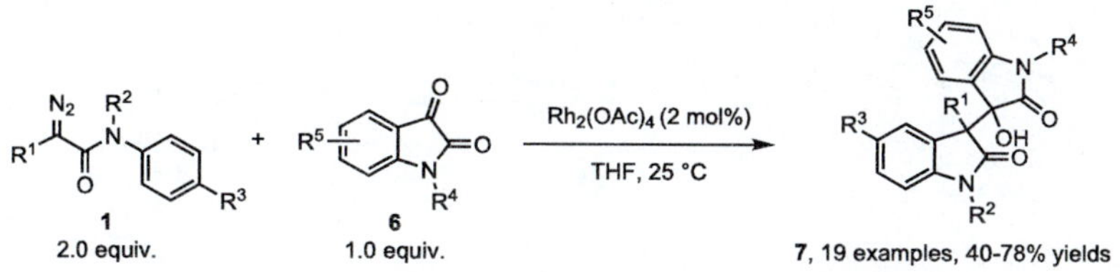

Scheme 7 Aldol-type interception of zwitterionic intermediates with isatins

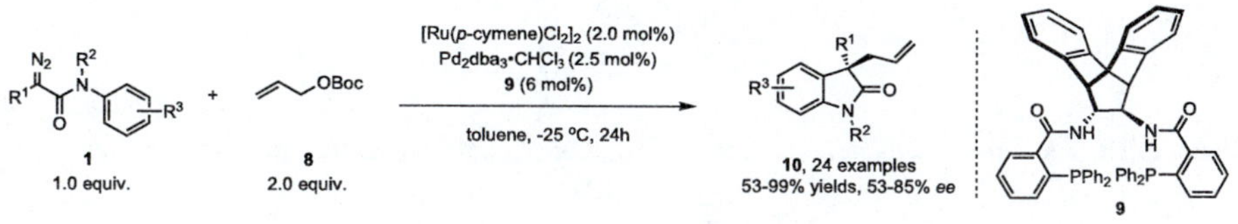

Scheme 8 The Ru/Pd co-catalyzed asymmetric Michael-type interception of enolate species with allyl *tert*-butyl carbonate

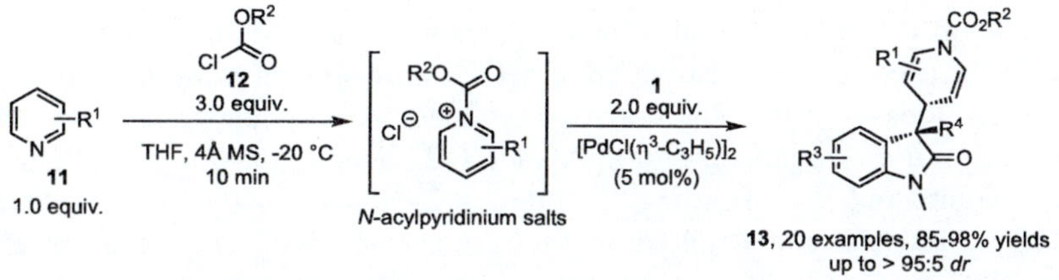

Scheme 9 Michael-type interception of zwitterionic intermediates with in-situ formed *N*-acylpyridinium electrophiles

Apart from imines, different electrophiles were employed to intercept these active zwitterionic intermediates in-situ generated from diazoacetamides. In the same year, Hu group described that isatins **6** could also serve as the electrophiles to catch the transient zwitterionic species for the rapid construction of 3,3′-bioxindole scaffolds (Scheme 7), which are prevalent skeletons existing in various alkaloids [34].

Later in 2016, Lautens' group depicted a novel Ru(II)/Pd (0)-catalyzed asymmetric one-pot reaction through interception of nucleophilic intermediates with allylic palladium species derived from allyl *tert*-butyl carbonate **8** [35]. The reaction initially generated the oxindole enolate species through ruthenium-catalyzed C-H functionalization of aryl α-diazoamides **1**, followed by palladium-catalyzed asymmetric allylic alkylation in the presence of chiral ligand **9**, providing a library of 3-allyl-3-aryl oxindole derivatives **10** in good yields with up to 85% *ee* (Scheme 8).

Beyond using of stable electrophiles, unstable and in-situ generated electrophiles could also be introduced to this interception process. In 2017, *N*-acylpyridinium salts, generated in-situ from pyridine **11** and chloroformate **12**, were used to effectively capture the zwitterionic intermediates by Hu group (Scheme 9) [36]. They

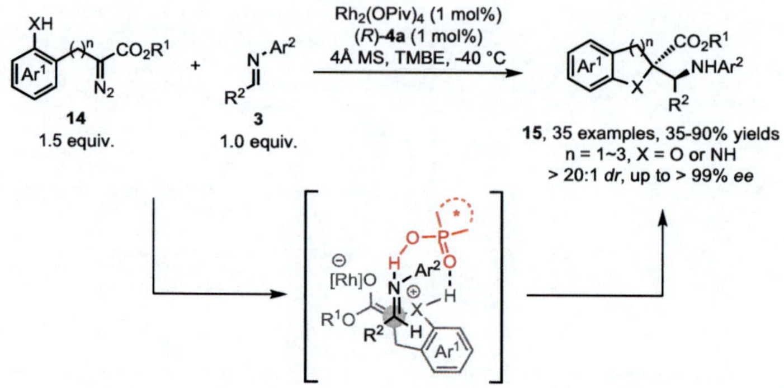

Scheme 10 Asymmetric Mannich-type interception of phenolic oxonium ylides with imines

illustrated a novel palladium-catalyzed three-component reaction of diazoamides **1** with pyridine **11** and chloroformate **12**, leading to a series of C-2 functionalized 2*H*-chromene derivatives **13** in good yields with moderate to good diastereoselectivities.

In addition to the diazoamides, other nucleophile-imbedded diazo compounds could also be used for the construction of heterocyclic structures through analogous cascade reactions. In 2013, Zhou's group reported a chiral copper-catalyzed intramolecular insertion of carbenes into phenolic O-H bonds of *ortho*-phenol tethered diazo compounds, affording chiral 2-carboxy dihydrobenzofurans, dihydrobenzopyrans, and tetrahydrobenzooxepines in high yields with excellent enantioselectivities [37]. Inspired by Zhou's work, Hu and coworkers described an intriguing enantioselective Rh(II)/CPA co-catalyzed intermolecular Mannich-type reaction of phenolic oxonium ylides with imines **3** for the direct construction of enantioenriched 2,2-disubstituted dihydrobenzofurans **15** with excellent enantioselectivity (Scheme 10) [38]. The mechanistic insight from this investigation was that the dual *H*-bonding of CPA circumvented the competitive intramolecular phenolic O-H bond insertion. Moreover, the results illustrated that isatins could also intercept these phenolic oxonium ylides in the presence of $Rh_2(OAc)_4$ to synthesize dihydrobenzofuran and 3-hydroxyoxindole hybrid products, which exhibited good anticancer potency against human colon cancer cells (HCT116 cells) [39].

Hydroxy group tethered diazo compounds **16** [40] reported by Hu and coworkers formed oxonium ylides under catalysis by copper, followed by aldol-type addition with electron-deficient aldehydes **17** or isatins **6**, leading to rapid assembly of *O*-containing heterocyclic skeletons bearing the 3-substituted 1,4-dioxan-2-one moiety in decent yields and good diastereoselectivities (Scheme 11).

In 2019, Hu group reported that metal carbenes, formed under rhodium (II) catalysis of *ortho*-amide group tethered diazo compounds **20**, were subsequently trapped by the carbonyl group intramolecularly to generate zwitterionic intermediates [41]. The Mannich-type interception of aforementioned zwitterionic intermediates with imines **3** provided high enantioselectivity in the formation of benzoxazine derivatives **21** under the cooperative catalysis of Rh(II)/CPA. In contrast, the Rh

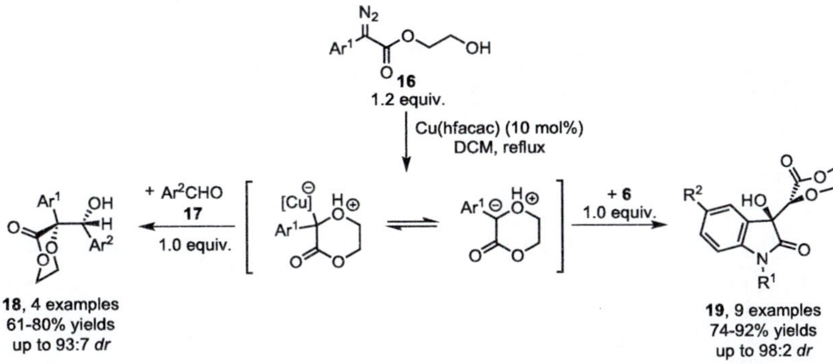

Scheme 11 Aldol-type interception of oxonium ylides derived from hydroxy group tethered diazo compounds **16**

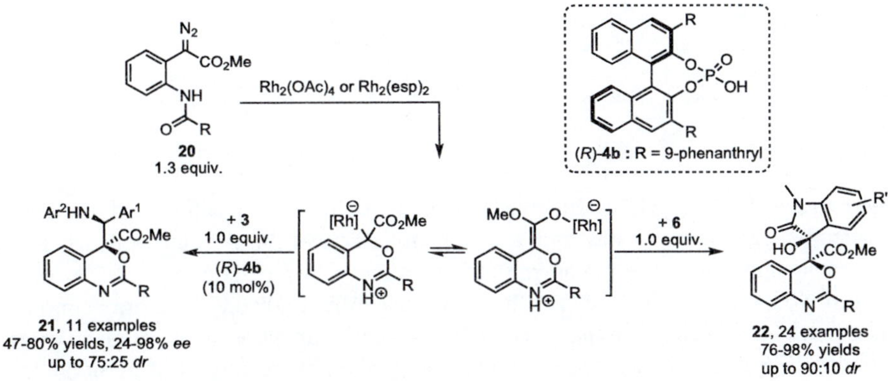

Scheme 12 The Mannich- and aldol-type interception of zwitterionic intermediates derived from *ortho*-amide group tethered diazo compounds **20**

(II)-catalyzed Aldol-type interception with isatins **6** afforded benzoxazine derivatives **22** in high yields with moderate diastereoselectivities (Scheme 12).

2.1.2 Alkynes as Carbene Precursors

Alkynes are common and readily available materials for generation of heterocyclic and carbocyclic frameworks through gold-catalyzed oxidative cyclization via carbene intermediates [42]. This approach can be traced to a 2008 report by Toste of an intermolecular [3 + 4]-cycloaddition of propargyl eaters **23** with α, β-unsaturated imines **24** [43]. With the assistance of a gold(I) catalyst, propargyl eaters **23** generated gold-stabilized vinylcarbenoid species through a catalytic 1,2-rearrangement, which subsequently went through a stepwise cyclization with imines **24** via allylgold intermediates, offering a series of azepines **25** in generally

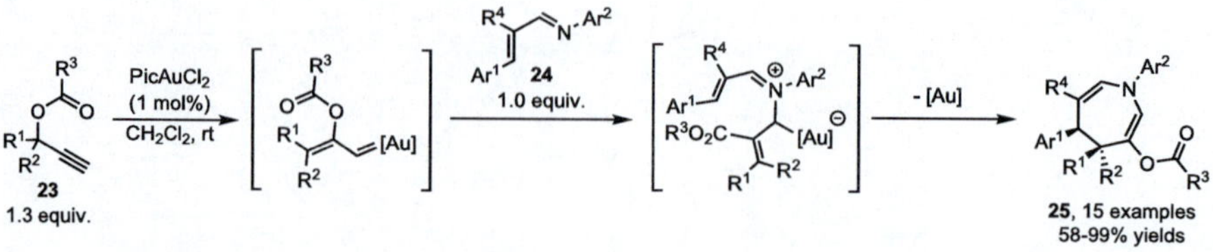

Scheme 13 Gold-catalyzed [3 + 4]-cycloaddition of propargyl esters with α,β-unsaturated imines

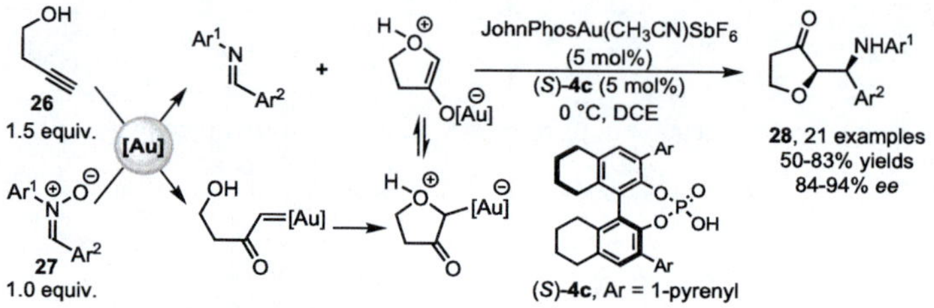

Scheme 14 Asymmetric Mannich-type reaction using 3-butynol with nitrones as carbene precursors

good yields (Scheme 13). However, given that the high reactivity of transient vinylcarbenoid intermediates and the lack of effective chiral gold catalysts, it is still challenging to realize the asymmetric version for this reaction.

Ten years later, Xu and Hu have successfully realized an atom-economy enantioselective Mannich-type interception of oxonium ylide or its enolate using 3-butynol **26** as carbene precursor under the cooperative catalysis of gold(I)/CPA (Scheme 14) [44]. Initially, gold-catalyzed alkyne oxidation formed the gold carbene species in the presence of nitrones **27**. This was followed by intramolecular nucleophilic attack with the tethered hydroxy group to form the ylide or its gold enolate species, then reassembly with the imine leaving fragment in the presence of CPA, leading to dihydrofuran-3-one derivatives **28** in good yields with excellent enantio- and diastereoselectivities. Notably, Liu's group disclosed a racemic version of this transformation using 2-ethynylphenols, acyclic 4-yn-1-ols, and phenoxyethynes as the carbene precursors under gold-catalysis at the same time [45].

Apart from the Mannich-type interception of above cascade reaction, Xu's group also illustrated an effective aldol-type addition of ylide intermediates derived from homopropargyl alcohols with isatins, which constructed 3-hydroxyoxindole derivatives with two vicinal stereocenters in good yields and excellent diastereoselectivities [46].

In 2021, Xu and Hu showed that homopropargyl sulfonamides **29** could serve as carbene precursors through a gold-catalyzed oxidative transformation [47]. Mechanistically, nucleophilic addition of the in-situ generated carbene intermediate with the tethered amino group formed a gold enamine species, followed by Mannich-type addition with imines in the presence of CPA, providing a direct access to chiral

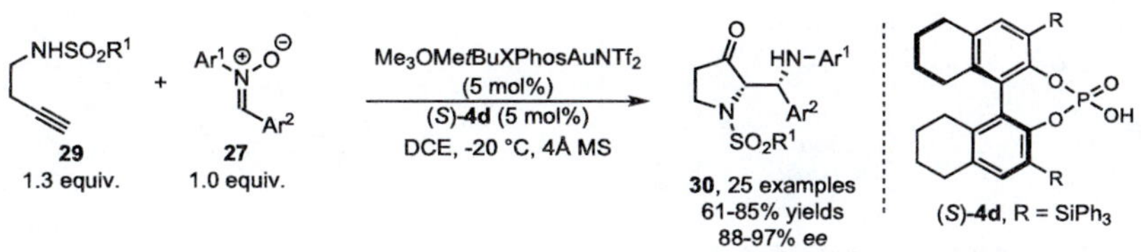

Scheme 15 Asymmetric Mannich-type reaction using homopropargyl sulfonamides as carbene precursors

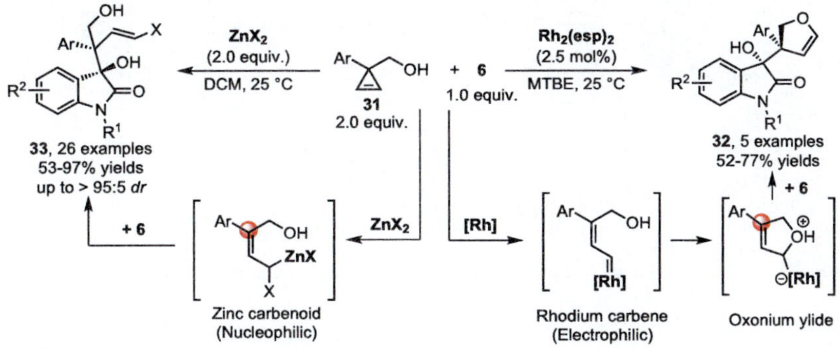

Scheme 16 Metal catalyst-dependent umpolung reactivity of cyclopropene alcohols

pyrrolidin-3-one derivatives **30** in good to high yields with high to excellent stereoselectivities with 100% atom economy (Scheme 15).

2.1.3 Cyclopropenes as Carbene Precursors

Cyclopropene, which is one of the highly strained carbocyclic structures, can serve as a carbene precursor through a catalytic ring-opening process [48]. In 2019, Hu group reported an effective carbene cascade reaction through aldol-type interception using cyclopropene alcohols **31** as carbene precursors [49]. Interestingly, the in-situ formed vinyl carbenes possess metal-dependent umpolung reactivity, which are beneficial for divergent syntheses (Scheme 16). When treated with $Rh_2(esp)_2$, cyclopropene alcohols **31** generated electrophilic carbenes, followed by intramolecular nucleophilic attack with the tethered hydroxyl group to form cyclic oxonium ylides, and finally trapped by isatins **6** at the vinylogous site to give butanolide derivatives **32** with 52-77% yields. In contrast, cyclopropene alcohols **31** formed nucleophilic zinc carbenoids under catalysis by Zn(II) halide, which were subsequently captured by isatins **6** to provide oxindole derivatives **33** in good yields and up to >95:5 *dr*.

An effective $Rh_2(esp)_2$/chiral Brønsted acid co-catalyzed asymmetric Mannich-type interception process of in-situ formed carboxylic oxonium ylides with imines **3** was disclosed by Hu and coworkers in 2020 [50]. In this transformation,

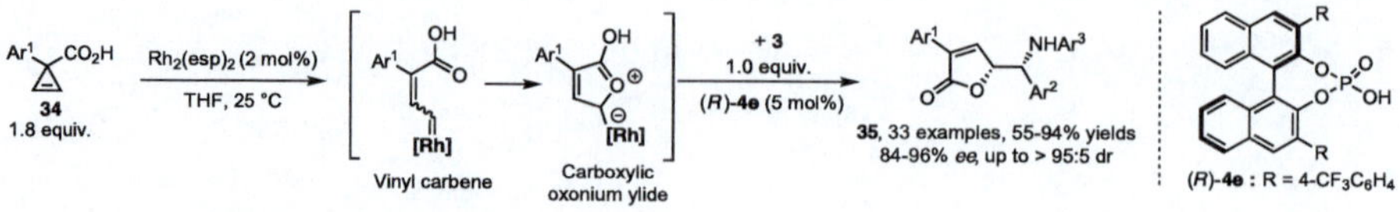

Scheme 17 Asymmetric Mannich-type reaction using cyclopropene carboxylic acids as carbene precursors

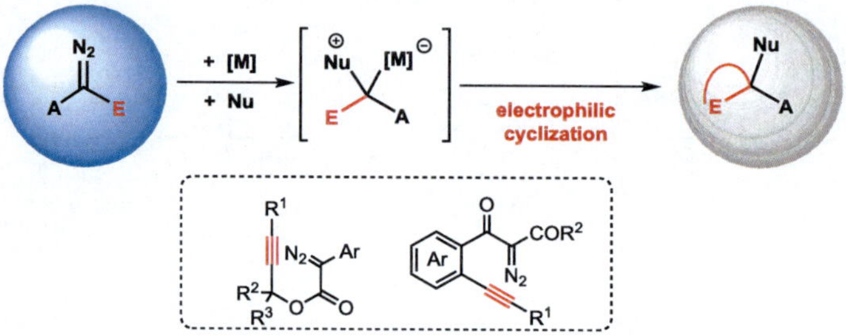

Scheme 18 Intramolecular electrophilic cyclization for the construction of heterocyclic skeleton

carboxylic oxonium ylides derived from cyclopropene carboxylic acids **34** underwent enantioselective Mannich-type addition with CPA-activated imines, providing chiral γ-butenolide derivatives **35** in good yields and stereoselectivities (Scheme 17). Moreover, the same group reported two analogous methods for the direct assembly of γ-butenolides using cyclopropene carboxylic acids **34** as carbene precursors - one was an Rh(II)-catalyzed sequential cycloisomerization/aldol-type addition with isatins **6** [51]; and the other was Michael-type interception of transient carboxylic oxonium ylides with α,β-unsaturated 2-acyl imidazoles [52].

2.2 *Heterocyclic Skeleton Construction via Electrophilic Cyclization*

Electrophilic functionalities tethered to diazo compounds can be used for the intramolecular interception of reactive intermediates that are derived in-situ from carbene species through addition with nucleophiles. This methodology provides an alternative approach for the assembly of multi-functional heterocycles (Scheme 18).

Zhang's group described a gold(I)-catalyzed cascade C–H functionalization/conia-ene type cascade reaction of *o*-alkynylaryl α-diazoesters with electron-rich aromatics in 2016, providing functionalized indene derivatives with high chemo- and site-selectivity [53]. Later, Xu's group developed an unprecedented gold-catalyzed water-mediated 6-*endo-dig*-yne carbocyclization with propargyl diazoacetates **36** for the direct construction of multi-substituted furans **37** in good yields (Scheme 19) [54]. In this transformation, the gold-associated oxonium ylides

Scheme 19 Gold-catalyzed cascade reaction with propargyl diazoacetates as carbene precursors

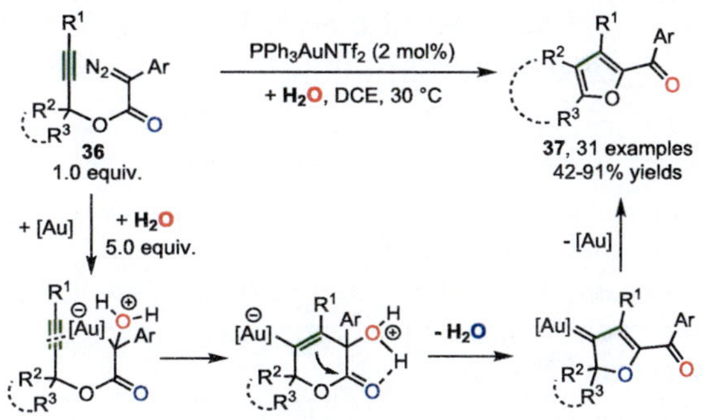

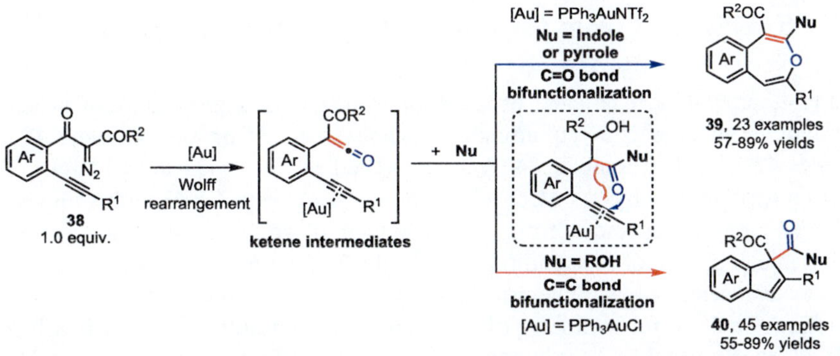

Scheme 20 Gold-catalyzed divergent ketene C=O or C=C bond bifunctionalization

generated from propargyl diazoacetates **36** and H_2O, followed by a sequential 6-*endo-dig*-yne carbocyclization/β-H elimination/protodeauration process, provided aromatized furans **37** in up to 91% yield.

Later in 2019, Xu disclosed a selective vinylogous reactivity of carbene intermediate derived via a gold-catalyzed alkyne carbocyclization to provide indenol derivatives in good yields with substrate generality [55]. Moreover, analogous approaches have also been applied to the construction of different carbocyclic structures [56–58]. Recently, the same group developed a gold-catalyzed C=O/C=C bifunctionalization strategy of ketene generated in-situ via Wolff rearrangement of alkyne-tethered diazoketones **38** (Scheme 20) [59]. The ketene intermediates went through a dual functionalization sequential approach via addition with nucleophiles and electrophiles. In the case with indoles/pyrroles as the nucleophiles, the reaction went through an *O*-7-*endo-dig* cyclization to give the 7-membered benzo[*d*]oxepines **39**; whereas, *C*-5-*endo-dig* carbocyclization occurred dominantly when alcohols were used as the nucleophiles, delivering mainly the indene products **40**.

3 Heterocyclic Skeleton Construction via Carbene *gem*-Difunctionalization

Metal carbene *gem*-difunctionalization through sequential addition with nucleophiles and electrophiles can be used for the direct construction of heterocycles via two different approaches: through carbene *gem*-difunctionalization with reagents containing both nucleophilic and electrophilic functionalities, or through one-pot carbene *gem*-difunctionalization and intramolecular cyclization cascade reactions (Scheme 21).

3.1 *Heterocycles from the Interception of Ammonium Ylides*

A novel and practical cascade process for synthesis of highly diastereoselective pyrrolidines via interception of ammonium ylides derived from diazoacetates **43** and anilines **41** with 4-*oxo*-enoates **42** was illustrated by Hu group in 2013 [60]. The reaction was initiated by *aza*-Michael addition of 4-*oxo*-enoates **42** with anilines **41**, followed by ylide generation from this adduct and diazoacetates **43**, and terminated by intramolecular aldol-type addition to afford the target products **44** in good yields and diastereoselectivities (Scheme 22). Later, Moody and coworkers established an analogous strategy for synthesis of functionalized pyrrolidines, which involved a N-H insertion and subsequent intramolecular aldol addition sequence under rhodium (II), copper(I), or iron(III) catalysis [61]. Remarkably, using β-nitroacrylates to trap

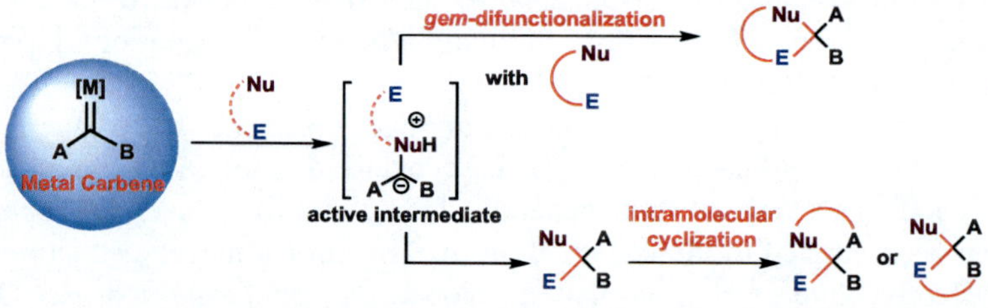

Scheme 21 Heterocyclic skeleton construction via carbene *gem*-difunctionalization cascade reactions

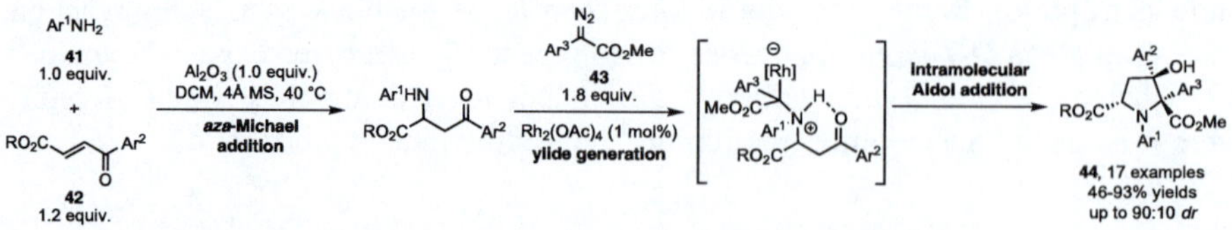

Scheme 22 Three-component reaction of diazoacetates with anilines and 4-*oxo*-enoates for the synthesis of pyrrolidines

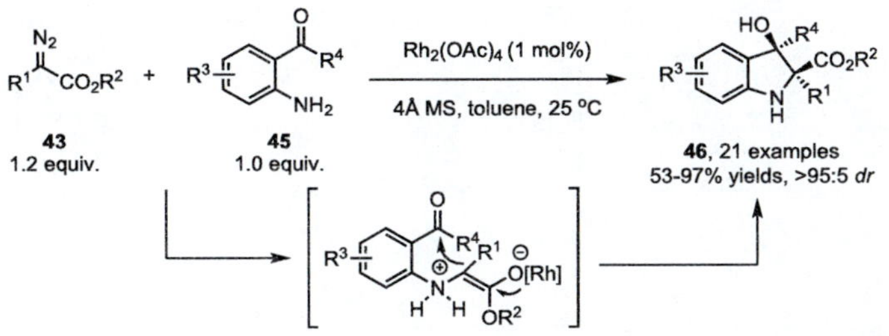

Scheme 23 Intramolecular aldol-type interception of ammonium ylides for the synthesis of indolines

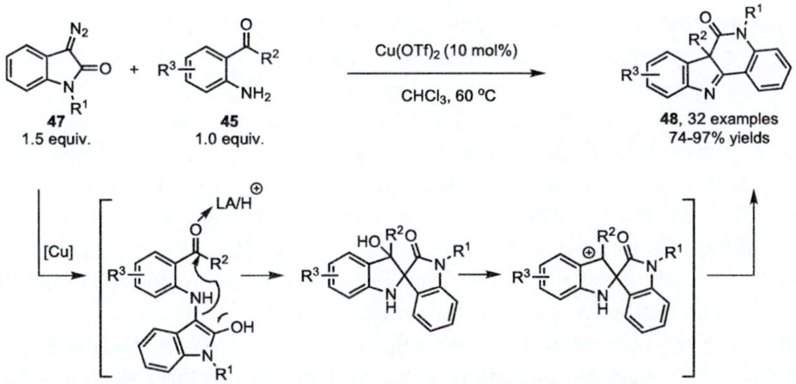

Scheme 24 Copper(II)-catalyzed N-H insertion/aldol-type interception/selective carbonyl migration cascade reaction

the aforementioned ammonium ylides, Hu group unveiled an efficient chiral Rh(I)-diene-catalyzed enantioselective three-component reaction to provide chiral γ-nitro-α-amino-succinates in good yields [62].

In 2014 Hu and coworkers reported an aldol-type addition of in-situ generated ammonium ylides using aryl diazoacetates **43** and 2-aminophenyl ketones **45** as materials (Scheme 23) [63]. The reaction provided a variety of 3-hydroxy-2,2,3-trisubstituted indolines **46** in good yields and excellent diastereoselectivities. Later, Rastogi's group presented an analogous rhodium carbene-mediated reaction of aryl diazoacetates with *N*-*o*-alkylamino benzoylbenzotriazoles in water, affording a direct access to pseudoindoxyl scaffolds in generally good yields [64].

Sekar's group disclosed a Cu(II)-catalyzed domino reaction of 2-aminochalcones **45** with 3-diazooxindoles **47** via a selective carbonyl migration process to produce indolo[3,2-*c*]quinolinones **48** with yields up to 97% (Scheme 24) in 2019 [65]. The mechanistic studies implied a stepwise process comprised of carbene N−H insertion, intramolecular aldol-type cyclization, and unprecedented ring expansion of the oxindole core through a C3-selective 1,2-carbonyl migration sequence.

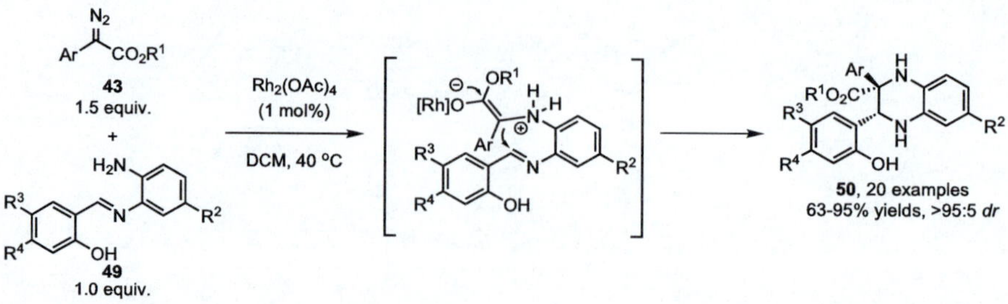

Scheme 25 Intramolecular Mannich-type interception of ammonium ylides for the synthesis of piperazines

Following the aldol-type addition, a similar approach via intramolecular Mannich-type addition of ammonium ylide reported by Hu and Liu via a formal, highly diastereoselective, [5 + 1]-cyclization reaction of aryl diazoacetates **43** with *ortho*-aminophenyl imine derivatives **49**. This transformation afforded trisubstituted tetrahydroquinoxalines **50** bearing a quaternary carbon center in moderate to good yields and excellent diastereoselectivities (Scheme 25) [66]. Mechanistic insights indicated that the reaction went through an intermolecular ammonium ylide formation and intramolecular Mannich-type addition sequence, and the presence of the *ortho*-hydroxyl group on the imines was essential for its elevated electrophilicity through intramolecular hydrogen bonding activation.

Anbarasan and coworkers unveiled a flexible route to construct various tri- and tetra-substituted indolines via diastereoselective palladium-catalyzed intramolecular interception of ammonium ylides with nonactivated C-C double bonds in 2017 [67]. Their experimental data and density functional theory (DFT) studies revealed that the ammonium ylides, derived from diazoacetates **43** and *o*-vinylanilines **51,** went through cyclization with tethered alkene species via a metallo−ene-type reaction involving a six-membered cyclic transition state. The reaction was terminated by protonation to give the desired indolines **52** and regenerate the palladium catalyst (Scheme 26).

In 2015, Sun and coworkers described a catalyst-controlled tandem annulation method for the construction of multi-substituted pyrrolidines with diazoacetates **43** and amino alkynes **53** (Scheme 27) [68]. With rhodium catalysis, diazoacetates **43** initially formed carbene intermediates, followed by N-H bond insertion with amino alkynes **53** via ammonium ylides. Then, intramolecular Conia-ene cyclization provided 3-methylene pyrrolidines **54** in moderate to good yields. Meanwhile, Nikolaev's group reported that $Rh_2(Oct)_4$-catalyzed reaction of diazomalonates with α,β-unsaturated δ-(*N*-aryl)amino esters went through the analogous transformation, affording the multi-functionalized pyrrolidines with yields of up to 82% and moderate diastereoselectivities [69].

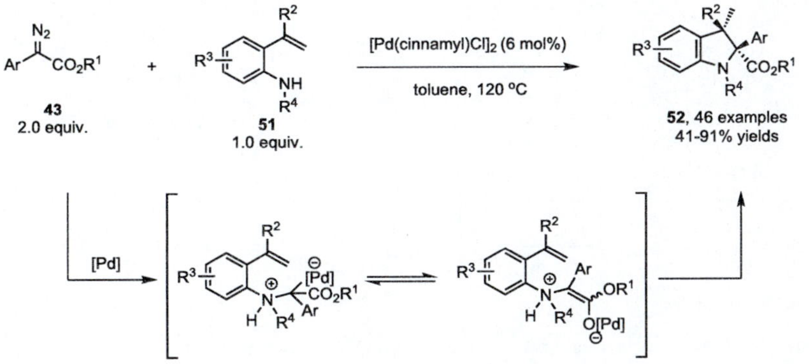

Scheme 26 Intramolecular interception of ammonium ylides with alkenes for the synthesis of indolines

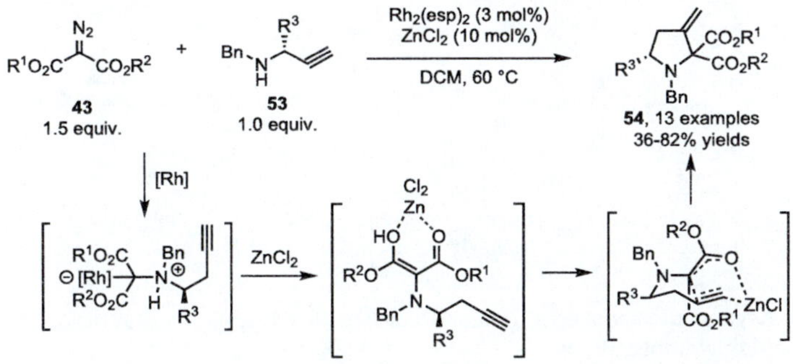

Scheme 27 Intramolecular formal 5-*exo-dig* cyclization for the synthesis of pyrrolidines

3.2 Heterocycles from the Interception of Oxonium Ylides

As disclosed independently by Moody [70] and Hu [71] (Scheme 28), highly diastereoselective multi-substituted tetrahydrofurans **56** can be generated from easily accessible β-hydroxyketones **55** and diazoacetates **43** via an aldol-type interception of in-situ generated oxonium ylide intermediates. Hatakeyama and coworkers also reported an Rh(II)/Zn(II)-catalyzed formal [4 + 1]-cycloaddition of diazo dicarbonyl compounds with homopropargyl alcohols, which provided substituted tetrahydrofurans in generally good yields [72]. In addition, the analogous stepwise approach, which involved O-H insertion and Michael-type addition of allyl alcohols with aryldiazoacetates was reported by Hu and coworkers in 2009, providing highly diastereoselective fully substituted tetrahydrofurans in one-pot [73].

Wang and coworkers reported a highly diastereo- and enantioselective approach for the synthesis of 2,3-dihydrobenzofurans **59** bearing tetrasubstituted carbon stereocenters through a copper-catalyzed intramolecular cyclization reaction of oxonium ylide in 2019 [74]. In this reaction, the copper(I)/bisoxazoline catalyst **58**

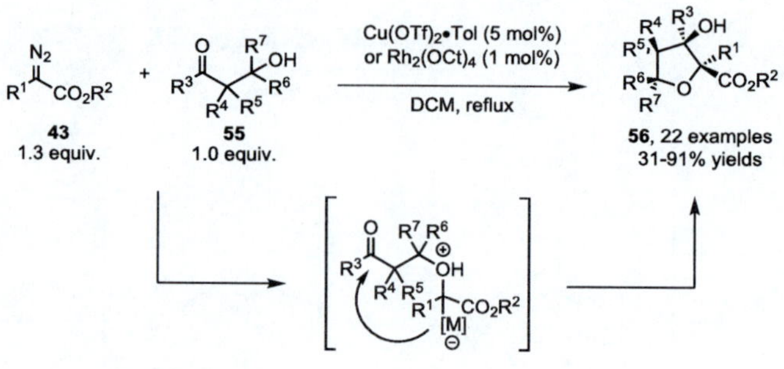

Scheme 28 Intramolecular aldol-type interception of oxonium ylides for the synthesis of tetrahydrofurans

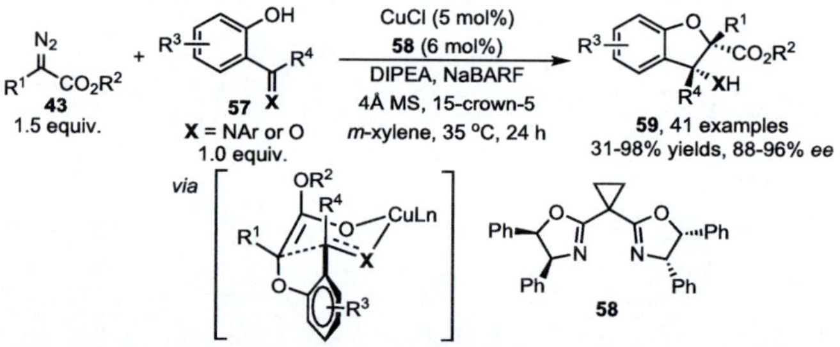

Scheme 29 Asymmetric annulation of diazoacetates with 2-iminyl 2-acyl-substituted phenols for the synthesis of dihydrobenzofurans

not only generated Cu(I)-carbene species with diazoacetates **43**, but also acted as a Lewis acid for the activation of the imine or ketone **57** for diastereo- and enantioselective cyclization with in-situ formed oxonium ylide (Scheme 29).

Recently, the Hu group assembled chiral 2,3-dihydropyrans **62** in good yields with excellent stereoselectivities through a chiral dirhodium(II)-catalyzed asymmetric annulation of α-hydroxyl ketones **61** with vinyl diazoesters **60** (Scheme 30) [75]. DFT studies indicated that the reaction proceeded with intramolecular asymmetric aldol-type interception of in-situ generated chiral Rh(II)-associated oxonium ylides to provide optical products **62** in good to excellent yields.

The synthesis of isochroman frameworks was accomplished by metal-catalyzed cyclization of alcohol-tethered enones **63** with diazo compounds by Hu group. With diazoindolinones **47** as carbene precursors, the transformation proceeded through asymmetric intramolecular Michael-type interception of oxonium ylide intermediates enabled by cooperative Rh(II)/CPA catalysis via a dual *H*-bonding activation model, affording chiral spirochroman-3,3-oxindoles **64** with vicinal quaternary and tertiary stereocenters (Scheme 31) [76]. Notably, 3,4-substituted isochromans could

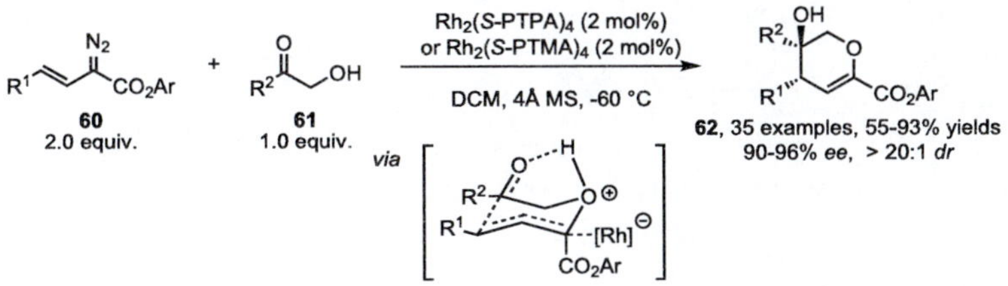

Scheme 30 Asymmetric aldol-type interception of oxonium ylides for the synthesis of dihydropyran

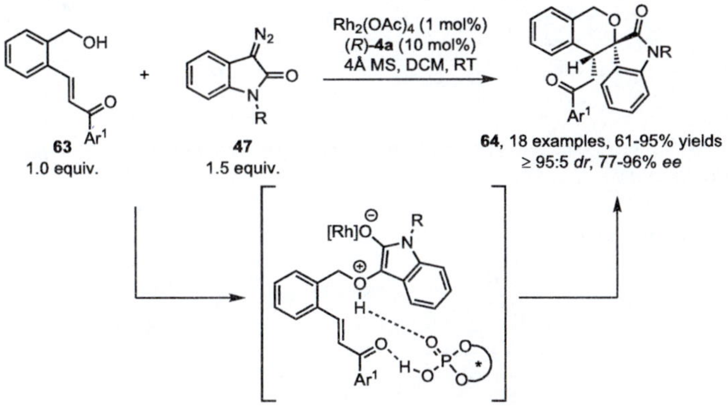

Scheme 31 Intramolecular Michael-type interception of oxonium ylides for the synthesis of spiro isochromans

be obtained selectively from diazoacetates and *o*-hydroxymethyl chalcones under the catalysis of an inexpensive copper(II) catalyst by an analogous approach [77].

Besides pre-synthesized electrophiles, in-situ generated isoquinolinium intermediates from 2-alkynylarylaldimines have also been employed to intercept oxonium ylides derived from diazoacetates **43** and water, providing 1,2-dihydroisoquinolines in high yields with moderate diastereoselectivities [78]. The Schneider group has reported an in-situ generated *ortho*-quinone methide (*o*-QM) for the trapping of active oxonium ylide under Rh(II)/CPA synergistic catalysis [79]. A remarkable feature of this reaction is using water as the nucleophile, which is a departing fragment in the formation of *ortho*-quinone methide (*o*-QM) under catalysis by CPA, to trap rhodium-carbene intermediates to furnish reactive oxonium ylides. The excellent diastereo- and enantioselectivities of target products **66** have been achieved with the assistance of CPA during the Michael-type interception of transient oxonium ylides with *o*-QM via *H*-bonding (Scheme 32).

In addition to C-C and C-X double bonds, Zhang and coworkers disclosed a practical approach to build the skeleton of 2,5-dihydrofurans **68** via gold(I)-catalyzed formal [4 + 1]-cycloaddition of α-diazoesters **43** with propargyl alcohols **67** using the alkyne unit as the acceptor [80]. Mechanistically, gold(I) carbenes

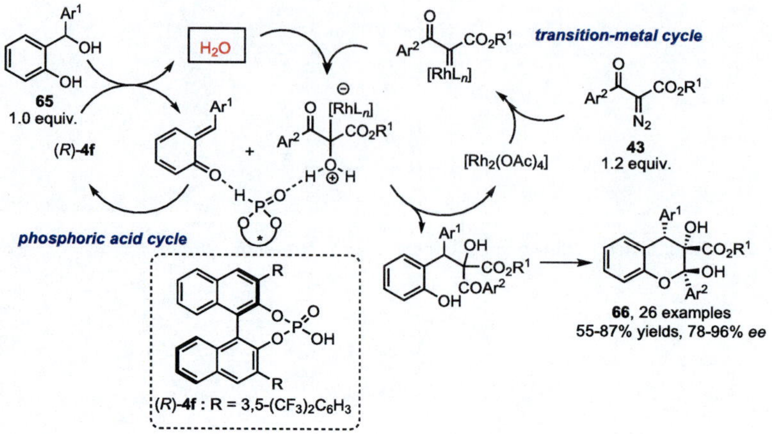

Scheme 32 Asymmetric Michael-type addition of reactive oxonium ylides for the synthesis of 1,2-dihydroisoquinolines

Scheme 33 Formal [4 + 1]-cycloaddition reaction of α-diazoesters with propargyl alcohols for the synthesis of dihydrofurans

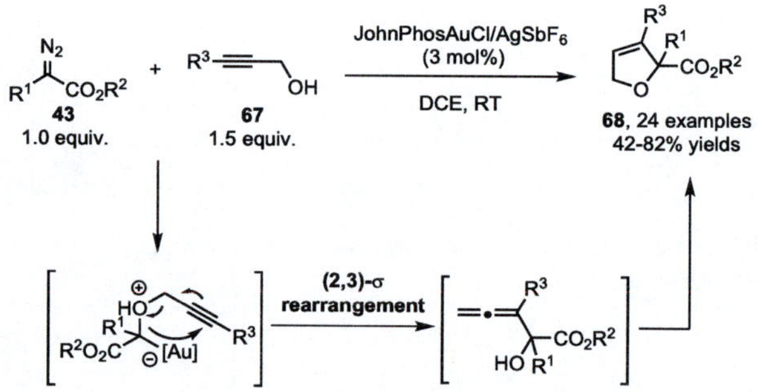

derived from α-diazoesters **43** combined with propargyl alcohols **67** to form oxonium ylides (or gold enolates), most likely followed by a [2,3]–σrearrangement to generate α-hydroxy allene intermediates, and terminated by 5-*endo-dig* cyclization to deliver 2,5-dihydrofurans **68** in 42-82% yields (Scheme 33). Hu group subsequently developed different catalytic systems for rapid access to 2,5-dihydrofurans **68**, such as Pd(II) [81] or Rh(II) catalysis [82] and Ag(I)/BINAP cooperative catalytic system [83].

3.3 Heterocycles from the Interception of Zwitterionic Intermediates

In 2010, Davies and coworkers disclosed an Rh(II)-catalyzed [3 + 2]-cycloaddition of styryl diazoacetates **60** with indoles **69**, affording cyclopenta[*b*]indoles **70** in good yields with high enantioselectivties [84]. Depending on the substitution pattern on

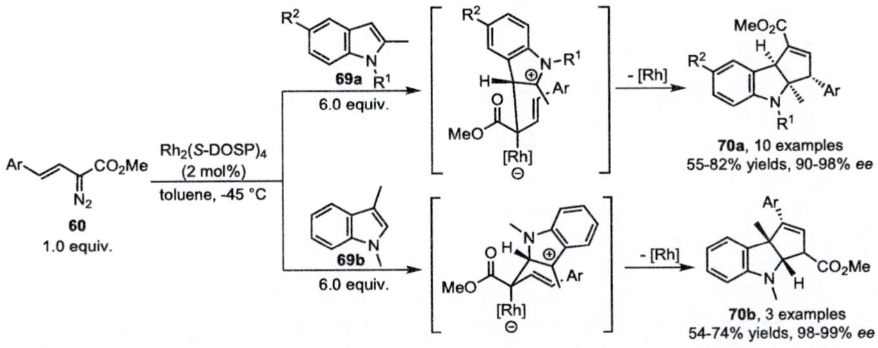

Scheme 34 Asymmetric [3 + 2]-Cycloaddition of vinyl diazoacetates with indoles

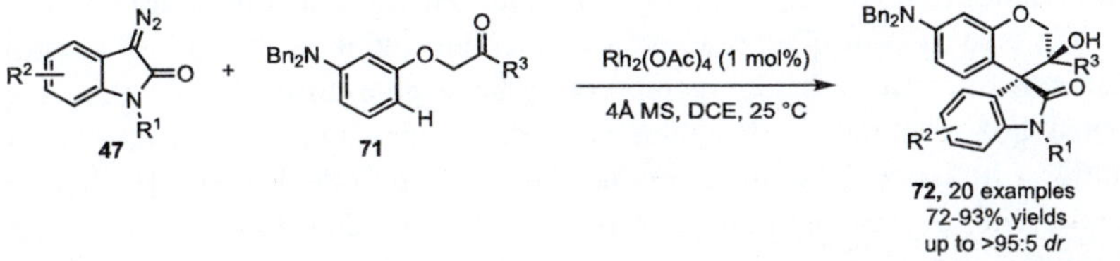

Scheme 35 Intramolecular aldol-type trapping zwitterionic intermediates for the synthesis of spirooxindoles

the indole unit, two distinct regioisomeric fused indolines could be obtained via intramolecular cyclization of in-situ formed zwitterionic intermediates (Scheme 34). In 2014, Katukojvala's group presented an analogous rhodium-catalyzed [4 + 2]-benzannulation of enaldiazo compounds with pyrroles, leading to substituted indoles in good yields [85].

In 2017, a straightforward method for the construction of spiro[chroman-4,3′-oxindole] skeleton with α-phenoxy ketones **71** and 3-diazooxindoles **47** was reported by Hu and coworkers [86]. In the presence of $Rh_2(OAc)_4$, the zwitterionic intermediates formed in-situ from 3-diazooxindoles **47** and α-phenoxy ketones **71**, underwent intramolecular aldol-type cyclization to offer spirooxindoles **72** in good yields and high diastereoselectivities (Scheme 35).

Recently, by utilizing *o*-nitroarylalkynes **73** as carbene precursors, an efficient gold(I)-catalyzed redox/nitroso aldol addition-cascade reaction was reported by Xu and coworkers [87]. The transformation was a 5-*exo-dig* cyclization of gold-activated alkyne with the *ortho* nitro group to form a gold carbene, followed by an intermolecular nucleophilic attack with indoles **69** to generate zwitterionic intermediates, which were trapped by the tethering nitroso group to afford 2-indolyl indolone *N*-oxides **74** in good yields (Scheme 36).

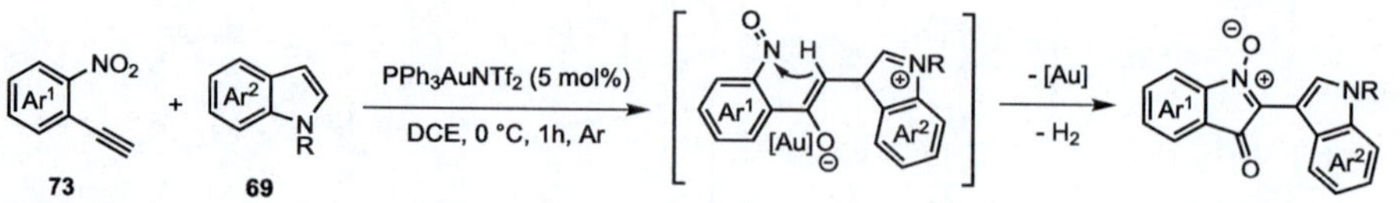

Scheme 36 Gold(I)-catalyzed redox/nitroso aldol addition cascade for the synthesis of indolone *N*-oxides

3.4 *Heterocycles from the Interception of Other Reactive Intermediates*

Besides ammonium ylides, oxonium ylides, and zwitterionic intermediates, complementary interception of other reactive intermediates with electrophiles has provided flexible construction of heterocycles via carbene *gem*-difunctionalization cascade reactions. For example, Padwa and coworkers have made great efforts for the assembly of heterocyclic natural products through a tandem carbonyl ylide formation and 1,3-dipolar cycloaddition process [9, 23]. In 2001, Doyle's group reported the stereospecific epoxidation via carbonyl ylides derived from diazoacetates and aldehydes in the presence of $Rh_2(OAc)_4$ [88]. Similarly, three-component [3 + 2]-cycloaddition of diazo compounds with aldehydes and dipolarophiles have been independently disclosed by Huisgen [89, 90], Nair [91], Fox [92], and Hu [93, 94]. In 2016, Hu and coworkers unveiled a facile synthesis of oxindole-fused spirotetrahydrofurochroman scaffolds via interception of carbonyl ylides derived from 3-diazooxindoles **47** and aldehydes **17** by nitroolefinenoates **75**, followed by base-catalyzed intramolecular Michael addition, generating the polycyclic spiro heterocycles **76a** and **76b** in good to excellent yields with excellent stereoselectivity (Scheme 37) [95]. The corresponding lactams were constructed by reduction of the nitro group and intramolecular amide formation, the products of which showed potential inhibitory activity against the enzyme protein tyrosine phosphatase 1B (PTP1B) in vitro [96].

Moreover, the same group disclosed an Rh(II)-catalyzed formal C=O bond carbene insertion reaction of (*E*)-2-hydroxycinnamaldehydes **77** with diazoacetates **43**, affording 2*H*-chromenes **78** in good yields with excellent diastereoselectivities (Scheme 38) [97]. In this transformation, the (*E*)-epoxide, initially generated from (*E*)-starting materials **77** and diazoacetates **43** in the presence of $Rh_2(esp)_2$, underwent a cascade epoxide ring-opening process for conversion to the (*Z*)-epoxide intermediate, which formed (*Z*)-products **78** with high diastereoselectivity.

At the same time, Hu and coworkers accomplished an intramolecular aldol-type trapping of sulfonium ylides derived from 3-diazooxindoles **47** and 2-mercaptophenyl ketones **79**, under co-catalysis of Rh(II)/CPA, giving chiral spirooxindole-fused thiaindans **80** in high yields with good diastereo- and enantioselectivities (Scheme 39) [98].

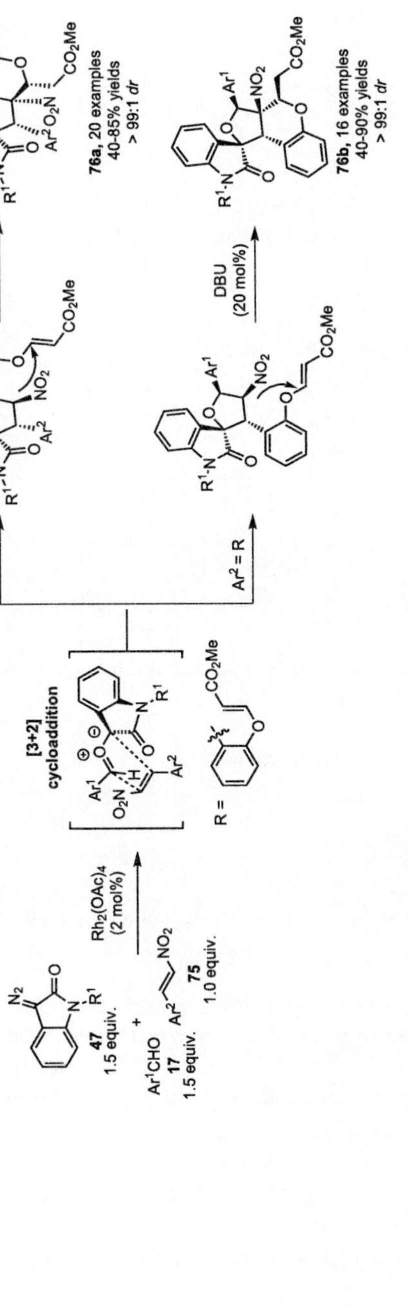

Scheme 37 One-pot three-component reaction of 3-diazooxindoles with aldehydes and nitroolefinenoates

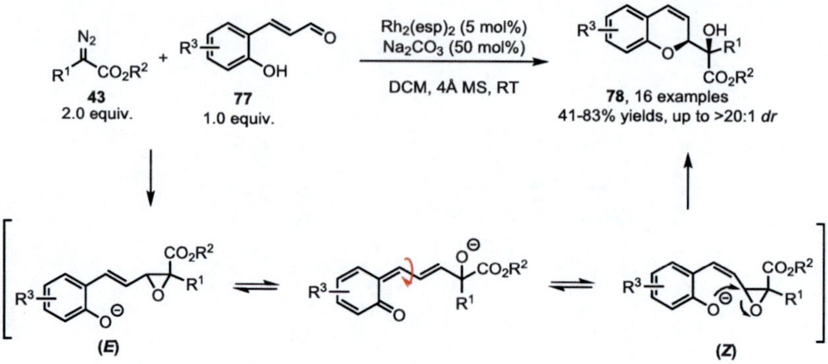

Scheme 38 Synthesis of 2*H*-chromenes involves epoxide intermediate

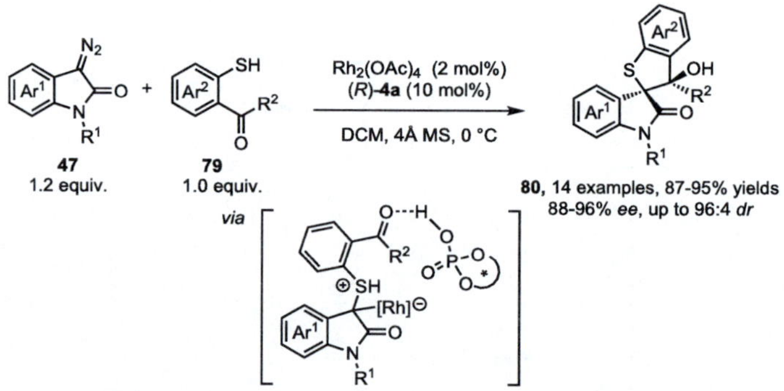

Scheme 39 Intramolecular aldol-type interception of sulfonium ylides

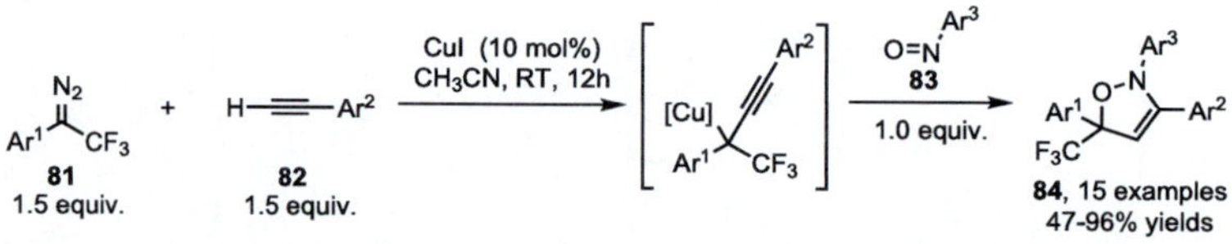

Scheme 40 Cu(I)-catalyzed three-component reaction for the synthesis of nitrosoarenes

In the same year, Hu group presented a novel Cu(I)-catalyzed [1 + 2 + 2]-cyclization reaction of trifluoromethyl diazo compounds **81** with terminal alkynes **82** and nitrosoarenes **83** [99]. A library of trifluoromethyl-substituted dihydroisoxazoles **84** was synthesized through interception of the copper propargyl intermediates by nitrosoarenes **83** (Scheme 40).

Enol intermediates [100, 101], derived from diazo pyruvates **85** and anilines **41** in the presence of ruthenium catalyst, were recently disclosed and confirmed by Hu group (Scheme 41). In addition to the enol intermediate, an imine ester intermediate generated by SET-mediated oxidation of enol was proposed as the key intermediate

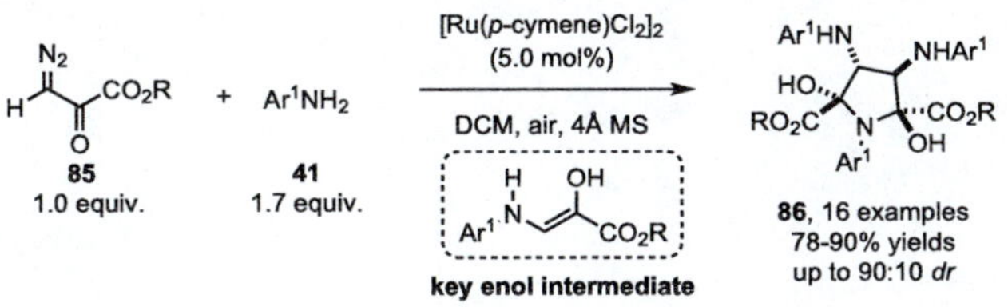

Scheme 41 Ru(II)-catalyzed two-component reaction for construction of fully substituted pyrrolidines

[100]. This strategy provided fully substituted pyrrolidines **86** in good yields with high diastereoselectivities.

3.5 *Heterocycles from One-Pot Cyclization Post the Carbene* gem-*Difunctionalization*

Instead of direct formation of heterocycles, intramolecular cyclization following metal carbene *gem*-difunctionalization also provided practical access to heterocyclic molecules with structural complexity. In 2009, Hu group developed an Rh(II)-catalyzed three-component tandem reaction of diazoacetates **43** with anilines **41** and β, γ-unsaturated α-ketoesters **87**, providing multi-substituted pyrrolidines **88** in moderate yields but with poor diastereoselectivities [102]. However, using diazo ketones **85** as carbene precursors, the same group established a highly diastereoselective tandem 1,4-conjugate addition–cyclization reaction with the same nucleophiles and electrophiles as above [103]. The carbene intermediates derived from $Rh_2(OAc)_4$-decomposed diazo ketones **85** reacted with anilines **41** to generate ammonium ylides, followed by 1,4-addition with β,γ-unsaturated α-ketoesters **87** and intramolecular cyclization of formed linear products, providing desired multi-substituted pyrrolidines **88** in good yields with high diastereoselectivities. Moreover, one-pot dehydration of **88** under acidic conditions afforded the corresponding 2,3-dihydropyrroles in excellent yields (Scheme 42).

Later, the same group accomplished a Pd(II)-catalyzed regiodivergent three-component reaction involving interception of ammonium ylides [104]. Readily available and easy-to-handle *N*-alkylquinolinium salts **89** were used to trap the transient ammonium ylides generated from diazoacetates **43** and anilines **41** via a 1,4-conjugate addition to provide 1,4-dihydroquinolines **90**. Then, protonation with the released HX formed iminium intermediates that by intramolecular nucleophilic cyclization afforded bridged tetrahydroquinolines **90** in high yields with moderate diasteroselectivities (Scheme 43).

At the same time, Hu and coworkers described an efficient Ru(II)/iminium co-catalyzed enantioselective formal [3 + 1 + 1]-cycloaddition reaction of diazo ketones **85** with anilines **41** and enals **91** for flexible assembly of optically active multi-substituted pyrrolidines **93**, which were produced in good yields with excellent

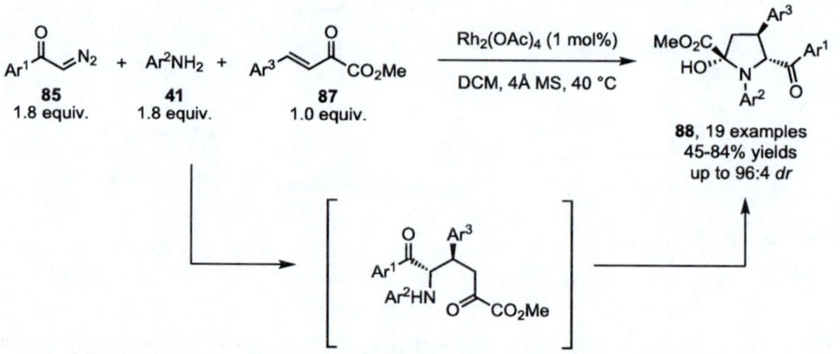

Scheme 42 Michael-type interception of ammonium ylides intramolecular cyclization for the synthesis of pyrrolidines

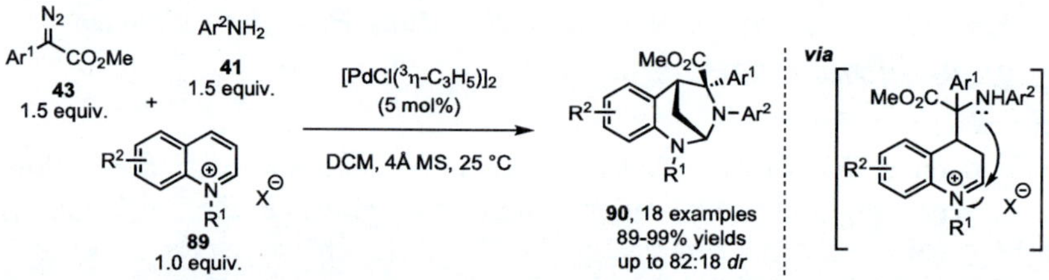

Scheme 43 Intermolecular interception of ammonium ylides with *N*-alkylquinolinium salts for the synthesis of bridged tetrahydroquinolines

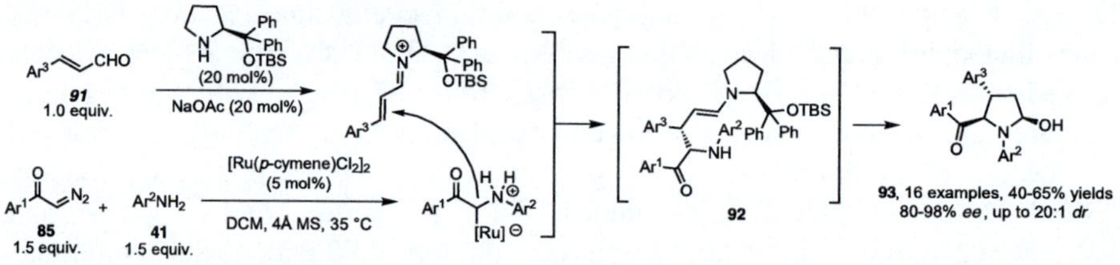

Scheme 44 Asymmetric formal [3 + 1 + 1]-cycloaddition reaction for construction of multi-substituted pyrrolidines

diastereo- and enantioselectivities. The reaction involved a Mannich-type addition of the two in-situ generated intermediates, ammonium ylide and iminium species, providing chiral linear products **92** which could be directly converted to the pyrrolidines **93** through an intramolecular cyclization (Scheme 44) [105].

In 2014, Katukojvala and coworkers reported a new class of enaldiazo compounds **94** [85, 106], which implemented a synergistic Rh(II)/BINOL phosphoric acid co-catalyzed three-component reaction with arylamines **41** and aryl aldehydes **17** for the direct synthesis of α-(3-pyrrolyl)benzylamines **96** (Scheme 45) [106]. The reaction was initiated by the formation of ammonium ylides from rhodium

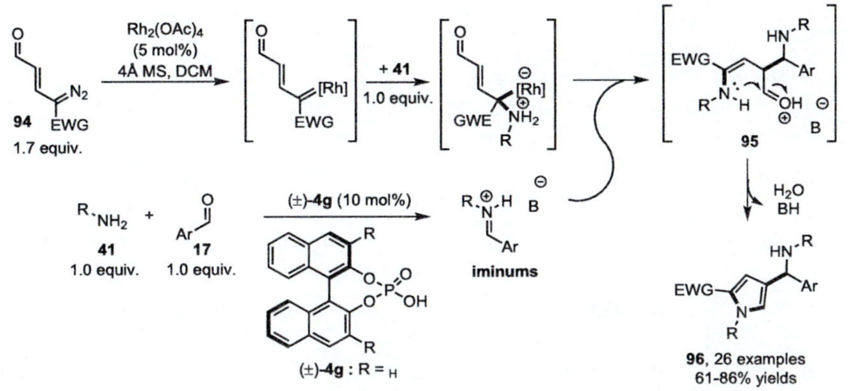

Scheme 45 Multi-component reaction for the synthesis of pyrroles involving ammonium ylide intermediates

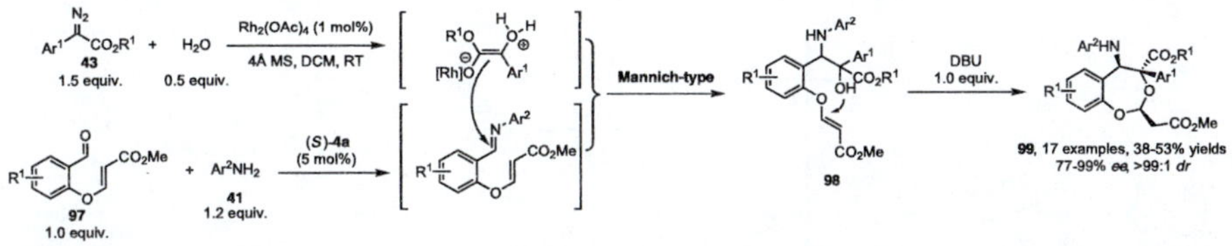

Scheme 46 Asymmetric four-component reaction of diazoacetates with water, anilines and methyl 3-(2-formylphenoxy)propenoates

enalcarbenes and arylamines **41**, followed by regioselective Mannich addition with in-situ generated iminiums derived from aldehydes **17** and arylamines **41**, and terminated by a cyclocondensation cascade, resulting in the target pyrrole products **96** in good yields.

Beyond the interception of ammonium ylides, analogous *O*-heterocycles could be formed using in-situ generated oxonium ylides. In 2014, Hu group designed an Rh (II)/CPA co-catalyzed four-component cascade reactions of aryl diazoesters **43** with water, anilines **41**, and methyl 3-(2-formylphenoxy)propenoates **97** to afford densely substituted seven-membered-ring *O*,*O*-acetals **99** in moderate yields with high diastereo- and enantioselectivities (Scheme 46) [107]. Initially, the Rh(II)-catalyzed the formation of oxonium ylides from diazoacetates **43** and water, which underwent Mannich-type interception with electrophiles generated in-situ from anilines **41** and methyl 3-(2-formylphenoxy)propenoates **97**. Subsequently, a DBU-catalyzed intramolecular *oxo*-Michael addition of linear products **98**, formed by the Mannich-type reaction, produced chiral cyclic acetals **99** with up to 53% yield and 99% *ee*. Later, the same group reported an asymmetric three-component reaction of diazo compounds with alcohols and aldimines/aldehydes under Rh(II)/Zr(IV)-BINOL co-catalysis, providing diversity-oriented syntheses of various chiral nitrogen- and/or oxygen-containing polyfunctional heterocycles [108].

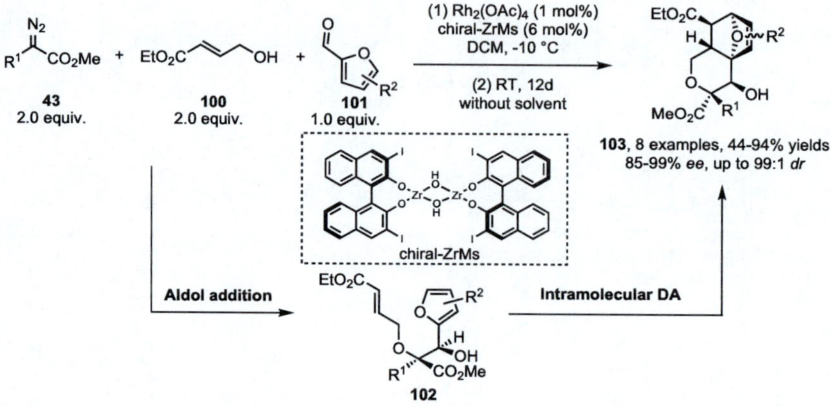

Scheme 47 One-pot three-component reaction for the synthesis of fused epoxyisochromenes

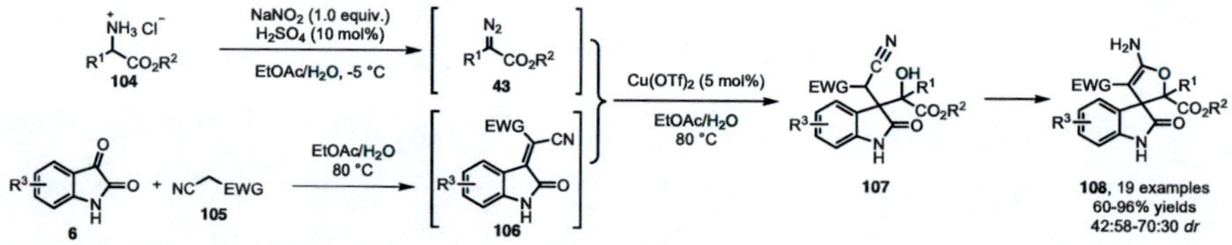

Scheme 48 Michael-type interception of oxonium ylides for the synthesis of spiro oxindoles

The same group reported in 2016 an efficient route for the rapid assembly of poly ring-fused chiral hydro-epoxyisochromene derivatives **103** through a one-pot $Rh_2(OAc)_4$/Zr-3-I-binol co-catalyzed enantioselective three-component cascade reactions of diazoacetates **43** with ethyl 4-hydroxybutenoate **100** and furfurals **101** (Scheme 47) [109]. DFT studies revealed that the reaction involved an oxonium ylide-trapping cascade reaction, followed by intramolecular Diels-Alder and epoxidation, providing the fused products **103** with high to excellent enantioselectivity.

In 2019 Hu group presented a green and step-economy access to spiro [2,3-dihydrofuran-3,3′-oxindole] derivatives **108** with in-situ generated α-diazo esters **43**, water, and in-situ-formed Michael acceptor **106**. The reaction went through a Michael-type interception of oxonium ylides that was followed by a formal [3 + 2]-cycloaddition under catalysis by $Cu(OTf)_2$ (Scheme 48) [110].

One year later a formal [3 + 2]-cycloaddition was disclosed by Hu and coworkers during the study of trapping of oxonium ylides derived from diazoketones **85** and alcohols **109** by azonaphthalenes **110**, providing *N*-substituted 1-amino-indole derivatives **113** in good yields when catalyzed by Rh(II)/PA co-catalysts (Scheme 49) [111].

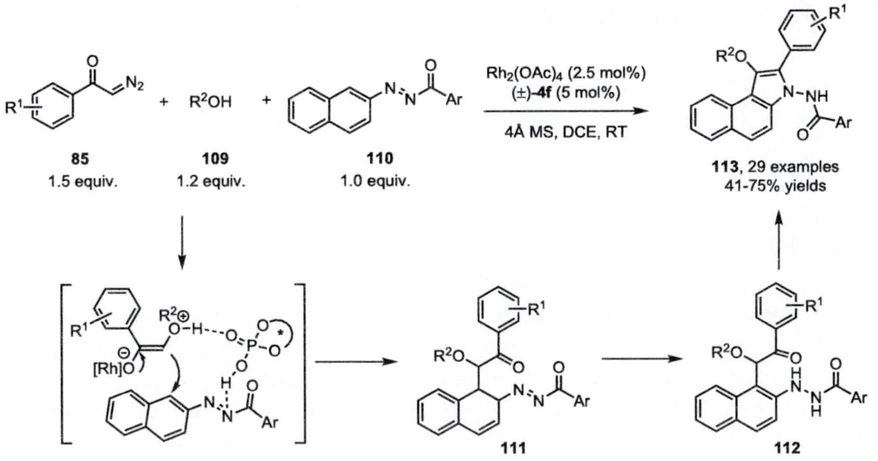

Scheme 49 Formal [3 + 2]-cycloaddition with diazoketones with alcohols and azonaphthalenes

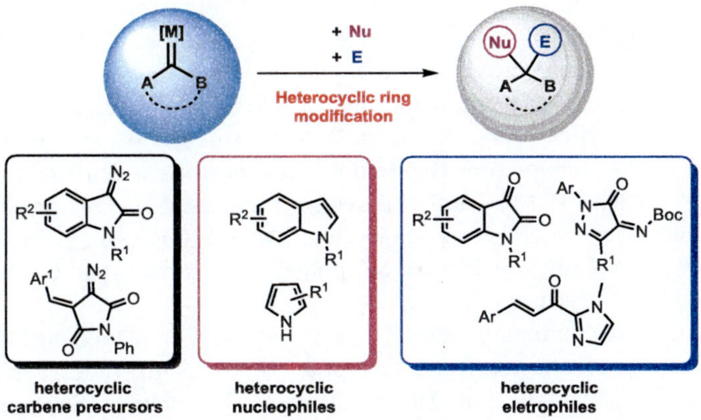

Scheme 50 Generally used heterocyclic reagents in carbene *gem*-difunctionalization

4 Heterocyclic Skeleton Modification via Carbene *gem*-Difunctionalization

Selective modification of heterocyclic skeletons via carbene *gem*-difunctionalization also plays an important role in heterocyclic chemistry. This process introduces the heterocyclic unit(s) from different reagents, including heterocyclic carbene precursors, heterocyclic nucleophiles, and heterocyclic electrophiles, leading to different heterocycles with structural complexity and diversity. The outcome of this multi-component reaction is a divergent heterocyclic ring modification protocol (Scheme 50).

4.1 Modification of Heterocyclic Carbene Precursors

Heterocyclic carbene precursors, mostly heterocyclic diazo compounds, are valuable synthons in heterocyclic chemistry. In this area, diazooxindoles **47** and vinyl diazosuccinimides **128**, are two heterocyclic carbene precursors, that react with nucleophiles to generate ammonium ylides or oxonium ylides, followed by their interception by different electrophiles to afford substituted heterocycles (Scheme 50, in black frame).

Gong and coworkers established in 2013 an efficient Rh(II)/CPA co-catalyzed aldol-type interception of ammonium ylides in-situ generated from 3-diazooxindoles **47** with amines **41** by glyoxylates, affording highly stereoselective 3-aminooxindoles with up to 99% *ee* and >20:1 *dr* [112]. One year later, Hu and coworkers reported an analogous approach for the synthesis of 3-amino-3-hydroxymethyloxindoles **116** in 42-98% yields through an Rh(II)-catalyzed aldol-type interception of analogous ammonium ylides by formaldehyde **17a**(Scheme 51, path **a**) [113]. Later, the same group successfully achieved Michael- and Mannich-type interception of corresponding ammonium ylides with (*E*)-1,4-enediones **114** [114] and isatin-derived ketimines **115** [115], respectively, producing functionalized 1,4-diketones **117** and vicinal diamine derivatives **118** with high diastereoselectivties (Scheme 51, path **b** and **c**).

Later in 2016, the asymmetric enantioselective version of Michael-type interception of ammonium ylides generated from 3-diazooxindoles **47** and amines **41** by nitroalkenes **75** was reported by Han and coworkers using a Ru(II)/chiral bifunctional organo catalyst **119** co-catalyzed system (Scheme 52). This transformation provided a variety of chiral 3-amino-3-alkyloxindoles **120** in good yields with high enantioselectivities [116], which could be applied to the formal synthesis of trimeric indole alkaloid (−)-psychotrimine.

Looking beyond ammonium ylides, Hu and coworkers employed the Rh(II)/CPA co-catalyst in the enantioselective interception of oxonium ylides derived from 3-diazooxindoles **47** and alcohols **109** by different electrophiles (Scheme 53). For example, the oxonium ylide intermediates could be trapped by *N*-benzhydryl-α-imino esters **121** under co-catalysis by Rh(II) and CPA, affording chiral β-tetrasubstituted α-amino acids **123** in good yields with good diastereo- and enantioselectivities (Scheme 53, path **a**) [117]. Moreover, α-propargylic indole iminium ions, which are derived from α-propargylic-3-indolymethanols **122**, were introduced as electrophiles for the interception of oxonium ylide intermediates through an enantioselective counteranion-directed propargylation process to produce the chiral propargylic indole derivatives **124** with good yields and enantioselectivities (Scheme 53, path **b**) [118]. Similarly, Shi's group reported that 2-indolymethanol could be used as an electrophile precursor for the trapping of oxonium ylides in a synergistic $Rh_2(OAc)_4$ and $(PhO)_2PO_2H$ catalyzed system, leading to 3-indolyl-3-alkoxy oxindole scaffolds in moderate to high yields [119].

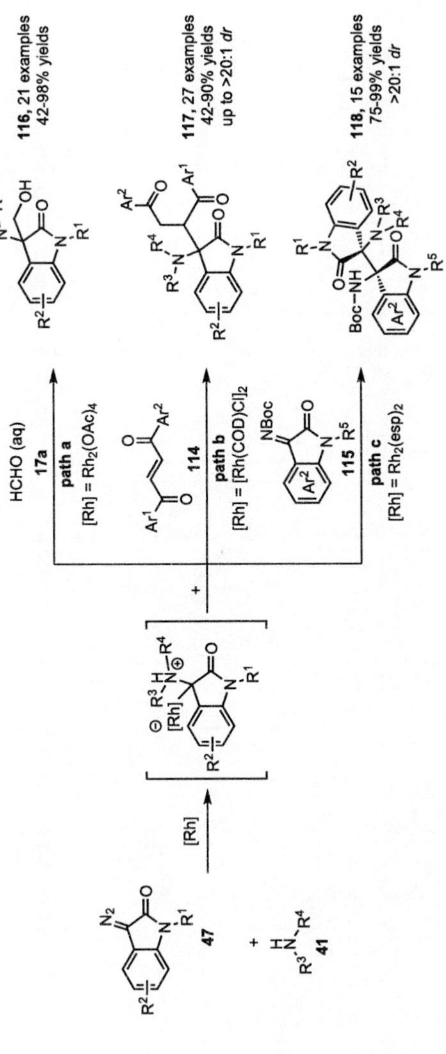

Scheme 51 Interception type of ammonium ylides derived from 3-diazooxindoles and anilines

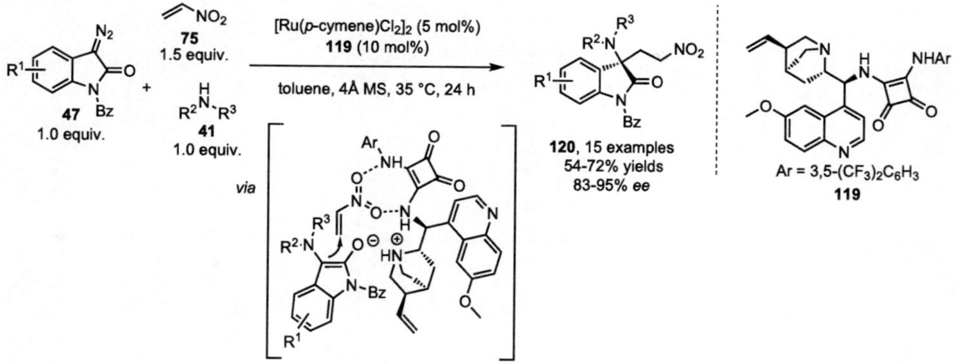

Scheme 52 The Ru(II)/chiral organo relay catalyzed three-component reaction of diazooxindoles, amines, and nitroalkene

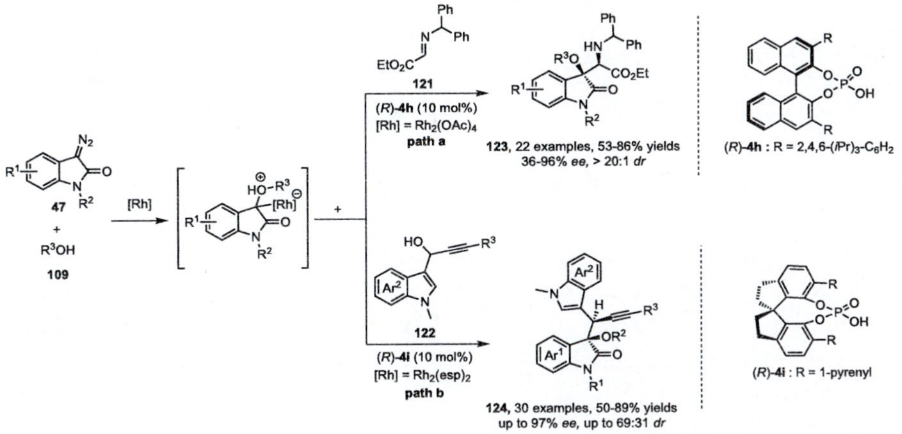

Scheme 53 Asymmetric Mannich-type interception of oxonium ylides derived from 3-diazooxindoles and alcohols

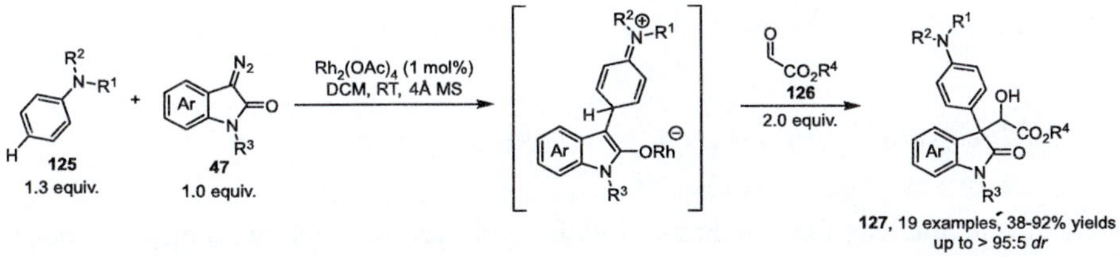

Scheme 54 Diastereoselective aldol-type interception of zwitterionic intermediates derived from diazooxindoles and *N,N*-disubstituted anilines

Besides the ammonium ylides and oxonium ylides, the Hu group also achieved highly diastereoselective interception of zwitterionic intermediates generated from 3-diazooxindoles **47** and *N,N*-disubstituted anilines **125** by glyoxylates **126** in the presence of $Rh_2(OAc)_4$ (Scheme 54) [120]. This three-component reaction provided a straightforward and convenient platform for the effective assembly of 3-aryl-3-substituted oxindole derivatives **127** in good yields.

In addition to 3-diazooxindoles **47**, Hu group recently illustrated a new type of heterocyclic diazo compound, vinyl diazosuccinimides **128** [121], which underwent an enantioselective three-component reaction with alcohols **109** and imines **3** to offer chiral 3,3-disubstituted succinimides **129** in good yields and with up to 97% *ee* (Scheme 55) [122]. The generated product with an alkenyl species could be easily converted to the chiral tricyclic structure **130** in 51% yield.

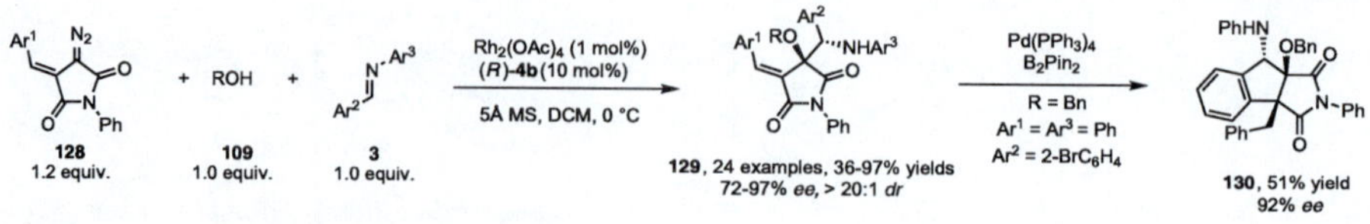

Scheme 55 Asymmetric Mannich-type interception of oxonium ylides derived from vinyl diazosuccinimides and alcohols

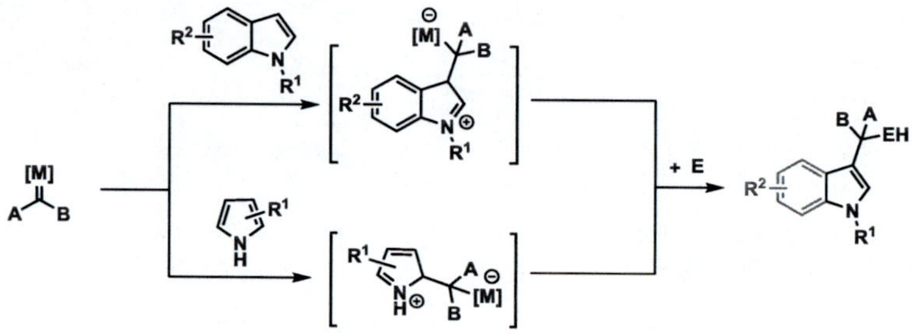

Scheme 56 Heterocyclic skeleton modification using heterocyclic nucleophiles

4.2 *Modification of Heterocyclic Nucleophiles*

Heterocyclic nucleophiles, including indoles and pyrroles, react with metal carbene species to afford zwitterionic intermediates, which can be trapped by different electrophiles to generate polyfunctionalized indole or pyrrole derivatives (Scheme 56).

In 2012 the Hu group successfully trapped the transient and highly reactive zwitterionic intermediates, formed in-situ from diazoacetates **43** and indoles **69** through CPA-activated imines, to afford polyfunctionalized indole derivates with excellent diastereo- and enantioselectivities [33]. Later, the same group established a highly enantioselective palladium(II) phosphate catalytic system to capture the zwitterionic intermediates generated from pyrrole **131** and diazoesters **47** with imines **3** (Scheme 57) [123]. All four possible stereoisomers of chiral pyrrole derivatives **132** were easily obtained in moderate yield with high control of diastereo- and enantioselectivities.

Shi and coworkers utilized isatin-derived ketimines in 2014 to trap zwitterionic intermediates generated from diazooxindoles **47** and indoles **69** in the presence of $Rh_2(OAc)_4$, providing functionalized 3,3′,3′′-trisindoles in generally high yields [124]. At the same time, the Hu group achieved an enantioselective four-component Mannich-type reaction by directly generating imines **3** in-situ with the corresponding aldehydes **126** and amines **41**. In a rare four-component reaction, 3-diazooxindoles **47**, indoles **69**, arylamines **41** and ethyl glyoxylate **126** with Rh(II)/CPA co-catalysis afforded chiral 3,3-disubstituted 3-indol-3′-yloxindoles **133** in good yields with

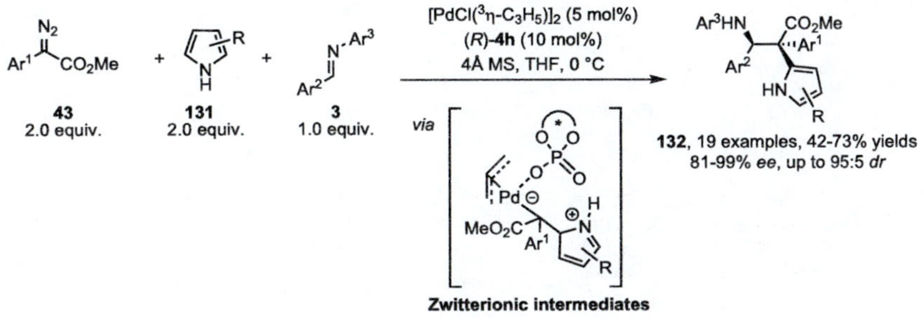

Scheme 57 Asymmetric Mannich-type interception of zwitterionic intermediates derived from diazoesters and pyrroles

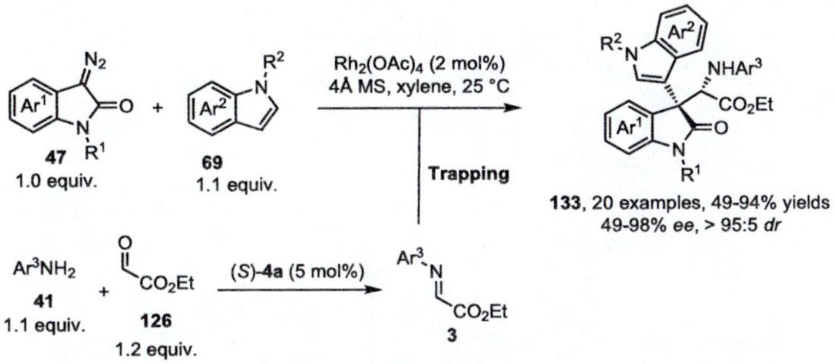

Scheme 58 Asymmetric Mannich-type interception of zwitterionic intermediates derived from diazooxindoles and indoles

excellent diastereo- and enantioselectivities through a Mannich-type assembly of two in-situ formed species, zwitterionic intermediates and imines **3** that derived from aldehydes **126** and amines **41** (Scheme 58) [125].

Moreover, the Hu group reported the aldol-type interception of the zwitterionic intermediates with ethyl glyoxylate **126** [126] or formalin **17a** [127], giving a series of mixed 3,3′-bisindoles **134** and **135** in good yields with excellent selectivities (Scheme 59). The target product could be potentially applied to the total synthesis of the natural alkaloid (±)-gliocladin C.

In a Michael-type interception, Gong and coworkers established an enantioselective sequential C-H functionalization/asymmetric Michael addition reaction of 3-diazooxindoles **47** with indoles **69** and nitroalkenes **75** under dual catalysis by ruthenium(II) complexes and quinine-derived squaramide, and chiral 3,3′-bis(indole) derivatives **136** were formed with moderate stereoselectivities (Scheme 60) [128]. In this transformation, 3-diazooxindoles **47** initially formed the ruthenium(II) carbene species followed by formation of zwitterionic intermediates with indoles **69**. The control experiments suggested that the zwitterionic intermediates underwent fast proton transfer to provide the insertion product rather than being directly trapped by activated nitroalkenes **75**. Later in 2015, the same group

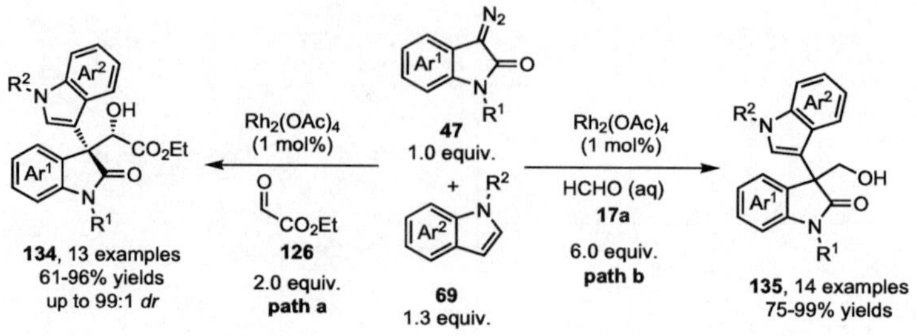

Scheme 59 Aldol-type interception of zwitterionic intermediates derived from diazooxindoles and indoles

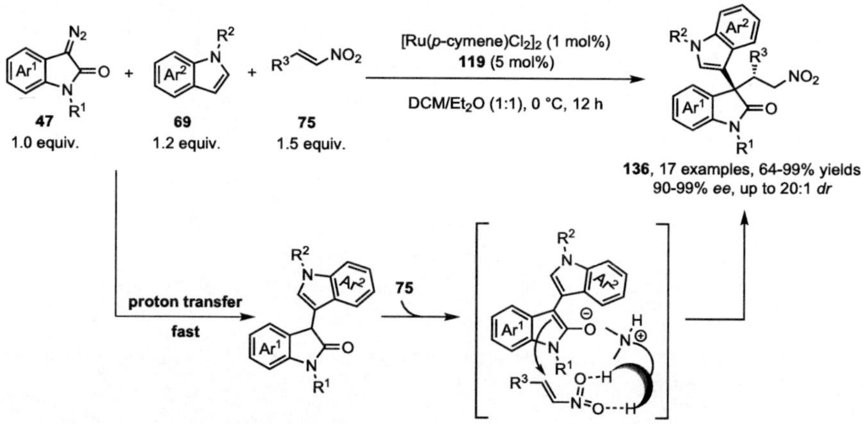

Scheme 60 Asymmetric sequential C-H functionalization/Michael-type addition for the synthesis of chiral 3,3′-bis(indole) derivatives

successfully trapped the zwitterionic intermediates by allenoates in the presence of $Rh_2(esp)_2$ and amino acid-derived chiral phosphines co-catalysts, affording a library of quaternary 3,3′-indolyloxindole derivatives with good yields and enantioselectivities [129]. Based on the observed positive effect of LiCl, they depicted an interaction model in which the lithium enolate generated a chiral thiourea-chloride complex through Li-Cl coordination to assist the stereochemical control. Recently, Hu group reported a highly efficient Rh(II)/chiral spiro phosphoric acid (SPA) co-catalyzed system, which enabled the asymmetric trapping of zwitterionic intermediates with in-situ generated chiral methylene iminium ions via asymmetric counter-anion-directed catalysis (ACDC) [130].

A facile iridium(I) and chiral secondary amine co-catalyzed enantioselective three-component reaction of diazoacetates **43** with indoles **69** and enals **91** was developed by Hu and Liu, providing a diverse range of 3-substituted indole derivatives **138** [131]. This transformation was proposed to go through an asymmetric interception of iridium(I)-associated zwitterionic intermediates with the chiral

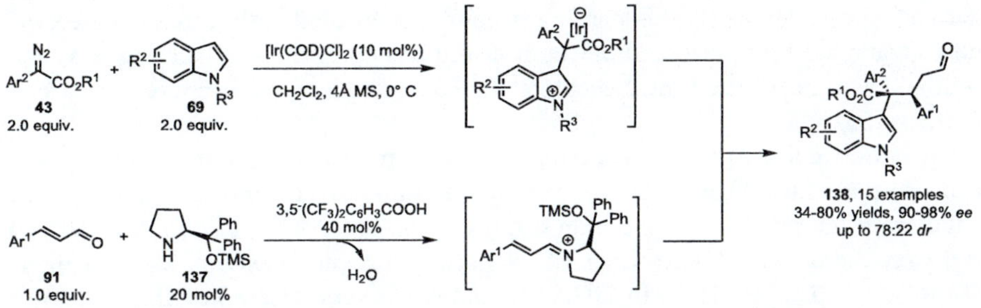

Scheme 61 Asymmetric Michael-type addition of zwitterionic intermediate derived from diazoacetates with indoles

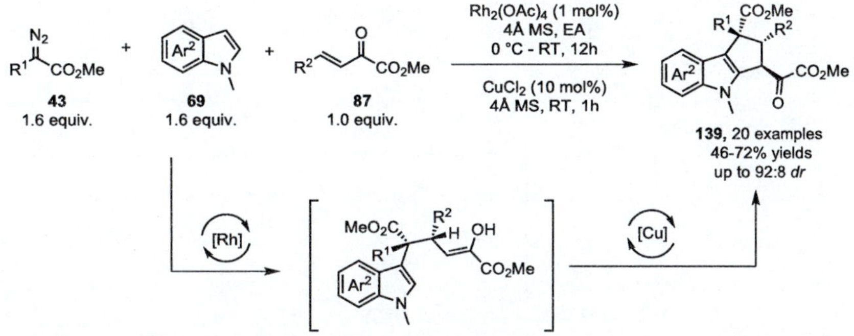

Scheme 62 The Rh(II)/Cu(II)-relay-catalyzed three-component reaction of diazoacetates with indoles and α-keto esters

amine-activated enals via a 1,4-addtition process (Scheme 61). With 1-sulfonyl-1,2,3-triazoles as carbene precursors, Hu and Xia presented an Rh(II)-catalyzed multi-component reaction through assembly of two in-situ formed species, α-amino enols and vinylimine ions [132].

Using a similar approach Hu and coworkers reported a sequential $Rh_2(OAc)_4$-catalyzed three-component reaction and $CuCl_2$-catalyzed aerobic dehydrogenative coupling process for the assembly of cyclopenta[*b*]indoles **139** in good to high yields (Scheme 62) [133]. Moreover, Xu's group reported a Cu(II)-catalyzed [4 + 1]-annulation of diazoacetates with 2-vinylindoles, giving polyfunctional cyclopenta [*b*]indoles in generally good yields [134].

4.3 Modification of Heterocyclic Electrophiles

Based on cascade reactions via carbene *gem*-difunctionalization, reactive heterocyclic electrophiles could be introduced for the in-situ interception of active intermediates derived from carbene precursors and nucleophiles for the modification of heterocyclic skeletons. In this part, electrophiles imbedded with different heterocycles, including cyclic azomethine imines, α,β-unsaturated 2-acyl imidazoles, and isatins have been used as acceptors for the different interception processes (Scheme 63).

In 2014, Reddy's group successfully captured transient oxonium ylides derived from diazoacetates **43** and alcohols **109** by isatin imines **115** through Mannich-type addition in the presence of rhodium catalyst. This process led to oxindole-derived α-alkoxy-β-amino acid derivatives **142** with two adjacent quaternary carbon centers (Scheme 64, path **a**) [135]. In 2020, Hu and coworkers intercepted the analogous oxonium ylides by pyrazolinone ketimines **140** to provide pyrazolone derivatives **143** in good yields with excellent diastereoselectivities (Scheme 64, path **b**) [136]. Recently, the same group achieved an efficient enantioselective Mannich-type interception of oxonium ylides by seven-membered imines **141** under the

Scheme 63 Heterocyclic skeleton modification using heterocyclic electrophiles

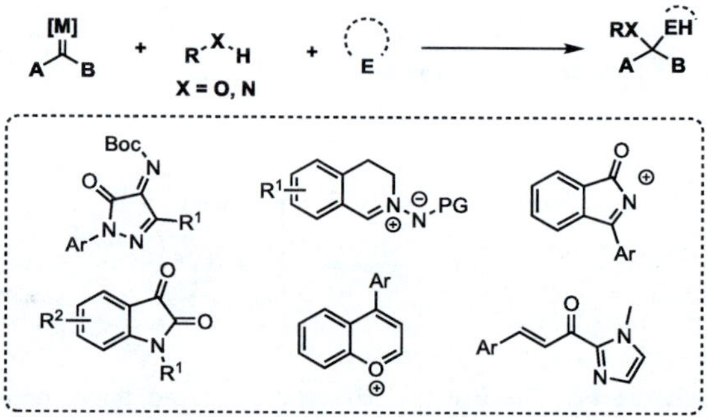

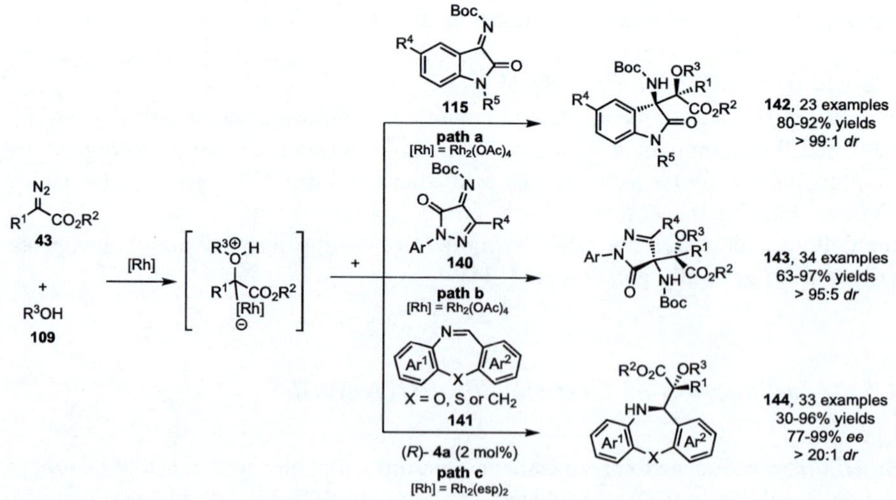

Scheme 64 Mannich-type interception of oxonium ylides with different heterocyclic imines

co-catalysis of Rh(II) and CPA, producing a series of chiral dibenzoazepine analogs **144** with excellent enantioselectivities (Scheme 64, path **c**, 91%-99% *ee*) [137].

In addition to imines, isatins, which are one of the most important heterocyclic carbonyl compounds, have been successfully employed for the interception of ammonium ylides and oxonium ylides. In 2017, the Hu group utilized isatins **6** to trap the in-situ formed ammonium ylide intermediates, providing non-protected β-hydroxy-α-aminoesters **146** with up to 77% yield and 62:38 *dr* (Scheme 65) [138].

Moreover, the same group also employed isatins **6** for the interception of transient oxonium ylides derived from diazoacetates **43** and water in the presence of $CuSO_4$, affording various oxindole derivatives **148** in good yields with moderate diastereoselectivities (Scheme 66, path **a**) [139]. Notably, water not only acted as a reactant, but was also used as the solvent in this process. Beyond isatins [139–141], indane-1,2,3-triones [142] could be used for the trapping of active oxonium ylides, offering polyfunctional 2-hydroxyindane-1,3-diones with moderate yields. Moreover, azetidine-2,3-diones **147** also exhibit good activity for the interception of oxonium ylides. Torres and coworkers reported that azetidine-2,3-diones **147** [143, 144] could also trap transient oxonium ylides for the rapid assembly of various β-lactam hybrids **149** with moderate yields and diastereoselectivities under the catalysis of $Rh_2(OAc)_4$ (Scheme 6, path **b**). Nevertheless, the enantioselective version of aldol-type interception of oxonium ylide intermediates by heterocyclic carbonyl compounds has not been accomplished thus far.

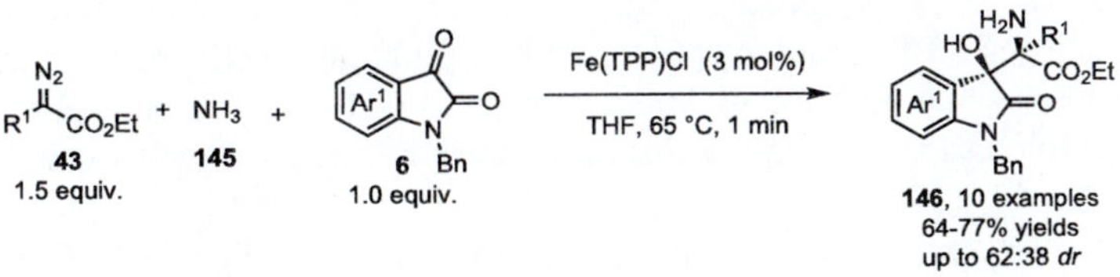

Scheme 65 Diastereoselective aldol-type interception of ammonium ylides with isatins

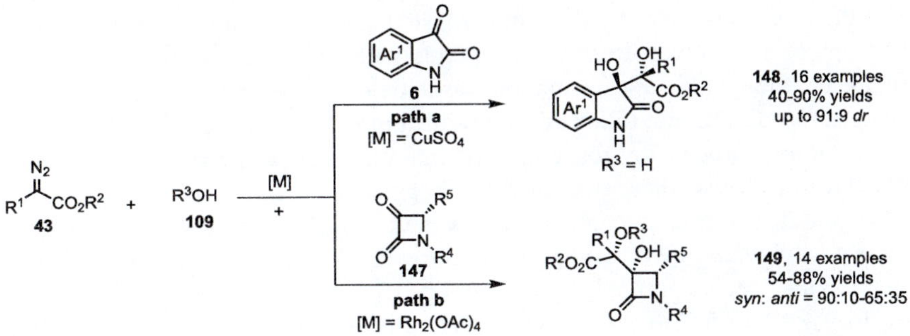

Scheme 66 Metal-catalyzed aldol-type interception of oxonium ylides with cyclic carbonyl compounds

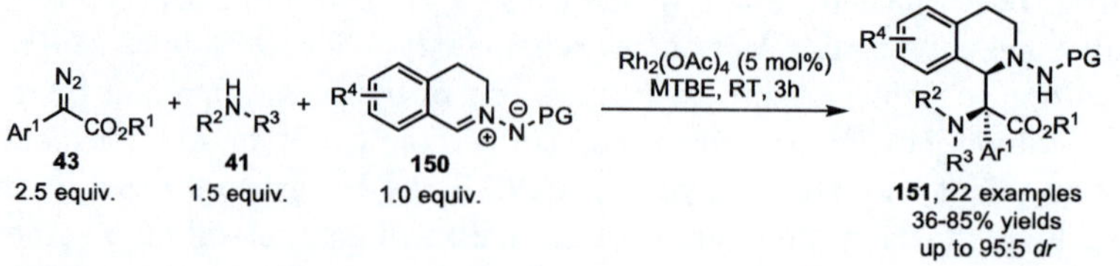

Scheme 67 Michael-type interception of ammonium ylides with cyclic azomethine imines

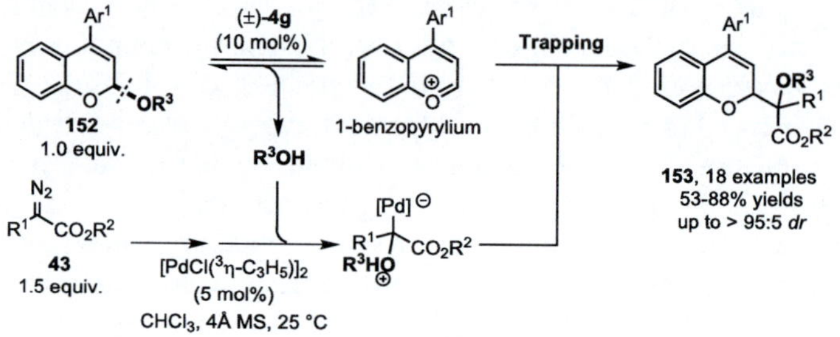

Scheme 68 Michael-type interception of oxonium ylides with 1-benzopyryliums

For the Michel-type interception of ammonium ylides or oxonium ylides, versatile heterocyclic Michael acceptors, either formed in-situ or stable reagents could be applied to the trapping process. In 2019, Hu and coworkers utilized cyclic azomethine imines **150** to intercept ammonium ylide intermediates generated from diazoacetates **43** and anilines **41** under catalysis by Rh(II), building pharmaceutically intriguing tetrahydroisoquinoline derivatives **151** in good yields with up to 95: 5 *dr* (Scheme 67) [145].

For the interception of oxonium ylides with heterocyclic Michael acceptors, excellent work has been accomplished by Hu [146–148], Schneider [79, 149] and others [150–152]. In 2015, Hu and coworkers established a novel strategy involving a bond cleavage, modification, and reassembly process [153]. Later, this protocol was applied for the interception of oxonium ylides by 1-benzopyryliums formed in-situ under co-catalysis of palladium/Brønsted acid, affording C-2 functionalized 2*H*-chromene derivatives **153** in good yields with moderate diastereoselectivities (Scheme 68) [154]. This transformation was initiated by the Brønsted acid-promoted cleavage of the C-O bond, followed by modification of the leaving alcohol fragments with palladium carbenes by forming ylide intermediates, and termination occurred through reassembly of these two modified fragments.

The enantioselective interception of oxonium ylides derived from diazo compounds **43** and water, by utilizing α,β-unsaturated 2-acyl imidazoles **154** as electrophiles, has also been reported by the Hu group [155]. With the co-catalytic system of $Rh_2(OAc)_4$, (*S*)-**155**-$Zn(OTf)_2$, and TsOH, this efficient transformation offered δ-hydroxyketone derivatives **156** containing a stereogenic quaternary carbon center in good yields with excellent enantioselectivities (Scheme 69).

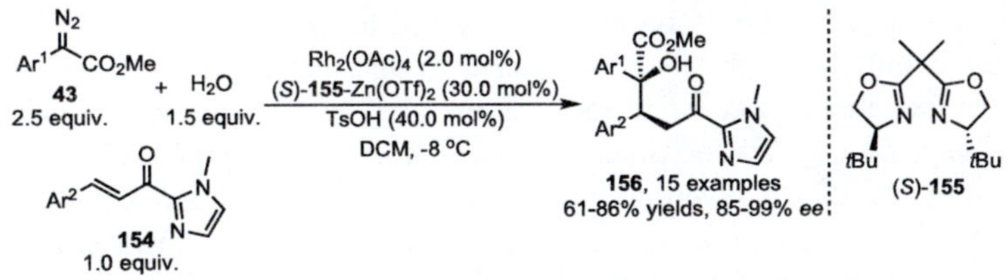

Scheme 69 Asymmetric Michael-type addition of oxonium ylides with α,β-unsaturated 2-acyl imidazoles

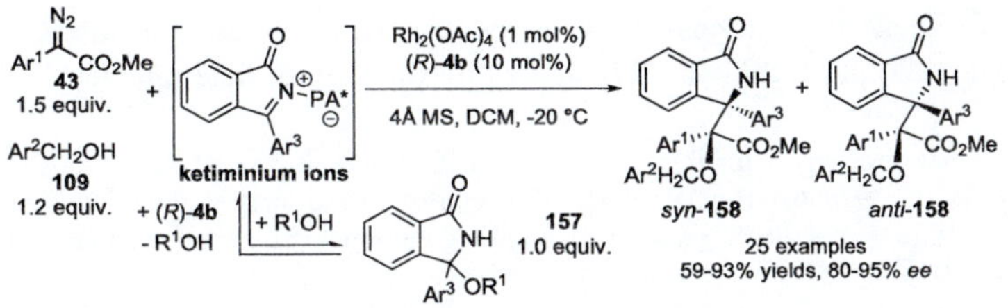

Scheme 70 Formal S_N1-type interception of oxonium ylides with cyclic ketiminium ions

Hu and coworkers also illustrated a formal S_N1 pathway for trapping oxonium ylide intermediates by using in-situ formed heterocyclic electrophiles [156, 157]. They employed ketiminium ions generated in-situ from 3-hydroxyisoindolinones **157** for the efficient capture of transient oxonium ylides, providing isoindolinone derivatives **158** with two contiguous quaternary stereogenic centers in moderate diastereoselectivities and excellent enantioselectivities (Scheme 70) [156]. Moreover, the same group described that Nicholas cations formed in-situ from propargylic alcohol–$Co_2(CO)_6$ complexes could intercept oxonium ylides through an Rh(II)/Ag(I)-cocatalyzed S_N1/S_N1'-type pathway, affording dicobalt hexacarbonyl-complexed 3,3-disubstituted oxindoles in good yields with moderate diastereoselectivities [157].

5 Conclusion

In this chapter, a variety of versatile cascade reactions via metal carbene *gem*-difunctionalization processes have been summarized for the expeditious synthesis of heterocycles with different substitution patterns and diverse frameworks. The key step of this general protocol involves an electrophilic interception of in-situ generated ylide/zwitterionic intermediates derived from carbene precursors with different electrophiles. The heterocyclic structures are formed in the ylide/zwitterionic intermediates formation step, at the electrophilic interception stage, through *gem*-difunctionalization or via a one-pot cyclization after the carbene *gem*-

difunctionalization process. Moreover, a series of heterocyclic carbene precursors, nucleophiles, and electrophiles are subjected to this method, providing a practical strategy for the heterocyclic ring modification. However, challenges remain in this area. (1) the source of carbene precursors is limited, generally because of access to diazo compounds, which severely restricted the structural diversity of the products; (2) the scarcity of effective catalytic systems for the construction of chiral heterocycles in multi-component reactions, including the chemo- and stereoselectivities; (3) synthetic applications of this unique carbene *gem*-difunctionalization reaction still under exploration, which might be served as a key step in natural product total synthesis and in medical chemistry for the generation of substantially potent bioactive molecules with novel structures. Overcoming these drawbacks with the development of selective and robust catalytic systems can open new vistas in synthetic chemistry for the practical construction or modification of heterocycles.

Acknowledgments Support for this research from the National Natural Science Foundation of China (92256301, 92056201, 21971262), Guangdong Basic and Applied Basic Research (2021A1515010384), and the Program for Guangdong Introducing Innovative and Entrepreneurial Teams (No. 2016ZT06Y337) is greatly acknowledged.

References

1. Xuan J, He XK, Xiao WJ (2020) Visible light-promoted ring-opening functionalization of three-membered carbo- and heterocycles. Chem Soc Rev 49:2546–2556
2. Zhang YC, Jiang F, Shi F (2020) Organocatalytic asymmetric synthesis of indole-based chiral heterocycles: strategies, reactions, and outreach. Acc Chem Res 53:425–446
3. Dongbang S, Confair DN, Ellman JA (2021) Rhodium-catalyzed C-H alkenylation/electrocyclization cascade provides dihydropyridines that serve as versatile intermediates to diverse nitrogen heterocycles. Acc Chem Res 54:1766–1778
4. Guo X, Hu W (2013) Novel multicomponent reactions *via* trapping of protic onium ylides with electrophiles. Acc Chem Res 46:2427–2440
5. Wang Y, Lu H, Xu PF (2015) Asymmetric catalytic cascade reactions for constructing diverse scaffolds and complex molecules. Acc Chem Res 48:1832–1844
6. Zhang D, Hu W (2017) Asymmetric multicomponent reactions based on trapping of active intermediates. Chem Rec 17:739–753
7. Kirmse W, Kapps M (1968) Reaktionen des diazomethans mit diallylsulfid und allyläthern unter kupfersalz-katalyse. Chem Ber 101:994–1003
8. Doyle MP, Tamblyn WH, Bagheri V (1981) Highly effective catalytic methods for ylide generation from diazo compounds. Mechanism of the rhodium- and copper-catalyzed reactions with allylic compounds. J Org Chem 46:5094–5102
9. Padwa A, Hornbuckle SF (1991) Ylide formation from the reaction of carbenes and carbenoids with heteroatom lone pairs. Chem Rev 91:263–309
10. Xia Y, Qiu D, Wang J (2017) Transition-metal-catalyzed cross-couplings through carbene migratory insertion. Chem Rev 117:13810–13889
11. Doyle MP, Ratnikov M, Liu Y (2011) Intramolecular catalytic asymmetric carbon-hydrogen insertion reactions. Synthetic advantages in total synthesis in comparison with alternative approaches. Org Biomol Chem 9:4007–4016
12. Nawrat CC, Moody CJ (2011) Natural products containing a diazo group. Nat Prod Rep 28:1426–1444

13. Cavitt MA, Phun LH, France S (2014) Intramolecular donor-acceptor cyclopropane ring-opening cyclizations. Chem Soc Rev 43:804–818
14. Caballero A, Perez PJ (2017) Dimensioning the term carbenoid. Chem 23:14389–14393
15. Pei C, Zhang C, Qian Y, Xu X (2018) Catalytic carbene/alkyne metathesis (CAM): a versatile strategy for alkyne bifunctionalization. Org Biomol Chem 16:8677–8685
16. Davies HML, Morton D (2011) Guiding principles for site selective and stereoselective intermolecular C–H functionalization by donor/acceptor rhodium carbenes. Chem Soc Rev 40:1857–1869
17. Zhu SF, Zhou QL (2012) Transition-metal-catalyzed enantioselective heteroatom-hydrogen bond insertion reactions. Acc Chem Res 45:1365–1377
18. Qian D, Zhang J (2014) Gold-catalyzed cyclopropanation reactions using a carbenoid precursor toolbox. Chem Soc Rev 44:677–698
19. Xu X, Doyle MP (2014) The [3+3]-cycloaddition alternative for heterocycle syntheses: catalytically generated metalloenolcarbenes as dipolar adducts. Acc Chem Res 47:1396–1405
20. Zhu D, Chen L, Fan H, Yao Q, Zhu S (2020) Recent progress on donor and donor-donor carbenes. Chem Soc Rev 49:908–950
21. Padwa A (2009) Domino reactions of rhodium(II) carbenoids for alkaloid synthesis. Chem Soc Rev 38:3072–3081
22. Xia Y, Zhang Y, Wang J (2013) Catalytic cascade reactions involving metal carbene migratory insertion. ACS Catal 3:2586–2598
23. Padwa A (1993) Synthetic applications of carbonyl ylides generated *via* the tandem cyclization cycloaddition reaction of rhodium carbenoids. J Chin Chem Soc 40:97–112
24. Xu X, Leszczynski JS, Mason SM, Zavalij PY, Doyle MP (2014) Expedient access to substituted 3-amino-2-cyclopentenones by dirhodium-catalyzed [3+2]-annulation of silylated ketene imines and enoldiazoacetates. Chem Commun 50:2462–2464
25. Deng Y, Yglesias MV, Arman H, Doyle MP (2016) Catalytic asymmetric synthesis of cyclopentyl β-amino esters by [3+2] cycloaddition of enecarbamates with electrophilic metalloenolcarbene intermediates. Angew Chem Int Ed 55:10108–10112
26. Xu X, Shabashov D, Zavalij PY, Doyle MP (2012) Unexpected catalytic reactions of silyl-protected enol diazoacetates with nitrile oxides that form 5-arylaminofuran-2(3*H*)-one-4-carboxylates. Org Lett 14:800–803
27. Davies HM, Xiang B, Kong N, Stafford DG (2001) Catalytic asymmetric synthesis of highly functionalized cyclopentenes by a [3 + 2] cycloaddition. J Am Chem Soc 123:7461–7462
28. Smith AG, Davies HM (2012) Rhodium-catalyzed enantioselective vinylogous addition of enol ethers to vinyldiazoacetates. J Am Chem Soc 134:18241–18244
29. Briones JF, Davies HM (2013) Enantioselective gold(I)-catalyzed vinylogous [3+2] cycloaddition between vinyldiazoacetates and enol ethers. J Am Chem Soc 135:13314–13317
30. Parr BT, Li Z, Davies HM (2011) Asymmetric synthesis of highly functionalized cyclopentanes by a rhodium- and scandium-catalyzed five-step domino sequence. Chem Sci 2:2378–2382
31. Lopez E, Lonzi G, Gonzalez J, Lopez LA (2016) Gold-catalyzed intermolecular formal (3+2) cycloaddition of stabilized vinyldiazo derivatives and electronically unbiased allenes. Chem Commun 52:9398–9401
32. Doyle MP, Shanklin MS, Pho HQ, Mahapatro SN (1988) Rhodium(II) acetate and nafion-H catalyzed decomposition of *N*-aryldiazoamides. Efficient synthesis of 2(3*H*)-indolinones. J Org Chem 53:1017–1022
33. Qiu H, Li M, Jiang LQ, Lv FP, Zan L, Zhai CW, Doyle MP, Hu WH (2012) Highly enantioselective trapping of zwitterionic intermediates by imines. Nat Chem 4:733–738
34. Li M, Zan L, Prajapati D, Hu W (2012) Rhodium(II) catalyzed diastereoselective reactions of diazoacetamides with isatins: an efficient approach to 3-hydroxy-3,3′-bioxindoles. Org Biomol Chem 10:8808–8813

35. Yamamoto K, Qureshi Z, Tsoung J, Pisella G, Lautens M (2016) Combining Ru-catalyzed C-H functionalization with Pd-catalyzed asymmetric allylic alkylation: synthesis of 3-allyl-3-aryl oxindole derivatives from aryl alpha-diazoamides. Org Lett 18:4954–4957
36. Zhang D, Kang Z, Hu W (2017) Trapping of transient zwitterionic intermediates by *N*-acylpyridinium salts: a palladium-catalyzed diastereoselective three-component reaction. J Org Chem 82:5952–5958
37. Song XG, Zhu SF, Xie XL, Zhou QL (2013) Enantioselective copper-catalyzed intramolecular phenolic O-H bond insertion: synthesis of chiral 2-carboxy dihydrobenzofurans, dihydrobenzopyrans, and tetrahydrobenzooxepines. Angew Chem Int Ed 52:2555–2558
38. Hong K, Dong S, Xu X, Zhang Z, Shi T, Yuan H, Xu X, Hu W (2021) Enantioselective intermolecular Mannich-type interception of phenolic oxonium ylide for the direct assembly of chiral 2,2-disubstituted dihydrobenzofurans. ACS Catal 11:6750–6756
39. Hong K, Yang X, Zhang Z, Xie X, Lv X, Xu X, Hu W (2022) Diastereoselective aldol-type interception of phenolic oxonium ylides for the direct assembly of 2,2-disubstituted dihydrobenzofurans. Org Biomol Chem 20:4635–4639
40. Guo X, Liu W, Hu W (2014) A facile access to polyfunctional oxygen-containing heterocycles *via* intramolecularly formed protic oxonium ylide trapping processes. Chem Asian J 9:117–120
41. Jia S, Yang X, Dong G, Ao C, Jiang X, Hu W (2019) Trapping of zwitterionic intermediates by isatins and imines: synthesis of benzoxazines bearing a C4-quaternary stereocenter. Org Lett 21:4014–4018
42. Li D, Zang W, Bird MJ, Hyland CJT, Shi M (2021) Gold-catalyzed conversion of highly strained compounds. Chem Rev 121:8685–8755
43. Shapiro ND, Toste FD (2008) Synthesis of azepines by a gold-catalyzed intermolecular [4+3]-annulation. J Am Chem Soc 130:9244–9245
44. Wei H, Bao M, Dong K, Qiu L, Wu B, Hu W, Xu X (2018) Enantioselective oxidative cyclization/Mannich addition enabled by gold(I)/chiral phosphoric acid cooperative catalysis. Angew Chem Int Ed 57:17200–17204
45. Sahani RL, Patil MD, Wagh SB, Liu RS (2018) Catalytic transformations of alkynes into either alpha-alkoxy or alpha-aryl enolates: Mannich reactions by cooperative catalysis and evidence for nucleophile-directed chemoselectivity. Angew Chem Int Ed 57:14878–14882
46. Cai J, Wang X, Qian Y, Qiu L, Hu W, Xu X (2019) Gold-catalyzed oxidative cyclization/aldol addition of homopropargyl alcohols with isatins. Org Lett 21:369–372
47. Zhou S, Xie X, Xu X, Dong S, Hu W, Xu X (2021) An asymmetric oxidative cyclization/Mannich-type addition cascade reaction for direct access to chiral pyrrolidin-3-ones. Chem Commun 57:12171–12174
48. Tichenor MS, Boger DL (2008) Yatakemycin: total synthesis, DNA alkylation, and biological properties. Nat Prod Rep 25:220–226
49. Zhang D, Kang ZH, Liu JW, Hu WH (2019) Metal-dependent umpolung reactivity of carbenes derived from cyclopropenes. iScience 14:292–300
50. Zhang D, Wang X, Zhang M, Kang Z, Xiao G, Xu X, Hu W (2020) Highly enantioselective trapping of carboxylic oxonium ylides with imines for direct assembly of enantioenriched γ-butenolides. CCS Chem 2:432–439
51. Zhang D, Wang X, Zhang M, Hu W (2020) Rhodium-catalyzed sequential cycloisomerization/aldol addition of cyclopropene carboxylic acids with isatins. Org Lett 22:5600–5604
52. Zhang M, Zhang T, Zhang D, Hu W (2020) Diastereoselective trapping of transient carboxylic oxonium ylides with α,β-unsaturated 2-acyl imidazoles. Adv Synth Catal 362:4662–4667
53. Ma B, Wu Z, Huang B, Liu L, Zhang J (2016) Gold-catalysed facile access to indene scaffolds *via* sequential C-H functionalization and 5-endo-dig carbocyclization. Chem Commun 52:9351–9354
54. Bao M, Qian Y, Su H, Wu B, Qiu L, Hu W, Xu X (2018) Gold(I)-catalyzed and H_2O-mediated carbene cascade reaction of propargyl diazoacetates: furan synthesis and mechanistic insights. Org Lett 20:5332–5335

55. Zhang C, Li H, Pei C, Qiu L, Hu W, Bao X, Xu X (2019) Selective vinylogous reactivity of carbene intermediate in gold-catalyzed alkyne carbocyclization: synthesis of indenols. ACS Catal 9:2440–2447
56. Zhang C, Hong K, Dong S, Pei C, Zhang X, He C, Hu W, Xu X (2019) Gold(I)-catalyzed aromatization: expeditious synthesis of polyfunctionalized naphthalenes. iScience 21:499–508
57. Zhang C, Hong K, Pei C, Zhou S, Hu W, Hashmi ASK, Xu X (2021) Gold(I)-catalyzed intramolecular cyclization/intermolecular cycloaddition cascade as a fast track to polycarbocycles and mechanistic insights. Nat Commun 12:1182–1193
58. Bao M, Xie X, Hu W, Xu X (2021) Gold-catalyzed carbocyclization/C=N bond formation cascade of alkyne-tethered diazo compounds with benzo[*c*]isoxazoles for the assembly of 4-iminonaphthalenones and indenes. Adv Synth Catal 363:4018–4023
59. Bao M, Chen J, Pei C, Zhang S, Lei J, Hu W, Xu X (2021) Gold-catalyzed ketene dual functionalization and mechanistic insights: divergent synthesis of indenes and benzo[*d*] oxepines. Sci China Chem 64:778–787
60. Jing C, Xing D, Qian Y, Shi T, Zhao Y, Hu W (2013) Diversity-oriented three-component reactions of diazo compounds with anilines and 4-oxo-enoates. Angew Chem Int Ed 52:9289–9292
61. Nicolle SM, Lewis W, Hayes CJ, Moody CJ (2016) Stereoselective synthesis of functionalized pyrrolidines by the diverted N-H insertion reaction of metallocarbenes with β-aminoketone derivatives. Angew Chem Int Ed 55:3749–3753
62. Ma X, Jiang J, Lv S, Yao W, Yang Y, Liu S, Xia F, Hu W (2014) An ylide transformation of rhodium(I) carbene: enantioselective three-component reaction through trapping of rhodium (I)-associated ammonium ylides by beta-nitroacrylates. Angew Chem Int Ed 53:13136–13139
63. Jing C, Xing D, Hu W (2014) Highly diastereoselective synthesis of 3-hydroxy-2,2,3-trisubstituted indolines *via* intramolecular trapping of ammonium ylides with ketones. Chem Commun 50:951–953
64. Devi L, Shukla R, Rastogi N (2018) Intramolecular trapping of ammonium ylides with *N*-benzoylbenzotriazoles in aqueous medium: direct access to the pseudoindoxyl scaffold. Org Biomol Chem 17:135–139
65. Arunprasath D, Sekar G (2019) Copper(II)-catalyzed domino synthesis of indolo[3,2-*c*] quinolinones *via* selective carbonyl migration. Org Lett 21:867–871
66. Guan XY, Tang M, Liu ZQ, Hu W (2019) A highly diastereoselective [5+1] annulation to 2,2,3-trisubstituted tetrahydroquinoxalines *via* intramolecular Mannich-type trapping of ammonium ylides. Chem Commun 55:9809–9812
67. Reddy ACS, Choutipalli VSK, Ghorai J, Subramanian V, Anbarasan P (2017) Stereoselective palladium-catalyzed synthesis of indolines *via* intramolecular trapping of *N*-ylides with alkenes. ACS Catal 7:6283–6288
68. Liu K, Zhu C, Min J, Peng S, Xu G, Sun J (2015) Stereodivergent synthesis of *N*-heterocycles by catalyst-controlled, activity-directed tandem annulation of diazo compounds with amino alkynes. Angew Chem Int Ed 54:12962–12967
69. Medvedev JJ, Galkina OS, Klinkova AA, Giera DS, Hennig L, Schneider C, Nikolaev VA (2015) Domino [4+1]-annulation of alpha,beta-unsaturated delta-amino esters with Rh(II)-carbenoids-a new approach towards multi-functionalized *N*-aryl pyrrolidines. Org Biomol Chem 13:2640–2651
70. Nicolle SM, Lewis W, Hayes CJ, Moody CJ (2015) Stereoselective synthesis of highly substituted tetrahydrofurans through diverted carbene O-H insertion reaction. Angew Chem Int Ed 54:8485–8489
71. Jing C, Xing D, Gao L, Li J, Hu W (2015) Divergent synthesis of multisubstituted tetrahydrofurans and pyrrolidines *via* intramolecular Aldol-type trapping of onium ylide intermediates. Chem 21:19202–19207
72. Urabe F, Miyamoto S, Takahashi K, Ishihara J, Hatakeyama S (2014) Formal [4+1]-cycloaddition of homopropargyl alcohols to diazo dicarbonyl compounds giving substituted tetrahydrofurans. Org Lett 16:1004–1007

73. Xu X, Han X, Yang L, Hu W (2009) Highly diastereoselective synthesis of fully substituted tetrahydrofurans by a one-pot cascade reaction of aryldiazoacetates with allyl alcohols. Chem 15:12604–12607
74. Liang XS, Li RD, Wang XC (2019) Copper-catalyzed asymmetric annulation reactions of carbenes with 2-iminyl- or 2-acyl-substituted phenols: convenient access to enantioenriched 2,3-dihydrobenzofurans. Angew Chem Int Ed 58:13885–13889
75. Xu A, Zhou X, Zheng R, Zhang Z, Yin X, Hong K, Ke Z, Qian Y, Hu W (2022) Chiral rhodium(II)-catalyzed asymmetric aldol-type interception of an oxonium ylide to assemble chiral 2,3-dihydropyrans. Sci China Chem 65:1607–1614
76. Gopi Krishna Reddy A, Niharika P, Zhou S, Jia SK, Shi T, Xu X, Qian Y, Hu W (2020) Bronsted acid catalyzed enantioselective assembly of spirochroman-3,3-oxindoles. Org Lett 22:2925–2930
77. Alavala GKR, Sajjad F, Shi T, Kang Z, Ma M, Xing D, Hu W (2018) Diastereoselective synthesis of isochromans *via* the Cu(ii)-catalysed intramolecular Michael-type trapping of oxonium ylides. Chem Commun 54:12650–12653
78. Guo Z, Cai M, Jiang J, Yang L, Hu W (2010) $Rh_2(OAc)_4$-AgOTf cooperative catalysis in cyclization/three-component reactions for concise synthesis of 1,2-dihydroisoquinolines. Org Lett 12:652–655
79. Alamsetti SK, Spanka M, Schneider C (2016) Synergistic rhodium/phosphoric acid catalysis for the enantioselective addition of oxonium ylides to *ortho*-quinone methides. Angew Chem Int Ed 55:2392–2396
80. Wang J, Yao X, Wang T, Han J, Zhang J, Zhang X, Wang P, Zhang Z (2015) Synthesis of 2,5-Dihydrofurans *via* a Gold(I)-catalyzed formal [4+1] cycloaddition of α-diazoesters and propargyl alcohols. Org Lett 17:5124–5127
81. Shi T, Guo X, Teng S, Hu W (2015) Pd(II)-catalyzed formal [4+1] cycloaddition reactions of diazoacetates and aryl propargyl alcohols to form 2,5-dihydrofurans. Chem Commun 51:15204–15207
82. Lv X, Yang H, Shi T, Xing D, Xu X, Hu W (2019) Rhodium(II)-catalyzed formal [4+1]-cycloaddition of pyridotriazoles and propargyl alcohols: synthesis of 2,5-dihydrofurans. Adv Synth Catal 361:1265–1270
83. Shi T, Teng S, Gopi Krishna Reddy A, Guo X, Zhang Y, Moore KT, Buckley T, Mason DJ, Wang W, Chapman E, Hu W (2019) Catalytic asymmetric synthesis of 2,5-dihydrofurans using synergistic bifunctional Ag catalysis. Org Biomol Chem 17:8737–8744
84. Lian Y, Davies HM (2010) Rhodium-catalyzed [3 + 2] annulation of indoles. J Am Chem Soc 132:440–441
85. Dawande SG, Kanchupalli V, Kalepu J, Chennamsetti H, Lad BS, Katukojvala S (2014) Rhodium enalcarbenoids: direct synthesis of indoles by rhodium(II)-catalyzed [4+2] benzannulation of pyrroles. Angew Chem Int Ed 53:4076–4080
86. Jia S, Lei Y, Song L, Krishna Reddy AG, Xing D, Hu W (2017) Diastereoselective intramolecular Aldol-type trapping of zwitterionic intermediates by ketones for the synthesis of spiro [chroman-4,3′-oxindole] derivatives. Adv Synth Catal 359:58–63
87. Zhou S, Liu Q, Bao M, Huang J, Wang J, Hu W, Xu X (2021) Gold(i)-catalyzed redox transformation of o-nitroalkynes with indoles for the synthesis of 2,3′-biindole derivatives. Org Chem Front 8:1808–1816
88. Doyle MP, Hu W, Timmons DJ (2001) Epoxides and aziridines from diazoacetates *via* ylide intermediates. Org Lett 3:933–935
89. de March P, Huisgen R (1982) Carbonyl ylides from aldehydes and carbenes. J Am Chem Soc 104:4952–4952
90. Huisgen R, De March P (1982) Three-component reactions of diazomalonic ester, benzaldehyde, and electrophilic olefins. J Am Chem Soc 104:4953–4954
91. Muthusamy S, Gunanathan C, Nethaji M (2004) Multicomponent reactions of diazoamides: diastereoselective synthesis of mono- and bis-spirofurooxindoles. J Org Chem 69:5631–5637
92. DeAngelis A, Taylor MT, Fox JM (2009) Unusually reactive and selective carbonyl ylides for three-component cycloaddition reactions. J Am Chem Soc 131:1101–1105

93. Lu CD, Chen ZY, Liu H, Hu WH, Mi AQ, Doyle MP (2004) A facile three-component one-pot synthesis of structurally constrained tetrahydrofurans that are t-RNA synthetase inhibitor analogues. J Org Chem 69:4856–4859
94. Qiu L, Guo X, Zhou J, Liu S, Yang L, Wu X, Hu W (2013) A stereoselective synthesis of fully substituted tetrahydrofurans through 1,3-dipolar cycloaddition with cinnamaldehydes: an easy access to chroman derivatives. RSC Adv 3:20065–20070
95. Qiu L, Wang D, Lei Y, Gao L, Liu S, Li J, Hu W (2016) Diastereoselective three-component cascade reaction to construct oxindole-fused spirotetrahydro-furochroman scaffolds for drug discovery. Eur J Org Chem 2016:2671–2680
96. Dong S, Lei Y, Jia S, Gao L, Li J, Zhu T, Liu S, Hu W (2017) Discovery of core-structurally novel PTP1B inhibitors with specific selectivity containing oxindole-fused spirotetrahydrofurochroman by one-pot reaction. Bioorg Med Chem Lett 27:1105–1108
97. Li M, Chu R, Zhao Y, Hu W, Liu S (2018) Formal carbene insertion into C-O double bond: a facile approach to the synthesis of 2*H*-chromenes. Tetrahedron 74:4551–4557
98. Xiao G, Chen T, Ma C, Xing D, Hu W (2018) Rh(II)/chiral phosphoric acid-cocatalyzed enantioselective synthesis of spirooxindole-fused thiaindans. Org Lett 20:4531–4535
99. Lv X, Kang Z, Xing D, Hu W (2018) Cu(I)-catalyzed three-component reaction of diazo compound with terminal alkyne and nitrosobenzene for the synthesis of trifluoromethyl dihydroisoxazoles. Org Lett 20:4843–4847
100. Sajjad F, Krishna Reddy AG, Che J, Hu W, Xing D (2020) Ruthenium-catalyzed diastereoselective synthesis of fully substituted pyrrolidines from anilines and diazo pyruvates. Org Lett 22:3094–3098
101. Sajjad F, Chen Y, Tian X, Dong S, Gopi Krishna Reddy A, Hu W, Xing D (2021) Facile synthesis of 1,4-oxazines by ruthenium-catalyzed tandem N-H insertion/cyclization of alpha-arylamino ketones and diazo pyruvates. Org Biomol Chem 19:1769–1772
102. Zhu Y, Zhai C, Yue Y, Yang L, Hu W (2009) One-pot three-component tandem reaction of diazo compounds with anilines and unsaturated ketoesters: a novel synthesis of 2,3-dihydropyrrole derivatives. Chem Commun 45:1362–1364
103. Zhang X, Ji J, Zhu Y, Jing C, Li M, Hu W (2012) A highly diastereoselective three-component tandem 1,4-conjugated addition-cyclization reaction to multisubstituted pyrrolidines. Org Biomol Chem 10:2133–2138
104. Kang Z, Zhang D, Hu W (2017) Regio- and diastereoselective three-component reactions *via* trapping of ammonium ylides with *N*-alkylquinolinium salts: synthesis of multisubstituted tetra- and dihydroquinoline derivatives. Org Lett 19:3783–3786
105. Li M, Chu R, Chen J, Wu X, Zhao Y, Liu S, Hu W (2017) Enantioselective formal [3+1+1] cycloaddition reaction by Ru(II)/Iminium cocatalysis for construction of multisubstituted pyrrolidines. Org Lett 19:1290–1293
106. Dawande SG, Kanchupalli V, Lad BS, Rai J, Katukojvala S (2014) Synergistic rhodium(II) carboxylate and bronsted acid catalyzed multicomponent reactions of enalcarbenoids: direct synthesis of alpha-pyrrolylbenzylamines. Org Lett 16:3700–3703
107. Qiu L, Guo X, Ma C, Qiu H, Liu S, Yang L, Hu W (2014) Efficient synthesis of chiral cyclic acetals by metal and Bronsted acid co-catalyzed enantioselective four-component cascade reactions. Chem Commun 50:2196–2198
108. Tang M, Xing D, Huang H, Hu W (2015) Divergent synthesis of chiral heterocycles *via* sequencing of enantioselective three-component reactions and one-pot subsequent cyclization reactions. Chem Commun 51:10612–10615
109. Tang M, Wu Y, Liu Y, Cai M, Xia F, Liu S, Hu W (2016) One-pot enantioselective multi-component cascade reactions for synthesis of chiral functionalized hydro-epoxyisochromenes: a rapid access to molecular complexity. Acta Chim Sin 74:54–60
110. Shi T, Teng S, Wei Y, Guo X, Hu W (2019) Synthesis of spiro[2,3-dihydrofuran-3,3′-oxindole] derivatives *via* a multi-component cascade reaction of α-diazo esters, water, isatins and malononitrile/ethyl cyanoacetate. Green Chem 21:4936–4940

111. Yin X, Xu A, Hu J, Bao M, Hu W, Qian Y (2020) A Rh(ii)/phosphoric acid co-catalyzed three-component reaction of diazo-ketones with alcohols and azonaphthalenes: access to indole derivatives *via* a formal [3+2]-cycloaddition. Org Biomol Chem 18:9805–9809
112. Ren L, Lian XL, Gong LZ (2013) Bronsted acid/rhodium(II) cooperative catalytic asymmetric three-component aldol-type reaction for the synthesis of 3-amino oxindoles. Chem 19:3315–3318
113. Wang C, Xing D, Wang D, Wu X, Hu W (2014) Synthesis of 3-amino-3-hydroxymethyloxindoles and 3-hydroxy-3-hydroxymethyloxindoles by $Rh_2(OAc)_4$-catalyzed three-component reactions of 3-diazooxindoles with formaldehyde and anilines or water. J Org Chem 79:3908–3916
114. Wei Q, Li Z, Yang X, Chen J, Liu X, Hu W, Liu S (2019) Rh(I)/$Sc(OTf)_3$-co-catalyzed Michael addition of ammonium ylide to (*E*)-1,4-enediones: synthesis of functionalized 1,4-diketones. Mol Divers 23:997–1010
115. Zhou K, Bao M, Sha H, Dong G, Hong K, Xu X, Hu W (2021) Highly diastereoselective synthesis of vicinal diamines *via* a Rh-catalyzed three-component reaction of diazo compounds with diarylmethanimines and ketimines. Org Chem Front 8:2997–3003
116. Lian XL, Meng J, Han ZY (2016) Ru(II)/organo relay catalytic three-component reaction of 3-diazooxindoles, amines, and nitroalkene: formal synthesis of (−)-psychotrimine. Org Lett 18:4270–4273
117. Jia S-K, Lei Y-B, Song L-L, Liu S-Y, Hu W-H (2017) Enantioselective trapping of oxonium ylide intermediates by *N*-benzhydryl-α-imino ester: synthesis of β-tetrasubstituted α-amino acids. Chin Chem Lett 28:213–217
118. Liu X, Tian X, Huang J, Qian Y, Xu X, Kang Z, Hu W (2022) Enantioselective propargylation of oxonium ylide with alpha-propargylic-3-indolymethanol: access to chiral propargylic indoles. Org Lett 24:1027–1032
119. Ma C, Zhou JY, Zhang YZ, Jiao Y, Mei GJ, Shi F (2018) Synergistic-catalysis-enabled reaction of 2-indolymethanols with oxonium ylides for the construction of 3-indolyl-3-alkoxy oxindole frameworks. Chem Asian J 13:2549–2558
120. Jia SK, Song LL, Lei YB, Gopi Krishna Reddy A, Xing D, Hu WH (2016) A Rh(ii)-catalyzed three-component reaction of 3-diazooxindoles with *N,N*-disubstituted anilines and glyoxylates for the synthesis of 3-aryl-3-substituted oxindoles. Org Biomol Chem 14:10157–10160
121. Chupakhin EG, Kantin GP, Dar'in DV, Krasavin M (2021) Convenient preparation of (*E*)-3-arylidene-4-diazopyrrolidine-2,5-diones in array format. Mendeleev Commun 31:36–38
122. Yuan H, Hong K, Liu X, Qian Y, Xu X, Hu W (2021) Enantioselective assembly of 3,3-disubstituted succinimides via three-component reaction of vinyl diazosuccinimides with alcohols and imines. Chem Commun 57:8043–8046
123. Zhang D, Qiu H, Jiang L, Lv F, Ma C, Hu W (2013) Enantioselective palladium(II) phosphate catalyzed three-component reactions of pyrrole, diazoesters, and imines. Angew Chem Int Ed 52:13356–13360
124. Mei LY, Tang XY, Shi M (2014) Rhodium-catalyzed three-component reaction of 3-diazooxindoles with indoles and isatin-derived ketimines: a facile and versatile approach to functionalized 3,3′,3″-trisindoles. Org Biomol Chem 12:1149–1156
125. Jing C, Xing D, Hu W (2015) Catalytic asymmetric four-component reaction for the rapid construction of 3,3-disubstituted 3-indol-3′-yloxindoles. Org Lett 17:4336–4339
126. Xing D, Jing C, Li X, Qiu H, Hu W (2013) Highly efficient synthesis of mixed 3,3′-bisindoles *via* Rh(II)-catalyzed three-component reaction of 3-diazooxindoles with indoles and ethyl glyoxylate. Org Lett 15:3578–3581
127. Jing C, Xing D, Wang C, Hu W (2015) Synthesis of 3-(hydroxymethyl)-3-indol-3′-yloxindoles *via* Rh(II)-catalyzed three-component reaction of 3-diazooxindoles, indoles and formalin. Tetrahedron 71:3597–3602
128. Chen DF, Zhao F, Hu Y, Gong LZ (2014) C-H Functionalization/asymmetric Michael addition cascade enabled by relay catalysis: metal carbenoid used for C-C bond formation. Angew Chem Int Ed 53:10763–10767

129. Chen DF, Zhang C-l, Hu Y, Han ZY, Gong LZ (2015) Catalytic enantioselective synthesis of quaternary 3,3′-indolyloxindoles by combination of Rh(ii) complexes and chiral phosphines. Org Chem Front 2:956–960
130. Kang Z, Zhang D, Xu X, Hu W (2019) Privilege-structure-oriented three-component asymmetric aminomethylation: assembly of chiral 3-aminomethyl indolones. Org Lett 21:9878–9883
131. Li M, Guo X, Jin W, Zheng Q, Liu S, Hu W (2016) An enantioselective three-component reaction of diazoacetates with indoles and enals by iridium/iminium co-catalysis. Chem Commun 52:2736–2739
132. Liu S, Yao W, Liu Y, Wei Q, Chen J, Wu X, Xia F, Hu W (2017) A Rh(II)-catalyzed multicomponent reaction by trapping an alpha-amino enol intermediate in a traditional two-component reaction pathway. Sci Adv 3:e1602467
133. Jiang L, Jin W, Hu W (2016) Double C–H functionalization of indoles *via* three-component reactions/$CuCl_2$-catalyzed aerobic dehydrogenative coupling for the synthesis of polyfunctional cyclopenta[*b*]indoles. ACS Catal 6:6146–6150
134. Dong K, Pei C, Zeng Q, Qiu L, Hu W, Qian Y, Xu X (2019) Copper-catalyzed [4+1]-annulation of 2-alkenylindoles with diazoacetates: a facile access to dihydrocyclopenta[*b*] indoles. Chem Commun 55:6393–6396
135. Rajasekaran T, Karthik G, Sridhar B, Kumar SK, Reddy BVS (2014) Highly diastereoselective reaction of α-diazoesters with aryl alcohols and isatin imines: rapid access to oxindole-derived α-alkoxy-β-amino acid derivatives with two adjacent quaternary carbon centers. Eur J Org Chem 2014:2221–2224
136. Ao C, Huang J, Xu X, Jia S, Kang Z, Hu W (2020) A Rh-catalyzed three-component reaction for the diastereoselective synthesis of pyrazolone derivatives with contiguous quaternary stereocenters. Org Biomol Chem 18:3466–3470
137. Xie X, Bao M, Chen K, Xu X, Hu W (2022) An asymmetric three-component reaction of a diazo compound with an alcohol and a seven-membered imine. Org Chem Front 9:2102–2108
138. Ma C, Chen J, Xing D, Sheng Y, Hu W (2017) Iron catalyzed efficient synthesis of polyfunctional primary amines *via* the direct use of ammonia. Chem Commun 53:2854–2857
139. Jing C, Shi T, Xing D, Guo X, Hu W-H (2013) $CuSO_4$-catalyzed three-component reaction of α-diazo ester, water and isatin: an efficient approach to oxindole derivatives. Green Chem 15:620–624
140. Guo X, Huang H, Yang L, Hu W (2007) Trapping of oxonium ylide with isatins: efficient and stereoselective construction of adjacent quaternary carbon centers. Org Lett 9:4721–4723
141. Ji J, Zhang X, Jiang L, Hu W (2012) $InBr_3$ catalyzed three-component reactions of an aryldiazoacetate, an alcohol, and a carbonyl compound. Tetrahedron Lett 53:182–185
142. Guo X, Wang J, Yang L, Hu W (2010) Rh(II) catalyzed three-component reactions of diazoacetates with benzenemethanol and indane-1, 2, 3-triones. Lett Org Chem 7:106–109
143. Alcaide B, Almendros P, Aragoncillo C, Callejo R, Ruiz MP, Torres MR (2009) Rhodium-catalyzed synthesis of 3-hydroxy-beta-lactams *via* oxonium ylide generation: three-component reaction between azetidine-2,3-diones, ethyl diazoacetate, and alcohols. J Org Chem 74:8421–8424
144. Alcaide B, Almendros P, Aragoncillo C, Callejo R, Ruiz MP, Torres MR (2012) Diastereoselective synthesis of β-lactam-oxindole hybrids through a three-component reaction of azetidine-2,3-diones, α-diazo-oxindoles, and alcohols catalyzed by $[Rh_2(OAc)_4]$. Eur J Org Chem 2012:2359–2366
145. Zhang D, Liu J, Kang Z, Qiu H, Hu W (2019) A rhodium-catalysed three-component reaction to access C1-substituted tetrahydroisoquinolines. Org Biomol Chem 17:9844–9848
146. Liu G, Yu S, Hu W, Qiu H (2019) A gold(i)-catalysed three-component reaction *via* trapping oxonium ylides with allenamides. Chem Commun 55:12675–12678
147. Yu S, Chen J, Liu G, Lei J, Hu W, Qiu H (2020) A gold(i)-catalysed chemoselective three-component reaction between phenols, alpha-diazocarbonyl compounds and allenamides. Chem Commun 56:1649–1652

148. Hu W, Jiang L, Han X, Gan M, Qiu H, Ji J, Zhang X (2011) Trapping of oxonium ylides with Michael acceptors: highly diastereoselective three-component reactions of diazo compounds with alcohols and benzylidene meldrum's acids/4-oxo-enoates. Synlett 2011:1717–1722
149. Suneja A, Loui HJ, Schneider C (2020) Cooperative catalysis for the highly diastereo- and enantioselective [4+3]-cycloannulation of *ortho*-quinone methides and carbonyl ylides. Angew Chem Int Ed 59:5536–5540
150. Harada S, Sakai C, Tanikawa K, Nemoto T (2019) Gold-catalyzed chemoselective formal (3 +2)-annulation reaction between β-naphthols and methyl aryldiazoacetate. Tetrahedron 75:3650–3656
151. Xu G, Tang S, Shao Y, Sun J (2019) $B(C_6F_5)_3$-Catalyzed formal (4+1)-annulation of *ortho*-quinone methides with diazoacetates: access to 2,3-dihydrobenzofurans. Chem Commun 55:9096–9099
152. Yadagiri D, Chaitanya M, Reddy ACS, Anbarasan P (2018) Rhodium catalyzed synthesis of benzopyrans *via* transannulation of *N*-sulfonyl-1,2,3-triazoles with 2-hydroxybenzyl alcohols. Org Lett 20:3762–3765
153. Zhang D, Zhou J, Xia F, Kang Z, Hu W (2015) Bond cleavage, fragment modification and reassembly in enantioselective three-component reactions. Nat Commun 6:5801–5809
154. Zhang D, Hu W (2018) Formal carbene insertion into C−O or C−N bond: an efficient strategy for the synthesis of 2-substituted 2*H*-chromene derivatives from chromene acetals or hemiaminal ethers. Adv Synth Catal 360:2446–2452
155. Guan XY, Yang LP, Hu W (2010) Cooperative catalysis in multicomponent reactions: highly enantioselective synthesis of γ-hydroxyketones with a quaternary carbon stereocenter. Angew Chem Int Ed 49:2190–2192
156. Kang Z, Zhang D, Shou J, Hu W (2018) Enantioselective trapping of oxonium ylides by 3-hydroxyisoindolinones via a formal S_N1 pathway for construction of contiguous quaternary stereocenters. Org Lett 20:983–986
157. Kang Z, Shou J, Xing D, Hu W (2020) Rh(II)/Ag(I)-cocatalyzed three-component reaction *via* S_N1/S_N1'-type trapping of oxonium ylide with the nicholas intermediate. J Org Chem 85:9850–9862

Heterocycles from Cyclopropanation of Five-Membered Heteroarenes

Kathrin Strunk and Oliver Reiser

Contents

Abstract The cyclopropantion of five-membered heteroarenes provides a facile entry to a broad variety of heterocycles, capitalizing on the high reactivity of carbenes to initially break the aromaticity of the starting materials as well as on the possibility for selective cleavage of any of the bonds in the cyclopropane moiety driven by the ring strain to allow skeletal rearrangements.

Keywords Cyclopropanation · Donor–acceptor cyclopropanes · Furans · Indole · Natural products · Pyran · Pyridine · Pyrroles · Ring expansion · γ-butyrolactone

1 Introduction

Five-membered heteroarenes are formidable starting materials for synthesizing heterocycles with great diversity: they are stable, inexpensive, and readily available, often produced from renewable resources on a ton scale. Prominent representatives are furans, which are considered platform chemicals derived from carbohydrates,

K. Strunk and O. Reiser (✉)
Institut für Organische Chemie, Universität Regensburg, Regensburg, Germany
e-mail: Kathrin.Strunk@ur.de; Oliver.Reiser@ur.de

Fig. 1 Aromatic stabilization energy of representative heteroarenes in comparison to carbocyclic arenes. [a] Ref. [1] [b] Ref. [2]

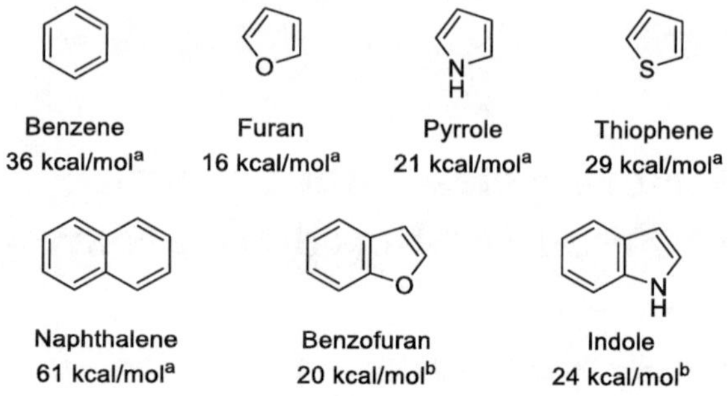

providing a facile entry to pyrroles as well. Alternatively, the condensation of 1,4-diketones with ammonia (Paal–Knorr synthesis) allows the versatile construction of pyrroles, while the analogous condensation with sulfidation reagents gives access to thiophenes (Paal–Knorr thiophene synthesis). Indoles are produced in the industry from anilines or nitroarenes. Depending on the electronic nature of such heteroarenes, the derivatization by electrophilic or nucleophilic substitution can generate new derivatives in great variety, however, maintaining the core structure of the heteroarene.

While five-membered heteroarenes are aromatic and consequently undergo archetypical transformations such as substitutions or cross-coupling reactions that retain their aromaticity, their aromatic stabilization energy is significantly lower than their carbocyclic analogs (Fig. 1). Consequently, it is comparatively easier to break their aromaticity, and consequently, reactions with carbenes leading to cyclopropanated heteroarenes have provided various opportunities for postmodification toward new heterocyclic derivatives.

Most commonly for cyclopropanation reactions of heteroarenes, diazo compounds have been employed as carbene precursors, provided the latter are stabilized by at least one conjugated substituent. Diazoacetates are predominantly applied since they can be synthesized and stored safely on a large scale. A very notable alternative is emerging with the use of sulfonylhydrazones as precursors, from which diazo compounds can be liberated upon base-induced elimination and converted in situ by a catalyst present to the desired products.

Cyclopropanations of heteroarenes with diazo compounds have been achieved by three conceptually different approaches (Scheme 1): (1) the irradiation of diazo compounds has been found effective for – presumably singlet – carbene transfer to alkenes in the combination of aryldiazoacetates and visible light; (2) decomposition in presence of Cu(I) or Rh(II) complexes to form closed shell metalcarbenoids which undergo cyclopropanation reactions following a concerted pathway; and (3) decomposition in the presence of Co(II) complexes which form open shell cobalt(III)-radical intermediates which add to alkenes in a stepwise radical type addition.

Scheme 1 Carbenes, metalcarbenoids, and metalloradicals as precursors for cyclopropanation reactions of alkenes

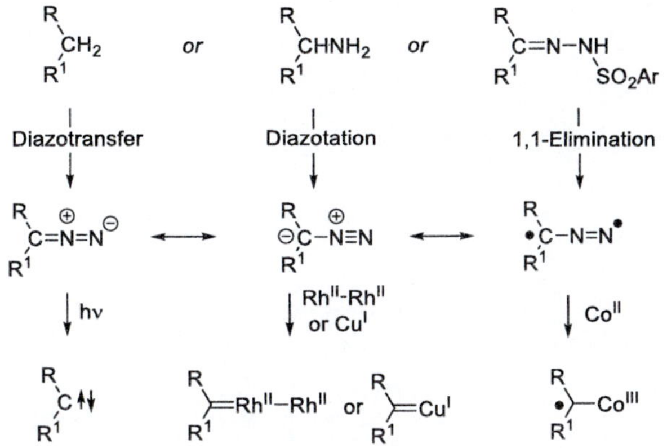

2 Cyclopropanation of Furans

The addition of carbenes to furans has a long history, dating back over 60 years (Scheme 2). Early reports on the thermal- and photoinduced decomposition of diazoesters that lead to cyclopropanated furan **2b** [3–5] were followed by the copper-bronze mediated reaction of diazomethane to give rise to **2a** [6, 7]. The seminal work by Wenkert and coworkers demonstrated the cyclopropanation of furans by catalysis with $Rh_2(OAc)_4$ to give cyclopropanated adducts **2b**–**2e** along with rearranged or ring-opened products, which demonstrated already the high intrinsic reactivity of such adducts for subsequent transformations [8–10]. Alternatively, Cu(I) has been identified as a capable catalyst for cyclopropanation with diazoacetates, which proved to be especially suitable for the synthesis of **2e** and **2f**, being crystalline, shelf-stable adducts. The cyclopropanation occurs regioselectively at the less substituted, presumably more electron-rich double bond and diastereoselectivity, mandated by the orientation of the ester moiety on the convex face of the bicycle. In combination with chiral bis(oxazoline) [11, 12] or aza (bisoxazoline) [13] ligands, these products can be prepared on multigram scale with high enantioselectivity [14–17]. The thermal decomposition (100°C) of *t*Butyl diazoacetate in the presence of a nickel(II)-catalyst **cat-6** was also shown to be suitable, leading to furan **2f** [18]. Aryldiazoacetates have also been applied, giving cyclopropanated furans **2g**–**2i** either by (asymmetric) rhodium(II) catalysis or catalyst-free photolysis with excellent results [21, 22]. Noteworthy, close to 100,000 turnover numbers have been achieved with rhodium(II) catalysts for these transformations, making them the most active metal complexes for the decomposition of diazo compounds to date.

The cycloadducts **2** offer a chameleon-like reactivity, allowing the construction into new heterocyclic scaffold in just a few steps (Scheme 3). Considering the furan oxygen as a donor and the ester moiety as an acceptor (i.e., donor–acceptor substituted cyclopropane [23–25]), the exocyclic cyclopropane bond in the

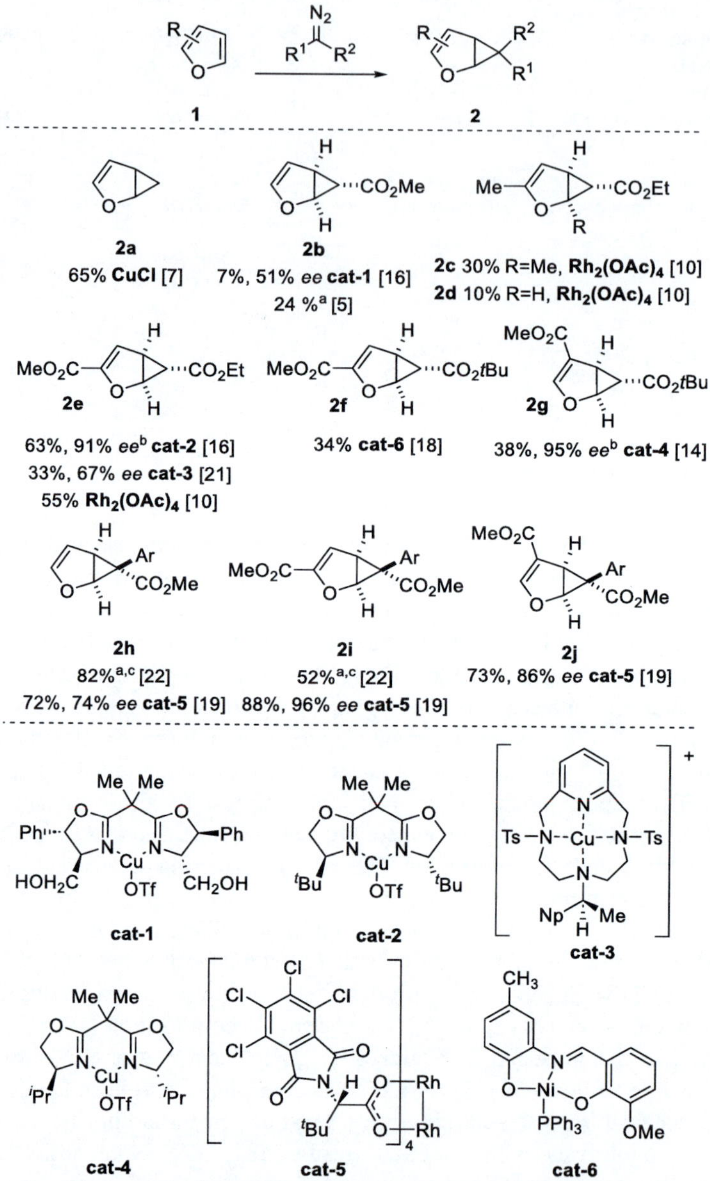

Scheme 2 Cyclopropanation of furans with diazo compounds. [a] Photochemical decomposition. [b] 99% ee after single crystallization. [c] in continuous flow [19, 20].

trajectory of these functionalities can easily be cleaved, which has been applied to the synthesis of furolactones **4**.

Such intermediates were also accessed via the cyclopropanation of 2,3-dihydrofurans [25, 26] or by radical additions to butenolides [26], which have

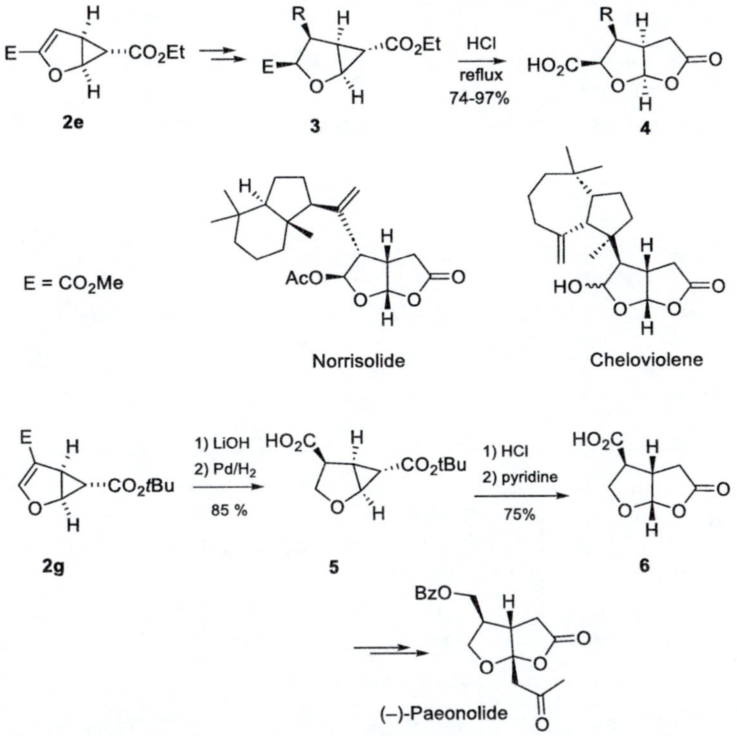

Scheme 3 Acid-catalyzed ring-opening/lactonization of cyclopropanated furans **2e** and **2g** toward furolactone natural products

been utilized to synthesize natural products such as Norrisolide [27] or Cheloviolene [26] or bicyclic peroxides [28] displaying antimalarial activity.

Alternatively, the donor–acceptor cyclopropane moiety can be utilized via cleaving the double bond by ozonolysis to give rise to cyclopropylcarbaldehydes **7**, which undergo upon addition of nucleophiles a retroaldol/cyclization cascade to yield γ-butyrolactones **8** (Scheme 4). This strategy has been applied to the synthesis of paraconic acids **9** [16, 21, 29] and various sesquiterpenes such as Xanthatin **10** [30] or Arglabin™ **11** [31].

Besides the exocyclic cyclopropane bond, the corresponding *endo*-cyclic bond in **2** can also be cleaved, thus resulting in the ring expansion of the furan to a pyran ring system (Scheme 5). To overwrite the inherent driving force for exocyclic bond cleavage due to the donor–acceptor relationship of the ring oxygen and the ester group in **2**, either a better acceptor center via a carbenium center or a formal anion masked as an organopalladium intermediate in 4-position of the tetrahydrofuran was found to be suitable to achieve this reaction mode. Thus, (oxidative) Heck-couplings with aryl iodides or aryl boronic acids with cyclopropanated furan adduct **2g** result in pyran derivatives **14** that are formed via a β-carbon elimination of intermediate **12** [32, 33]. Alternatively, **2h** could be converted to the mesylate **15**, which undergoes

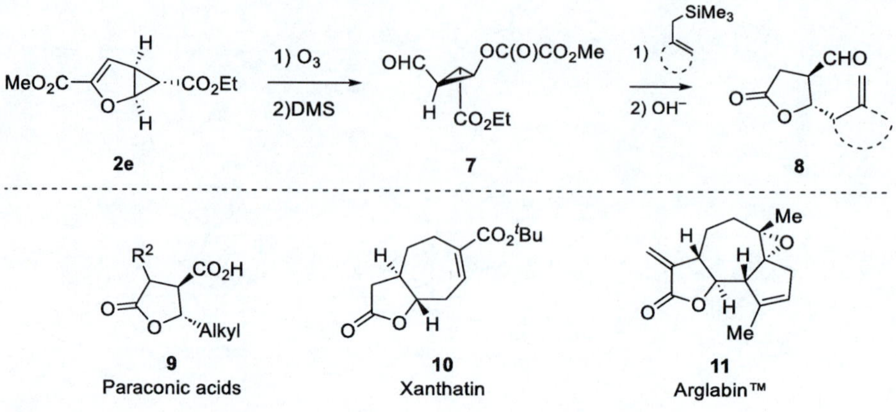

Scheme 4 Cyclopropanated furans as precursors for γ-butyrolactone natural products

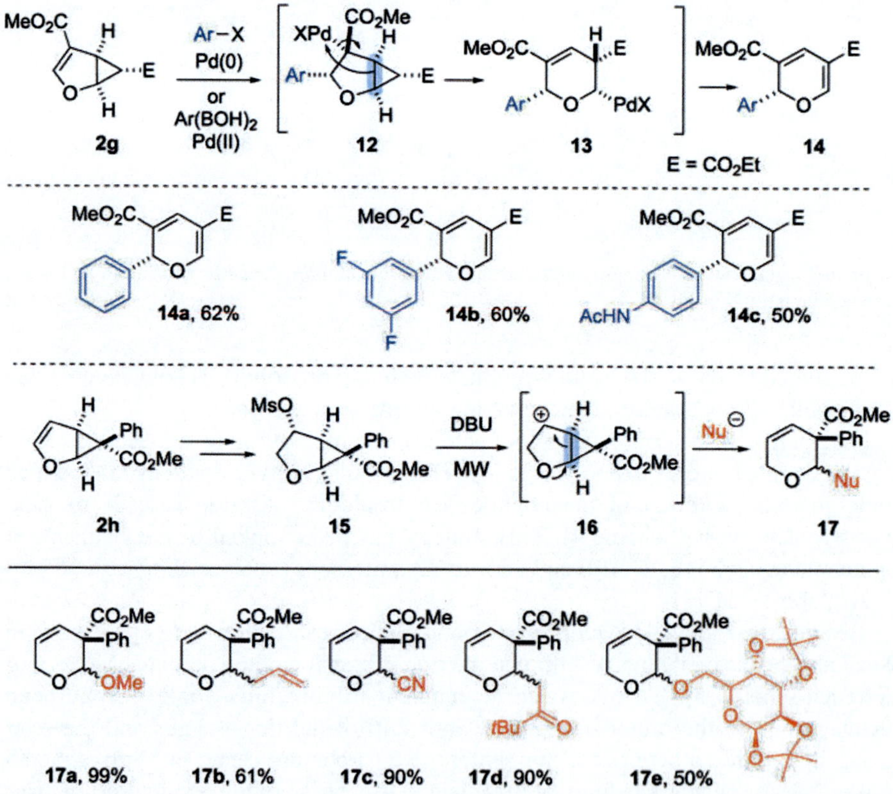

Scheme 5 Heterocycle formation via endocyclic ring-opening reactions of cyclopropanated furans

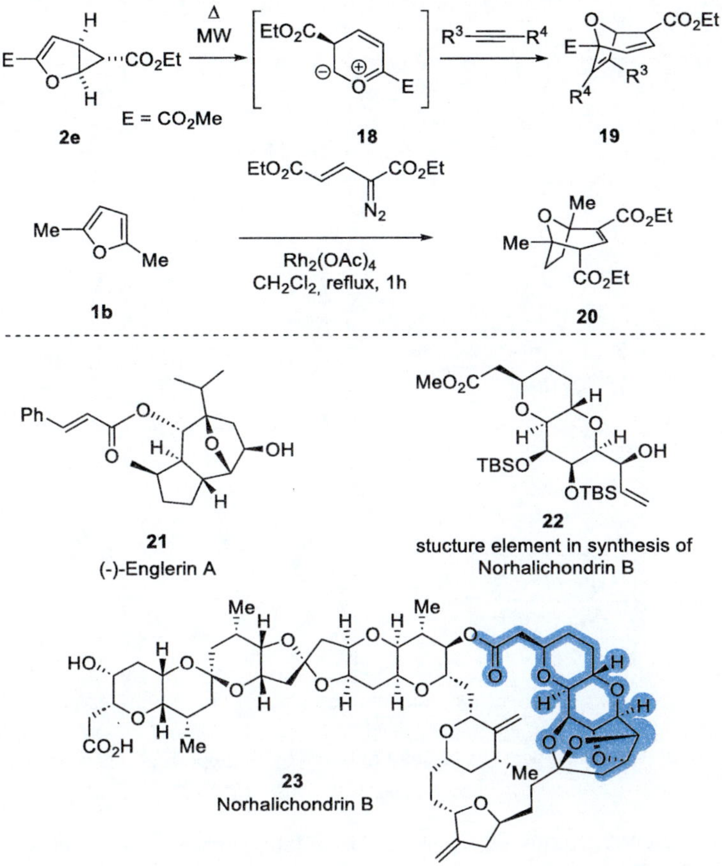

Scheme 6 Synthesis of oxo-[3.2.1]-frameworks via cyclopropanation of furans followed by electrocyclic ring-opening or Cope-rearrangements

ring expansion to **17** in the presence of various nucleophiles upon microwave-assisted heating [34].

Another possibility for endocyclic bond cleavage was found for derivative **2e**, which undergoes [3 + 2]-cycloadditions to [3.2.1]-scaffolds **19** with alkenes or alkynes via a 6π-electrocyclic ring-opening intermediate **15** (Scheme 6) [35]. This ring system can also be accessed by a tandem cyclopropanation/Cope rearrangement between 2,5-dimethylfuran (**1b**) and vinyldiazoacetates [36–40], as elegantly demonstrated by Davies et al. to achieve key intermediates relevant for the synthesis of natural products such as Englerin A **21** [41, 42] or Norhalichondrin **23** [43].

While the monocyclopropanated products can be obtained selectively by employing an excess of the furan, also the twofold cyclopropanation of furan **1a** giving rise to **21** was demonstrated with high *anti*-selectivity for the second cycloaddition, being possible either by rhodium- [44] or copper catalysis (Scheme 7) [45–47]. This was elegantly utilized by Werz et al. to synthesize ion-channel like

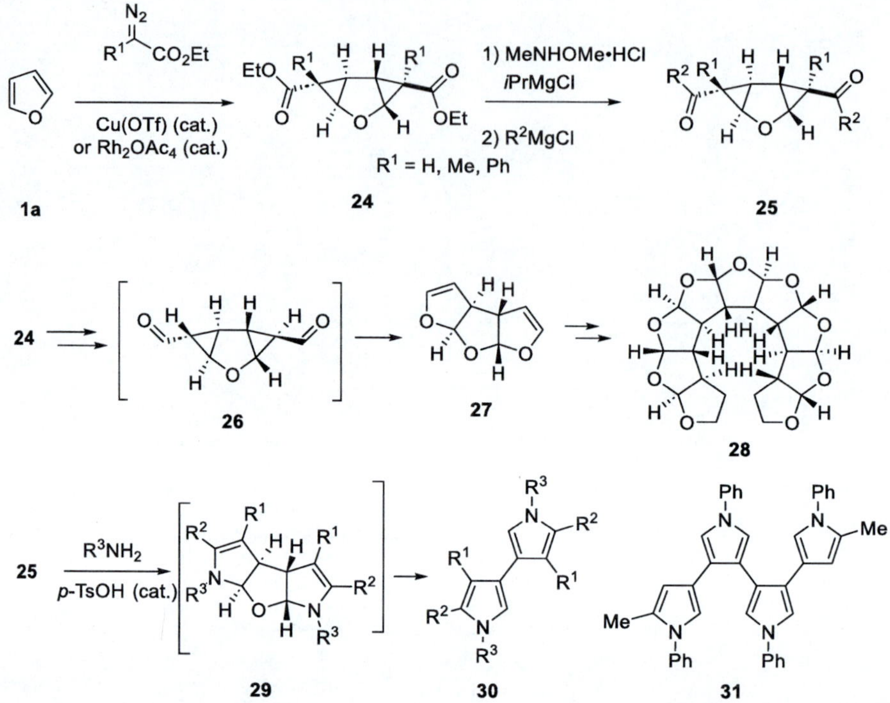

Scheme 7 Twofold cyclopropanation of furan followed by ring-opening/cyclization cascades

polycyclic tetrahydrofurans **28** (Scheme 7). Notably, upon conversion of the ester moieties in **24** ($R^1 = H$) to formyl groups, ring-opening/cyclization of **26** occurs to **27** without the need for (Lewis) acid activation, reflecting the better acceptor quality of an aldehyde compared to an ester. Subsequent cyclopropanation of the 2,3-dihydrofuran in **27** allowed the connection of up to 9 tetrahydrofuran moieties iteratively [43–45].

Alternatively, the ester moieties in **24** can be converted via the corresponding Weinreb amides to ketones **25**, which could be engaged with amines in a similar acid-catalyzed ring-opening cyclization to **29**, which further gave rise to bispyrroles **30** and in extension tetrapyrrole **31** applying an iterative strategy.

3 Cyclopropanation of Pyrroles

The higher aromatic stabilization energy of pyrroles compared to furans makes it more challenging to carry out dearomative transformations. Nevertheless, provided that an electron-withdrawing protecting group is attached to nitrogen, pyrroles readily undergo carbene addition to yield stable adducts (Scheme 8).

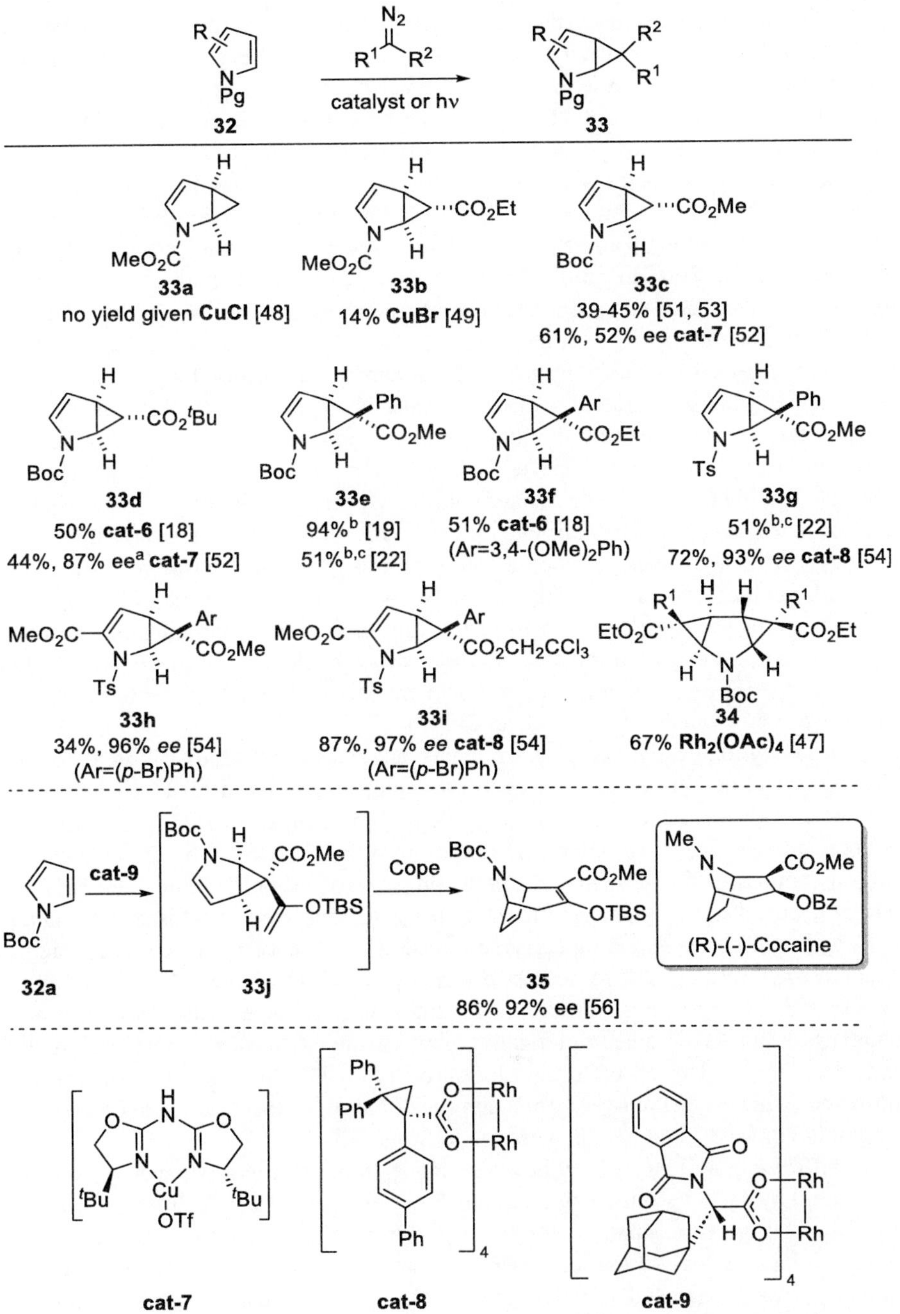

Scheme 8 Cyclopropanations of pyrroles: [a]99% *ee* after single crystallization. [b]Photochemical decomposition. [c] in continuous flow.

Pioneered by Fowler, *N*-carbamate substituted pyrroles **32** upon decomposition of diazomethane [48] or diazoacetates [49] with CuCl undergo cyclopropanation, contrasting *N*-methyl pyrrole which only gives substitution products [50]. It was subsequently shown that Cu(OTf), which can be generated in situ from $Cu(OTf)_2$/ $PhNHNH_2$, is a more active catalyst for the cyclopropanation of *N-Boc*-pyrrole with diazoacetates, allowing the synthesis of monocyclopropanated derivatives **33c** on multigram scale [51]. In combination with chiral bis(oxazoline) or aza(bisoxazoline) ligands, the cyclopropanation can be rendered asymmetric, giving rise to **33d** in up to 99%ee after recrystallization [52]. Alternatively, separating both enantiomers of (±)-**26c** by simulated moving bed (SMB) chromatography is highly efficient [53]. Comparatively few substituted pyrroles have been explored for this transformation, but ester-substituted pyrroles are amenable substrates for such [2 + 1]-cycloadditions as exemplified with the synthesis of **33h-i** [54]. Aryl-substituted diazoacetates require (chiral) rhodium catalysts, and moreover, *N*-tosyl protected pyrrole was found to give higher yields than *N*-Boc-pyrrole in the cyclopropanation reaction, leading to **33g-I** in high enantiopurity. Most impressively, vinyl-substituted diazoacetates undergo a tandem cyclopropanation/Cope rearrangement directly to azabicyclo[3.2.1]-octanes of type **35**, representing the critical framework of many alkaloids such as cocaine [55, 56]. While the twofold cyclopropanation of pyrrole was observed as a byproduct in the synthesis of monocyclopropanated pyrroles [51, 57], Werz et al. achieved the synthesis of such tricyclic adducts **34** with the aid of copper(I) or rhodium(II) catalysis in moderate to good yields by using an excess (3–5 equivalents) of diazoacetates [47].

Various synthetically valuable strategies have been reported that involve the cleavage of either the endo- or exocyclic cyclopropane bond in **33** (Scheme 9). Fowler already described endocyclic cyclopropane ring cleavage by the thermal isomerization at 285°C of **33a** to the dihydropyridine **36** [48] or by the endocyclic ring-opening at 100°C of **33a** or **33b** in the presence of a dipolarophile to give rise to **34** [58], however, no yields for these transformations were disclosed. The latter strategy could be expanded by carrying out the [3 + 2]-cycloaddition under microwave irradiation at 150°C to access to azabicyclo[3.2.1]-frameworks **31b** in great variety [35]. The reactions proceed with high diastereoselectivity; moreover, the chiral information of enantiomerically pure starting materials is retained in the products. The value of building blocks of type **31** was shown with various postmodifications. Moreover, this approach allows the facile synthesis of isoquinuclidines **39** in a two-step sequence from **38d**.

A different possibility of endocyclic ring-opening of cyclopropanated pyrroles was found by carrying out Heck-reactions between aryl- or vinylhalides or -triflates and **26c**, giving rise to substituted dihydropyridines **40** en route to a broad diversity of piperidines such as **42** (Scheme 10). Alternatively, dihydropyridines **40** could be directly dehydrogenated to pyridines **41** by 2,3-dichloro-5,6-dicyano-1,4-benzoquione (DDQ) in a one-flask protocol. Especially, 2-aryl substituted pyridines **41** become accessible this way, which are difficult to prepare by cross-coupling methodology due to the lack of readily available pyridine precursors [32, 33].

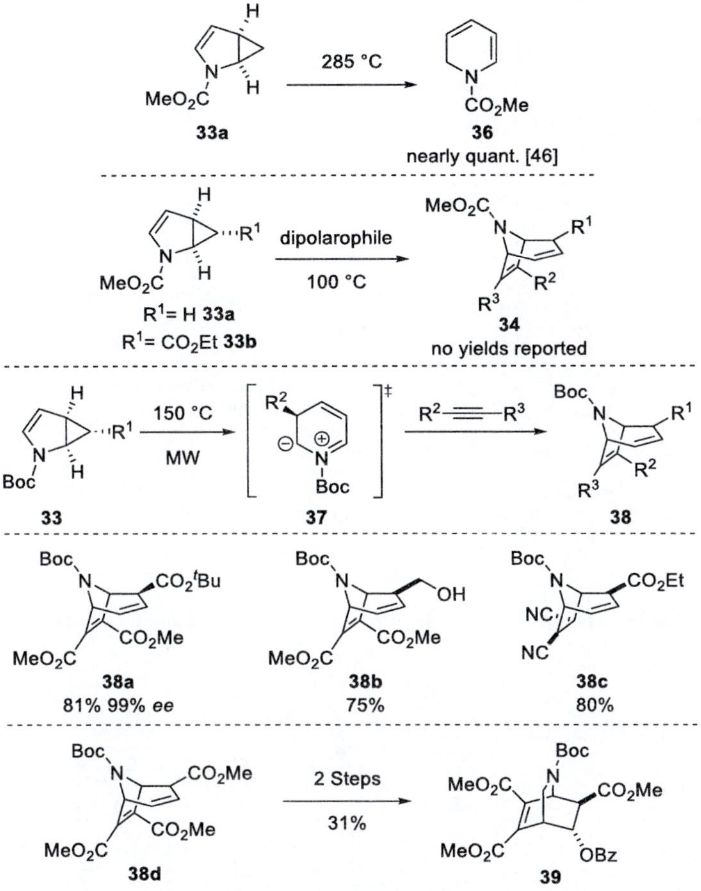

Scheme 9 Thermal endocyclic ring-opening of cyclopropanated pyrroles

Alternatively, the endocyclic ring expansion can be triggered from **33e** via the mesylate **43** under thermal conditions or via the epoxide **45** by Lewis acid catalysis, giving rise to tetrahydropyridines **44** or dihydropyridines **46**, respectively [34] (Scheme 11).

While the cyclopropanated furans are stable and only undergo ring-opening via the exocyclic cyclopropane bond upon activation by acids, the corresponding adducts derived from *N*-unprotected are not stable, reflecting the better donor properties of nitrogen compared to oxygen. Thus, cyclopropanation of *N*-unprotected pyrroles **47** with dichlorocarbene, typically generated under harsh conditions from chloroform/NaOH [59] or organomercury precursors [60], directly yields 3-chloropyridines **49** (Ciamician–Dennstedt rearrangement, Scheme 12) via the corresponding cyclopropane adducts **48** followed by endocyclic ring-opening. Given the harsh basic conditions that are required for the formation of dichlorocarbene from chloroform, yields however are generally quite low, especially

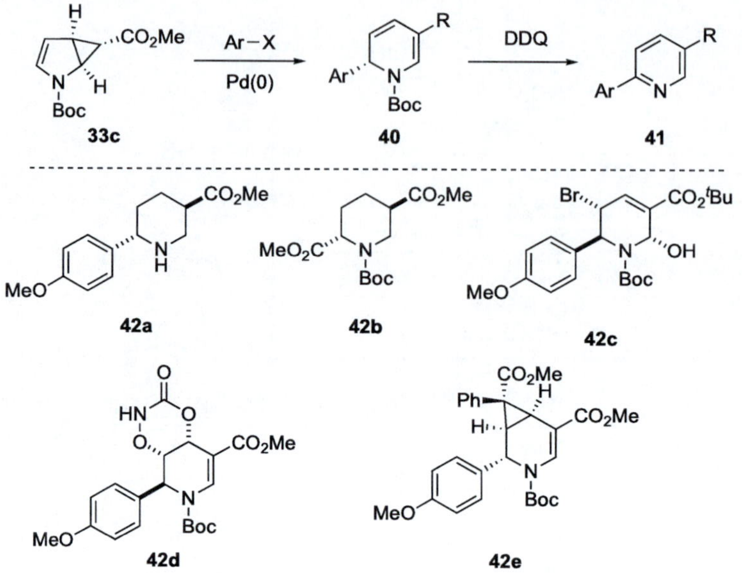

Scheme 10 Heck-couplings with concurrent ring expansion of cyclopropanated pyrrole **33c**

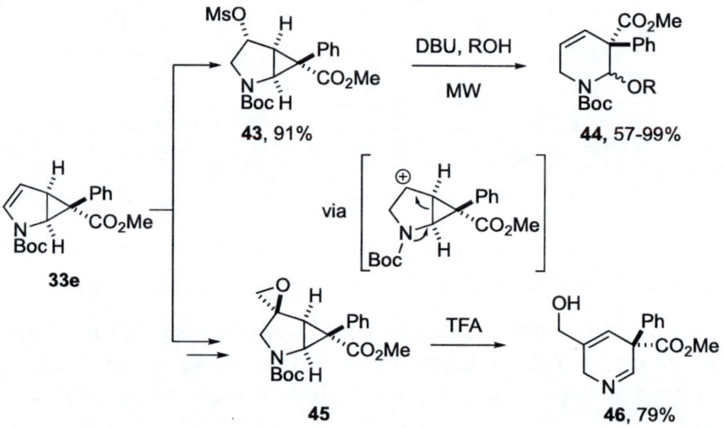

Scheme 11 Endocyclic ring-opening of cyclopropanated pyrrole **33e** via cationic intermediates of type **49**

since the Reimer–Tiemann formylation of pyrrole is a competing pathway. An improved protocol was reported by using the sodium salt of trichloroacetic acid instead of chloroform under phase transfer conditions [61], which has found a most elegant application in the total synthesis of neurotrophic alkaloids Complanadine A and Lycodine [62], allowing the transformation of **47a** into **49** in respectable yields as the key step.

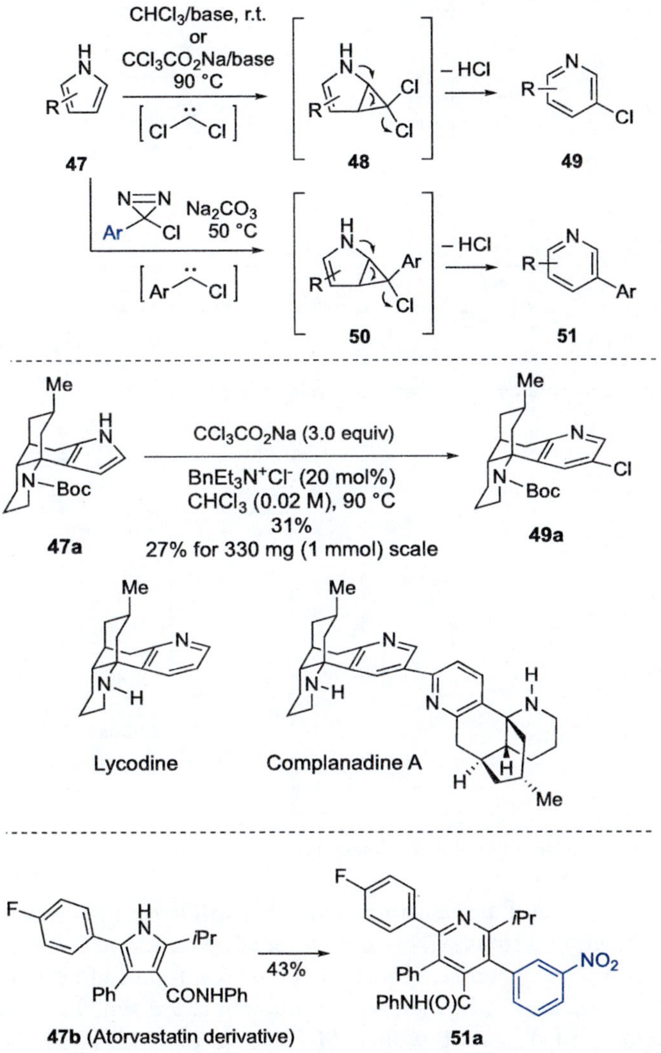

Scheme 12 Ciamician–Dennstedt rearrangement and variants for the synthesis of pyridines

A significant extension of the Ciamician–Dennstedt rearrangement was reported by using readily available 3-chloro-3-aryl-3*H*-diazirines that give rise to arylchlorocarbene under mild thermal and basic conditions. This way, pyrroles **47** – and by extension indoles (vide infra) – can be converted to 3-arylpyridines **51** in up to 85% yield. Moreover, complex structures with various functional groups such as the Atorvastatin derivative **47b** can be transformed to the pyridine **51a** [63], demonstrating the potential of this methodology for late-stage transformations.

Capitalizing on the donor–acceptor relationship of the ring-nitrogen and the ester group, the exocyclic ring-opening of cyclopropanated pyrroles **33** is facile.

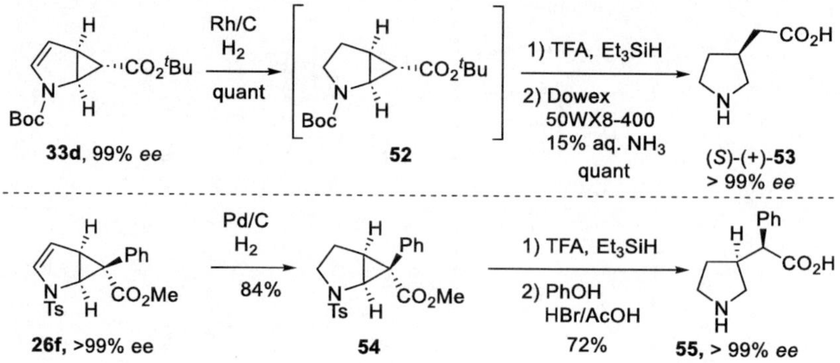

Scheme 13 Synthesis of homo-β-prolines from cyclopropanated pyrroles **33d** and **33f**

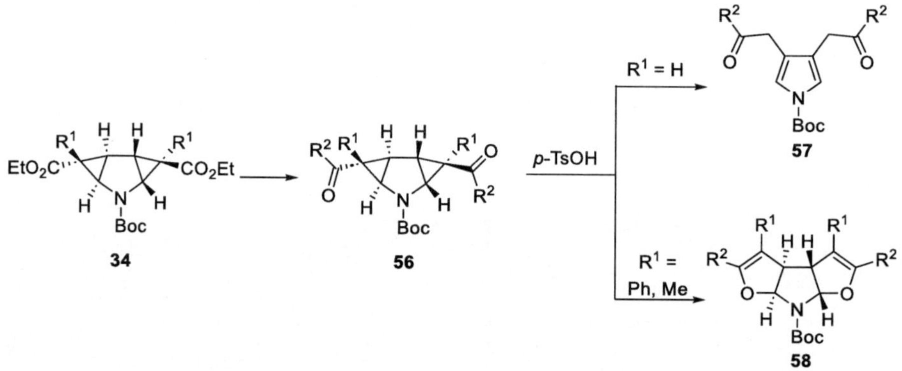

Scheme 14 Rearrangement of twofold cyclopropanated pyrrole **34**

Preparatively most useful are transformations that will trigger rearomatization to a pyrrole and therefore, removing the double bond in **33** typically precedes ring-opening reactions. Thus, simple hydrogenation of **33e** followed by treatment with either trifluoroacetic acid or BF_3 allows the straightforward synthesis of the GABA uptake inhibitor (*S*)-(+)-homo-β-proline **47** (Scheme 13) [52]. A similar sequence starting from **33f** can be applied for the synthesis of homo-β-proline derivatives such as **55** [54].

Werz and coworkers converted the twofold cyclopropanation adducts **34** via their corresponding Weinreb amides to **50**, setting the stage to a divergent acid-catalyzed rearrangement to either 3,4-disubstituted pyrroles **57** or tricyclic *N,O*-acetals **58** depending on the substitution pattern on the cyclopropane moiety (Scheme 14).

The in situ generation and conversion of diazo compounds is a most attractive strategy from an operational point of view, especially with a view for large-scale applications. Sulfonyl hydrazones, being shelf-stable, generally crystalline, and readily available from their corresponding aldehydes or ketones have been recognized as suitable starting materials for this purpose. Calling for a base to initiate the 1,1-elimination to generate the diazo compound must be compatible with conditions

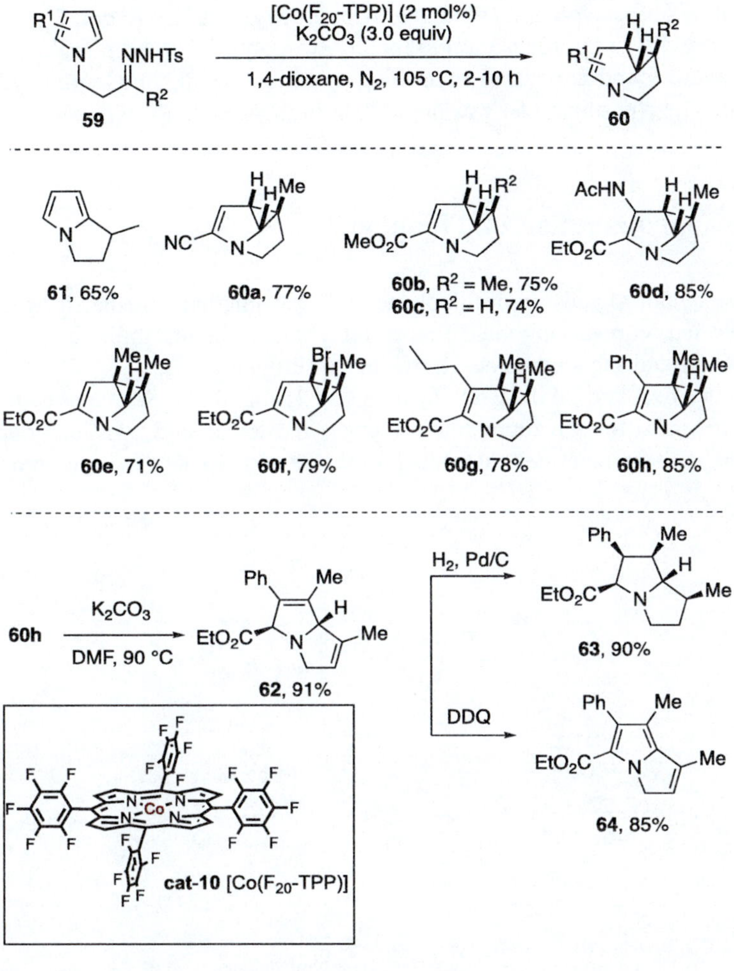

Scheme 15 Co(II) catalyzed intramolecular cyclopropanation of pyrroles with in situ generated diazo compounds from tosylhydrazones.

to trigger its decomposition in a productive fashion subsequently. The combination of an inorganic base such as K_2CO_3 in a polar, non-protic solvent with catalytic amounts of Co(II)-porphyrin complexes is compatible, which has been most elegantly applied in intramolecular cyclopropanations of pyrroles **59** (Scheme 15) [64]. In the presence of the perfluorinated Co(II)-catalyst **cat-10**, the diazo compounds generated from **59** upon elimination with base would give the highly congested cyclopropane adducts **60** in very good yields. An electron-withdrawing substituent in the 2-position of **59** was required to overcome the competing C, H-insertion to **61**. Other catalysts based on copper, ruthenium, iridium, or rhodium gave greatly inferior results. Cyclopropane **60** can be readily converted into heterocyclic scaffolds relevant for medicinal chemistry. For example, the cyclopropane ring in **60h** was cleaved upon treatment with base, giving rise to dihydropyrrolizine **62**. Stereoselective hydrogenation to the *all-exo* polysubstituted pyrrolizidine **63**, a ring structure found in naturally occurring pyrrolizidine alkaloids, or dehydrogenation to the 3*H*-pyrrolizine **64** was possible in high yield.

4 Cyclopropanation of Thiophene

The monocyclopropanation of thiophene with diazomethane mediated by CuCl has been reported with varying yields, most likely due to the instability of the resulting adduct **66a** requiring storage at −80°C under nitrogen [6, 7].

With the discovery of the CNS medicine LY2140023 (**71**, Scheme 16), evaluated in oral treatment for schizophrenia in a phase II clinical study, an important target molecule has been identified that called for the development of the cyclopropanation

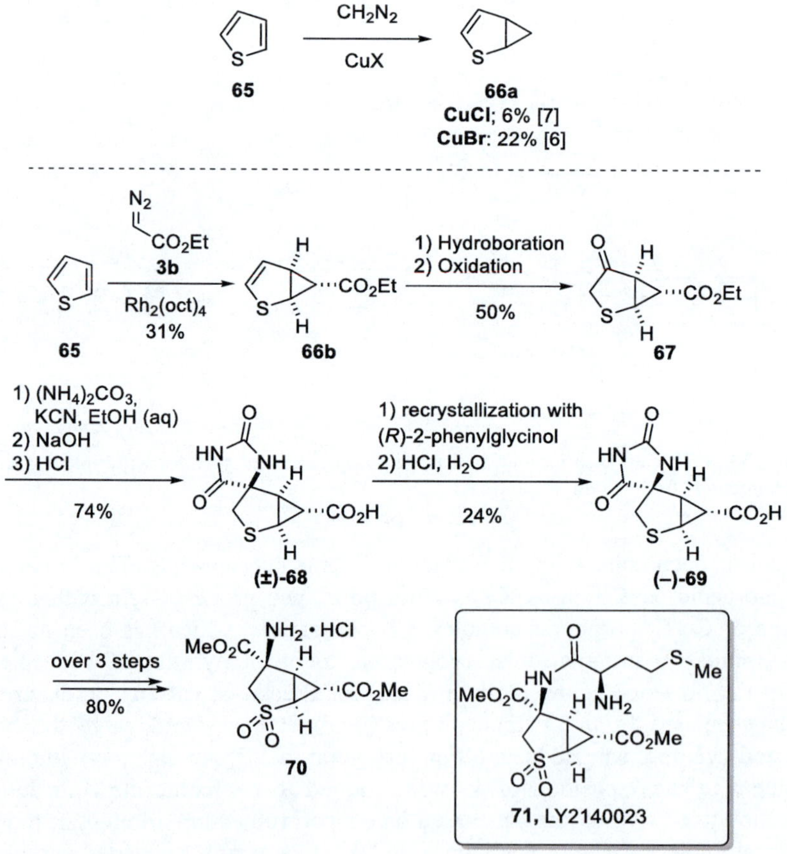

Scheme 16 Synthesis of LY2140023 (**71**)

of thiophene. In contrast to furan and pyrrole, thiophene proved to be a more challenging substrate, probably reflecting that sulfur acts as a catalyst poison for many transition metals and, moreover, has a high tendency to form ylides with carbenes. Indeed, applying copper catalysts does not result in the desired [2 + 1]-cycloaddition, while Rh(II), and especially Rh(II)octanoate, gives rise to **66b** in moderate yield but allows a low catalyst loading of 0.05 mol% [65].

Asymmetric variants employing chiral rhodium(II) catalysts have not succeeded yet. The cyclopropane adduct **66b** is remarkably stable, allowing several subsequent transformations necessary toward the target. A hydroboration/oxidation sequence led to ketone **67**, which was converted to hydantoin (±)-**68** on a 300 g scale. At this stage the separation of the hydantoin enantiomers (±)-**68** by recrystallization with (*R*)-2-phenylglycinol was possible, and the desired enantiomer (−)-**68** could be converted to **69** by oxidation to the sulfone and cleavage hydantoin cleavage. Coupling with methionine completed the synthesis, which could be upscaled to give 22 kg of the final product **71**.

Werz and coworkers disclosed that neither direct twofold cyclopropanation of thiophene nor the sequential process via the monocyclopropanated adduct **66b** was possible. Moreover, an analogous rearrangement, which was successful with cyclopropanated furans (Scheme 6) or pyrroles (Scheme 13), of **66b** to an *O*,*S*-acetal was not possible [47].

5 Cyclopropanation of Benzofuran and Benzothiophene

The overall aromatic stabilization energy of benzofuran (**72**) is similar to that of furan. However, the C-C-double bond in the heterocycle also shows reactivity like that found in an alkene, given that after its functionalization a benzene moiety is retained, thus overall increasing the aromatic character. For example, benzofuran can be formylated in a Vilsmeier–Haack reaction as expected from an electron-rich arene but undergoes addition of bromine rather than substitution. Along the lines discussed for furan, cyclopropanation with diazoacetates is facile under rhodium(II)-catalysis [21, 66, 67] or catalyst-free by initiation of blue-LED light [22], resulting in cyclopropanated adducts **73** in high yields (Scheme 17). The reaction is highly diastereoselective, generally giving rise to the adducts in which the ester group is found on the *exo*-face, but the rhodium(I)-catalyst **cat-11** has been discovered that gives rise to the endo-diastereomer **73b** exclusively [67]. Besides the remarkable switch in diastereoselectivity, it should be noted that in this case a rhodium(I) rather than the typical rhodium(II) catalysts promote this transformation. To render this transformation asymmetric, chiral rhodium catalysts have been investigated, however, **73d** could only be obtained in up to 25%*ee*, greatly contrasting the high enantioselectivities that could be achieved with furans or pyrroles under the same conditions. However, based on wild-type myoglobin (Mb), an iron-heme enzyme complex has been identified which allowed the synthesis of **73c** with perfect diastereo- and enantioselectivity [68]. Besides the spectacular chemical

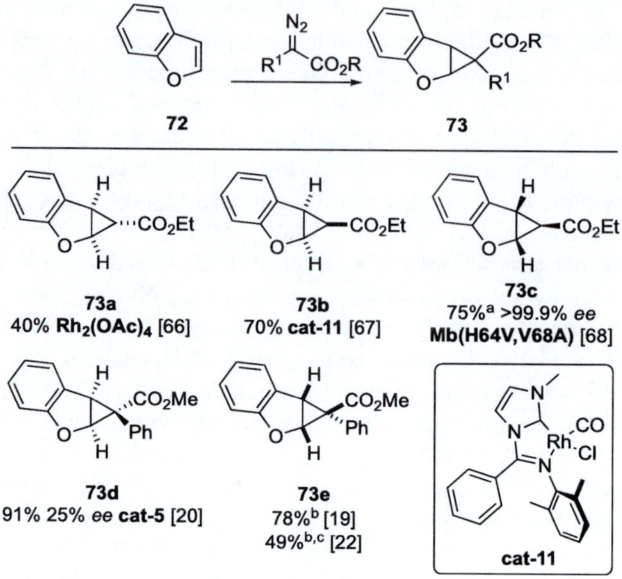

Scheme 17 Cyclopropanation of benzofuran with diazoacetates: [a]Mb(H64V,V68A?), an iron-heme enzyme catalyst. [b]Photochemical decomposition. [c]In continuous flow [20]

transformation achieved, this result is remarkable because neither Fe(TPP) nor hemin catalyzed this reaction, emphasizing the critical interplay of the protein matrix and the protein-embedded iron-porphyrin for the cyclopropanation reaction.

In a conceptually intriguing approach, Echavarren and coworkers demonstrated that gold(I) carbenes can be generated by tandem cyclization and 1,5-alkoxy migration of 1,6-enynes, providing a notable alternative to carbene generation from diazo compounds. The so-generated **75** undergoes cycloaddition to benzofuran or thiobenzofuran, giving rise to **79** or **80** with excellent relative, notably *endo*-stereocontrol (Scheme 18). Extending this transformation to indole does not lead to the corresponding cyclopropane **81** but rather to the C-3-alkylation adduct **82** [69].

6 Cyclopropanation of Indoles

One of the most prominent scaffolds in medicinal chemistry is the indole moiety, being found in numerous natural products and drugs. Therefore, its functionalization by adding carbenes is promising to generate diverse functionalized derivatives. In contrast to pyrrole (see Sect. 3), in indole the lone pair of nitrogen is conjugated to an aromatic ring, thus making it less nucleophilic. Consequently, cyclopropanation of indole does not require an electron-withdrawing protecting group but is also possible with *N*-alkylated or even with the unprotected compound; however, in the latter

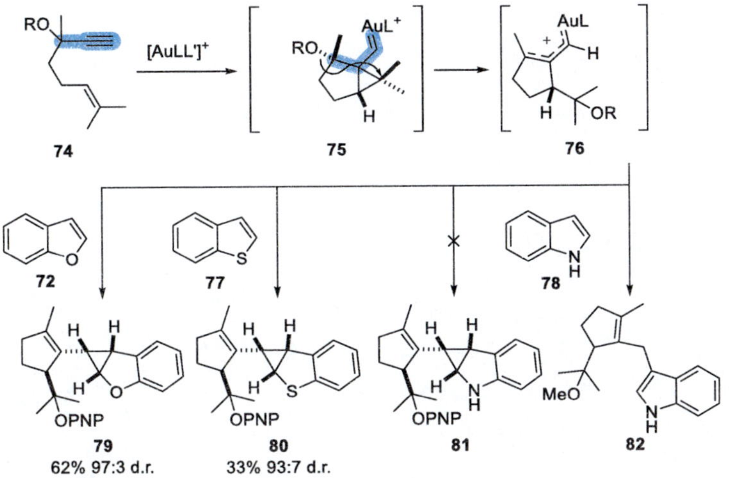

Scheme 18 Cyclopropanation of benzofuran and thiobenzofuran by gold(I) carbenes

cases, further ring-opening is challenging to prevent. This can be turned into an advantage by designing tandem cyclopropanation/ring-opening sequences as was elegantly applied to the synthesis of 3-aryl substituted quinolines from indoles [63] (analogously to the synthesis of **51**, cf. Scheme 14), quinoline-3-carboxylates **84** [70] or **86** [71] from **78** or **85** (Scheme 19). Noteworthy, by employing halodiazoacetates rather than cleaving the exocyclic cyclopropane bond in **83,** being in a donor–acceptor relationship between the indole nitrogen and the ester group, the endocyclic bond cleaves with bromide as the leaving group. This strategy was applied for the efficient synthesis of norfloxacin hydrochloride from **86b**, being a potent broad-spectrum quinolone antibiotic.

Substituted indoles can be synthesized from N-methylindole (**88**) via the Cu-catalyzed cyclopropanation with ethyl diazoacetate (Scheme 20). While the cyclopropanated adduct **89** is stable enough to confirm its formation by NMR, upon treatment with silica exocyclic opening along the donor–acceptor cyclopropane bond with concurrent rearomatization takes place, giving rise to **90** in excellent yield [72]. Alternatively, regioselective cuprate addition onto C-2 of **89** is possible, giving rise to disubstituted indolines **91**, an essential structural element found in plant growth hormones [73–75] and herbicides [76].

With the synthesis of (±)-Communesin F, belonging to a family of polycyclic indole alkaloids isolated from the marine fungal strain of *Penicillium* sp. displaying significant cytotoxicity [77], a most impressive application of this strategy was demonstrated (Scheme 21). Starting from the diazoacetate **92**, the intramolecular cyclopropanation catalyzed by low quantities of Cu(OTf) led to the stable adduct **93**, however, with low diastereoselectivity. Reduction of the azide to an amine triggered the nucleophilic ring-opening of the cyclopropane ring and kinetic protonation of the enolate intermediate. This way, **94** was obtained as a single diastereomer in high yield, which was further converted to the target compound **95** [78].

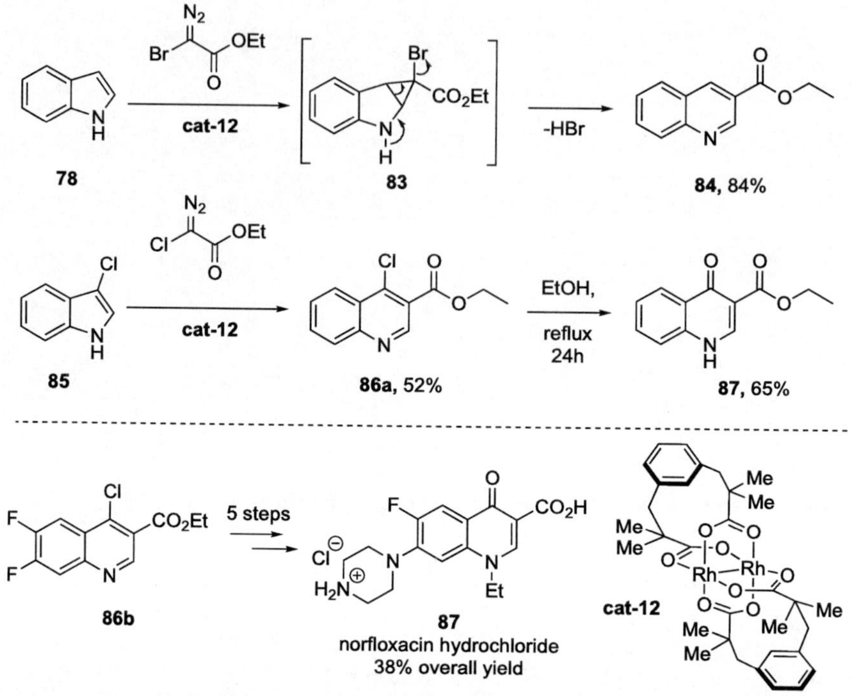

Scheme 19 Synthesis of quinolines and quinolones by tandem cyclopropanation/ring-opening of indoles

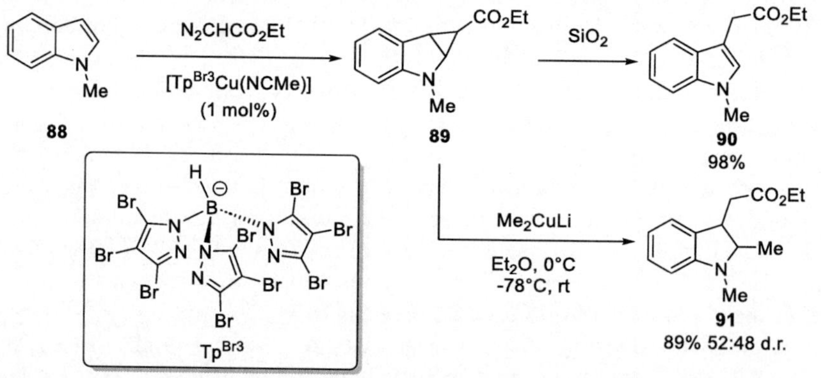

Scheme 20 Synthesis of substituted indoles and indolines by cyclopropanation of *N*-methylindole (**88**)

Although not strictly necessary, the cyclopropanation works exceptionally well with indoles being protected on nitrogen with electron-withdrawing substituents, presumably taming the nucleophilicity of nitrogen to trigger ring-oping reactions.

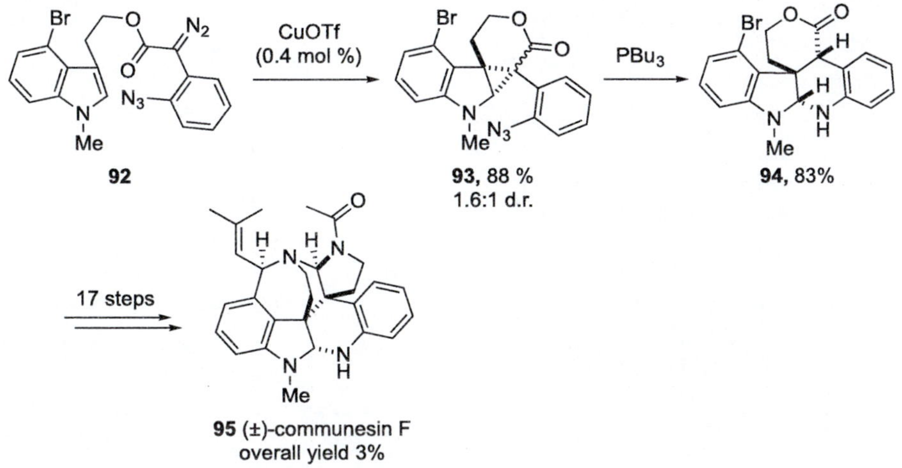

Scheme 21 Synthesis of (±)-Communesin F featuring the intramolecular cyclopropanation of **92**

Wenkert and coworkers reported that indoles **96a–c** undergo cyclopropanation with ethyl diazoacetate upon catalysis with copper bronze in moderate yields (Scheme 22). Notably, in the course of these studies, it was discovered that besides the expected *exo*-diastereomers **97a–c**, also the *endo*-diastereomers **98a–c** are formed, contrasting the exclusive formation of *exo*-diastereomers in the similar reactions with furans (cf Sect. 2) or pyrroles (cf Sect. 3) [79]. Using the combination of $Cu(OTf)_2/PhNHNH_2$ and methyl diazoacetate, the cyclopropanated adducts of *N*-Boc (Boc = *tert*-butyloxycarbonyl) protected indoles **97d/98d** and **97e/98e** were obtained with improved yields, especially, when a large excess of the diazo compound was used [80]. Again, the adducts formed as a 2–3:1 *exo/endo* mixture, which could be readily separated. Subsequent oxidative degradation revealed a very different behavior of the diastereomers: while ozonolysis of **97e** led to the diketone **99**, which was subsequently converted to the cyclopropyl amino acid **100**, the ozonolysis of **98e** stopped at the partially oxidized alkene **101**. Aryl-substituted diazoacetates under blue light irradiation have also been shown to generate suitable carbene precursors for the addition to indole **96f**, giving rise to cyclopropane **97f** with exclusive *exo*-selectivity with respect to the ester substituent [22]. Alternatively, the thermal decomposition of diazo carbonyl compounds in the presence of nickel catalyst **cat-6** proved to be amenable for the cyclopropanation of indoles and, notably, in this variant the products such as **97g** were obtained with excellent *exo*-selectivity.

Using a carbohydrate-modified bis(oxazoline) **ligand-1** in combination with Cu (OTf), the cyclopropanation of indole **96f** with ethyl diazoacetate could be rendered asymmetric, giving rise to **97d** in good yields and appreciable enantioselectivity (Scheme 23). Likewise, cyclopropanation of **96g** followed by acid-catalyzed ring-opening led to **102**, which was further converted to the anticholinergic drug (–)-desoxyeseroline (**104**). As key step, the ring-opening/cyclization to **103**,

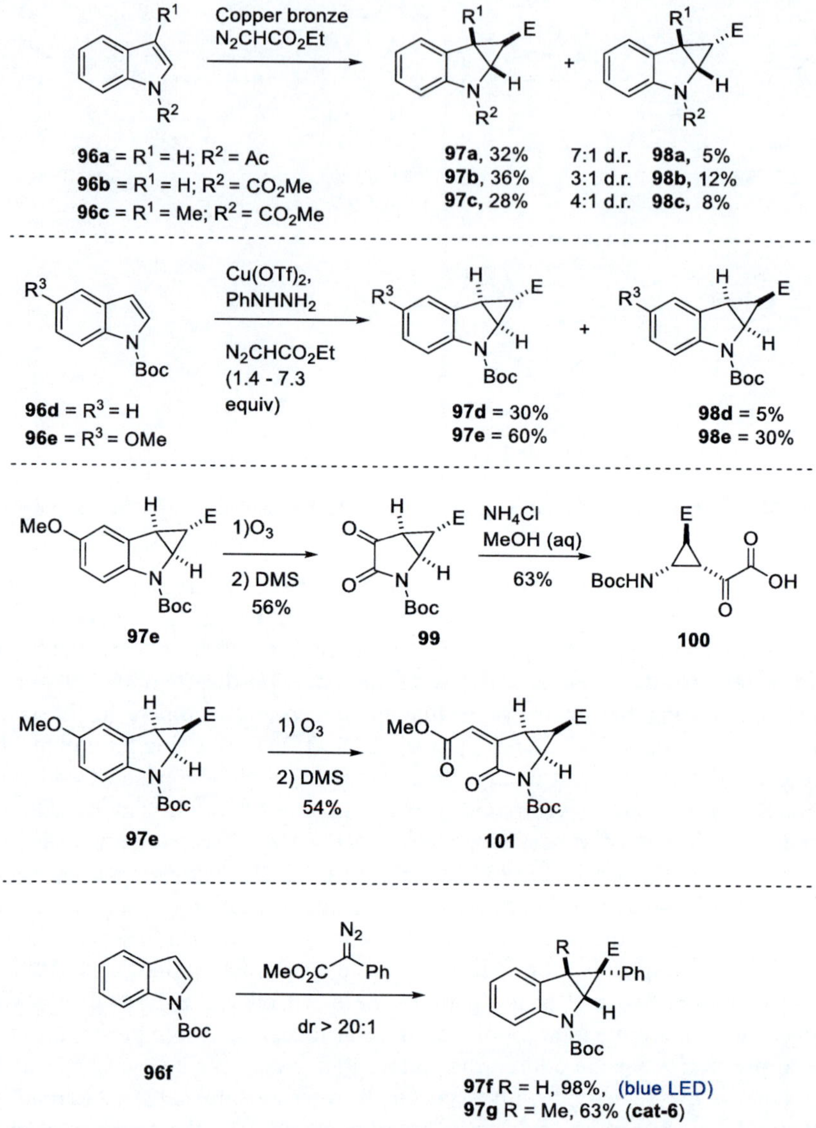

Scheme 22 Cyclopropanation of *N*-acetyl and *N*-carbamate protected indoles. E = CO_2Me

representing overall a [3 + 2]-heteroannulation onto indole, gave facile access to the core structure **104** [81]. Conceptually related, the reaction of vinyl carbenes derived from the corresponding vinyldiazoacetates with indole leads to its carbocyclic [3 + 2]-annulation [82–84], however, in these cases vinylcyclopropanes are not implicated as intermediates, contrasting the reaction course that was found for pyrroles (cf. Scheme 7).

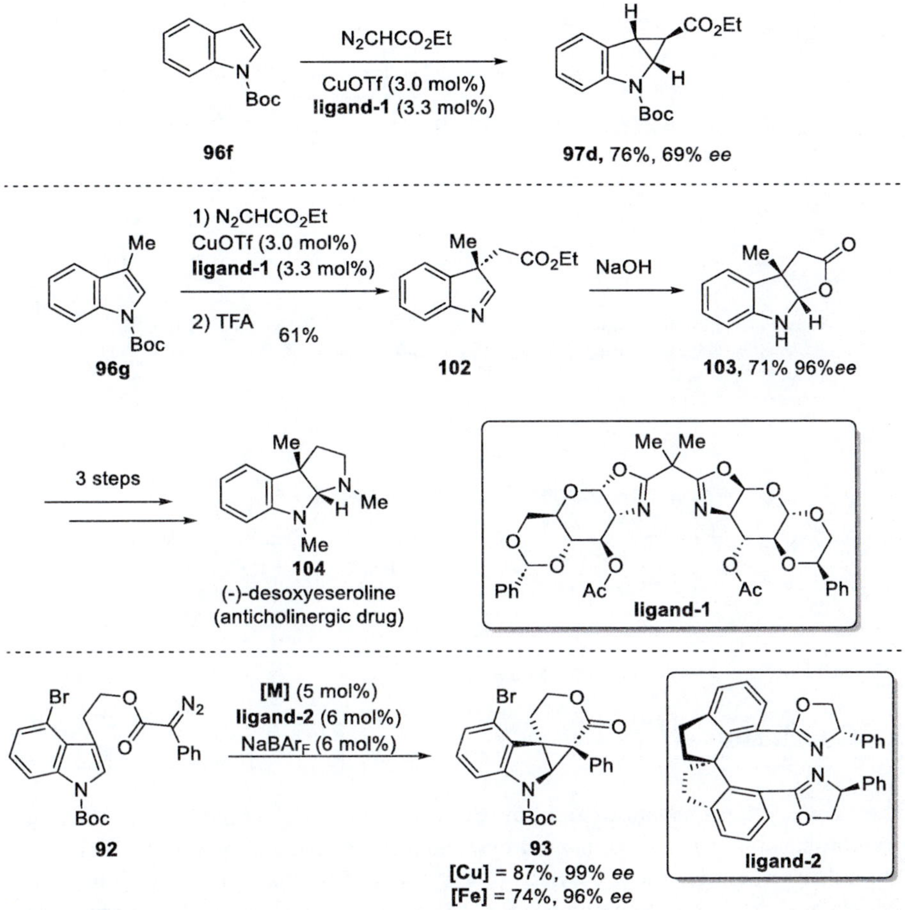

Scheme 23 Asymmetric cyclopropanation of indoles

In a related approach as described for the synthesis of (±)-Communesin F (Scheme 19), the intramolecular cyclopropanation of **92** using bis(oxazoline ligand) **ligand-1** proceeded with very high diastereo- and enantioselectivity. Notably, besides Cu(I), Fe(II) was also found to be suitable in this reaction [85].

Along the lines discussed for pyrrole derivatives (cf. Scheme 15), the Co(II)-catalyzed intramolecular cyclopropanation of indoles has been achieved [59]. The transformation shows a remarkable scope, tolerating a wide range of electronic steric variations of the starting materials **105**. Ring-opening of the strained cyclopropane ring in **106** is facile, e.g. by photoredox-catalyzed oxidation, presumably via the amino radical **107,** allows the synthesis of 2,3-dihydro-1*H*-pyrrolo[1,2a]indole core **108**, which is widely found in bioactive molecules (Scheme 24).

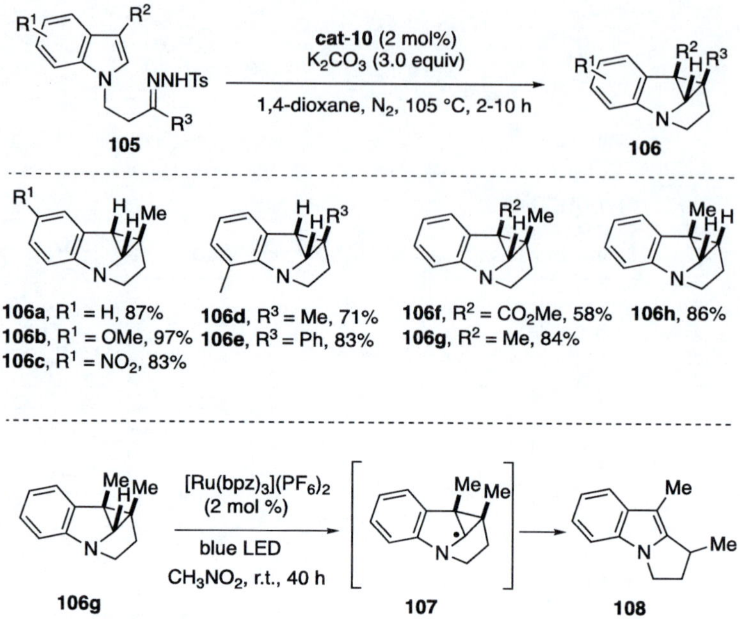

Scheme 24 Intramolecular cyclopropanation of indoles

7 Conclusion

The monocyclopropanation and, in extension but less explored, the twofold cyclopropanation of five-membered heteroarenes has been exploited to synthesize a broad variety of heterocycles. Such [2 + 1]-cycloadditions go along with breaking the aromaticity of the heteroarene, for which metal-catalyzed and photomediated variants have been developed, mainly using of diazocarbonyl compounds as carbene precursors. The resulting cycloadducts have a distinct curved shape, making subsequent functionalization of this core highly stereoselective from the convex face (Scheme 24). In combination with enantioselective variants of the cyclopropanation, it is possible in this way to assemble 5 contiguous stereocenters with high diastereo- and enantiocontrol in just two steps. The exocyclic cyclopropane bonds can be readily cleaved utilizing the donor–acceptor relationship (R^3 or R^4 = ketone, ester), but can also be triggered by oxidation of the ring-heteroatom. In turn, the endocyclic cyclopropane bond can be opened thermally in an electrocyclic ring-opening, by generating a positive charge in the β-position of the five-membered ring (cyclopropylcarbinyl rearrangement) or an organopalladium intermediate that triggers a carbon elimination (Scheme 25).

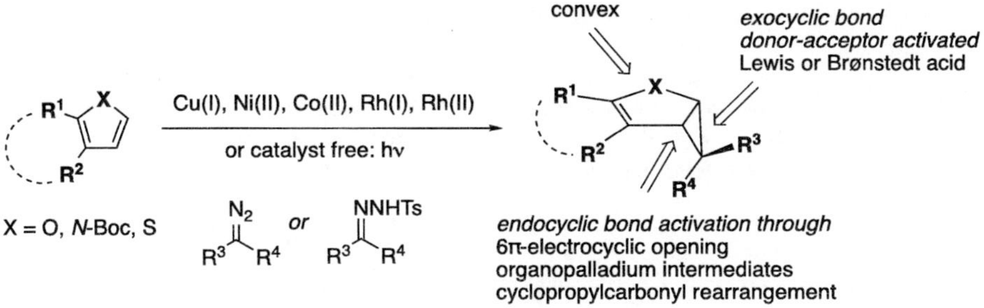

Scheme 25 Cyclopropanated heteroarenes as building blocks for heterocycles

References

1. Zhussupova GE, Zhussupova AI (2015) Impact of "organic chemistry" course on subsequent courses, studying biologically active compounds. Procedia Soc Behav Sci 191:1247–1254
2. Dewar JS, Harget AJ, Trinajstić N, Worley SD (1970) Ground states of conjugated molecules–XXI. Tetrahedron 26:4505–4516
3. Novak J, Sorm F (1957) Chem Listy:1693
4. Schenck GO, Steinmetz R (1958) Strahlenchemische umsetzungen mit diazoessigester. Angew Chem 70:504–505
5. Schenck GO, Steinmetz R (1963) Photochemische bildungsweisen und umlagerungen von thiopheno- und furano-cyclopropancarbonsäureestern.cis,trans-muconaldehydsäureester durch valenzisomerisierung von 2-oxa-bicyclo[3.1.0]hexen-(3)-carbonsäure-(6)-ester!! Eur J Org Chem 668:19–30
6. Müller E, Kessler H, Fricke E, Suhr H (1963) Homologisierung von 5-ringaromaten mit diazomethan. Tetrahedron Lett 4:1047–1049
7. Herges R, Ugi I (1986) Improved preparation of 2-oxabicyclo[3.1.0]hex-3-ene (homofuran) and 2-thiabicyclo[3.1.0]hex-3-ene (homothiophene). Synthesis 1986:1059
8. Wenkert E, Khatuya H, Klein PS (1999) Reactions of ethyl diazoacetate with β-methylfurans. Tetrahedron Lett 40:5171–5174
9. Wenkert E, Khatuya H (1999) The effect of substituents of α-alkyl sidechains on furan-diazoester interactions. Tetrahedron Lett 40:5439–5442
10. Wenkert E, Rodolfo Lavilla MG, Poter B, Ramachandran K, Sheu J-H (1990) Polyene synthesis. Ready construction of retinol-carotene fragments,(.+–.)-6 (E)-LTB3 leukotrienes, and corticrocin. J Org Chem 55:6203–6214
11. Evans DA, Woerpel KA, Hinman MM, Faul MM (1991) Bis(oxazolines) as chiral ligands in metal-catalyzed asymmetric reactions. Catalytic, asymmetric cyclopropanation of olefins. J Am Chem Soc 113:726–728
12. Evans DA, Peterson GS, Johnson JS, Barnes DM, Campos KR, Woerpel KA (1998) An improved procedure for the preparation of 2,2-bis[2-[4(*S*)-*tert*-butyl-1,3-oxazolinyl]]propane [(*S,S*)-*tert*-butylbis(oxazoline)] and derived copper(II) complexes. J Org Chem 63:4541–4544
13. Werner H, Vicha R, Gissibl A, Reiser O (2003) Improved synthesis of aza-bis(oxazoline) ligands. J Org Chem 68:10166–10168
14. Harrar K, Reiser O (2012) Enantioselective synthesis of (−)-paeonilide. Chem Commun (Camb) 48:3457–3459
15. Werner H, Herrerías CI, Glos M, Gissibl A, Fraile JM, Pérez I, Mayoral JA, Reiser O (2006) Synthesis of polymer bound azabis(oxazoline) ligands and their application in asymmetric cyclopropanations. Adv Synth Catal 348:125–132
16. Chhor RB, Nosse B, Sörgel S, Böhm C, Seitz M, Reiser O (2003) Enantioselective synthesis of paraconic acids. Chemistry 9:260–270

17. Schinnerl M, Böhm C, Seitz M, Reiser O (2003) New bis(oxazoline) ligands with secondary binding sites for the asymmetric cyclopropanation of furans. Tetrahedron Asymmetry 14:765–771
18. Nag E, Gorantla SMNVT, Arumugam S, Kulkarni A, Mondal KC, Roy S (2020) Tridentate nickel(II)-catalyzed chemodivergent C-H functionalization and cyclopropanation: regioselective and diastereoselective access to substituted aromatic heterocycles. Org Lett 22: 6313–6318
19. Lehner V, Davies HML, Reiser O (2017) Rh(II)-catalyzed cyclopropanation of furans and its application to the total synthesis of natural product derivatives. Org Lett 19:4722–4725
20. Jurberg ID, Davies HML (2018) Blue light-promoted photolysis of aryldiazoacetates. Chem Sci 9:5112–5118
21. Castano B, Gallo E, Cole-Hamilton DJ, Dal Santo V, Psaro R, Caselli A (2014) Continuous flow asymmetric cyclopropanation reactions using Cu(I) complexes of Pc-L* ligands supported on silica as catalysts with carbon dioxide as a carrier. Green Chem 16:3202–3209
22. Klöpfer V, Eckl R, Floß J, Roth PMC, Reiser O, Barham JP (2021) Catalyst-free, scalable heterocyclic flow photocyclopropanation. Green Chem 23:6366–6372
23. Schneider TF, Kaschel J, Werz DB (2014) A new golden age for donor-acceptor cyclopropanes. Angew Chem Int Ed 53:5504–5523
24. Werz DB, Biju AT (2020) Uncovering the neglected similarities of arynes and donor-acceptor cyclopropanes. Angew Chem Int Ed 59:3385–3398
25. Reissig H-U, Zimmer R (2003) Donor-acceptor-substituted cyclopropane derivatives and their application in organic synthesis. Chem Rev 103:1151–1196
26. Garnsey MR, Slutskyy Y, Jamison CR, Zhao P, Lee J, Rhee YH, Overman LE (2018) Short enantioselective total syntheses of cheloviolenes A and B and dendrillolide C *via* convergent fragment coupling using a tertiary carbon radical. J Org Chem 83:6958–6976
27. Brady TP, Kim SH, Wen K, Theodorakis EA (2004) Stereoselective total synthesis of (+)-norrisolide. Angew Chem Int Ed 43:739–742
28. Budde S, Goerdeler F, Floß J, Kreitmeier P, Hicks EF, Moscovitz O, Seeberger PH, Davies HML, Reiser O (2020) Visible-light mediated oxidative ring expansion of anellated cyclopropanes to fused endoperoxides with antimalarial activity. Org Chem Front 7:1789–1795
29. Bandichhor R, Nosse B, Reiser O (2005) Paraconic acids – the natural products from lichen symbiont. In: Mulzer J (ed) Natural product synthesis I, vol 243. Springer, Berlin, pp 43–72
30. Bergmann A, Reiser O (2014) Enantioselective synthesis of xanthatin. Chemistry 20:7613–7615
31. Kalidindi S, Jeong WB, Schall A, Bandichhor R, Nosse B, Reiser O (2007) Enantioselective synthesis of arglabin. Angew Chem Int Ed 46:6361–6363
32. Yedoyan J, Wurzer N, Klimczak U, Ertl T, Reiser O (2019) Regio- and stereoselective synthesis of functionalized dihydropyridines, pyridines, and 2*H*-pyrans: Heck coupling of monocyclopropanated heterocycles. Angew Chem Int Ed 58:3594–3598
33. Wurzer N, Klimczak U, Babl T, Fischer S, Angnes RA, Rehbein J, Reiser O (2021) Heck-type coupling of fused bicyclic vinylcyclopropanes: Synthesis of 1,2-dihydropyridines, 2,3-dihydro-1*H*-azepines, 1,4-cyclohexadienes, 2*H*-pyrans, and 1,3-butadienes
34. Eckl R, Fischer S, Sonnleitner CM, Schmidhuber D, Rehbein J, Reiser O (2022) Stereoselective synthesis of biologically relevant tetrahydropyridines and dihydro-2*H*-pyrans *via* ring-expansion of monocyclopropanated heterocycles. ACS Org Inorg Au 2:169–174
35. Sonnleitner CM, Park S, Eckl R, Ertl T, Reiser O (2020) Stereoselective synthesis of tropanes *via* a 6π-electrocyclic ring-opening/Huisgen 3+2-cycloaddition cascade of monocyclopropanated heterocycles. Angew Chem Int Ed 59:18110–18115
36. Davies HM (1993) Tandem cyclopropanation/Cope rearrangement: a general method for the construction of seven-membered rings. Tetrahedron 49:5203–5223
37. Davies HM, Clark DM, Alligood DB, Eiband GR (1987) Mechanistic aspects of formal [3 + 4] cycloadditions between vinylcarbenoids and furans. Tetrahedron 43:4265–4270

38. Davies HML, Ahmed G, Churchill MR (1996) Asymmetric synthesis of highly functionalized 8-oxabicyclo[3.2.1]octene derivatives. J Am Chem Soc 118:10774–10782
39. Davies HML, McAfee MJ, Oldenburg CEM (1989) Scope and stereochemistry of the tandem intramolecular cyclopropanation/Cope rearrangement sequence. J Org Chem 54:930–936
40. Davies HM, Clark DM, Smith TK (1985) [3 + 4] cycloaddition reactions of vinyl carbenoids with furans. Tetrahedron Lett 26:5659–5662
41. Xu J, Caro-Diaz EJE, Theodorakis EA (2010) Enantioselective formal synthesis of (−)-englerin A *via* a Rh-catalyzed 4 + 3 cycloaddition reaction. Org Lett 12:3708–3711
42. Xu J, Caro-Diaz EJE, Batova A, Sullivan SDE, Theodorakis EA (2012) Formal synthesis of (−)-englerin A and cytotoxicity studies of truncated englerins. Chem Asian J 7:1052–1060
43. Jackson KL, Henderson JA, Motoyoshi H, Phillips AJ (2009) A total synthesis of norhalichondrin B. Angew Chem 121:2382–2386
44. Wenkert E, Khatuya H (1998) Facile synthesis of furan-3,4-diacetates. Helv Chim Acta 81: 2370–2374
45. Schneider TF, Kaschel J, Dittrich B, Werz DB (2009) Anti-oligoannelated THF moieties: synthesis *via* push-pull-substituted cyclopropanes. Org Lett 11:2317–2320
46. Schneider TF, Kaschel J, Awan SI, Dittrich B, Werz DB (2010) From furan to molecular stairs: syntheses, structural properties, and theoretical investigations of oligocyclic oligoacetals. Chemistry 16:11276–11288
47. Kaschel J, Schneider TF, Schirmer P, Maaß C, Stalke D, Werz DB (2013) Rearrangements of furan-, thiophene- and *N*-Boc-pyrrole-derived donor-acceptor cyclopropanes: scope and limitations. Eur J Org Chem 2013:4539–4551
48. Fowler FW (1971) Synthesis and thermolysis of *N*-methoxycarbonyl-2,3-homopyrrole. Angew Chem Int Ed 10:135
49. Fowler FW (1969) The 2,3-homopyrrole ring system. J Chem Soc D:1358–1359
50. Rees CW, Smithen CE (1964) The reactions of heterocyclic compounds with carbenes. In: Advances in heterocyclic chemistry, vol 3. Elsevier, pp 57–78
51. Beumer R, Bubert C, Cabrele C, Vielhauer O, Pietzsch M, Reiser O (2000) The synthesis of diastereo- and enantiomerically pure beta-aminocyclopropanecarboxylic acids. J Org Chem 65: 8960–8969
52. Pilsl LKA, Ertl T, Reiser O (2017) Enantioselective three-step synthesis of homo-β-proline: a donor-acceptor cyclopropane as key intermediate. Org Lett 19:2754–2757
53. Gheorghe A, Schulte M, Reiser O (2006) Synthesis of functionalized pyrrolidin-2-ones and (*S*)-Vigabatrin from pyrrole. J Org Chem 71:2173–2176
54. Fu J, Wurzer N, Lehner V, Reiser O, Davies HML (2019) Rh(II)-catalyzed monocyclopropanation of pyrroles and its application to the synthesis pharmaceutically relevant compounds. Org Lett 21:6102–6106
55. Davies HM, Young WB, Smith H (1989) Novel entry to the tropane system by reaction of rhodium(II) acetate stabilized vinylcarbenoids with pyrroles. Tetrahedron Lett 30:4653–4656
56. Reddy RP, Davies HML (2007) Asymmetric synthesis of tropanes by rhodium-catalyzed 4 + 3 cycloaddition. J Am Chem Soc 129:10312–10313
57. Hedley SJ, Ventura DL, Dominiak PM, Nygren CL, Davies HML (2006) Investigation into factors influencing stereoselectivity in the reactions of heterocycles with donor-acceptor-substituted rhodium carbenoids. J Org Chem 71:5349–5356
58. Fowler FW (1971) Cycloadditions ofN-Methoxycarbonyl-2,3-homopyrrole. Angew Chem Int Ed 10:135–136
59. Ciamician GL, Dennstedt M (1881) Ueber die einwirkung des chloroforms auf die kaliumverbindung pyrrols. Ber Dtsch Chem Ges 14:1153–1163
60. Dhanak D, Reese CB (1987) Synthesis of [6](2,4)pyridinophanes. J Chem Soc Perkin Trans 1: 2829
61. Raheem IT, Thiara PS, Jacobsen EN (2008) Regio- and enantioselective catalytic cyclization of pyrroles onto N-acyliminium ions. Org Lett 10:1577–1580

62. Ma D, Martin BS, Gallagher KS, Saito T, Dai M (2021) One-carbon insertion and polarity inversion enabled a pyrrole strategy to the total syntheses of pyridine-containing lycopodium alkaloids: complanadine a and lycodine. J Am Chem Soc 143:16383–16387
63. Dherange BD, Kelly PQ, Liles JP, Sigman MS, Levin MD (2021) Carbon atom insertion into pyrroles and indoles promoted by chlorodiazirines. J Am Chem Soc 143:11337–11344
64. Reddy AR, Hao F, Wu K, Zhou C-Y, Che C-M (2016) Cobalt(II) porphyrin-catalyzed intramolecular cyclopropanation of *N*-alkyl indoles/pyrroles with alkylcarbene: efficient synthesis of polycyclic *N*-heterocycles. Angew Chem Int Ed 55:1810–1815
65. Waser M, Moher ED, Borders SSK, Hansen MM, Hoard DW, Laurila ME, LeTourneau ME, Miller RD, Phillips ML, Sullivan KA, Ward JA, Xie C, Bye CA, Leitner T, Herzog-Krimbacher B, Kordian M, Müllner M (2011) Process development for a key synthetic intermediate of LY2140023, a clinical candidate for the treatment of schizophrenia. Org Process Res Dev 15:1266–1274
66. Matlin SA, Chan L, Catherwood B (1990) 6-Diazopenicillanates. Part 1. Reactions with furans. J Chem Soc Perkin Trans 1:89
67. Rosenberg ML, Vlašaná K, Sen Gupta N, Wragg D, Tilset M (2011) Highly cis-selective Rh(I)-catalyzed cyclopropanation reactions. J Org Chem 76:2465–2470
68. Vargas DA, Khade RL, Zhang Y, Fasan R (2019) Biocatalytic strategy for highly diastereo- and enantioselective synthesis of 2,3-dihydrobenzofuran-based tricyclic scaffolds. Angew Chem Int Ed 58:10148–10152
69. Calleja P, Pablo Ó, Ranieri B, Gaydou M, Pitaval A, Moreno M, Raducan M, Echavarren AM (2016) α,β-Unsaturated gold(I) carbenes by tandem cyclization and 1,5-alkoxy migration of 1,6-enynes: mechanisms and applications. Chemistry 22:13613–13618
70. Mortén M, Hennum M, Bonge-Hansen T (2015) Synthesis of quinoline-3-carboxylates by a Rh (II)-catalyzed cyclopropanation-ring expansion reaction of indoles with halodiazoacetates. Beilstein J Org Chem 11:1944–1949
71. Peeters S, Berntsen LN, Rongved P, Bonge-Hansen T (2019) Cyclopropanation-ring expansion of 3-chloroindoles with α-halodiazoacetates: novel synthesis of 4-quinolone-3-carboxylic acid and norfloxacin. Beilstein J Org Chem 15:2156–2160
72. Delgado-Rebollo M, Prieto A, Pérez PJ (2014) Catalytic functionalization of indoles by copper-mediated carbene transfer. ChemCatChem 6:2047–2052
73. Teale WD, Paponov IA, Palme K (2006) Auxin in action: signalling, transport and the control of plant growth and development. Nat Rev Mol Cell Biol 7:847–859
74. Duca DR, Glick BR (2020) Indole-3-acetic acid biosynthesis and its regulation in plant-associated bacteria. Appl Microbiol Biotechnol 104:8607–8619
75. Chen D, Chen Y, Ma Z, Zou L, Li J, Liu Y (2018) One-pot synthesis of indole-3-acetic acid derivatives through the cascade Tsuji-Trost reaction and Heck coupling. J Org Chem 83:6805–6814
76. Menon N, Richmond D, Rahman MR, Menon BRK (2022) Versatile and facile one-pot biosynthesis for amides and carboxylic acids in E. coli by engineering auxin pathways of plant microbiomes. ACS Catal 12:2309–2319
77. Numata A, Takahashi C, Ito Y, Takada T, Kawai K, Usami Y, Matsumura E, Imachi M, Ito T, Hasegawa T (1993) Communesins, cytotoxic metabolites of a fungus isolated from a marine alga. Tetrahedron Lett 34:2355–2358
78. Yang J, Wu H, Shen L, Qin Y (2007) Total synthesis of (+/−)-communesin F. J Am Chem Soc 129:13794–13795
79. Wenkert E, Alonso ME, Gottlieb HE, Sanchez EL, Pellicciari R, Cogolli P (1977) Reactions of ethyl diazoacetate with thianaphthene, indoles, and benzofuran. J Org Chem 42:3945–3949
80. Gnad F, Poleschak M, Reiser O (2004) Stereoselective synthesis of novel conformationally restricted β- and γ-amino acids. Tetrahedron Lett 45:4277–4280
81. Özüduru G, Schubach T, Boysen MMK (2012) Enantioselective cyclopropanation of indoles: construction of all-carbon quaternary stereocenters. Org Lett 14:4990–4993

82. Barluenga J, Tudela E, Ballesteros A, Tomás M (2009) Asymmetric C2-C3 cyclopentannulation of the indole ring. J Am Chem Soc 131:2096–2097
83. Lian Y, Davies HML (2010) Rhodium-catalyzed 3 + 2 annulation of indoles. J Am Chem Soc 132:440–441
84. Jing C, Cheng Q-Q, Deng Y, Arman H, Doyle MP (2016) Highly regio- and enantioselective formal 3 + 2-annulation of indoles with electrophilic enol carbene intermediates. Org Lett 18: 4550–4553
85. Xu H, Li Y-P, Cai Y, Wang G-P, Zhu S-F, Zhou QL (2017) Highly enantioselective copper- and iron-catalyzed intramolecular cyclopropanation of indoles. J Am Chem Soc 139:7697–7700

Heterocycles from Cycloaddition Reactions

Kostiantyn O. Marichev

Contents

Abstract The synthesis of heterocyclic compounds via catalytic cycloaddition reactions of metallocarbenes is robust, atom-economic, and often allows access to chiral heterocycles with high levels of stereocontrol. This methodology has demonstrated advantages over other metal- and organo-catalytic approaches towards heterocycles (e.g., lower catalyst loading, use of inexpensive metal catalysts, mild reaction conditions, excellent stereoselectivities, etc.). This chapter is comprehensive, and covers all available literature reports on cycloaddition using metallocarbenes to produce different ring size heterocycles: from small strained tri-membered to medium-size eight-membered rings.

Keywords Asymmetric reactions · Catalysis · Cycloaddition · Diazo compounds · Donor-acceptor cyclopropenes · Metal carbenes

K. O. Marichev (✉)
Department of Chemistry, Georgia State University, Atlanta, GA, USA
e-mail: kmarichev@gsu.edu

1 Introduction

Since the discovery of the Diels–Alder reaction, cycloaddition reactions have been regarded as one of the most versatile and useful chemical reactions for the construction of carbo- and heterocyclic ring systems [1, 2]. *Cycloaddition* is defined as a stepwise or concerted "reaction in which two or more unsaturated molecules (or parts of the same molecule) combine with the formation of a cyclic adduct in which there is a net reduction of the bond multiplicity" [3]. Among various efficient chemical processes, the central importance of cycloaddition for the synthesis of heterocyclic compounds, which exhibit a wide range of pharmaceutical properties [4, 5], is in its stereospecificity, ability to generate complex molecules under mild reaction conditions and with atom economy.

Metallocarbenes [6] are reactive compounds or intermediates that undergo cycloaddition reactions with a variety of dipolar species, and they provide convenient access to different sized ring heterocyclic compounds often with high levels of stereocontrol via asymmetric catalysis [7–9]. The resulting cycloaddition products obtained from metallocarbenes contain at least one carbon atom in a heterocyclic ring. This chapter is focused on advances in the synthesis of different sized ring heterocycles using cycloaddition reactions with various metal carbene or carbenoid surrogates. The reaction scope includes only cycloadditions that form new heterocyclic rings. Literature reports on derivatizations of already existing heterocyclic rings will not be part of this chapter. Furthermore, cycloaddition reactions of metallocarbenes which form carbocyclic compounds will not be discussed.

In these cycloaddition reactions, which occur via a stepwise mechanism, metallocarbenes or carbenoids can act as "one-atom" ([*n* + 1]-), "two-atom" ([*n* + 2]-), or "three-atom" ([*n* + 3]-cycloaddition) surrogates. The mechanism for [*n* + 1]- and [*n* + 2]-cycloaddition generally involves an initial nucleophilic attack onto the "carbenic" carbon followed by ring closure (Scheme 1); however, the

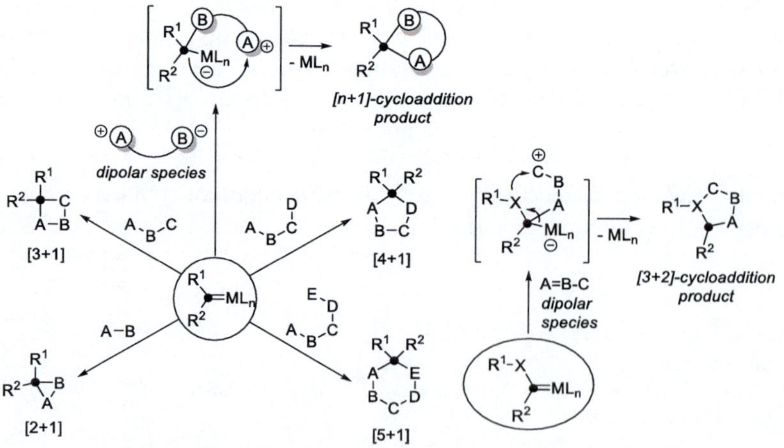

Scheme 1 [*n* + 1]- and [*n* + 2]-Cycloaddition with metallocarbenes

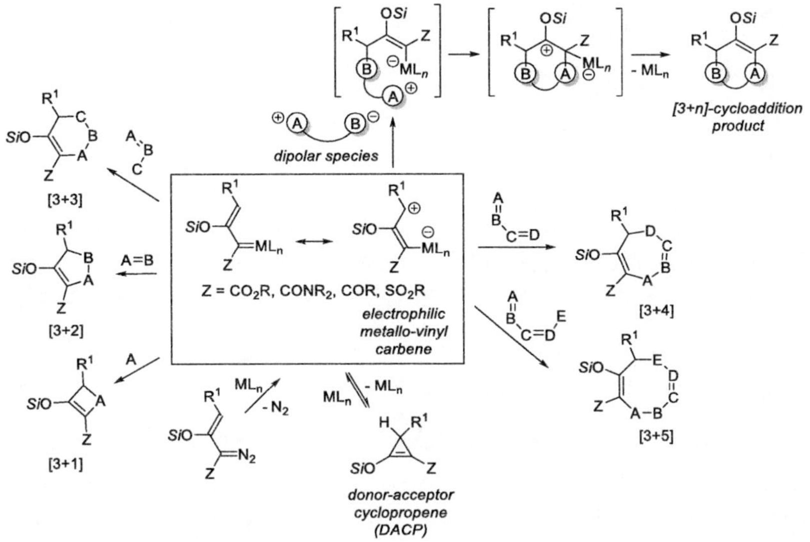

Scheme 2 [3 + *n*]-Cycloaddition with silyloxy-substituted metallo-vinylcarbenes and generation of donor-acceptor cyclopropenes (DACPs)

[*n* + 3]- (or [3 + *n*]-) cycloadditions mainly occur with metallo-vinyl carbenes: the initial nucleophilic attack at the "vinylogous" carbon followed by ring closure and elimination of a metal catalyst (Scheme 2).

In the case of substituted metallo-vinyl carbenes as "three-atom" surrogates for [3 + *n*]-cycloaddition, their extended conjugation increases the stability of metallocarbenes and facilitates nucleophilic attack of a dipolar molecule at the vinylogous position through the formation of a vinylmetal intermediate that is susceptible to electrophilic attack and ring closure followed by the elimination of the metal ML_n (Scheme 2). 3-Silyloxy-substituted metallo-vinyl carbenes are generated from the corresponding silyl-protected enoldiazo compounds (-acetates, -acetamides, -ketones, and -sulfones) by their reaction with a metal catalyst (ML_n) accompanied by dinitrogen extrusion. Notably, the nature of metal, its ligands, substituents on the "carbenic" carbon, and the type of dipolar species determine the rate of cycloaddition and whether nucleophilic reaction occurs at the vinylogous position or the carbene center ([3 + *n*] vs. [2 + 3]). An important feature of 3-silyloxy-substituted metallo-vinyl carbenes is their ability to generate donor–acceptor cyclopropenes (DACPs), which have been known as a resting state of metallo-vinylcarbenes (Scheme 2) [10]. Recently established connectivity between DACPs formed from γ-substituted enoldiazo compounds by dinitrogen extrusion, and *Z*- and *E*-geometrical isomers of metallo-enolcarbenes identified them as an effective source for vinyldiazo compounds in catalytic asymmetric [3 + *n*]-cycloaddition reactions [11].

Special attention in this chapter will be given to formal cycloaddition reactions for the construction of heterocyclic ring systems. These reactions do not follow the

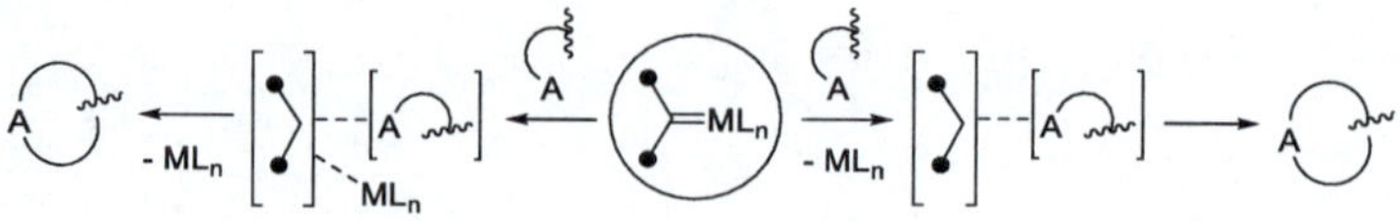

Scheme 3 Different pathways for formal cycloaddition with metallocarbenes

above-mentioned mechanisms, but the net process appears as a cycloaddition. The process involves a metallocarbene in a simple cascade or complex rearrangement cascade reactions with the elimination of a metal catalyst, which can occur in any step (Scheme 3).

2 Three-Membered Heterocycles: [2 + 1]-Cycloaddition

The most common and important three-membered heterocyclic compounds are small-ring systems such as oxiranes (epoxides) and aziridines, although the construction of these rings via cycloaddition using carbenes or carbenoids is not a traditional method for their synthesis. Among other methods, classic approaches for the synthesis of oxiranes include the Prilezhaev reaction [12] – epoxidation of alkenes using organic peroxycarboxylic acids – and intramolecular Williamson ether synthesis from halohydrins using a strong base. The important asymmetric versions of epoxidation were later discovered by Sharpless [13–15], Corey [16], and Jacobsen [17, 18].

However, examples of the synthesis of epoxides that involve carbenes or metallocarbenes are rare. An attractive method for their synthesis via formal [2 + 1]-cycloadditions of aldehydes and ketones with carbenoid compounds, dimethyloxosulfonium and dimethylsulfonium methylides, was proposed by Corey and Chaykovsky (Scheme 4a) [19]. Catalytic asymmetric sulfur ylide-mediated epoxidation of carbonyl compounds using chiral sulfur ylides (obtained from the reaction of corresponding sulfide with benzyl bromide, and further treatment with a base) was first discovered by Furukawa [20] and later developed by several research groups [21–29]. Model epoxide **2** was obtained using stoichiometric amounts of chiral sulfur ylides in moderate to excellent yields with up to 98% *ee* and >98:2 *dr* (Scheme 4b). An alternative to this process was proposed by Aggarwal with the use of a transition metal benzylidene (PhCH = [M]) generated from (diazomethyl)-benzene and only catalytic amounts of a chiral sulfide (Scheme 4c) [30, 31]. High yields and stereoselectivities were obtained with the use of 5 mol% of copper (II) acetylacetonate or 1 mol% of dirhodium(II) tetraacetate and 20 mol% of chiral sulfide **7**. (Diazomethyl)benzene is an unstable compound that was generated from benzaldehyde tosylhydrazone sodium salt and 10 mol% of benzyl triethylammonium chloride – a phase transfer catalyst. This reaction is a good example of the synthesis of oxiranes via formal [2 + 1]-cycloaddition that involves metallocarbenes, and it was optimized for a broad scope of chiral oxiranes. Further

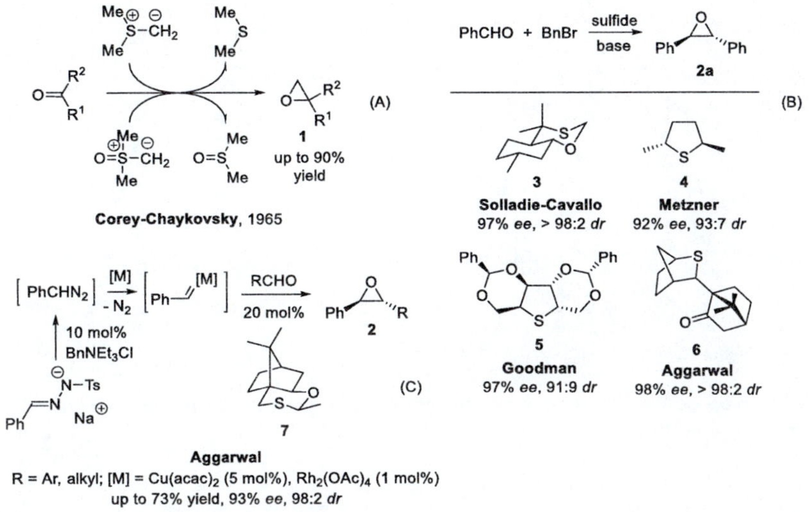

Scheme 4 (**a**) First example of epoxide synthesis using sulfur ylides as carbenoids. (**b**, **c**) Development of catalytic asymmetric synthesis of oxiranes using chiral sulfur ylides and metallocarbenes

optimizations allowed lower chiral sulfide loading to 5 mol% when sulfide **6** was utilized [32]. The main advantages of this method are: (1) the reaction requires only catalytic amounts of a sulfide; (2) occurs under neutral conditions; (3) metallocarbenes are more reactive than alkyl halides, which makes possible the employment of less reactive sulfides in this reaction, and therefore, the expansion of the scope of chiral sulfides that can be used.

A proposed mechanism for asymmetric synthesis of oxiranes with stochiometric amounts of a chiral sulfide involves the reaction of a sulfide with an alkyl halide (R^2CH_2X) to give the corresponding sulfonium salt, which then undergoes deprotonation by hydroxide or another base to afford the corresponding sulfur ylide, which upon the reaction with an aldehyde (R^1CHO) affords the epoxide (Scheme 5a). An alternative method for the generation of sulfur ylides involves the reaction of metallocarbenes with catalytic amounts of a chiral sulfide. The mechanism includes a catalytic cycle that is developed by the reaction of a chiral sulfide with a diazo compound (R^2CHN_2) in the presence of a transition metal catalyst [M] to yield a chiral sulfur ylide, which then undergoes the reaction with an aldehyde (R^1CHO) to afford the desired oxirane, returning the chiral sulfide to the catalytic cycle. (Scheme 5b).

Similar to oxiranes, the pioneering work by Corey and Chaykovsky on reaction of sulfur ylides with imines is an old approach for the construction of simple aziridine heterocycles [19]. However, since the discovery of copper-catalyzed formal [2 + 1]-cycloaddition of alkenes with in situ generated nitrenes by Evans in 1991 [33], this reaction became a traditional method for the synthesis of the aziridine ring and a synonym to "aziridination" [34–41]. Other ways to construct aziridine rings include cyclization of imines with organolithium compounds [42], ammonium ylides [43],

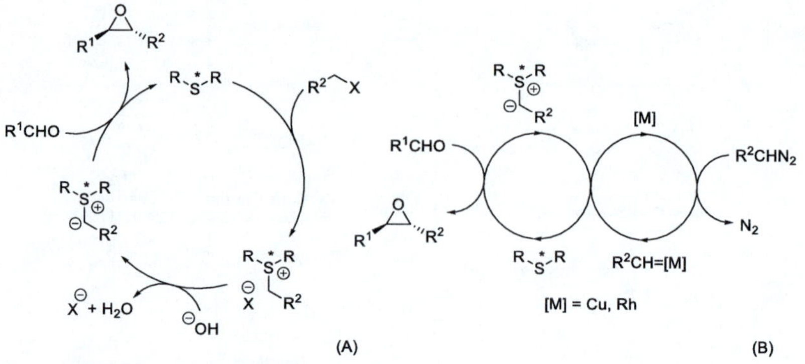

Scheme 5 Proposed mechanism for catalytic asymmetric synthesis of oxiranes using chiral sulfur ylides and metallocarbenes

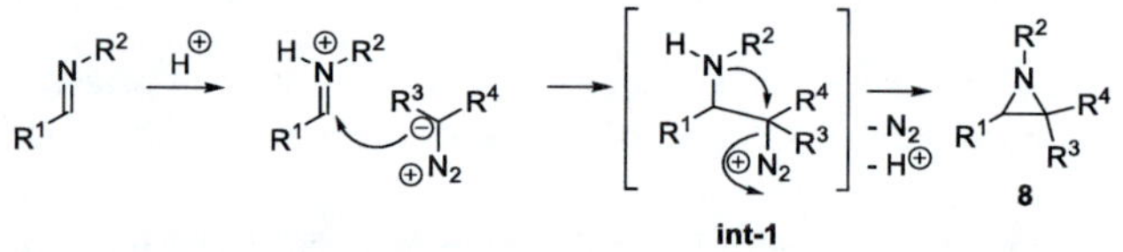

Scheme 6 General scheme for the acid catalyzed formal [2 + 1]-cycloaddition of imines with diazo compounds to form aziridines

and a well-documented modification of Corey–Chaykovsky reaction using imines (two-atom component) and diazo compounds as a carbenoid reactant (one-atom component) (Scheme 6) [44–47]. Formally, this is a [2 + 1]-cycloaddition, but the process does not involve the generation of a carbene or metallocarbene, and, analogous to sulfur ylides, it is a stepwise addition of the nucleophilic carbon of a diazo compound to the electrophilic sp^2 carbon of an imine with further S_N2 displacement of dinitrogen to close the aziridine ring. In order to activate the sp^2 carbon of the imine this reaction is typically catalyzed by Brønsted or Lewis acids. In Zhong's work, the employment of a chiral BINOL-derived phosphoric acid allowed access to chiral aziridines in high yields and with excellent stereocontrol [48].

The only reported method for asymmetric aziridination with the use of metallocarbenes is Aggarwal's [2 + 1]-cycloaddition of an imine with in situ generated (diazomethyl)benzene in the presence of dirhodium(II) tetraacetate (1 mol%) and chiral sulfide **6** (20 mol%) (Scheme 7) [49]. Aziridine **9** was obtained in 52% yield with 98% *ee* and 8:1 *dr*, and successfully transformed to compound **10**, a side chain of Taxol, with complete retention of optical purity.

In a recent report [50], the construction of 1,2,2-trisubstituted aziridines was successfully performed via [2 + 1]-cycloaddition reaction of free singlet carbenes (generated in situ from α-diazo esters under blue LED irradiation) with hexahydro-1,3,5-triazines (Scheme 8). The authors demonstrated solvent-dependent divergent reactivity of the reactants: the reaction in DMSO resulted in the formation of

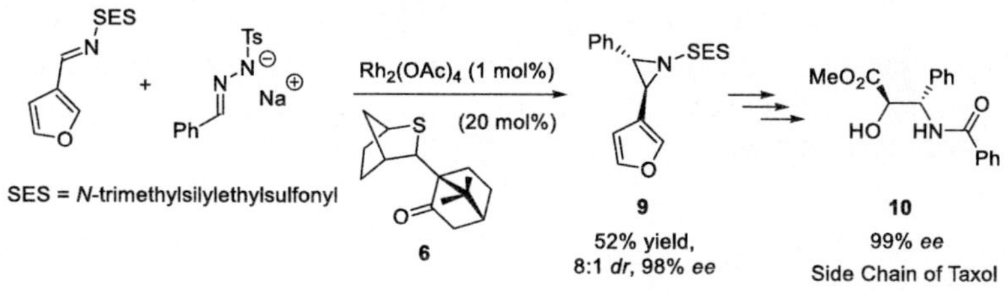

Scheme 7 Catalytic asymmetric aziridination using metallocarbenes: a route to the Taxol side chain

Scheme 8 Aziridines from [2 + 1]-cycloaddition of hexahydro-1,3,5-triazines with free singlet carbenes generated from α-diazo esters under blue LED irradiation: reaction scope and mechanistic studies

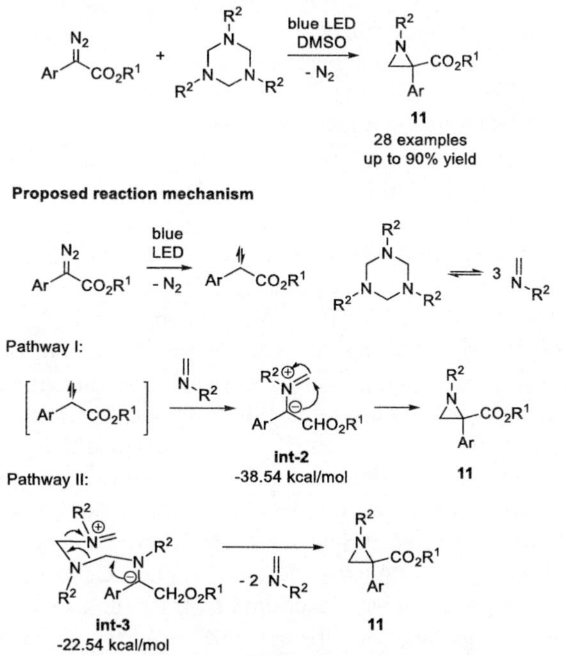

aziridines via [2 + 1]-cycloaddition, and in DCM imidazolidines were produced via [4 + 1]-cycloaddition, which will be later discussed in this chapter. DFT calculations indicate that the process is stepwise rather than concerted (Scheme 8), and intermediate **int-2** in pathway I has lower energy than **int-3** in pathway II by 16 kcal/mol, which favors dissociation of a hexahydro-1,3,5-triazine to the corresponding monomer, and then its reaction with a free singlet carbene. This results in [2 + 1]- and not [4 + 1]-cycloaddition that is favored by pathway II.

In summary, three-membered heterocyclic compounds are commonly obtained via [2 + 1]-cycloaddition; however, a few methods involve free carbenes and metallocarbenes. The most common carbenoid compounds for the synthesis of oxiranes and aziridines are sulfur ylides and diazo compounds that provide a

C-atom as the one-atom component for [2 + 1]-cycloaddition. To my knowledge, there are no reports on the synthesis of other important classes of three-membered heterocycles (e.g., thiiranes, azirines, oxirenes, and thiirenes) via [2 + 1]-cycloaddition of carbenes or metallocarbenes.

3 Four-Membered Heterocycles: [3 + 1]-Cycloaddition

Among typical representatives of four-membered heterocycles (azetidines, azetines, oxetanes, and thietanes), azetidine and 2-azetidinone structural motif is the most abundant in naturally occurring compounds (e.g., Mugineic Acids, Penaresidins, Penicillin, and Cephalosporin) and synthetic drug molecules (e.g., Melagatran) [51–54]. Therefore, the construction of azetidine ring systems, including their asymmetric synthesis, deserves special attention.

Traditional approaches to access azetidines are cyclization of 1,3-bis(halogeno)- or 1,3-bis(tosylato)-propanes with amines in basic media [55, 56], intramolecular cyclization of organoboronates [57], three-membered ring expansions [58, 59], strain release of azabicyclo[1.1.0]butane [60], and a recent visible-light-mediated [2 + 2]-photocycloaddition [61]. 2-Azetines as unsaturated analogs of azetidines have received scarce attention, and are typically obtained via elimination reactions of azetidine derivatives [62, 63], or intermolecular [2 + 2]-cycloaddition/isomerization of allenyl imides [64]. It should be noted that none of these methods for the synthesis of these strained heterocycles involve metallocarbenes. The first [3 + 1]-cycloaddition for the synthesis of azetidine derivatives was realized in 2016 using sulfur ylides [65], and the first formal [3 + 1]-cycloaddition involving metallocarbenes was reported for the synthesis of 2-azetine derivatives using copper(I) catalysis (Scheme 9a) [66]. However, this process is sequential: [2 + 1]-cycloaddition with a nitrene source to form aziridine-substituted diazo compounds **int-4**, and their further copper (I)-catalyzed ring expansion to azetines **12**.

Direct [3 + 1]-cycloaddition requires three- and one-atom surrogates to construct the azetine ring. The first example of this type of cycloaddition was reported in 2017 by Doyle and co-workers for highly enantioselective synthesis of cyclobutenes using a copper-vinylcarbene (three-atom component) generated from silyl-protected enoldiazoacetates and α-acyl sulfur ylides (one-atom component, carbene donor) [67].

Inspired by this discovery, a nitrene version of the [3 + 1]-cycloaddition was realized for the highly enantioselective synthesis of donor–acceptor 2-azetine derivatives **13** via reaction of silyl-protected enoldiazoacetates with *N*-aryl-imido sulfur ylides (Scheme 9b) [68, 69]. The reaction is metal catalyst sensitive and occurs only with cationic copper(I) salts. Furthermore, simple bisoxazoline (Box) ligands were not suitable, providing low enantiocontrol, but modified double-sidearmed bisoxazoline (Sabox) ligand **L1** was optimal to achieve up to 95% yield and 99% ee. Notably, unlike in the [3 + 1]-cycloaddition to form chiral cyclobutenes, this cycloaddition reaction occurred with (*Z*)- but not with (*E*)-enoldiazoacetates, and the

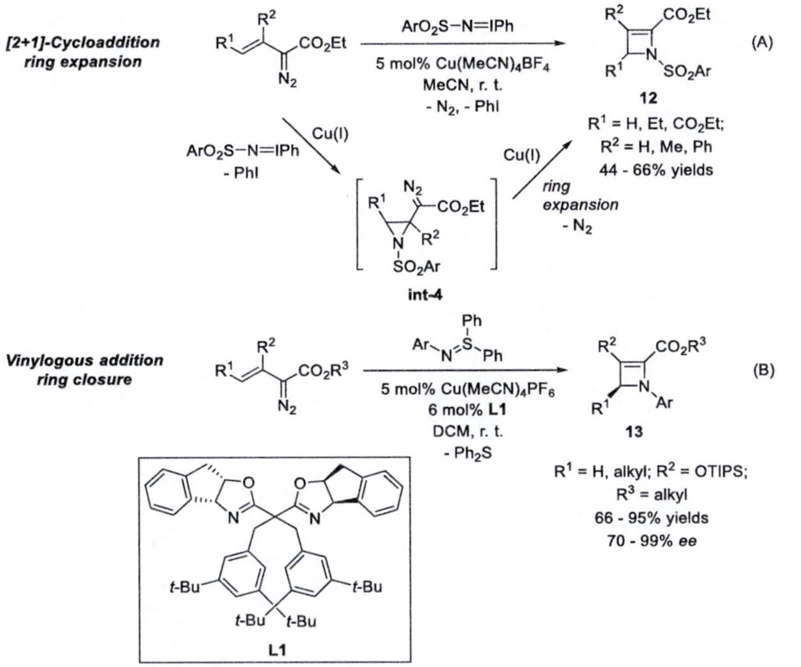

Scheme 9 (**a**) Copper(I)-catalyzed synthesis of 2-azetines via [2 + 1]-cycloaddition/ring expansion and (**b**) highly enantioselective [3 + 1]-cycloaddition (vinylogous addition/ring closure)

corresponding donor–acceptor cyclopropene (DACP) formed from the enoldiazoacetate was not reactive with *N*-aryl-imido sulfur ylides [11]. Furthermore, in the synthesis of donor–acceptor azetines, sulfur ylide reactivities were variable: *N*-aryl-imido sulfur ylides reacted with silyl-protected enoldiazoacetates at room temperature, but α-acyl-imido sulfur ylides were completely unreactive even at elevated temperatures.

A proposed mechanism for this asymmetric [3 + 1]-cycloaddition involves nucleophilic addition of the *N*-aryl-imido sulfur ylides (one-atom surrogate) to the vinylogous electrophilic carbon of the Cu(I)-carbene intermediate (three-atom surrogate) (Scheme 10). The intramolecular cyclization of **int-5** occurs through the elimination of diphenylsulfide Ph_2S, aided by electron donation from the oxygen of the siloxy group, via apparent S_N2-like backside displacement to form **int-6** [68]. Further dissociation of the ligated Cu(I) results in the formation of the four-membered ring product and regeneration of the reaction catalyst.

The chiral donor−acceptor azetines obtained are reported to be stable at room temperature but reactive with nucleophiles, including water that is a weak nucleophile. An important application of these strained ring systems has been reported in the synthesis of chiral unnatural *N*-aryl-substituted amino acid derivatives via *retro*-Claisen nucleophilic ring opening with complete retention of enantiopurity

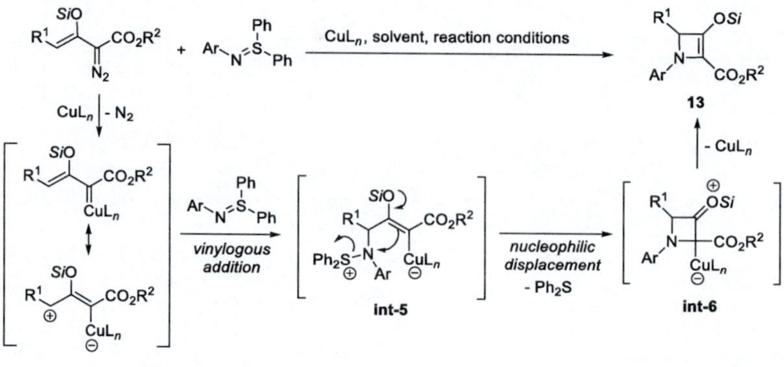

Scheme 10 Proposed mechanism for copper(I)-catalyzed [3 + 1]-cycloaddition of silyl-protected enoldiazo acetates with *N*-aryl-imido sulfur ylides

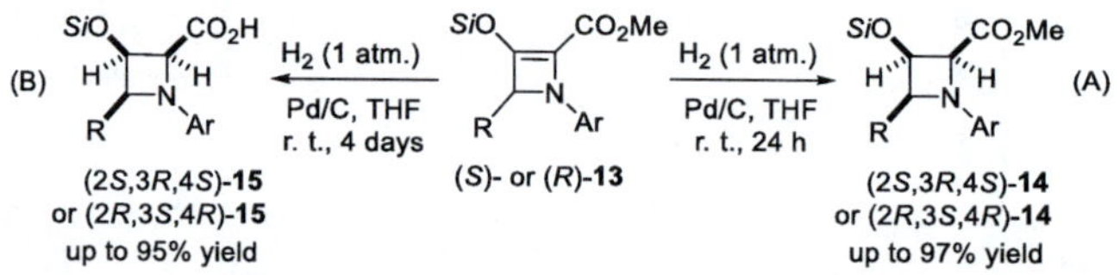

Scheme 11 Synthesis of chiral tetrasubstituted azetidine-2-carboxylates **14** azetidine-2-carboxylic acids **15** via hydrogenation of chiral azetine-2-carboxylates **13**

demonstrating enormous potential with a broad scope of nucleophiles (amines, alcohols, thiols, water, hydrazines, hydroxylamines, etc.) [69].

Considering the reactivity of chiral donor–acceptor azetines, they were subjected to classic (Pd/C)-catalyzed *cis*-hydrogenation of the double bond to stabilize the four-membered heterocycle by releasing ring strain, which provided easy access to chiral tetrasubstituted methyl 2-azetidine-carboxylates that possess three chiral centers (Scheme 11a) [68]. Notably, *cis*-hydrogenation occurs from the site opposite to substituent R at the 4-position of azetidine ring in high yields (up to 97%) with exceptional diastereocontrol (*dr* > 20:1) in most cases and complete retention of enantiopurity. Hydrogenation of *p*-methoxybenzyl (PMB) azetine-2-carboxylates performed under the same reaction conditions as their methyl analogs, but with longer reaction times, led to azetidine-2-carboxylic acids (Scheme 11b) in high yields (up to 95%) and exceptional diastereocontrol (*dr* > 20:1). The obtained chiral azetidine-2-carboxylic acids were easily converted to azetidine iminosugar derivatives (structural unit of naturally occurring azetidine compounds) via reduction by lithium aluminum hydride.

In summary, the success of the [3 + 1]-cycloaddition of metallo-enolcarbenes with *N*-aryl-imido sulfur ylides to form chiral azetine- and azetidine-2-carboxylates and possibilities of further functionalizations open an underexplored and broad potential of [3 + 1]-cycloaddition reactions for the synthesis of other four-membered heterocycles. Moreover, the application of other sulfur ylides, phosphorus, or nitrogen ylides as alternatives for one-atom surrogates in [3 + 1]-cycloaddition is yet to be discovered.

4 Five-Membered Heterocycles

4.1 [3 + 2]-Cycloaddition

Construction of five-membered heterocycles via [3 + 2]-cycloaddition has a long history [70, 71], and one of the classic and impactful transformations is the azide–alkyne Huisgen reaction – a simple [3 + 2]-cycloaddition between an azide (1,3-dipole) and a terminal or internal alkyne (1,2-dipole) to give a 1,2,3-triazole [72]. The reaction was later named "click chemistry" by Sharpless [73], and received enormous attention in medicinal chemistry and drug delivery [74–76].

[3 + 2]-Cycloaddition with the use of metallo-vinyl carbenes (three-atom surrogate) is one of the most common and straightforward methods for rapid construction of various five-membered heterocycles under mild reaction conditions. The majority of catalysts employed for [3 + 2]-cycloaddition have been those of dirhodium (II) carboxylates; however, cationic silver and gold catalysts are also suitable for the synthesis of aromatic nitrogen containing heterocycles. A plausible mechanism for [3 + 2]-cycloaddition includes initial nucleophilic addition to the electrophilic carbene center or to the vinylogous position of a metallo-vinyl carbene followed by ring closure with regeneration of the metal catalyst.

Styryldiazoacetates and γ-alkyl vinyldiazoacetates have been known to selectively undergo a nucleophilic attack at the carbene position, rather than at the vinylogous position. One of the examples of their heterocyclization via [3 + 2]-cycloaddition is the $Rh_2(OAc)_4$-catalyzed reaction with imines to form dihydropyrroles **16** through a zwitterionic intermediate (**int-7**) (Scheme 9a) [77]. Notably, the cyclization of **int-7** occurs prior to or concurrent with dissociation of the rhodium(II) catalyst. Another example of [3 + 2]-cycloaddition is the rhodium (II)-catalyzed reaction of a styryldiazoacetate with oximes to produce fully substituted oxazoles **17** (Scheme 12a) [78]. The initial nucleophilic attack by nitrogen of the oxime occurs at the carbene position (**int-8**) with further cyclization to complete the [3 + 2]-cycloaddition, which after elimination of water and regeneration of the rhodium(II)-catalyst produced oxazoles **17**. Interestingly, the cyclization occurred with the ester carbonyl oxygen, and the styryl group remained intact.

In contrast, nucleophilic attack occurred at the vinylogous position when sterically hindered chiral dirhodium(II) catalysts were applied in [3 + 2]-cycloaddition with nitrones to form chiral tetrasubstituted dihydroisoxazoles **18** (Scheme 12b) [79]. In this case, the cyclization of **int-9** occurs prior to dissociation of the rhodium catalyst. This is the first example of highly enantioselective [3 + 2]-heterocyclization of metallo-vinylcarbenes (up to 99% *ee*).

An unexpected cascade process with massive rearrangement occurred in the rhodium(II) catalyzed reaction of the unsubstituted vinyldiazoacetate with nitrones,

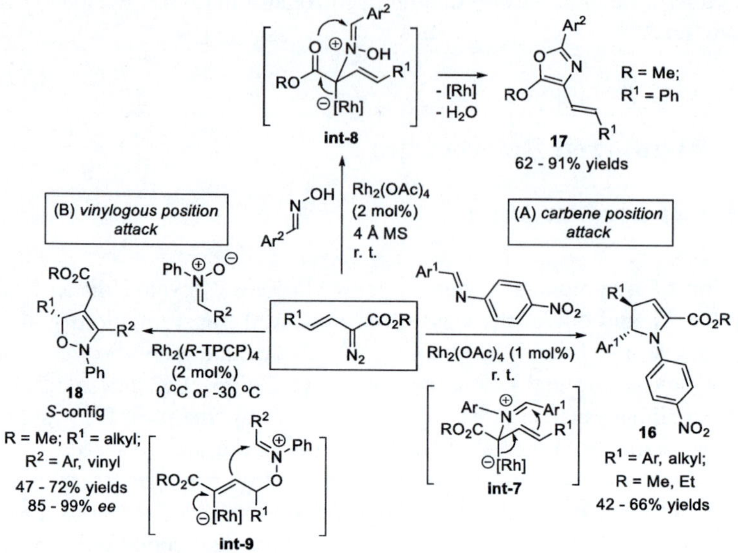

Scheme 12 Synthesis of dihydropyrroles **16**, oxazoles **17**, and chiral dihydroisoxazoles **17** via rhodium(II)-catalyzed [3 + 2]-cycloaddition of vinyldiazoacetates

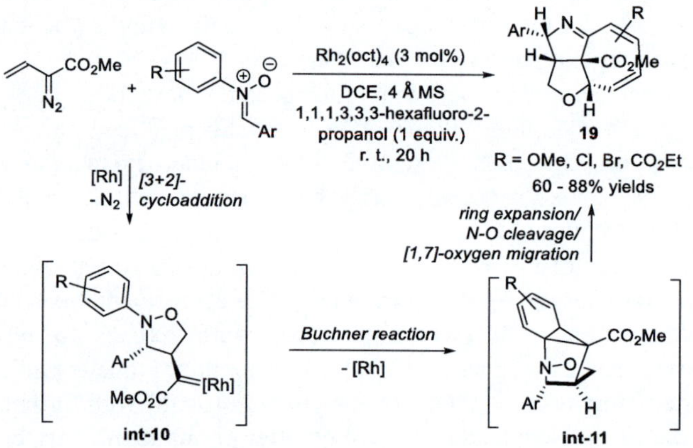

Scheme 13 Tandem rhodium(II)-catalyzed [3 + 2]-cycloaddition – cyclopropanation/rearrangement of 2-diazo-3-butenoate to form tricyclic compound **19**

and resulted in tricyclic derivative **19** (Scheme 13) [80]. The initial step in the reaction mechanism is [3 + 2]-cycloaddition, which starts with nucleophilic attack of nitrone at the vinylogous position of dirhodium(II)-vinylcarbene to generate intermediate **int-10**. The cascade process starts with an intramolecular Buchner reaction to form **int-11**, which then undergoes a 6π-electrocyclic ring expansion with subsequent N–O cleavage and [1,7]-oxygen migration to form tricyclic product

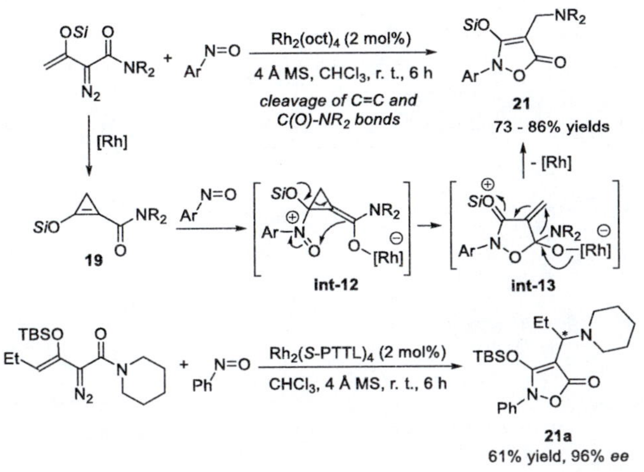

Scheme 14 Rhodium(II) carboxylate catalyzed formal [3 + 2]-cycloaddition reaction between enoldiazoacetamides and nitrosoarenes with extensive rearrangement

19. The proposed mechanistic pathway for this cascade process was later confirmed by computational studies [81].

As part of the study on divergent behavior of metal catalysts, an interesting example of formal [3 + 2]-cycloaddition with extensive rearrangement was reported in a rhodium(II) octanoate catalyzed reaction of enoldiazoacetamides and nitrosoarenes (Scheme 14) [82]. The reaction starts with rhodium(II)-catalyzed generation of stable DACP **20**, which upon reaction with the nucleophilic nitrogen of a nitrosoarene forms cyclopropane intermediate **int-12** that undergoes ring expansion to generate **int-13**. Rearrangement via cleavage of C=C and C(O)NR_2 bonds in **int-13.** produces multifunctionalized 5-isoxazolones **21** in high yields. An asymmetric version of this cycloaddition was realized using chiral rhodium(II) catalyst Rh_2(*S*-PTTL)$_4$ in the reaction of a 4-ethyl-substituted enoldiazoacetamide with nitrosobenzene to afford 5-isoxazolone **21a** in 61% yield with 96% *ee*.

The synthesis of substituted pyrazole heterocycles has been reported via rhodium (II)/Lewis acid catalyzed formal [3 + 2]-cycloaddition of donor–acceptor substituted hydrazones with enoldiazoacetates (Scheme 15a) [83]. This regiospecific cascade process starts from rhodium(II)-catalyzed reaction of enol diazoacetates and substituted donor–acceptor hydrazones to form stable isomeric acyclic products **22** or **23** (with a C–N or N–N double bond) after C–H carbene insertion. Subsequent Sc(OTf)$_3$ (Lewis acid) promoted cyclization, desilylation, and aromatization to afford pyrazoles **24** can be performed either with compounds **22** or **23**, or as a sequential one-pot reaction.

An interesting ligand-controlled discrimination has been reported between two diazo compounds (aryl- and styryl-diazoacetates) via formal [3 + 2]-cycloaddition that leads to isomeric trisubstituted pyrazoles **25** and **26** with the use of gold (I) catalysis (Scheme 15b) [84]. The reaction outcome depends on the nature and

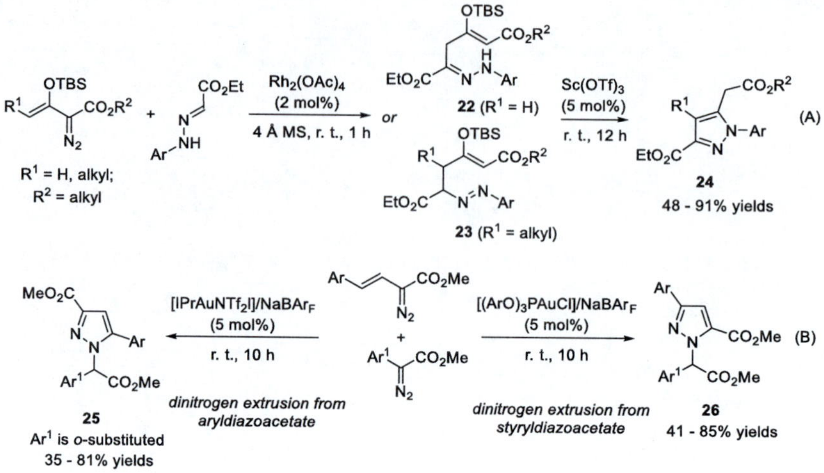

Scheme 15 Synthesis of substituted pyrazoles **24–26** via [3 + 2]-cycloaddition that involves metallocarbenes

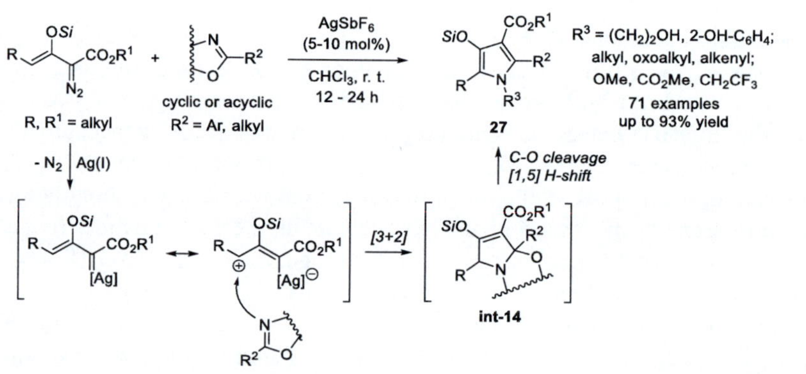

Scheme 16 Synthesis of highly functionalized pyrroles via cascade silver(I)-catalyzed [3 + 2]-cycloaddition of γ-substituted enoldiazoacetates and cyclic or acyclic imino ethers with subsequent C–O cleavage and [1, 5] *H*-shift

electronic structure of the ligand and steric hindrance of the arydiazoacetate. Pyrazoles **25** are formed with a NHC ligated Au(I) catalyst and only with *o*-substituted arydiazoacetates (dinitrogen extrusion occurs from arydiazoacetate, but does not generate the metallo-vinylcarbene), whereas a phosphite ligated Au (I) catalyst forms pyrazoles **26** in moderate to high yields (dinitrogen extrusion occurs from the styrydiazoacetate and the metallo-vinylcarbene is generated).

In a recent silver(I)-catalyzed cascade transformation of silyl-protected γ-substituted enoldiazoacetates with imino ethers that forms fully substituted and highly functionalized pyrroles (Scheme 16) [85, 86], the initial step is [3 + 2]-cycloaddition of a metallo-vinylcarbene (three-atom component) with the C=N unit of an imino ether (two-atom component) to generate **int-14**, which then undergoes C–O cleavage and [1,5]-proton transfer to produce fully substituted pyrroles **27**. The reaction tolerates a very broad scope of substrates and affords pyrrole products in good to high yields (up to 93%) under mild reaction conditions.

4.2 [4 + 1]-Cycloaddition

[4 + 1]-Cycloaddition with metallo-carbenes is typically a cascade process that involves the initial nucleophilic attack at the carbene center with subsequent 1,5-heterocyclization on the same carbon atom to produce a five-membered heterocycle (Scheme 17). To favor cyclization on the same carbon, this type of cycloaddition often requires structurally rigid metallo-carbenes (e.g., cyclic diazoamides). The [4 + 1]-cycloaddition cascade reaction is known to be catalyzed by Rh(II) [87, 88], and Cu(I) [89, 90], but examples of Pd(II) catalysis have also been reported [91]. Scheme 17 illustrates a plausible mechanism for the [4 + 1]-cycloaddition with 3-diazo-1-methylindolin-2-one derivatives and the scope of Rh(II)-catalyzed reactions. Cu(I)- and Pd(II)-catalyzed reactions are summarized in Scheme 18.

The synthesis of imidazolidine heterocycles has been reported via Au(I)-catalyzed formal [4 + 1]-cycloaddition of 2,4,6-triaryl-hexahydro-1,3,5-triazines

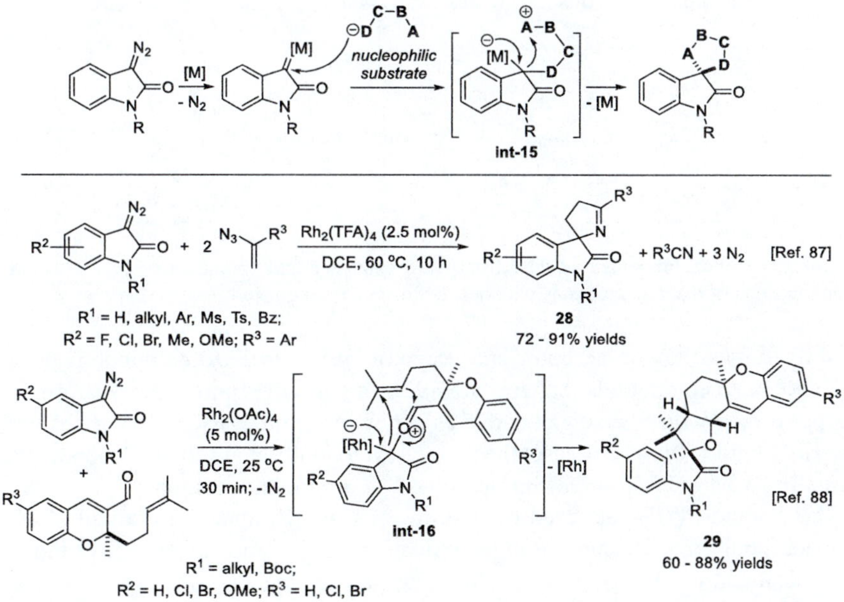

Scheme 17 Plausible mechanism of the [4 + 1]-cycloaddition with 3-diazo-1-methylindolin-2-one derivatives and the scope of Rh(II)-catalyzed reactions

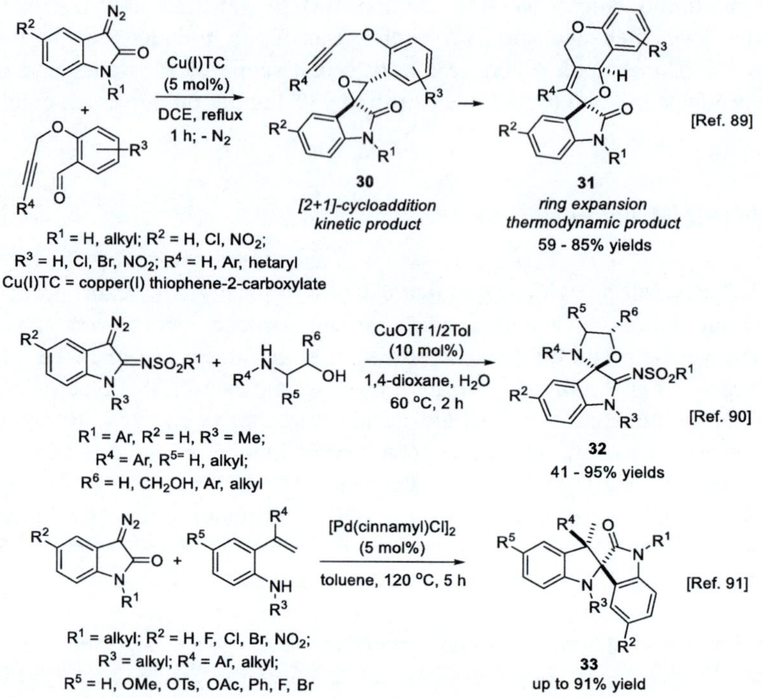

Scheme 18 The scope of Cu(I)- and Pd(II)-catalyzed [4 + 1]-cycloaddition reactions

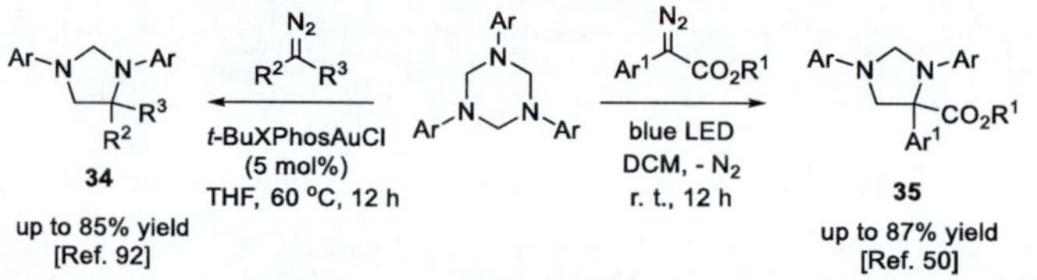

Scheme 19 Imidazolidine heterocycles from Au(I)-catalyzed and photoinduced [4 + 1]-cycloaddition reactions of diazo compounds with hexahydro-1,3,5-triazines

with a variety of donor–acceptor and acceptor–acceptor diazo compounds but not with enoldiazo compounds (Scheme 19) [92]. In a recent report, this transformation was performed with free singlet carbenes generated from α-aryldiazoacetates under blue LED light [50]. In this case, the [4 + 1]-cycloaddition is solvent dependent and occurs in DCM but not in DMSO that favors [2 + 1]-cycloaddition.

Rh(III)- and Pd(II)-catalyzed transformations, which appear as formal [4 + 1]-cycloadditions and occur through oxidative C–H and N–H activation via cyclometallation [93–95], do not follow the general mechanism for cycloaddition and will not be discussed in this chapter.

5 Six-Membered Heterocycles

5.1 [3 + 3]-Cycloaddition

In order for [3 + 3]-cycloaddition to occur with metallo-carbenes, they must react as three-atom surrogates. Metallo-vinylcarbenes, and especially those substituted at the β-position with an electron-donating group, are great reactants for this transformation. Silyl-protected enoldiazo compounds, upon reaction with a metal catalyst, generate metallo-vinylcarbenes that are well set for [3 + 3]-cycloaddition by reacting exclusively at the vinylogous position due to stabilization by a siloxy group. The role of DACP and its equilibrium with the metallo-vinylcarbene is crucial for the outcome of the [3 + 3]-cycloaddition reaction and its stereoselectivity.

A variety of dipolarophiles have been applied in catalytic [3 + 3]-cycloaddition with silyl-protected enoldiazo compounds, and the asymmetric reactions of nitrones are the most extensively studied. Nitrones are common and the most widely used 1,3-dipoles for different types of cycloaddition reactions to form six-membered heterocycles [96].

The first reported example of asymmetric [3 + 3]-cycloaddition enoldiazoacetates with nitrones is rhodium(II)-catalyzed reaction with a phthalimide-alanine ligated catalyst [Rh_2(*S*-PTA)$_4$], which afforded trisubstituted 3,6-dihydro-2*H*-1,2-oxazines **36** in moderate to high yields and enantioselectivities (up to 93% *ee*) (Scheme 20) [97]. An analogous dirhodium(II) acetate catalyzed [3 + 3]-cycloaddition of silyl-protected enoldiazoacetates but without enantiocontrol was later applied in the synthesis of cyclic nitroso acetals – *N*-silyloxy-3,6-dihydro-2*H*-1,2-oxazines [98].

The successful outcome of this dirhodium(II)-catalyzed [3 + 3]-cycloaddition was encouraging, but limited scope and the high cost of rhodium catalysts prompted the search for alternative metal catalysts and ligands that would be general for the cycloaddition of enoldiazo compounds with nitrones. The investigation of reactivities and stereoselectivities of enoldiazoacetamides with nitrones under copper (I) catalysis revealed that the asymmetric [3 + 3]-cycloaddition to form **37** occurred with higher yields and exceptional enantiocontrol with the use of chiral Box ligand **L2** (Scheme 20) [99]. In a further search for an optimal and general catalytic system for [3 + 3]-cycloaddition of enoldiazocarbonyl compounds with nitrones, Box ligand optimization was performed for reactions with enoldiazoketones. The results indicated that in order to achieve high yields and enantiocontrol in the synthesis of 3,6-dihydro-2*H*-1,2-oxazines **38**, the use of Sabox ligand with copper(I) catalyst was necessary, and Sabox ligand **L3** was optimal (up to 99% *ee*) (Scheme 20) [100]. Notably, these conditions were suitable to achieve excellent enantioselectivities for enoldiazo-acetates and -acetamides making the Cu $(MeCN)_4BF_4$-**L3** catalytic system general for highly enantioselective [3 + 3]-cycloaddition of enoldiazocarbonyl compounds with nitrones. Chiral dirhodium (II) catalysts were also tested in the [3 + 3]-cycloaddition of enoldiazoketones with nitrones but did not provide high yields of [3 + 3]-cycloaddition products **38** due to a competing intramolecular reaction of the enoldiazoketone to form

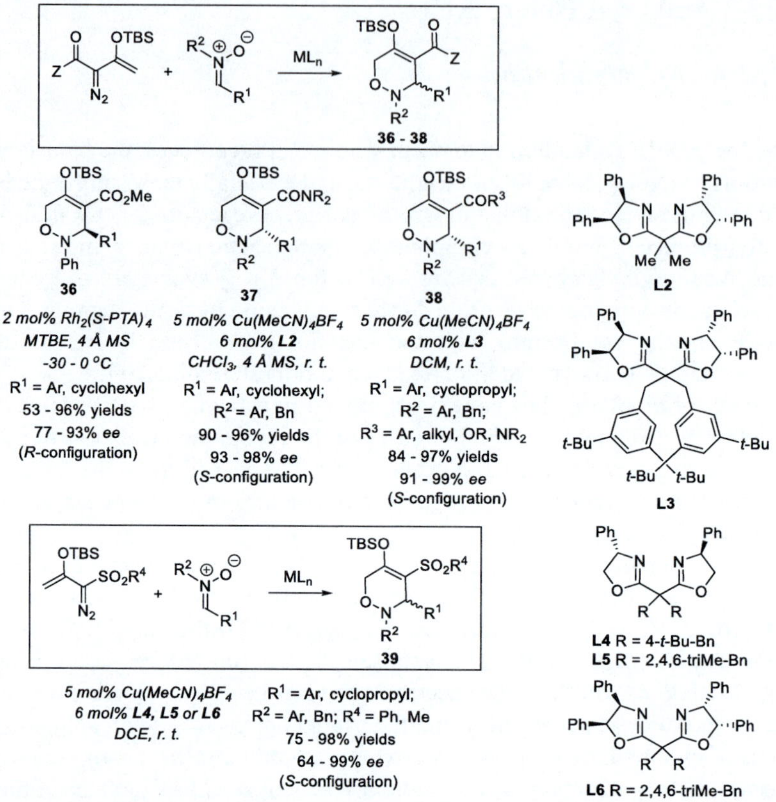

Scheme 20 Summary of dirhodium(II)- and copper(I)-catalyzed asymmetric [3 + 3]-cycloaddition of enoldiazocarbonyl compounds and enoldiazo sulfones with nitrones

2,5-disubstituted furans [101]. The results from chiral rhodium(II) vs. copper (I) catalysis in [3 + 3]-cycloaddition of enoldiazoacetates with nitrones indicate that the (*R*)-absolute configuration of 3,6-dihydro-2*H*-1,2-oxazines **36** was produced by chiral dirhodium(II) with (*S*)-carboxylate ligands, but copper(I) with (*R*)-Sabox ligands allowed access to the opposite enantiomer (*S*)-**36**.

Further studies of [3 + 3]-cycloaddition of enoldiazo compounds with nitrones demonstrated the suitability of enoldiazosulfones to achieve high yields and enantiocontrol in the synthesis of 3,6-dihydro-2*H*-1,2-oxazines **39** under copper (I) catalysis with optimized Sabox ligands **L4**–**L6** (Scheme 20) [102]. Unlike with enoldiazocarbonyl compounds, rhodium(II) catalysis was completely unsuitable for [3 + 3]-cycloaddition of enoldiazosulfones because of the enhanced stability of sulfonyl-substituted DACPs, which underwent a [3 + 2]-cycloaddition with nitrones to form a bicyclic product.

The above-mentioned [3 + 3]-cycloaddition reactions with nitrones (Scheme 20) describe reactivity and selectivity of enoldiazo compounds without a substituent at the 4-position. Early work on (*Z*)-4-phenyl-substituted enoldiazoacetates showed

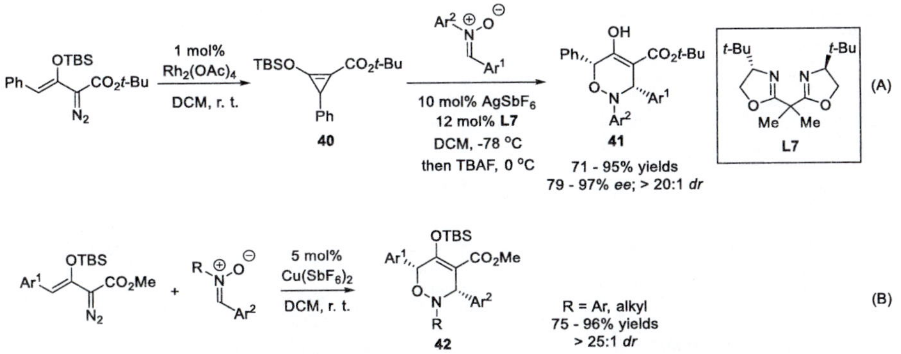

Scheme 21 (**a**) Silver(I)-catalyzed asymmetric formal [3 + 3]-cycloaddition of DACP **40** with nitrones. (**b**) Copper(I)-catalyzed direct vinylogous [3 + 3]-cycloaddition of silyl-protected 4-aryl-substituted enoldiazoacetates with nitrones

that they were not reactive with nitrones under standard conditions of copper(I) or dirhodium(II) catalysis. An alternative two-step pathway for their [3 + 3]-cycloaddition with nitrones was discovered with the use of rhodium(II)-silver(I) catalysis (Scheme 21a) [103]. Stable DACP **40** was generated in situ with dirhodium (II) tetraacetate and then treated with a silver(I) salt ligated with classic Box ligand **L7**. Silver(I)-vinylcarbene regenerated from DACP **40** then undergoes the [3 + 3]-cycloaddition with nitrones to afford 3,6-dihydro-2*H*-1,2-oxazines **41** after the silyl group removal with TBAF.

A general approach for the synthesis of silyl-protected 3,6-dihydro-2*H*-1,2-oxazines **42** has been reported via copper(I)-catalyzed vinylogous [3 + 3]-cycloaddition of silyl-protected 4-aryl-substituted enoldiazoacetates with various nitrones (Scheme 21b) [104]. Although this reaction is not asymmetric, it occurs with high diastereocontrol (>25:1 *dr*) to produce *cis*-**42** in high yields.

Years later an asymmetric version of copper(I)-catalyzed [3 + 3]-cycloaddition of silyl-protected 4-alkyl-substituted enoldiazoacetates with nitrones was reported using Box ligands. This reaction is potentially challenging due to the fact that 4-substituted enoldiazoacetates exist as (Z)- and (*E*)-geometrical isomers, which are different in reactivity and stereoselectivity. However, the comparison of [3 + 3]-cycloaddition reactions of nitrones with (Z)- and (*E*)-isomers, as well as with their corresponding DACPs, indicated no difference in their reactivities and selectivities, which confirms the central role of DACPs in the cycloaddition process [11]. The [3 + 3]-cycloaddition of (Z)-4-alkyl-substituted enoldiazoacetates with nitrones occurs either via a (Z)-copper(I) vinylcarbene directly formed from the (Z)-4-alkyl-substituted enoldiazoacetate or from the DACP (resting state of the carbene) after the elimination of Cu(I). However, (*E*)-copper(I) vinylcarbene can only be formed directly from (*E*)-4-alkyl-substituted enoldiazoacetates and not from the corresponding DACP (Scheme 22) [105]. A highly enantioselective synthesis of alkyl-substituted 3,6-dihydro-2*H*-1,2-oxazines **43** has been reported from the [3 + 3]-cycloaddition of (Z)-4-alkyl-substituted enoldiazoacetates with nitrones

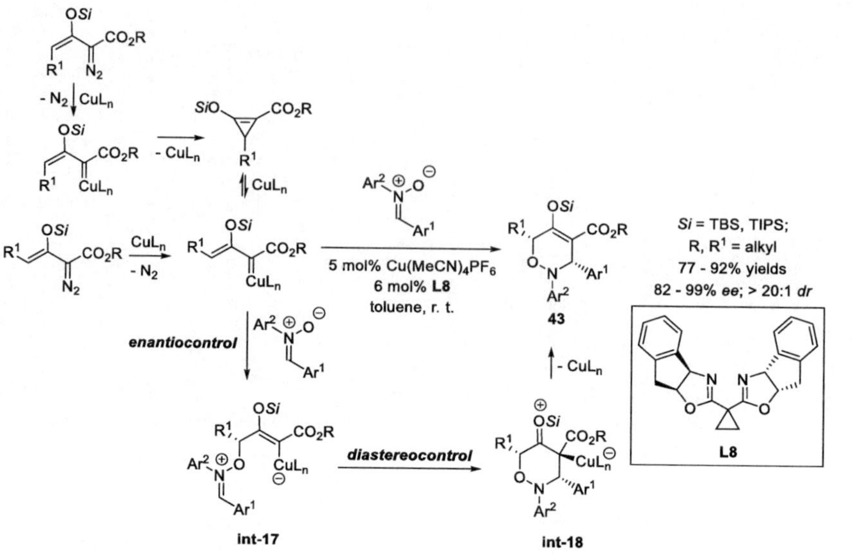

Scheme 22 Copper(I)-catalyzed asymmetric [3 + 3]-cycloaddition of 4-alkyl-substituted enoldiazoacetates with nitrones

using $Cu(MeCN)_4PF_6$ and optimal indeno-Box ligand **L8** (Scheme 22) [105]. Unlike with unsubstituted enoldiazoacetates, [3 + 3]-cycloaddition reactions with 4-substituted occur with high enantiocontrol established in the vinylogous addition step, and exclusive *cis*-diastereocontrol – in the ring closing step. Detailed mechanistic information on the effect of the chiral catalyst on stereoselectivities requires further theoretical studies.

The effect of an electron-donating group (EDG) at the 3-position of vinyldiazoacetates in [3 + 3]-cycloaddition with nitrones was further studied with silyloxy group being replaced by other EDGs such as aryl and alkyl. 4-Alkyl-substituted vinyl- and styryl-diazoacetates have been reported to exclusively undergo [3 + 2]-cycloaddition with nitrones under chiral dirhodium catalysis [79], however, 3-alkyl- and 3-aryl-substituted vinyldiazoacetates underwent asymmetric [3 + 3]-cycloaddition using the copper(I) catalyst $Cu(MeCN)_4BF_4$ with indeno-Box ligand **L8** being optimal (Scheme 23) [106].

Yields and enantioselectivities of aryl(alkyl)-3,6-dihydro-2*H*-1,2-oxazines **44** are moderate to high, and generally lower than those reported for silyloxy analogs. Moreover, alkyl-substituted heterocycles **44** were obtained in much lower yields and enantioselectivities than corresponding aryl-substituted compounds. These results indicate that a stronger EDG in the 3-position of vinyldiazoacetate in the reaction with nitrones favors [3 + 3]-cycloaddition with high enantiocontrol. Based on this observation, other strong EDGs analogous to the silyloxy group (e.g., OR, OAc, halide) in vinyldiazoacetates are anticipated to facilitate copper(I)-catalyzed [3 + 3]-cycloaddition reactions with high yields and stereoselectivities, but these transformations have not yet been reported. Other metal catalysts such as dirhodium

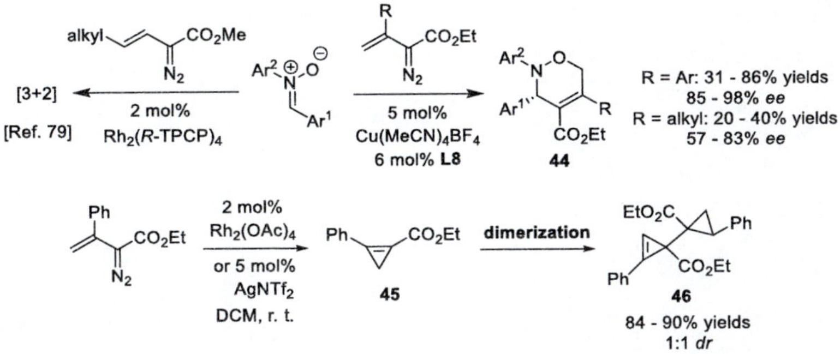

Scheme 23 Asymmetric copper(I)-catalyzed [3 + 3]-cycloaddition of 3-alkyl and aryl-substituted vinyldiazoacetates with nitrones

(II) tetraacetate and silver bis(trifluoromethanesulfonyl)imide were not suitable for the synthesis of heterocycles **44** due to a competing reaction – the facile self-dimerization of DACP **45** to form compound **46** (Scheme 23).

As described above, nitrones have been extensively studied in [3 + 3]-cycloaddition with vinyldiazo compounds, and those reactions follow the general mechanism: vinylogous addition to a metallo-vinyl carbene followed by cyclization. Other reported examples of [3 + 3]-cycloaddition of enoldiazo compounds that follow this mechanism include dearomative cycloaddition reactions of pyridinium ylides to form bi- or tricyclic structures. Although asymmetric [3 + 3]-cycloaddition reactions of pyridinium ylides with various dipolarophiles have been reported [107, 108], there are few reports on their reactions with metallo-vinyl carbenes. In the first report of dearomatizing [3 + 3]-cycloaddition of enoldiazoacetates with *N*-acylimino-pyridinium or -quinolinium ylides, chiral dirhodium(II) catalysts with phthalimide-amino acid ligands [$Rh_2(S\text{-PTTL})_4$ and $Rh_2(S\text{-PTAD})_4$] were employed to achieve high yields and enantioselectivities in the synthesis of chiral bi- and tricyclic diazines **47** at reduced temperatures (Scheme 21a) [109]. Later investigations of chiral copper (I) catalysis in this [3 + 3]-cycloaddition revealed that optimal Sabox ligand **L9** used with $Cu(MeCN)_4BF_4$ achieved high yields and enantioselectivities in the synthesis of chiral bi- and tricyclic diazines **49** at room temperature, and this catalyst was superior to the corresponding chiral rhodium(II) catalysts. Notably, compounds **49** are enantiomers of **47**, and can be obtained using readily available Sabox ligands which are derivatives of natural amino acids. The $Cu(MeCN)_4BF_4$/**L9** catalytic system was found to be the best and general in [3 + 3]-cycloadditions of *N*-acylimino-pyridinium or -quinolinium ylides not only with enoldiazoacetates but with enoldiazo-ketones and -acetamides to form chiral diazine derivatives **48** and **50**, respectively (Scheme 24a) [100].

The [3 + 3]-cycloaddition of enoldiazoacetates with pyridinium and isoquinolinium dicyano- or dicarbomethoxy-methylides (Scheme 24b) has been reported to form chiral bi- or tricyclic azines **51** in moderate to high yields and stereoselectivities using the chiral rhodium(II) catalyst $Rh_2(S\text{-PTIL})_4$

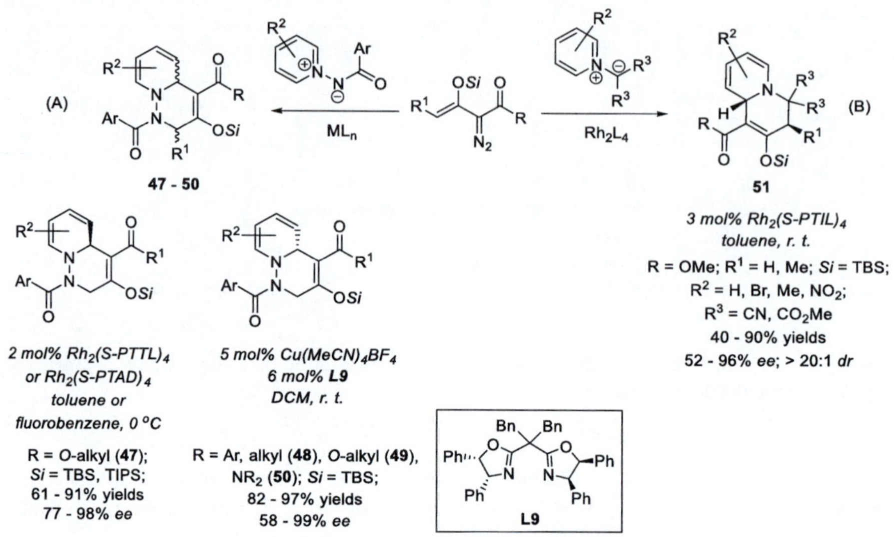

Scheme 24 Asymmetric [3 + 3]-cycloaddition of silyl-protected enoldiazo compounds with pyridinium ylides

[110]. However, copper(I) catalysts were ineffective, and their use did not lead to the formation of **51**. Interestingly, to achieve high yields and stereoselectivities of **51**, at least 3 mol% of Rh_2(*S*-PTIL)$_4$ was required. Lowering the Rh(II) catalyst loading led to mixtures of desired [3 + 3]-cycloaddition product **51** and products of competing [3 + 2]-cycloaddition with the corresponding DACP formed from the enoldiazoacetate. This result was one of the first discoveries of DACPs acting as a resting state of metallo-vinyl carbenes and named in literature as a "boomerang effect" [110]. A detailed mechanism, selectivity, and ligand effect in these competing [3 + 3]- and [3 + 2]-cycloaddition reactions were further investigated by DFT calculations [111].

It is worth mentioning that [3 + 3]-cycloaddition of silyl-protected enoldiazoacetates with achiral dipolarophiles follow the general vinylogous addition/ring closure mechanism. They form bi- and tricyclic heterocyclic systems in high yields and diastereocontrol without use of a chiral catalyst. For example, azomethine imines, which are commonly used in the synthesis of five-membered heterocycles via [3 + 2]-cycloaddition [112], were utilized in the synthesis of pyrazoline-fused six-membered heterocycles via [3 + 3]-cycloaddition with trimethylenemethane and propargyl esters under palladium [113] and gold catalysis [114], respectively. Only one example of the [3 + 3]-cycloaddition of enoldiazoacetates with azomethine imines has been reported; using dirhodium (II) tetraacetate to form bicyclic pyrazolines **52** this cycloaddition occurred with high diastereocontrol toward the *cis*-stereoisomer (Scheme 25a) [115]. According to computational studies [116], the origin of the *cis*-diastereocontrol is in the combination of steric and electronic effects that reduce the repulsion between the R^3 and

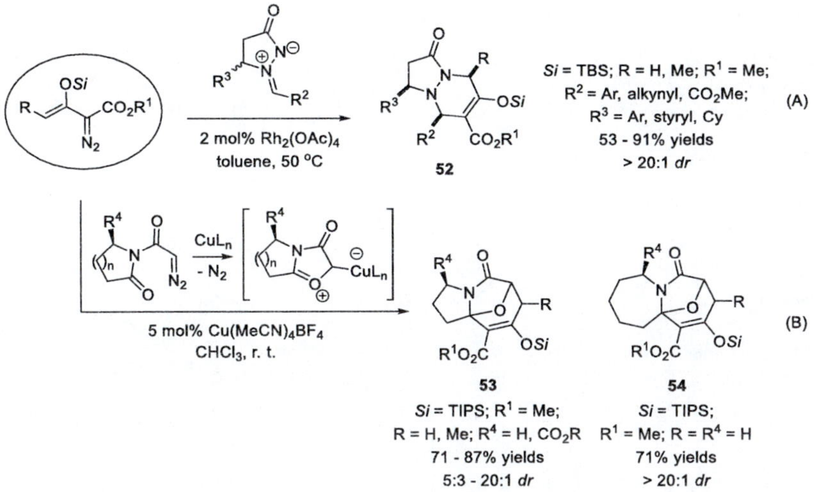

Scheme 25 Diastereoselective synthesis of bi- and tri-cyclic heterocycles via [3 + 3]-cycloaddition of silyl-protected enoldiazoacetates with achiral dipolarophiles

silyloxy groups in the rhodium(II)-vinylcarbene complex and transition state for cyclization.

Highly substituted azepine tricyclic systems **53** and **54** have been constructed via [3 + 3]-cycloaddition of enoldiazoacetates with cyclic carbonyl ylides, which were generated from five- or seven-membered α-diazocarboximides using copper (I) catalyst $Cu(MeCN)_4BF_4$ (Scheme 25b) [117]. The reaction occurs in high yields and with good to excellent diastereocontrol when R = H (up to >20:1 *dr*), however the diastereoselectivity of this transformation is very poor when R = Me (5:3 *dr*). The introduction of substituent R^4 in the reactant α-diazocarboximides generates a chiral center and allows formation of the enantiopure heterocycles **53** as the result of highly diastereoselective [3 + 3]-cycloaddition.

An important type of reaction for the construction of heterocycles that involves metal carbenes is "formal" cycloaddition. The [3 + 3]-cycloaddition of metallo-vinylcarbenes follows the general vinylogous addition/ring closure mechanism, and often occurs through formation of stable intermediates, but the net sum of atoms in the cycle mimics the [3 + 3]-cycloaddition process.

A formal chiral rhodium(II)-catalyzed [3 + 3]-cycloaddition of silyl-protected enoldiazoacetates with phenylhydrazones has been reported as a two-step process [118]. The first step involves the enantioselective N-H insertion at the vinylogous position, presumably via formation of an ammonium ylide followed by proton induced catalyst expulsion to form phenylhydrazone derivative **55** (Scheme 26a). Rh_2(*R*-PTL$)_4$ was found to be optimal chiral catalyst to achieve high yields and enantioselectivities in the N-H insertion step (up to 97% *ee*). Further heterocyclization is a separate step catalyzed by a Lewis acid $Sc(OTf)_3$, and affords chiral pyridazines **56** by Mannich addition/cyclization and a one-pot silyl group

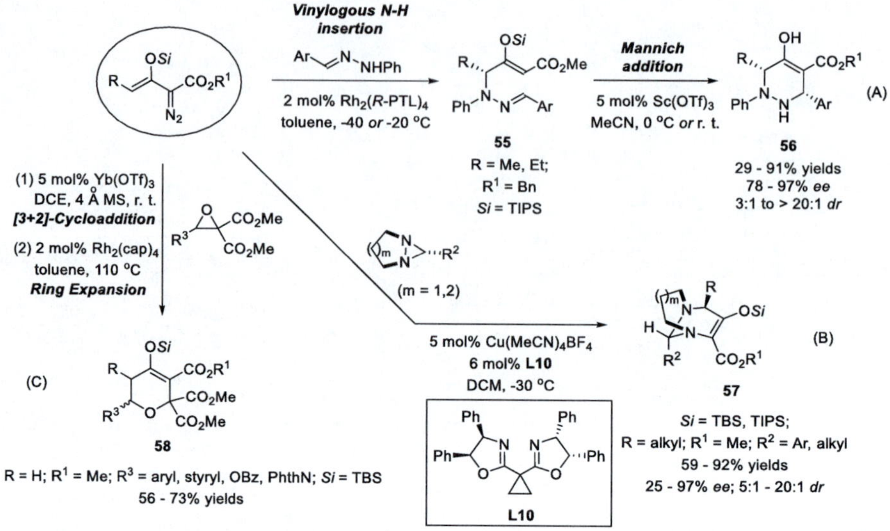

Scheme 26 Examples of formal [3 + 3]-cycloaddition of silyl-protected enoldiazoacetates to form nitrogen (**56**, **57**) and oxygen (**58**) containing heterocycles

removal with high *cis*-diastereocontrol (up to >20:1 *dr*) and complete retention of enantiopurity.

A chiral copper(I)-catalyzed formal [3 + 3]-cycloaddition of (*Z*)-4-alkyl-substituted enoldiazoacetates with symmetric diaziridines undergoes desymmetrization of the reactant diaziridines and subsequent N–N bond cleavage to produce 1,5-diazabicyclo[*n*.3.1]non-2-ene derivatives **43** in high yields and stereocontrol that is optimal with Box ligand **L10** (Scheme 26b) [119]. The reaction is considered to be a formal [3 + 3]-cycloaddition because after the initial vinylogous addition, N–N bond cleavage is required to complete ring closure. Notably, the assessment of reactivity and selectivity of (*Z*)- and (*E*)-4-alkyl-substituted enoldiazoacetates, as well as the corresponding DACP in this asymmetric [3 + 3]-cycloaddition indicated high reactivity and stereoselectivity of the (*Z*)-isomer and the DACP. However, the (*E*)-isomer demonstrated slow conversion and low stereoselectivity. This outcome is different from the asymmetric [3 + 3]-cycloaddition of 4-alkyl-substituted enoldiazoacetates with nitrones, which indicated no effect of *Z*- or *E*-geometrical isomers to yields and stereoselectivities of the [3 + 3]-cycloaddition [11, 105].

Another type of formal [3 + 3]-cycloaddition to form heterocycles is a sequential [3 + 2]-cycloaddition/ring expansion of enoldiazoacetates with strained activated oxiranes to form highly substituted tetrahydropyrans **58** (Scheme 26c) [120]. The first step is Lewis acid catalyzed, which favors reactivity of the C=C bond of the enoldiazoacetate with the oxirane ring, and affords a [3 + 2]-cycloaddition product with the diazo group being intact (was isolated and characterized). Subsequent dinitrogen extrusion with dirhodium(II) tetracaprolactamate $Rh_2(cap)_4$ generates a

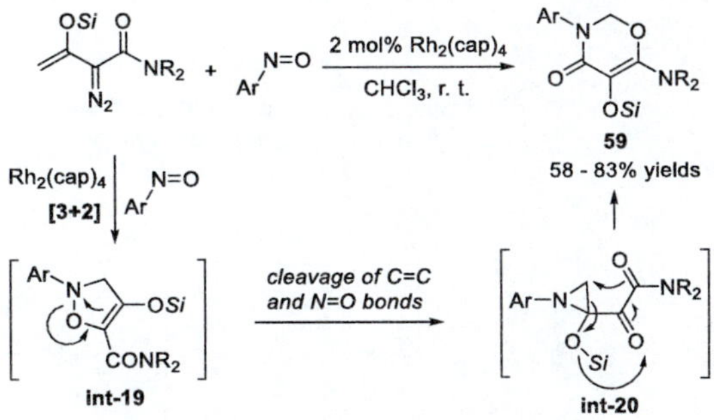

Scheme 27 Synthesis of 2,3-dihydro-4*H*-1,3-oxazin-4-ones **59** via rhodium(II)-catalyzed formal [3 + 3]-cycloaddition of enoldiazoacetamides with nitrosoarenes: a cascade of [3 + 2]-cycloaddition and rearrangements

rhodium(II) carbene, which then initiates ring expansion to form highly substituted tetrahydropyrans **58** and complete formal [3 + 3]-cycloaddition (see, Scheme 9a for the similar mechanism). This formal [3 + 3]-cycloaddition was also carried out in one pot via sequential addition of $Yb(OTf)_3$ and $Rh_2(cap)_4$ catalysts to a mixture of reactants.

An elegant synthesis of substituted 2,3-dihydro-4*H*-1,3-oxazin-4-ones **59** has been reported via formal [3 + 3]-cycloaddition of silyl-protected enoldiazoacetamides with nitrosoarenes using $Rh_2(cap)_4$ catalyst (Scheme 27) [82]. The process involves a complex cascade of [3 + 2]-cycloaddition, rearrangements with the cleavage of C=C and N=O bonds, silyl group transfer, and heterocyclization through proposed intermediates **int-19** and **int-20**. In contrast, a more Lewis acidic $Rh_2(oct)_4$ catalyst used with the same reactants leads to five-membered heterocycle **21** through a different complex rearrangement cascade (see, Scheme 14).

5.2 [4 + 2]-Cycloaddition

The synthesis of six-membered heterocycles via [4 + 2]-cycloaddition is well-known and documented in hetero-Diels–Alder reactions that occur between heteroatom-containing diene and ene surrogates. This transformation demonstrated broad applications in the synthesis of biologically relevant compounds and natural products [121, 122]. However, to the best of my knowledge, the construction of six-membered heterocycles via direct [4 + 2]-cycloaddition that involves metal carbenes has not yet been reported.

An example of "formal" [4 + 2]-cycloaddition has been reported in the synthesis of highly substituted indolizidinone derivatives **60** via copper(I)-catalyzed cascade reaction of enol-substituted carbonyl ylides (generated in situ from enoldiazoimides) and sulfur ylides (Scheme 28) [123]. The reaction occurs through a copper(I)-vinylcarbene (**int-21**) that generates proposed enol-substituted carbonyl ylides (**int-22** and **int-23**) with subsequent nucleophilic attack by the sulfur ylide to form **int-24** followed by a cascade of rearrangements (C–O bond cleavage, silyl and acetyl

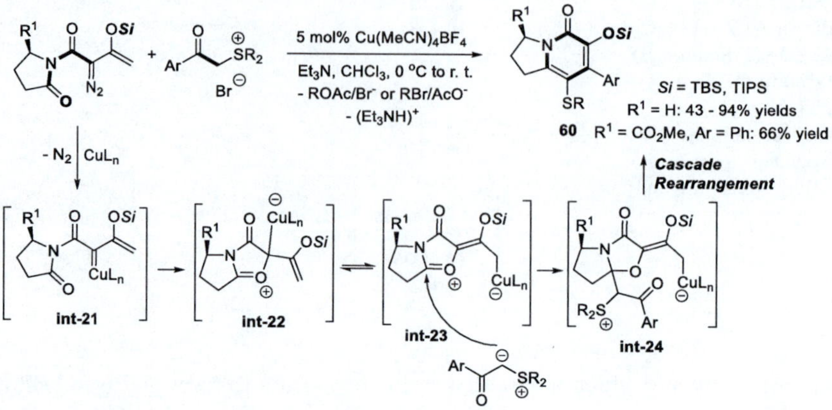

Scheme 28 Synthesis of highly substituted indolizidinone derivatives **60** via copper(I)-catalyzed formal [4 + 2]-cycloaddition of enoldiazoimides with sulfur ylides

group migration, and C–S bond cleavage) to produce uniquely substituted indolizidinones **60** not available by other synthetic methods. One example of the synthesis of a chiral indolizidinone has been demonstrated by the introduction of an asymmetric center to the reactant enoldiazoimide ($R^1 = CO_2Me$).

5.3 [5 + 1]-Cycloaddition

[5 + 1]-Cycloaddition reactions with metallo-carbenes that lead to six-membered heterocycles are even more scarce in the literature than are [4 + 1]-cycloadditions (see, Scheme 17) due to a higher entropy for the intramolecular cyclization. Similarly, the [5 + 1]-cycloaddition is a cascade process that involves initial nucleophilic attack at the carbene center with subsequent 1,6-heterocyclization on the same carbon atom to produce a six-membered heterocycle. Two examples of this cycloaddition have been reported for the synthesis of spirocyclic oxygen-nitrogen containing heterocycles (Scheme 29) [124, 125]. A robust synthesis of the spiro [chroman-4,3′-oxindole] derivatives **61** was realized via a sequential aromatic C–H insertion of activated α-phenoxy ketones to a rhodium(II) carbene (**int-25**) generated from 3-diazooxindoles with subsequent spirocyclization of **int-27** that involves intramolecular aldol-type trapping, hydrogen transfer, and elimination of the rhodium(II) catalyst (Scheme 29a) [124]. This $Rh_2(OAc)_4$-catalyzed cascade reaction is a formal [5 + 1]-cycloaddition that occurs with moderate to high diastereocontrol (5:1–20:1 *dr*) and provides access to rare heterocyclic spiro[chroman-4,3′-oxindoles] **61** in high yields. Similar 3-diazoindolin-2-imine reactants have been reported to undergo $Rh_2(esp)_2$-catalyzed cascade reaction with benzo[*d*]isoxazoles to produce substituted spiro[benzo[*e*][1,3]oxazine-2,3′-indolin]-2-imine derivatives **62** in moderate yields (Scheme 29b) [125]. This formal [5 + 1]-cycloaddition starts with a

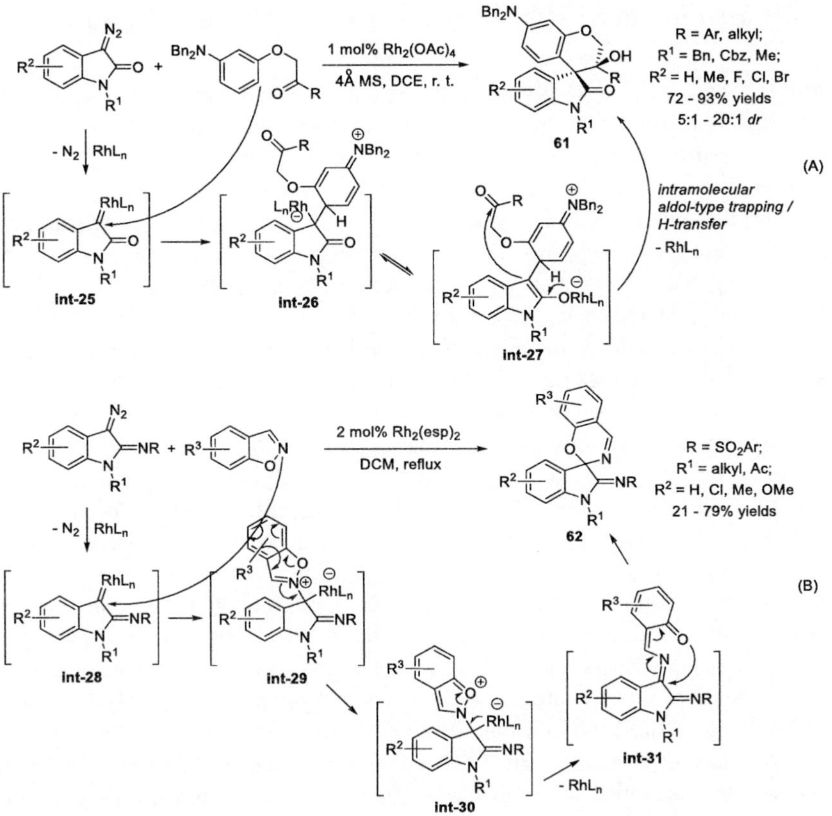

Scheme 29 Representative examples of the synthesis of spirocyclic compounds via Rh(II)-catalyzed formal [5 + 1]-cycloaddition

traditional nucleophilic attack of the *N*-atom of a benzo[*d*]isoxazole onto the Rh(II)-carbenic center of **int-28** to generate **int-29** and **int-30**. Subsequent benzo[*d*]isoxazole ring opening accompanied by elimination of Rh(II)-catalyst forms **int-31**, which then undergoes an uncatalyzed recyclization to spiro[benzo[*e*][1,3]oxazine-2,3′-indolin]-2-imine **62**. In addition, the reaction of a 3-diazoindolin-2-imine with a simple isoxazole did not form the desired [5 + 1]-cycloaddition product, which indicates that the extended conjugation of the benzenoid moiety in benzo[d]isoxazoles favors ring opening and recyclization to complete the formal [5 + 1]-cycloaddition.

6 Seven-Membered Heterocycles: [3 + 4]-Cycloaddition

Most of the reported [3 + 4]-cycloaddition reactions that involve metal carbenes are cascade cyclopropanation/Cope rearrangement transformations. In fact, a metal carbene is involved only in the cyclopropanation step – formation of three-membered carbocycles, which are not part of this chapter. Nevertheless, it is worth mentioning that this formal [3 + 4]-cycloaddition is an important and versatile tool for the construction of seven-membered heterocycles, which are common structural units of many natural products [126], including tropane alkaloids [127]. In this cascade process, stereocontrol is established in the cyclopropanation step, while the Cope rearrangement is stereospecific. One the most important formal [3 + 4]-cycloaddition that involves metal carbenes is the chiral rhodium(II)-catalyzed reaction of silyl-protected enoldiazoacetates with substituted pyrroles to construct a tropane heterocyclic system in a stereoselective manner. The latter have been successfully employed in total syntheses of naturally occurring isostemofoline [128] and several batzelladines [129, 130]. Examples of the synthesis of substituted 4,5-dihydrooxepine heterocycles via catalytic formal [3 + 4]-cycloaddition have also been documented [131].

[3 + 4]-Cycloaddition reactions which occur through a zwitterionic intermediate are rare due to kinetic and thermodynamic factors of the addition and cyclization processes. To my knowledge, there is one report on the synthesis of seven-membered heterocycles via direct [3 + 4]-cycloaddition: a gold(I)-catalyzed reaction of silyl-protected enoldiazo compounds (enoldiazo-acetates and -acetamides) with hexahydro-*s*-triazines (Scheme 30) [92]. The requirement for the [3 + 4]-cycloaddition is that the hexahydro-*s*-triazines must react directly through **int-32**, and not as dissociated formaldimine monomers, which would undergo [3 + 2]-cycloaddition. The Au[^{t}BuXPhos]Cl-catalyzed [3 + 4]-cycloaddition allowed access to pentasubstituted tetrahydro-1*H*-1,3-diazepines **63** in moderate yields. In contrast, gold-vinylcarbenes formed from β-unsubstituted vinyldiazoacetates undergo [4 + 1]-cycloaddition under the same reaction conditions to form five-membered heterocycles (see, Scheme 19).

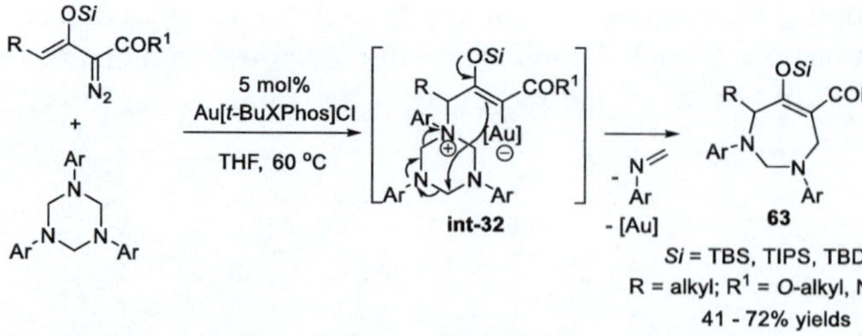

Scheme 30 Synthesis of pentasubstituted tetrahydro-1*H*-1,3-diazepines **63** via gold(I)-catalyzed [3 + 4]-cycloaddition of enoldiazo compounds with hexahydro-*s*-triazines

7 Eight-Membered Heterocycles: [3 + 5]-Cycloaddition

The construction of eight-membered heterocycles via cycloaddition reactions has been challenging, and [3 + 5]-cycloadditions are very rare. One of the recently developed methods for generation of a 1,5-dipole is palladium-catalyzed decarboxylative ring opening of vinylethylene carbonates (VECs), which form zwitterionic allylpalladium intermediates suitable for [3 + 5]-cycloaddition with 1,3-dipolarophiles [132, 133]. The first report on [3 + 5]-cycloaddition of 1,5-dipolarophiles with metallo-vinylcarbenes is the rhodium(II)-catalyzed reaction of pyridinium zwitterions with TBS protected enoldiazoacetates to form substituted diazocine derivatives **64** in moderate 43–73% yields (Scheme 31a) [134]. Interestingly, this dearomative cycloaddition follows the general vinylogous addition/ring closure mechanism, despite the formed zwitterionic intermediate (**int-33**) being remote. One asymmetric example of this rhodium(II)-catalyzed cycloaddition was reported to occur with high enantiocontrol (63% yield, 90% *ee* for **65**) using the chiral dirhodium(II) carboxylate catalyst $Rh_2(S\text{-PTAD})_4$, but the absolute configuration of **65** was not determined. This initial result of asymmetric

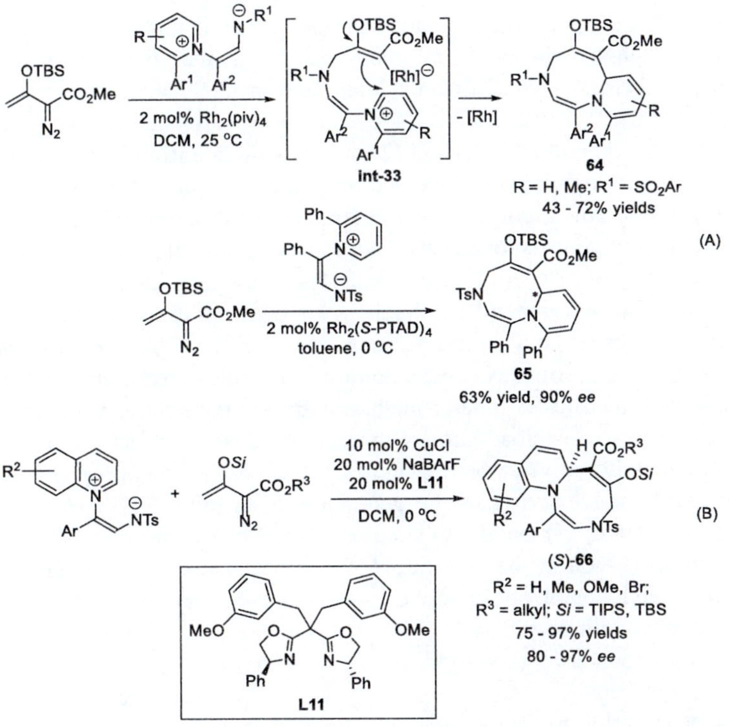

Scheme 31 Dearomative [3 + 5]-cycloaddition of pyridinium and quinolinium zwitterions with electrophilic rhodium(II)- and copper(I)-vinylcarbenes via vinylogous addition/ring closure

[3 + 5]-cycloaddition demonstrates its potential for remote asymmetric control via chiral dirhodium(II) catalysis.

In a very recent report, the same research group discovered a chiral copper(I)-catalyzed version of this dearomative [3 + 5]-cycloaddition, however, while quinolinium zwitterions were suitable substrates, pyridinium zwitterions were unreactive (Scheme 31b) [135]. Catalyst and ligand optimization revealed that 10 mol% of CuCl, 20 mol% of NaBArF, and 20 mol% of chiral Sabox ligand **L11** are required to achieve high yields and enantioselectivities (up to 97% *ee*) in the synthesis of chiral diazocines (*S*)-**66** at 0°C.

8 Conclusion

The synthesis of heterocyclic compounds via catalytic cycloaddition reactions of metallocarbenes is robust, atom-economic, and often allows access to chiral heterocycles with high levels of stereocontrol. This methodology has demonstrated advantages over other metal- and organo-catalytic approaches toward heterocycles (e.g., lower catalyst loading, use of inexpensive metal catalysts, mild reaction conditions, excellent stereoselectivities, etc.). This chapter is comprehensive and covers all available literature reports on cycloaddition using metallocarbenes to produce different ring size heterocycles: from small strained tri-membered to medium-size eight-membered rings. Two distinct cycloaddition mechanisms with metal carbenes have been described: (1) nucleophilic addition at the "carbenic" carbon followed by ring closure ([*n* + 1] and [3 + 2]) and (2) nucleophilic addition at the "vinylogous" carbon of metallo-vinylcarbenes followed by ring closure and elimination of a metal catalyst ([3 + *n*]). An important method for the construction of structurally very unique heterocyclic rings is formal cycloaddition with metallo-vinyl carbenes, and it includes examples of [3 + 2]-, [3 + 3]-, and [4 + 2]-cycloadditions. These reactions do not follow the two general mechanisms of cycloaddition, and occur through sequences of cascades and rearrangements. A significant part of this chapter is devoted to catalytic asymmetric cycloaddition of metallo-vinyl carbenes (especially 3-silyloxy-substituted) as a general methodology for the synthesis of heterocyclic compounds via [3 + *n*]-cycloaddition. The established connection between enoldiazo compounds, metallo-vinyl carbenes, and donor–acceptor cyclopropenes (DACPs) makes DACPs suitable surrogates for generation of metallo-vinyl carbenes and use in asymmetric [3 + *n*]-cycloaddition reactions. Notably, the presence of an electron-donating group (EDG) at the 3-position of a vinyldiazo compound facilitates the "vinylogous" nucleophilic addition of a 1,3-dipolarophile to a metallo-vinyl carbene and selectively completes the [3 + 3]-cycloaddition, while 3-unsubstituted vinyldiazo compounds (e.g., styryldiazo compounds) in metal catalyzed reactions with a 1,3-dipolarophile undergo [3 + 2]-cycloaddition. The role of an EDG at the 3-position of a vinyldiazo compound is evident, and studies on [3 + *n*]-cycloadditions with other EDGs (besides reported silyloxy, aryl, or alkyl) are open for future discoveries. As seen in all sections of this chapter, various transition metals have been utilized as catalysts for asymmetric cycloaddition with diazo compounds or sulfur ylides, but cycloaddition reactions catalyzed by chiral dirhodium(II) and, more recently, by copper(I) with chiral Box ligands are the most explored and have the broadest applications. Chiral copper(I) as a less costly alternative to chiral dirhodium(II) catalysts has demonstrated higher levels of stereoselectivities in a variety of cycloadditions. Future studies on cycloaddition reactions using other transition metals, ligand design, and structural variations of metallocarbene precursors and dipolarophiles open enormous possibilities for the synthesis of new heterocyclic compounds not yet available by other synthetic methods.

References

1. Carruthers W (1990) Cycloaddition reactions in organic synthesis. Pergamon Press, Oxford
2. Kobayashi S, Jorgensen K (2003) Cycloaddition reactions in organic synthesis. Wiley VCH, Weinheim
3. IUPAC Gold Book. http://goldbook.iupac.org
4. Katritzky AR (2004) Introduction: heterocycles. Chem Rev 104:2125–2126
5. Kabir E, Uzzaman M (2022) A review on biological and medicinal impact of heterocyclic compounds. Res Chem 4:100606
6. Moss RA, Doyle MP (eds) (2014) Contemporary carbene chemistry. Wiley, New York
7. Xu X, Doyle MP (2014) The [3 + 3]-cycloaddition alternative for heterocycle syntheses: catalytically generated metalloenolcarbenes as dipolar adducts. Acc Chem Res 47:1396–1405
8. Deng Y, Cheng Q-Q, Doyle MP (2017) Asymmetric [3+3] cycloaddition for heterocycle synthesis. Synlett 28:1695–1706
9. Marichev KO, Doyle MP (2019) Catalytic asymmetric cycloaddition reactions of enoldiazo compounds. Org Biomol Chem 17:4183–4195
10. Müller P, Pautex N, Doyle MP, Bagheri V (1990) Rh(II) catalyzed isomerizations of cyclopropenes. Evidence for Rh(II)-complexed vinylcarbene intermediates. Helv Chim Acta 73:1233–1241
11. Dong K, Marichev KO, Doyle MP (2019) The role of donor–acceptor cyclopropenes in metal carbene reactions. Conversion of *E*-substituted enoldiazoacetates to *Z*-substituted metallo-enolcarbenes. Organometallics 38:4043–4050
12. Prileschajew N (1909) Oxydation ungesättigter Verbindungen mittels organischer superoxyde [oxidation of unsaturated compounds by means of organic peroxides]. Ber Dtsch Chem Ges (in German) 42:4811–4815
13. Rossiter B, Katsuki T, Sharpless KB (1981) Asymmetric epoxidation provides shortest routes to four chiral epoxy alcohols which are key intermediates in syntheses of methymycin, erythromycin, leukotriene C-1, and disparlure. J Am Chem Soc 103:464–465
14. Martin V, Woodard S, Katsuki T, Yamada Y, Ikeda M, Sharpless KB (1981) Kinetic resolution of racemic allylic alcohols by enantioselective epoxidation. A route to substances of absolute enantiomeric purity? J Am Chem Soc 103:6237–6240
15. Finn MG, Sharpless KB (1991) Mechanism of asymmetric epoxidation. 2. Catalyst structure. J Am Chem Soc 113:113–126
16. Corey EJ, Shibata S, Bakshi RK (1988) An efficient and catalytically enantioselective route to (*S*)-(−)-phenyloxirane. J Org Chem 53:2861–2863
17. Zhang W, Loebach JL, Wilson SR, Jacobsen EN (1990) Enantioselective epoxidation of unfunctionalized olefins catalyzed by salen manganese complexes. J Am Chem Soc 112: 2801–2803

18. Jacobsen EN, Zhang W, Muci AR, Ecker JR, Deng L (1991) Highly enantioselective epoxidation catalysts derived from 1,2-diaminocyclohexane. J Am Chem Soc 113:7063–7064
19. Corey EJ, Chaykovsky M (1965) Dimethyloxosulfonium methylide ($(CH_3)_2SOCH_2$) and dimethylsulfonium methylide ($(CH_3)_2SCH_2$). Formation and application to organic synthesis. J Am Chem Soc 87:1353–1364
20. Furukawa N, Sugihara Y, Fujihara H (1989) Camphoryl sulfide as a chiral auxiliary and a mediator for one-step synthesis of optically active 1,2-diaryloxiranes. J Org Chem 54:4222–4224
21. Li A-H, Dai L-X, Aggarwal VK (1997) Asymmetric ylide reactions: epoxidation, cyclopropanation, aziridination, olefination, and rearrangement. Chem Rev 97:2341–2372
22. Aggarwal VK, Richardson J (2003) The complexity of catalysis: origins of enantio- and diastereocontrol in sulfur ylide mediated epoxidation reactions. Chem Commun:2644–2651
23. Aggarwal VK, Winn CL (2004) Catalytic, asymmetric sulfur ylide-mediated epoxidation of carbonyl compounds: scope, selectivity, and applications in synthesis. Acc Chem Res 37:611–620
24. Julienne K, Metzner P, Henyron V (1999) Asymmetric synthesis of epoxides from aldehydes mediated by (+)-(2*R*,5*R*)-2,5-dimethylthiolane. J Chem Soc Perkin Trans 1:731–735
25. Zanardi J, Leriverend C, Aubert D, Julienne K, Metzner P (2001) Catalytic cycle for the asymmetric synthesis of epoxides using sulfur ylides. J Org Chem 66:5620–5623
26. Winn CL, Bellanie B, Goodman JM (2002) A highly enantioselective one-pot sulfur ylide epoxidation reaction. Tetrahedron Lett 43:5427–5430
27. Solladie-Cavallo A, Diep-Vohuule A, Sunjic V, Vinkovic V (1996) A two-step asymmetric synthesis of pure *trans*-(*R*,*R*)-diaryl-epoxides. Tetrahedron Asymmetry 7:1783–1788
28. Solladie-Cavallo A, Roje M, Isarno T, Sunjic V, Vinkovic V (2000) Pyridyl and furyl epoxides of more than 99% enantiomeric purities: the use of a phosphazene base. Eur J Org Chem:1077–1080
29. Breau L, Ogilvie WW, Durst T (1990) Preparation of optically active stilbene oxides *via* sulfonium salts derived from C2 symmetric thiolanes. Tetrahedron Lett 31:35–38
30. Aggarwal VK, Ford JG, Fonquerna S, Adams H, Jones RVH, Fieldhouse R (1998) Catalytic asymmetric epoxidation of aldehydes. Optimization, mechanism and discovery of stereoelectronic control involving a combination of anomeric and Cieplak effects in sulfur ylide epoxidations with chiral 1,3-oxathianes. J Am Chem Soc 120:8328–8339
31. Aggarwal VK (1998) Catalytic asymmetric epoxidation and aziridination mediated by sulfur ylides. Evolution of a project. Synlett 4:329–336
32. Aggarwal VK, Alonso E, Hynd G, Lydon KM, Palmer MJ, Porcelloni M, Studley JR (2001) Catalytic asymmetric synthesis of epoxides from aldehydes using sulfur ylides with *in situ* generation of diazocompounds. Angew Chem Int Ed 40:1430–1433
33. Evans DA, Faul MM, Bilodeau MT (1991) Copper-catalyzed aziridination of olefins by (*N*-(*p*-toluenesulfonyl)imino)phenyliodinane. J Org Chem 56:6744–6746
34. Catino AJ, Nichols JM, Forslund RE, Doyle MP (2005) Efficient aziridination of olefins catalyzed by mixed-valent dirhodium(II,III) caprolactamate. Org Lett 7:2787–2790
35. Chang JWW, Ton TMU, Chan PWH (2011) Transition-metal-catalyzed aminations and aziridinations of C–H and C–C bonds with iminoiodinanes. Chem Rec 11:331–357
36. Driver TG (2010) Recent advances in transition metal-catalyzed *N*-atom transfer reactions of azides. Org Biomol Chem 8:3831–3846
37. Cramer SA, Jenkins DM (2011) Synthesis of aziridines from alkenes and aryl azides with a reusable macrocyclic tetracarbene iron catalyst. J Am Chem Soc 133:19342–19345
38. Sabir S, Pandey CB, Yadav AK, Tiwari B, Jat JL (2018) Direct N-H/N-Me aziridination of unactivated olefins using *O*-(sulfonyl)hydroxylamines as aminating agents. J Org Chem 83: 12255–12260
39. Shimbayashi T, Sasakura K, Eguchi A, Okamoto K, Ohe K (2019) Recent progress on cyclic nitrenoid precursors in transition-metal-catalyzed nitrene-transfer reactions. Chem A Eur J 25: 3156–3180

40. Deng T, Mazumdar W, Yoshinaga Y, Patel PB, Malo D, Malo T, Wink DJ, Driver TG (2021) Rh_2(II)-catalyzed intermolecular *N*-aryl aziridination of olefins using nonactivated N atom precursors. J Am Chem Soc 143:19149–19159
41. Tan H, Samanta S, Maity A, Roychowdhury P, Powers DC (2022) *N*-Aminopyridinium reagents as traceless activating groups in the synthesis of *N*-aryl aziridines. Nat Commun 13:3341
42. Concellón JM, Rodríguez-Solla H, Simal C (2008) Addition reactions of iodomethyllithium to imines. A direct and efficient synthesis of aziridines and enantiopure amino aziridines. Org Lett 10:4457–4460
43. Yadav LDS, Kapoor R, Garima (2009) Organocatalytic stereoselective aziridination of imines *via* ammonium ylides. Synlett 19:3123–3126
44. Aggarwal VK, Ferrara M (2000) Highly selective aziridination of imines using trimethylsilyldiazomethane and applications of C-silylaziridines in synthesis. Org Lett 2: 4107–4110
45. Yadav JS, Reddy BVS, Reddy PN, Shesha Rao M (2003) Bi(OTf)3-[Bmim]PF6: a novel and reusable catalytic system for the synthesis of *cis*-aziridine carboxylates. Synthesis 9:1387–1390
46. Borkin D, Carlson A, Török B (2010) K-10-catalyzed highly diastereoselective synthesis of aziridines. Synlett 5:745–748
47. Williams AL, Johnston JN (2004) The Brønsted acid-catalyzed direct Aza-Darzens synthesis of *N*-alkyl *cis*-aziridines. J Am Chem Soc 126:1612–1613
48. Zeng X, Zeng X, Xu Z, Lu M, Zhong G (2009) Highly efficient asymmetric *trans*-selective aziridination of diazoacetamides and *N*-Boc-imines catalyzed by chiral Brønsted acids. Org Lett 11:3036–3039
49. Aggarwal VK, Vasse J-L (2003) Asymmetric sulfur ylide mediated aziridination: application in the synthesis of the side chain of Taxol. Org Lett 5:3987–3990
50. Cheng X, Cai B-G, Mao H, Lu J, Li L, Wang K, Xuan J (2021) Divergent synthesis of aziridine and imidazolidine frameworks under blue LED irradiation. Org Lett 23:4109–4114
51. Yoda H, Takahashi M, Sengoku T (2011) Majumdar KC, Chattopadhyay SK (eds) Heterocycles in natural product synthesis1st edn. Wiley-VCH, Weinheim, pp 41–62
52. Brandi A, Cicchi S, Cordero FM (2008) Novel syntheses of azetidines and azetidinones. Chem Rev 108:3988–4035
53. Antermite D, Degennaro L, Luisi R (2017) Recent advances in the chemistry of metallated azetidines. Org Biomol Chem 15:34–50
54. West FG, Bott TM (2012) Preparation and synthetic applications of azetidines. Heterocycles 84:223–264
55. Ju Y, Varma RS (2006) Aqueous *N*-heterocyclization of primary amines and hydrazines with dihalides: microwave-assisted syntheses of *N*-azacycloalkanes, isoindole, pyrazole, pyrazolidine, and phthalazine derivatives. J Org Chem 71:135–141
56. Hillier MC, Chen C-Y (2006) A one-pot preparation of 1,3-disubstituted azetidines. J Org Chem 71:7885–7887
57. Xu P, Zhang M, Ingoglia B, Allais C, Dechert-Schmitt A-MR, Singer RA, Morken JP (2021) Construction of azacycles by intramolecular amination of organoboronates and organobis (boronates). Org Lett 23:3379–3383
58. Malik S, Nadir UK (2008) A facile synthesis of 1-arenesulfonylazetidines through reaction of 1-arenesulfonylaziridines with dimethylsulfoxonium methylide generated under microwave irradiation. Synlett 1:108–110
59. Han J-Q, Zhang H-H, Xu P-F, Luo Y-C (2016) Lewis acid and (hypo)iodite relay catalysis allows a strategy for the synthesis of polysubstituted azetidines and tetrahydroquinolines. Org Lett 18:5212–5215
60. Fawcett A, Murtaza A, Gregson CHU, Aggarwal VK (2019) Strain-release-driven homologation of boronic esters: application to the modular synthesis of azetidines. J Am Chem Soc 141: 4573–4578

61. Becker MR, Wearing ER, Schindler CS (2020) Synthesis of azetidines *via* visible-light-mediated intermolecular [2+2] photocycloadditions. Nat Chem 12:898–905
62. Hodgson DM, Pearson CI, Kazmi M (2014) Generation and electrophile trapping of *N*-Boc-2-lithio-2-azetine: synthesis of 2-substituted 2-azetines. Org Lett 16:856–859
63. Baumann AN, Eisold M, Music A, Haas G, Kiw YM, Didier D (2017) Methods for the synthesis of substituted azetines. Org Lett 19:5681–5684
64. Pang S, Yang X, Cao Z-H, Zhang Y-L, Zhao Y, Huang Y-Y (2018) Intermolecular [2 + 2] cycloaddition/isomerization of allenyl imides and unactivated imines for the synthesis of 1-azadienes catalyzed by a $Ni(ClO_4)_2{\cdot}6H_2O$ Lewis acid. ACS Catal 8:5193–5199
65. Li C, Jiang K, Ouyang Q, Liu T-Y, Chen Y-C (2016) [3+1]- and [3+2]-cycloadditions of azaoxyallyl cations and sulfur ylides. Org Lett 18:2738–2741
66. Barluenga J, Riesgo L, Lonzi G, Tomás M, López LA (2012) Copper(I)-catalyzed [3+1] cycloaddition of alkenyldiazoacetates and iminoiodinanes: easy access to substituted 2-azetines. Chem A Eur J 18:9221–9224
67. Deng Y, Massey LA, Zavalij PY, Doyle MP (2017) Catalytic asymmetric [3+1]-cycloaddition reaction of ylides with electrophilic metallo-enolcarbene intermediates. Angew Chem Int Ed 56:7479–7483
68. Marichev KO, Wang K, Dong K, Greco N, Massey LA, Deng Y, Arman H, Doyle MP (2019) Synthesis of chiral tetrasubstituted azetidines from donor–acceptor azetines *via* asymmetric copper(I)-catalyzed imido-ylide [3+1]-cycloaddition with metallo-enolcarbenes. Angew Chem Int Ed 58:16188–16192
69. Marichev KO, Dong K, Massey LA, Deng Y, De Angelis L, Wang K, Arman H, Doyle MP (2019) Chiral donor−acceptor azetines as powerful reactants for synthesis of amino acid derivatives. Nat Commun 10:5328
70. Gothelf KV, Jørgensen KA (1998) Asymmetric 1,3-dipolar cycloaddition reactions. Chem Rev 98:863–910
71. Breugst M, Reissig HU (2020) The Huisgen reaction: milestones of the 1,3-dipolar cycloaddition. Angew Chem Int Ed 59:12293–12307
72. Huisgen R (1961) Centenary lecture – 1,3-dipolar cycloadditions. Proc Chem Soc:357–396
73. Kolb HC, Finn MG, Sharpless KB (2001) Click chemistry: diverse chemical function from a few good reactions. Angew Chem Int Ed 40:2004–2021
74. Kolb HC, Sharpless BK (2003) The growing impact of click chemistry on drug discovery. Drug Discov Today 8:1128–1137
75. Jiang X, Hao X, Jing L, Wu G, Kang D, Liu X, Zhan P (2019) Recent applications of click chemistry in drug discovery. Expert Opin Drug Discovery 14:779–789
76. Tron GC, Rirali T, Billington RA, Canonico PL, Sorba G, Genazzani AA (2008) Click chemistry reactions in medicinal chemistry: applications of the 1,3-dipolar cycloaddition between azides and alkynes. Med Res Rev 28:278–308
77. Doyle MP, Yan M, Hu W, Gronenberg LS (2003) Highly selective catalyst-directed pathways to dihydropyrroles from vinyldiazoacetates and imines. J Am Chem Soc 125:4692–4693
78. Xu X, Zavalij PY, Hu W, Doyle MP (2012) Efficient synthesis of oxazoles by dirhodium(II)-catalyzed reactions of styryl diazoacetate with oximes. Chem Commun 48:11522–11524
79. Qin C, Davies HML (2013) Rh_2(*R*-TPCP)$_4$-catalyzed enantioselective [3+2]-cycloaddition between nitrones and vinyldiazoacetates. J Am Chem Soc 135:14516–14519
80. Wang XC, Abrahams QM, Zavalij PY, Doyle MP (2012) Highly regio- and stereoselective dirhodium vinylcarbene induced nitrone cycloaddition with subsequent cascade carbenoid aromatic cycloaddition/N-O cleavage and rearrangement. Angew Chem Int Ed 51:5907–5910
81. Yang X, Xu P, Xue Y (2014) Mechanism and regioselectivity of the cycloaddition between nitrone and dirhodium vinylcarbene catalyzed by $Rh_2(O_2CH)_4$: a computational study. Theor Chem Accounts 133:1549
82. Cheng Q-Q, Lankelma M, Wherritt D, Arman H, Doyle MP (2017) Divergent rhodium-catalyzed cyclization reactions of enoldiazoacetamides with nitrosoarenes. J Am Chem Soc 139:9839–9842

83. Xu X, Zavalij PY, Hu W, Doyle MP (2013) Vinylogous reactivity of enol diazoacetates with donor–acceptor substituted hydrazones. Synthesis of substituted pyrazole derivatives. J Org Chem 78:1583–1588
84. Xu G, Zhu C, Gu W, Li J, Sun J (2015) Gold(I)-catalyzed diazo cross-coupling: a selective and ligand controlled denitrogenation/cyclization cascade. Angew Chem Int Ed 54:883–887
85. Dong K, Humeidi A, Griffith W, Arman H, Xu X, Doyle MP (2021) Ag^{I}-catalyzed reaction of enol diazoacetates and imino ethers: synthesis of highly functionalized pyrroles. Angew Chem Int Ed 60:13394–13400
86. Dong K, Zheng H, Su Y, Humeidi A, Arman H, Xu X, Doyle MP (2021) Catalyst-directed divergent catalytic approaches to expand structural and functional scaffold diversity *via* metallo-enolcarbene intermediates. ACS Catal 11:4712–4721
87. Yi R, Qian L, Wan B (2019) Synthesis of spiropyrrolidine oxindoles through Rh(II)-catalyzed olefination/cyclization of diazooxindoles and vinyl azides. Chinese J Catal 40:177–183
88. Subba Reddy BV, Pravardhan Reddy E, Sridharc B, Jayaprakash Rao Y (2016) Rhodium-catalyzed cycloaddition of carbonyl ylides for the synthesis of spiro[furo[2,3-*a*] xanthene-2,3′-indolin]-2′-one scaffolds. RSC Adv 6:50497–50499
89. Muthusamy S, Prabu A, Suresh E (2019) Copper-catalyzed synthesis of spiroindolofurobenzopyrans: tandem reactions of diazoamides and O-propargyl salicylaldehydes. Org Biomol Chem 17:8088–8093
90. Sheng G, Ma S, Bai S, Mao J, Lu P, Wang Y (2018) A copper-catalyzed reaction of 3-diazoindolin-2-imines with 2-(phenylamino)ethanols: convenient access to spiro[indoline-3,2′-oxazolidin]-2-imines. Chem Commun 54:1529–1532
91. Reddy ACS, Reddy PM, Anbarasan P (2020) Diastereoselective palladium catalyzed carbenylative amination of ortho-vinylanilines with 3-diazoindolin-2-ones. Adv Synth Catal 362:801–806
92. Zhu C, Xu G, Sun J (2016) Gold-catalyzed formal [4+1]/[4+3] cycloadditions of diazo esters with triazines. Angew Chem Int Ed 55:11867–11871
93. Hyster TK, Ruhl KE, Rovis T (2013) A coupling of benzamides and donor/acceptor diazo compounds to form γ-lactams *via* Rh(III)-catalyzed C−H activation. J Am Chem Soc 135: 5364–5367
94. Lam H-W, Man K-Y, Chan WW, Zhou Z, Yu W-Y (2014) Rhodium(III)-catalyzed formal oxidative [4 + 1] cycloaddition of benzohydroxamic acids and α-diazoesters. A facile synthesis of functionalized benzolactams. Org Biomol Chem 12:4112–4116
95. Ma B, Wu P, Wang X, Wang Z, Lin HX, Dai HX (2019) Efficient synthesis of spirooxindole pyrrolones by a rhodium(III)-catalyzed C–H activation/carbene insertion/Lossen rearrangement sequence. Angew Chem Int Ed 58:13335–13339
96. Anderson LL (2016) Diverse applications of nitrones for the synthesis of heterocyclic compounds. Asian J Org Chem 5:9–30
97. Wang X, Xu X, Zavalij PY, Doyle MP (2011) Asymmeric formal [3+3]-cycloaddition reactions of nitrones with electrophilic vinylcarbene intermediates. J Am Chem Soc 133: 16402–16405
98. Shved AS, Tabolin AA, Novikov RA, Nelyubina YV, Timofeev VP, Ioffe SL (2061) Six-membered cyclic nitroso acetals: synthesis and studies of the nitrogen inversion process of *N*-silyloxy-3,6-dihydro-2*H*-1,2-oxazines. Eur J Org Chem:5569–5578
99. Cheng Q-Q, Yedoyan J, Arman H, Doyle MP (2016) Copper-catalyzed divergent addition reactions of enoldiazoacetamides with nitrones. J Am Chem Soc 138:44–47
100. Marichev KO, Adly FG, Carranco AM, Garcia EC, Arman H, Doyle MP (2018) Catalyst choice for highly enantioselective [3+3]-cycloaddition of enoldiazocarbonyl compounds. ACS Catal 8:10392–10400
101. Marichev KO, Wang Y, Carranco AM, Garcia EC, Yu ZX, Doyle MP (2018) Rhodium(II)-catalysed generation of cycloprop-1-en-1-yl ketones and their rearrangement to 5-aryl-2-siloxyfurans. Chem Commun 54:9513–9516

102. Adly FG, Marichev KO, Jensen JA, Arman H, Doyle MP (2019) Enoldiazosulfones for highly enantioselective [3 + 3]-cycloaddition with nitrones catalyzed by copper(i) with chiral BOX ligands. Org Lett 21:40–44
103. Xu X, Zavalij PJ, Doyle MP (2013) A donor-acceptor cyclopropene as a dipole source for a silver(I) catalyzed asymmetric catalytic [3+3]-cycloaddition with nitrones. Chem Commun 49: 10287–10289
104. Qian Y, Xu X, Wang X, Zavalij PJ, Hu W, Doyle MP (2012) Rhodium(II)- and copper(II)-catalyzed reactions of enol diazoacetates with nitrones: metal carbene versus Lewis acid directed pathways. Angew Chem Int Ed 51:5900–5903
105. Dong K, Xu X, Doyle MP (2020) Copper(I)-catalyzed highly enantioselective [3+3]-cycloaddition of γ-alkyl enoldiazoacetates with nitrones. Org Chem Front 7:1653–1657
106. Zheng H, Faghihi I, Doyle MP (2021) Copper(I)-catalyzed highly enantioselective [3+3]-cycloaddition of β-aryl/alkyl vinyl diazoacetates with nitrones. Helv Chim Acta 104:e2100081
107. Chrzanowska M, Rozwadowska MD (2004) Asymmetric synthesis of isoquinoline alkaloids. Chem Rev 104:3341–3370
108. Guo C, Fleige M, Janssen-Müller D, Daniliuc CG, Glorius F (2015) Switchable selectivity in an NHC-catalysed dearomatizing annulation reaction. Nat Chem 7:842–847
109. Xu X, Zavalij PY, Doyle MP (2013) Highly enantioselective dearomatizing formal [3+3]-cycloaddition reactions of *N*-acyliminopyridinium ylides with electrophilic enolcarbene intermediates. Angew Chem Int Ed 52:12664–12668
110. Xu X, Zavalij PY, Doyle MP (2013) Catalytic asymmetric syntheses of quinolizidines by dirhodium-catalyzed dearomatization of isoquinolinium/pyridinium methylides – the role of catalyst and carbene source. J Am Chem Soc 135:12439–12447
111. Li S-J, Fang D-C (2018) DFT studies on the dirhodium-catalyzed [3 + 2] and [3 + 3] cycloaddition reactions of enol diazoacetates with isoquinolinium methylide: mechanism, selectivity, and ligand effect. Organometallics 37:1373–1380
112. Požgan F, Mamari HA, Grošelj U, Svete J, Štefane B (2018) Synthesis of non-racemic pyrazolines and pyrazolidines by [3+2] cycloadditions of azomethine imines. Molecules 23:3
113. Shintani R, Hayashi T (2006) Palladium-catalyzed [3+3] cycloaddition of trimethylenemethane with azomethine imines. J Am Chem Soc 128:6330–6331
114. Shapiro ND, Shi Y, Toste FD (2009) Gold-catalyzed [3+3]-annulation of azomethine imines with propargyl esters. J Am Chem Soc 131:11654–11655
115. Qian Y, Zavalij PJ, Hu W, Doyle MP (2013) Bicyclic pyrazolidinone derivatives from diastereoselective catalytic [3+3]-cycloaddition reactions of enoldiazoacetates with azomethine imines. Org Lett 15:1564–1567
116. Zhang Y, Yang Y, Zhao J, Xue Y (2018) Mechanism and diastereoselectivity of [3+3] cycloaddition between enol diazoacetate and azomethine imine catalyzed by dirhodium tetracarboxylate: a theoretical study. Eur J Org Chem:3086–3094
117. Deng Y, Massey LA, Rodriguez Núñez YA, Arman H, Doyle MP (2017) Catalytic divergent [3+3]- and [3+2]-cycloaddition by discrimination between diazo compounds. Angew Chem Int Ed 56:12292–12296
118. Xu X, Zavalij PY, Doyle MP (2012) Synthesis of tetrahydropyridazines by a metal-carbene directed highly enantioselective vinylogous N-H insertion/Lewis acid catalyzed diastereoselective Mannich addition. Angew Chem Int Ed 51:9829–9833
119. Zheng H, Doyle MP (2019) Catalytic desymmetric cycloaddition of diaziridines with metalloenolcarbenes. The role of donor-acceptor cyclopropenes. Angew Chem Int Ed 58: 12502–12506
120. Cheng Q-Q, Qian Y, Zavalij PY, Doyle MP (2015) Lewis acid/rhodium-catalyzed formal [3+3]-cycloaddition of enoldiazoacetates with donor–acceptor cyclopropanes. Org Lett 17: 3568–3571
121. Eschenbrenner-Lux V, Kumar K, Waldmann H (2014) The asymmetric hetero-Diels–Alder reaction in the syntheses of biologically relevant compounds. Angew Chem Int Ed 53:11146–11157

122. Heravi MM, Ahmadi T, Ghavidel M, Heidari B, Hamidi H (2015) Recent applications of the hetero Diels–Alder reaction in the total synthesis of natural products. RSC Adv 5:101999–102075
123. Cheng Q-Q, Massey LA, Willett BS, Deng Y, Arman H, Doyle MP (2018) Copper-catalyzed formal [4+2] cycloaddition of enoldiazoimides with sulfur ylides. Angew Chem Int Ed 57: 10343–10346
124. Jia S, Lei Y, Song L, Reddy AGK, Xing D, Hu W (2017) Diastereoselective intramolecular aldol-type trapping of zwitterionic intermediates by ketones for the synthesis of spiro [chroman-4,3′-oxindole] derivatives. Adv Synth Catal 359:58–63
125. Han C, Wu W, Chen Z, Pu S (2019) Rhodium-catalyzed [5+1]-cycloaddition reactions to spiro-benzo[*e*][1,3]oxazineindoline imines. Asian J Org Chem 8:1385–1389
126. Yin Z, He Y, Chiu P (2018) Application of (4+3) cycloaddition strategies in the synthesis of natural products. Chem Soc Rev 47:8881–8924
127. Fodor G, Dharanipragada R (1986) Tropane alkaloids. Nat Prod Rep 3:181–184
128. Kende AS, Smalley Jr TL, Huang H (1999) Total synthesis of (±)-isostemofoline. J Am Chem Soc 121:7431–7432
129. Parr BT, Economou C, Herzon SB (2015) A concise synthesis of (+)-batzelladine B from simple pyrrole-based starting materials. Nature 525:507–510
130. Economou C, Romaire JP, Scott TZ, Parr BT, Herzon SB (2018) A convergent approach to batzelladine alkaloids. Total syntheses of (+)-batzelladine E, (−)-dehydrobatzelladine C, and (+)-batzelladine K. Tetrahedron 74:3188–3197
131. Xu X, Hu W-H, Zavalij PY, Doyle MP (2011) Divergent outcomes of carbene transfer reactions from dirhodium and copper-based catalysts separately or in combination. Angew Chem Int Ed 50:11152–11155
132. Yuan C, Wu Y, Wang D, Zhang Z, Wang C, Zhou L, Zhang C, Song B, Guo H (2018) Formal [5+3] cycloaddition of zwitterionic allylpalladium intermediates with azomethine imines for construction of N,O-containing eight-membered heterocycles. Adv Synth Catal 360:652–658
133. Niu B, Wu X-Y, Wei Y, Shi M (2019) Palladium-catalyzed diastereoselective formal [5 + 3] cycloaddition for the construction of spirooxindoles fused with an eight-membered ring. Org Lett 21:4859–4863
134. Lee DJ, Ko D, Yoo EJ (2015) Rhodium(II)-catalyzed cycloaddition reactions of non-classical 1,5-dipoles for the formation of eight-membered heterocycles. Angew Chem Int Ed 54:13715–13718
135. Lee JY, Kumar Varshnaya R, Yoo EJ (2022) Synthesis of chiral diazocine derivatives *via* a copper-catalyzed dearomative [5+3] cycloaddition. Org Lett 24:3731–3735

Alkynes as Carbene Precursors for the Synthesis of Heterocycles

Long-Wu Ye

Contents

L.-W. Ye (✉)
Key Laboratory for Chemical Biology of Fujian Province, State Key Laboratory of Physical Chemistry of Solid Surfaces, and College of Chemistry and Chemical Engineering, Xiamen University, Xiamen, China
e-mail: longwuye@xmu.edu.cn

Abstract Catalytic metal carbene-involved transformations are considered as one of the most important aspects in homogeneous transition metal catalysis. The generation of gold carbenes from readily available alkynes represents a significant advance in metal carbene chemistry. This chapter summarizes the advances in the gold-catalyzed amination of alkynes with azides, nitrogen ylides, isoxazoles, and anthranils, and gold-catalyzed oxidation of alkynes with nitro compounds, nitrones, sulfoxides, and pyridine *N*-oxides, through the presumed α-imino gold carbene and α-oxo gold carbene intermediates. Importantly, these methods allow the rapid and efficient assembly of various structurally complex heterocycles. These gold-catalyzed protocols are presented by highlighting their product diversity, selectivity, applicability, as well as the mechanistic rationale.

Keywords Alkynes · Amination · Gold carbene · Heterocycles · Oxidation

1 Introduction

During the past decades, homogeneous gold catalysis has received considerable attention because of its high catalytic reactivity, mild reaction conditions, as well as good functional group compatibility [1–13]. Consequently, a wide range of efficient synthetic methodologies have been developed, and these methods have been successfully applied to the synthesis of numerous natural products, biologically active molecules, as well as pharmaceutical molecules [14–16]. Owing to the divergent reactivity of carbenes, the generation of gold carbenes is arguably the most important aspect of homogeneous gold catalysis [1–13]. Among reactions involving gold carbenes, the gold-catalyzed amination and oxidation of readily available alkynes have proven to be one of the most efficient and straightforward ways to produce gold carbene intermediates, without the requirement of hazardous and explosive reagents.

Catalytic amination reactions are considered as one of the most effective tools for construction of *N*-containing molecules [17–21]. In particular, the gold-catalyzed amination of alkynes has attracted particular attention because of its high efficiency in the synthesis of *N*-containing molecules, especially valuable *N*-heterocycles. During the past decades, an array of amination reagents in the field of gold catalysis, such as azides, nitrogen ylides, isoxazoles, and anthranils, have been discovered [22–24]. The typical reaction model of this gold-catalyzed amination of alkyne is shown in Scheme 1a. The nitrogen atom initially attacks the gold-activated triple bond to form the vinyl gold intermediate, triggering fragment of N–X bond to generate the key α-imino gold carbene intermediate which were rarely obtained from the corresponding α-imino diazo compounds by metal-catalyzed dediazotizations of diazo compounds [25, 26]. Benefiting from the unique reactivity

Scheme 1 Formation of gold carbenes through nitrogen and oxygen transfer reactions of alkynes

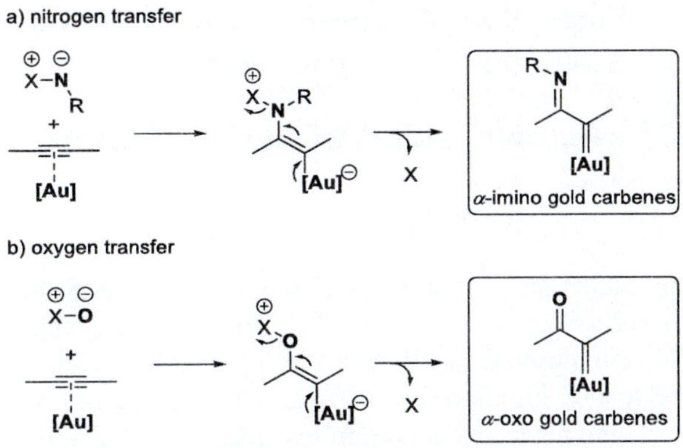

of the α-imino gold carbene intermediates, different carbene chemistries, such as 1,2-migration, X–H insertion, formal annulation, cyclopropanation and carbene metathesis, could be achieved for the efficient construction of important *N*-containing heterocycles.

Over the past decades, the gold-catalyzed oxidation of alkynes has been extensively explored, and a series of oxygen atom transfer reagents have been developed, including nitro compounds, nitrones, sulfoxides, and *N*-oxides [27–30]. Of note, carbene formation based on diazo compounds is not within the scope of this chapter, although they have been vigorously investigated [4, 9]. The reaction model is very similar to the gold-catalyzed amination of alkynes, as depicted in Scheme 1b. First, the oxygen atom of the oxygen transfer reagent attacks the gold-activated alkynes, delivering the vinyl gold intermediates. Subsequent O–X bond cleavage produces the pivotal α-oxo gold carbenes. It is noteworthy that the traditional generation of α-oxo gold carbene intermediates requires the use of not easily accessible α-diazo carbonyl compounds as precursors [4, 9]. Therefore, this protocol offers a safe and convenient methodology in organic synthesis. Similar to the α-imino gold carbenes, these highly versatile α-oxo gold carbene intermediates can undergo various valuable transformations including X–H insertion, 1,2-migration, formal annulation, cyclopropanation and rearrangement, enabling the divergent synthesis of heterocycles.

The aim of this chapter is to provide the latest and comprehensive introduction of gold-catalyzed carbene formation based on alkynes by highlighting the synthetic diversity, selectivity and applicability, and the mechanistic rationale is presented where possible. It should be pointed out that carbene formation based on alkynes catalyzed by other metals are not included in this chapter due to the limited space.

2 Formation of α-Imino Gold Carbenes Through Amination of Alkynes with Azides

2.1 Intramolecular Amination of Alkynes with Azides

2.1.1 1,2-Migration

In 2005, Toste and co-workers reported the first example for the use of azides acting as nitrene transfer reagents [31]. The cascade cyclization of homopropargyl azides **1** in the presence of 2.5 mol% of (dppm)Au_2Cl_2 and 5 mol% of $AgSbF_6$ as catalysts led to the formation of 1*H*-pyrroles **2** (Scheme 2). Various substituents such as alkyl, aryl, and heterocycle on homopropargyl azides were well tolerated in this reaction. This pioneering work opened a new field in the gold-catalyzed nitrene transfer reactions based on azides, providing an efficient and divergent way for the synthesis of a wide range of functionalized *N*-containing heterocycles, which are privileged structural motifs in bioactive molecules, pharmaceuticals, and natural products.

On the basis of experimental observations, the proposed mechanism of this acetylenic Schmidt reaction is shown in Scheme 3. The reaction begins with

Scheme 2 Gold-catalyzed acetylenic Schmidt reaction of homopropargyl azides **1**

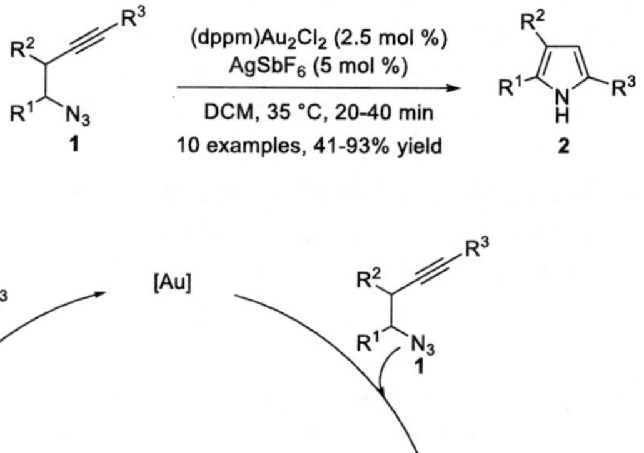

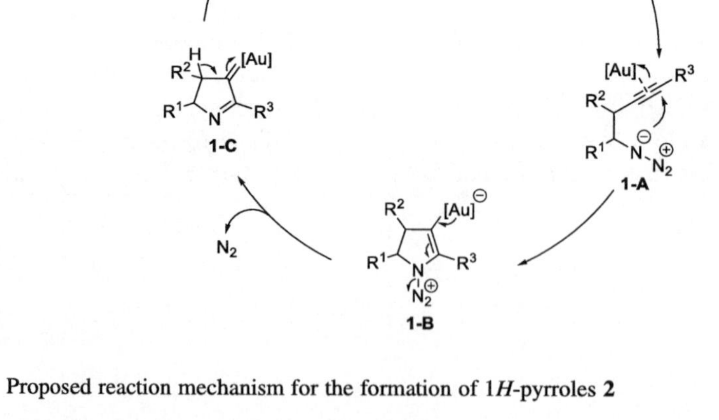

Scheme 3 Proposed reaction mechanism for the formation of 1*H*-pyrroles **2**

coordination of the gold catalyst to the alkyne followed by 5-*endo-dig* cyclization, delivering the vinyl gold intermediates **1-B**. Subsequently, the elimination of nitrogen forms the α-imino gold carbenes **1-C**, which undergoes formal a 1,2-*H* shift to produce 2*H*-pyrroles **1-D** via protodeauration with the regeneration of the gold catalyst. Finally, tautomerization of 2*H*-pyrroles affords the 1*H*-pyrroles **2**.

Other elegant protocols for gold-catalyzed azide–yne cyclization were then developed, affording a variety of functionalized polysubstituted quinolines [32] and 2*H*-1,3-oxazines [33].

2.1.2 Oxidation

The further oxidation of α-imino gold carbene intermediates was first realized by Toste group in 2007 [34]. As shown in Scheme 4, treatment of homopropargyl azide **3** with 10 mol% of IPrAuCl/$AgSbF_6$ (IPr = [1,3-bis(2,6-diisopropylphenyl)imidazo-2-ylidine]) as catalyst and 4 equiv. of diphenylsulfoxide as oxidant led to the formation of pyrrolone **4** in 73% yield together with a small amount of alkyl shift product **4′**. This protocol presumably involves the generation of an α-imino gold carbene intermediate **3-C**, followed by oxidation of α-imino gold carbene via intermolecular oxygen atom transfer from diphenylsulfoxide.

2.1.3 X–H Insertion

Transition-metal-catalyzed X–H insertion into metal carbenes is one of the most straightforward ways to construct the carbon–carbon or carbon–heteroatom bond. However, X–H insertion reactions into azide-incorporated gold carbenes were unknown until 2011 when Gagosz and co-workers described a gold-catalyzed tandem O–H insertion/Claisen rearrangement [35]. As shown in Scheme 5, cyclization of 2-alkynyl arylazides **5** in the presence of 4 mol% of $IAdAuNTf_2$ generates α-imino gold carbene intermediates **5-B**, which were further captured by a large excess of allylic alcohols via O–H insertion, yielding 3-allyloxyindoles **5-C**. Claisen

Scheme 4 Gold-catalyzed oxidation of α-imino gold carbene for the generation of pyrrolone **4**

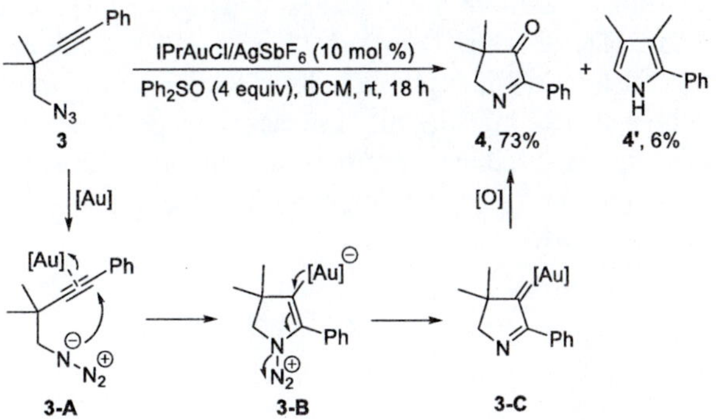

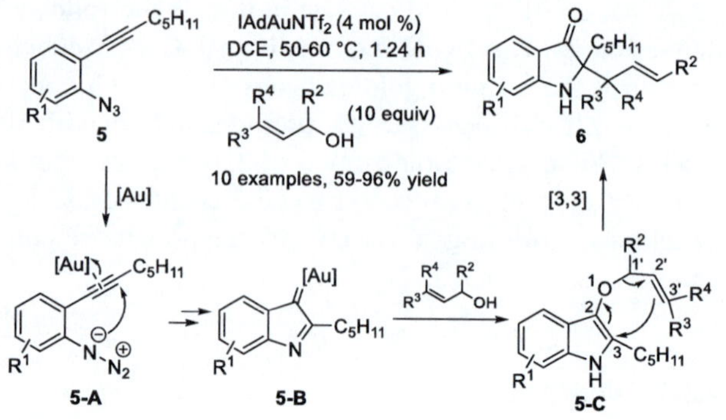

Scheme 5 Gold-catalyzed O–H insertion/claisen rearrangement of 2-alkynyl arylazides **5** for the construction of indolin-3-ones **6**

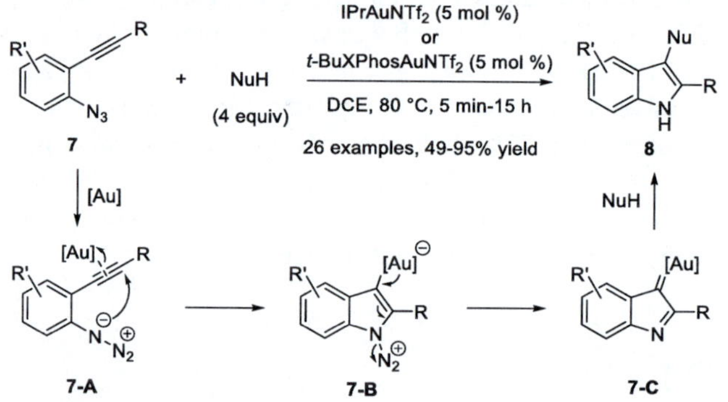

Scheme 6 Gold-catalyzed aminative X–H insertion of 2-alkynyl arylazides **7** for the synthesis of indoles **8**

rearrangement of **5-C** allowed the formation of indolin-3-ones **6** in moderate to excellent yields with moderate diastereoselectivity. A range of primary and secondary allylic alcohols and 2-alkynyl arylazides bearing electron-rich or electron-poor substituents served as appropriate substrates.

In the same year L. Zhang and co-workers reported a similar gold-catalyzed C–H and O–H insertion of α-imino gold carbenes [36]. As outlined in Scheme 6, the reaction of 2-alkynyl arylazides **7** with various nucleophiles in the presence of $IPrAuNTf_2$ or t-$BuXPhosAuNTf_2$ (5 mol%) as the catalyst in DCE at 80°C led to the construction of versatile indoles **8** in generally moderate to excellent yields. The carbon-nucleophiles in this case can be electron-rich arenes or heteroarenes, such as anisoles, naphthalenes, pyrroles, and indoles. Thus, this protocol provides an alternative approach to achieving umpolung reactivity of indoles at the three-position via gold catalysis.

2.1.4 Formal Annulation

In 2012, L. Zhang and co-workers discovered that the in situ generated α-imino gold carbenes undergo a gold-catalyzed formal [3 + 2]-annulation with nitriles, enabling the synthesis of bicyclic imidazoles (Scheme 7) [37]. Various azidoalkynes **9** and nitriles were transformed to substituted bicyclic imidazoles **10** in good yields at room temperature, thus constituting a bimolecular [2 + 2 + 1] cycloaddition. It is notable that the nitrile, a weak nucleophile, was used as the reaction solvent to improve the reaction efficiency. In addition, the use of MsOH (1.1 equiv) as an additive avoided the coordination of gold catalyst with basic imidazoles and triazoles, while the use of $AuCl_3$ as the catalyst could prohibit the competing intramolecular Huisgen reaction. The related formal annulations of alkynyl azides with alkynols and nitriles were also developed for the synthesis of polycyclic skeletons [38–40].

In 2017, Ohno and co-workers reported a gold-catalyzed formal [4 + 2] annulation of azido diynes **11** with pyrroles **12** for the construction of the pyrrolo[2,3-*c*] carbazoles **13** (Scheme 8) [41]. Importantly, this strategy has been successfully employed as the crucial step in the total synthesis of dictyodendrin B, F and formal synthesis of dictyodendrin C, E.

2.1.5 Cyclopropanation/Cyclopropenation

As the distinctive reactivity of carbenes, cascade cyclopropanation and cyclopropenation reactions have also been developed based on nitrogen transfer reactions of alkynes. In 2014, Ohno and Fujii reported a gold-catalyzed cascade cyclization of (azido)ynamides **15** for the efficient preparation of various indoloquinolines (Scheme 9) [42]. The tethered alkene moiety can serve as nucleophilic agent to trap the α-imino gold-carbenoid species, yielding the cyclopropane-fused indoloquinolines **16** in high yields. In addition, further studies revealed that

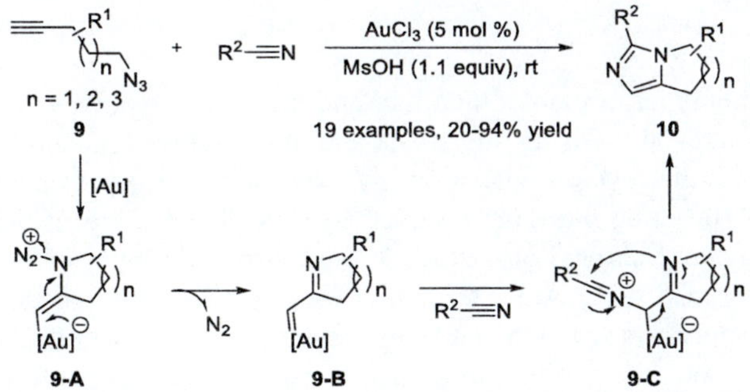

Scheme 7 Au(III)-catalyzed formal [3 + 2] annulation of azidoalkynes **9** and nitriles for the assembly of bicyclic imidazoles **10**

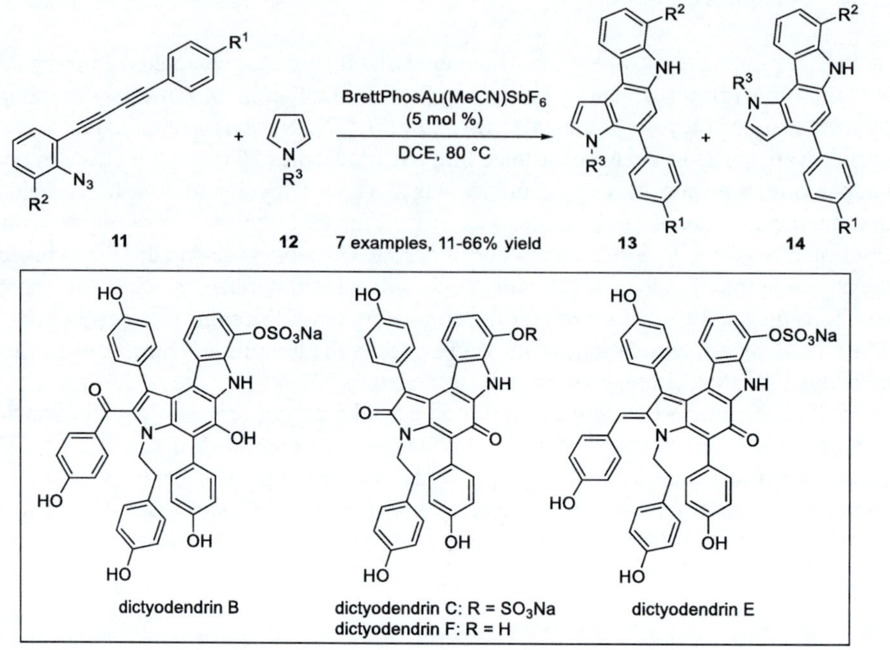

Scheme 8 Gold-catalyzed formal [4 + 2] annulation of azido diynes **11** with pyrroles **12**

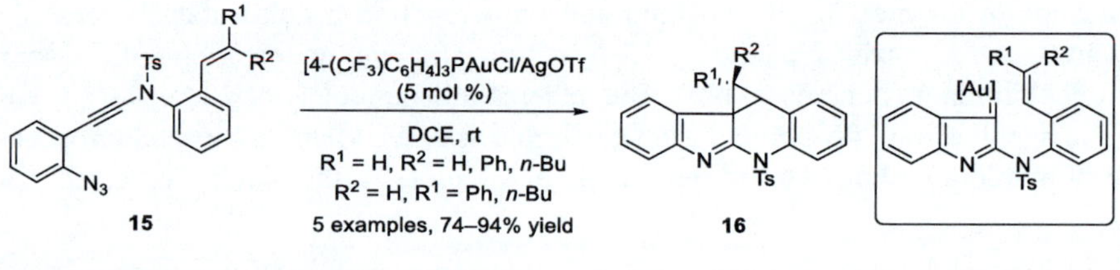

Scheme 9 Gold-catalyzed aminative cyclization of (azido)ynamides **15** for the construction of indoloquinolines **16**

syntheses of other types of indoloquinolines and indole-fused polycyclic compounds could also be achieved by using allylsilane and other tethered nucleophiles.

Alternatively, the cyclopropanation-involved cascade cyclization was also reported. In 2016, Gong and Han established a synthetic route of aryl-annulated carbazoles **19** from 2-alkynyl arylazides **17** and terminal alkynes **18** (Scheme 10) [43]. In the presence of $IPrAuNTf_2$ (5 mol%) as catalyst the reaction proceeded smoothly to produce a range of valuable aryl-annulated carbazoles **19** in moderate to good yields. Of note, the addition of benzoic acid and water could further improve the yield. The transformation showed good tolerance towards electron-donating, electron-withdrawing, and bulky substituents in azido alkynes and various terminal alkynes.

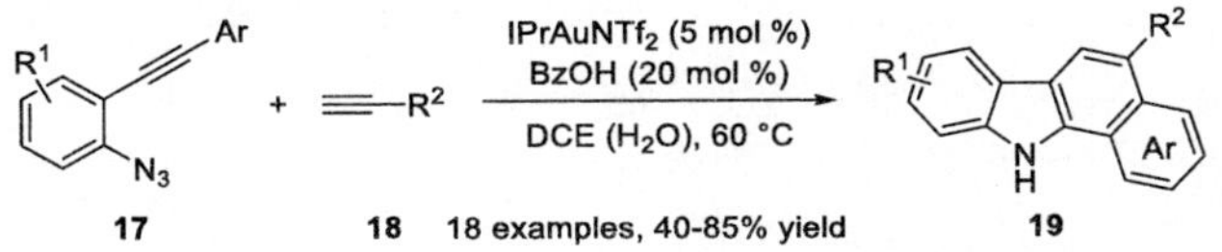

Scheme 10 Gold-catalyzed tandem cyclization of 2-alkynyl arylazides **17** with alkynes

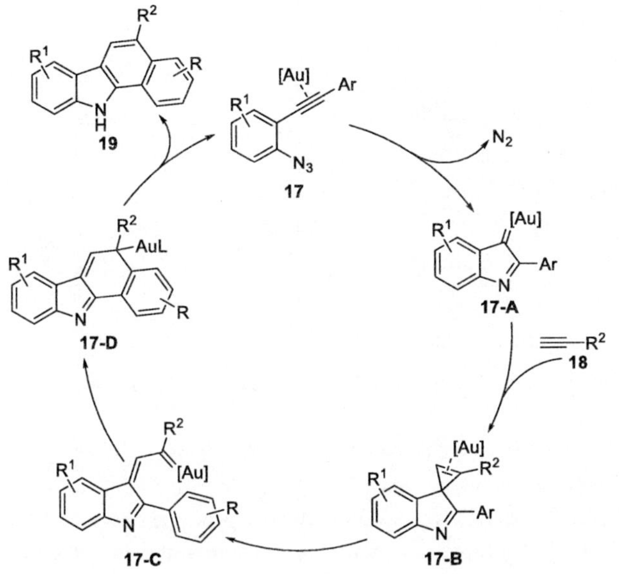

Scheme 11 Proposed reaction mechanism for the synthesis of carbazoles **19**

The proposed mechanism is shown in Scheme 11. First, the generated α-imino gold carbene intermediates **17-A** undergo a cyclopropenation with alkynes **18** affording cyclopropenes **17-B**, which are transformed into alkenyl gold carbene intermediates **17-C**. Subsequent intramolecular Friedel-Crafts type reaction, followed by protonation and tautomerization, leads to the final carbazoles **19**. The authors believe that water and benzoic acid, which accelerated the reaction and led to an improved yield, weaken coordination of imino intermediates (**17-A** – **17-D**) to the gold catalyst and facilitate the protonation and tautomerization of intermediates **17-D**.

2.1.6 4π Electrocyclization

By employing the similar nitrogen transfer approach, L. Zhang and co-workers reported a gold-catalyzed cascade cyclization of linear azidoenynes **20** for the efficient synthesis of 2,3-dihydro-1*H*-pyrrolizines **21** bearing electron-withdrawing

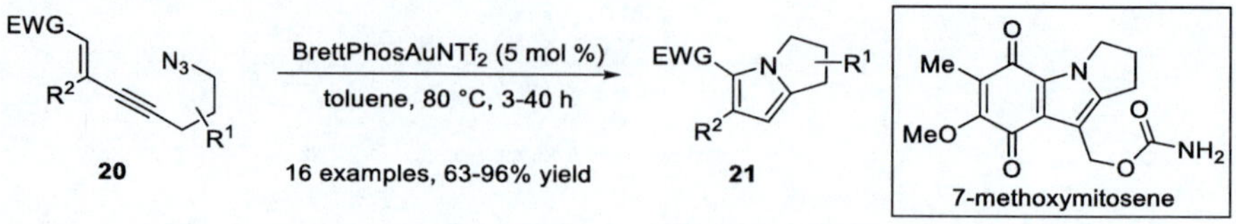

Scheme 12 Gold-catalyzed aminative cyclization of azidoenynes **20**

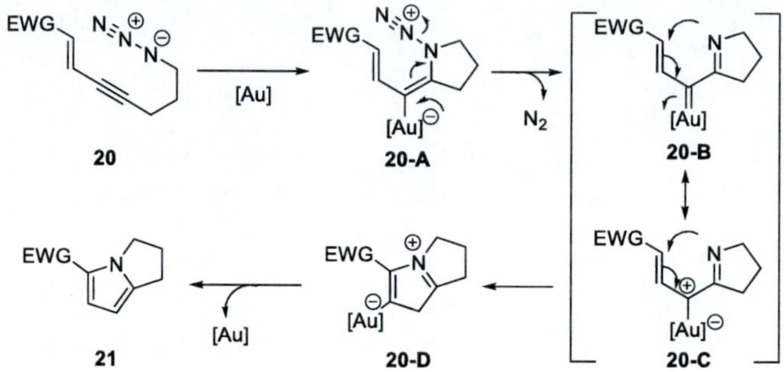

Scheme 13 Proposed reaction mechanism for the formation of 2,3-dihydro-1*H*-pyrrolizines **21**

groups at the five-position [44]. With bulky $BrettPhosAuNTf_2$ (5 mol%) as the catalyst, the reactions were compatible with various electron-withdrawing groups (EWG), such as acyl, ester as well as the sulfonyl group (Scheme 12). Moreover, the synthetic utility of this protocol was nicely demonstrated in a formal synthesis of 7-methoxymitosene.

According to experimental observations and their previous report, a plausible mechanism is described in Scheme 13. First, Au-activated azidoenynes **20** undergo 5-*exo-dig* cyclization followed by departure of N_2 to generate the α-imino gold carbenes **20-B**. Subsequent 4π electrocyclization occurs to eventually produce the desired 2,3-dihydro-1*H*-pyrrolizines **21**.

2.1.7 Carbene Metathesis

In 2017, Ye and co-workers disclosed a nitrene transfer initiated metal carbene metathesis by azido-diynes as substrates [45]. The treatment of azido-diynes **22** with $Ph_3PAuNTf_2$ (2 mol%) as the catalyst in CH_3NO_2 at room temperature allowed the efficient formation of pyrrolo[2,3-*b*]indoles **23** in generally good to excellent yields (Scheme 14). It is notable that, starting from the same substrates, the reaction led to the divergent synthesis of pyrrolo[3,4-*c*]quinolin-1-ones in the presence of Cu (I) as the catalyst and *N*-oxide as the external oxidant.

On the basis of experimental observations and DFT calculations, the proposed mechanism is depicted in Scheme 15. The reaction starts with 5-*endo-dig* cyclization of ynamides **22** and followed by elimination of N_2, generates the α-imino gold

Scheme 14 Gold-catalyzed aminative cyclization of azido-diynes **22**

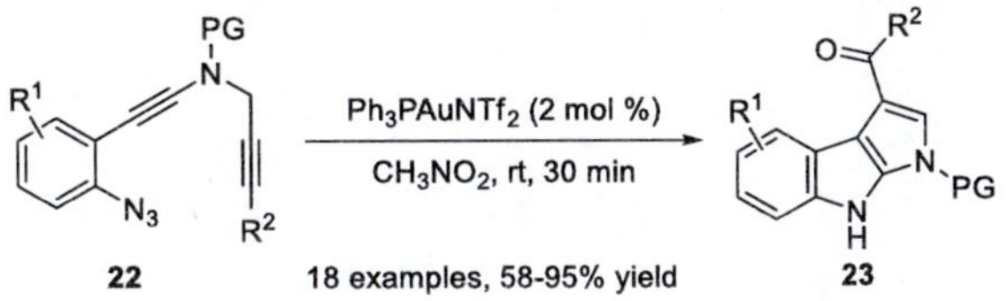

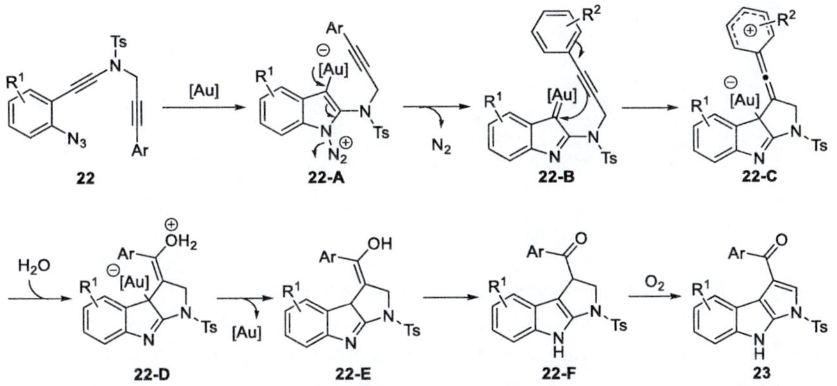

Scheme 15 Proposed reaction mechanism for the construction of pyrrolo[2,3-*b*]indoles **23**

carbenes **22-B**, which are further captured by another alkyne moiety to furnish enylium-cationic intermediates **22-C**. Then, intermediates **22-C** are proposed to be readily trapped by trace amounts of water in the reaction system, followed by protodeauration, delivering intermediates **22-E**. Subsequent enol tautomerism, aromatization and dehydrogenative oxidation by O_2 afford the final products **23**.

2.2 Intermolecular Amination of Alkynes with Azides

Due to the low nucleophilicity of azides, intermolecular nitrene transfer based on azides was challenging. However, by employing benzyl azides and ynamides as substrates Ye, Lu and co-workers reported the first intermolecular nitrene transfer reaction based on the reaction of ynamides with azides in 2015 [46]. By using 5 mol % of $IPrAuNTf_2$ as catalyst, 3-bromobenzyl azide was found to be the best nitrene transfer reagent to react with ynamides **24**, thus allowing the facile synthesis of various 2-aminoindoles **26** in mostly good yields via formal [3 + 2]-annulation (Scheme 16). It should be mentioned that other benzyl azides were also suitable nitrene transfer reagents, and the benzyl group in products could be easily removed and further transformed.

Notably, the reaction would undergo formal [4 + 2]-annulation instead of [3 + 2]-annulation by employing indolyl azides as nitrene transfer reagents (Scheme 17). In the presence of 5 mol% of $IPrAuNTf_2$ as catalyst and 1.1 equiv. of AgOAc as

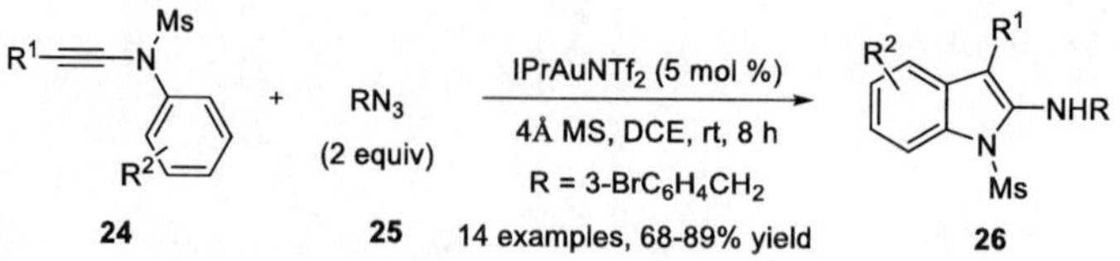

Scheme 16 Gold-catalyzed annulation of ynamides **24** with 3-bromobenzyl azide **25** for the synthesis of 2-aminoindoles **26**

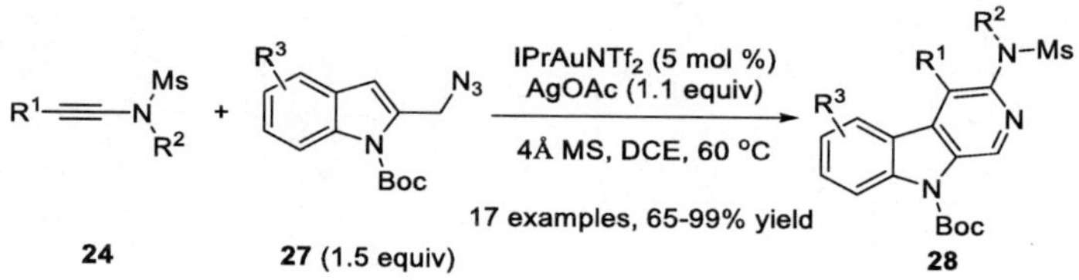

Scheme 17 Gold-catalyzed annulation of ynamides **24** with indolyl azides **27**

oxidant, the reaction of ynamides **24** with indolyl azides **27** led to the efficient formation of valuable 3-amino-*β*-carbolines **28**. A variety of functional groups, including aliphatic substituents, as well as electron-withdrawing and electron-donating groups on the aromatic ring, were readily tolerated in this tandem reaction.

The proposed mechanism for the formation of aminoindoles **26** and 3-amino-*β*-carbolines **28** was illustrated in Scheme 18. The reaction is initiated by the intermolecular addition of azido group onto gold-activated ynamides **24-A**, affording the vinyl gold complexes **24-B**. After departure of N_2, *α*-imino gold carbenes **24-C** are formed as the key intermediates, which would be further trapped by the *N*-phenyl or indolyl group leading to the corresponding intermediates **24-D1** and **24-D2**. The selectivity in two reactions is mainly attributed to the different nucleophilicities of phenyl and indolyl groups. Finally, proton transfer and ligand exchange occur to deliver the desired 2-aminoindoles **26** and intermediates **24-F2**, respectively, and the latter undergoes further dehydrogenative oxidation to afford 3-amino-*β*-carbolines **28**. When 3-indolyl azides were employed as nitrene transfer reagents, the reaction also led to the formation of the corresponding *β*-carbolines via 1,2-alkyl migration [47].

3 Formation of *α*-Imino Gold Carbenes Through Amination of Alkynes with Nitrogen Ylides

3.1 *Pyridine-Based Aza-Ylides*

Nitrogen ylides such as pyridine- and sulfur-based aza-ylides have also been widely employed as intermolecular nitrene transfer reagents in the gold-catalyzed nitrogen

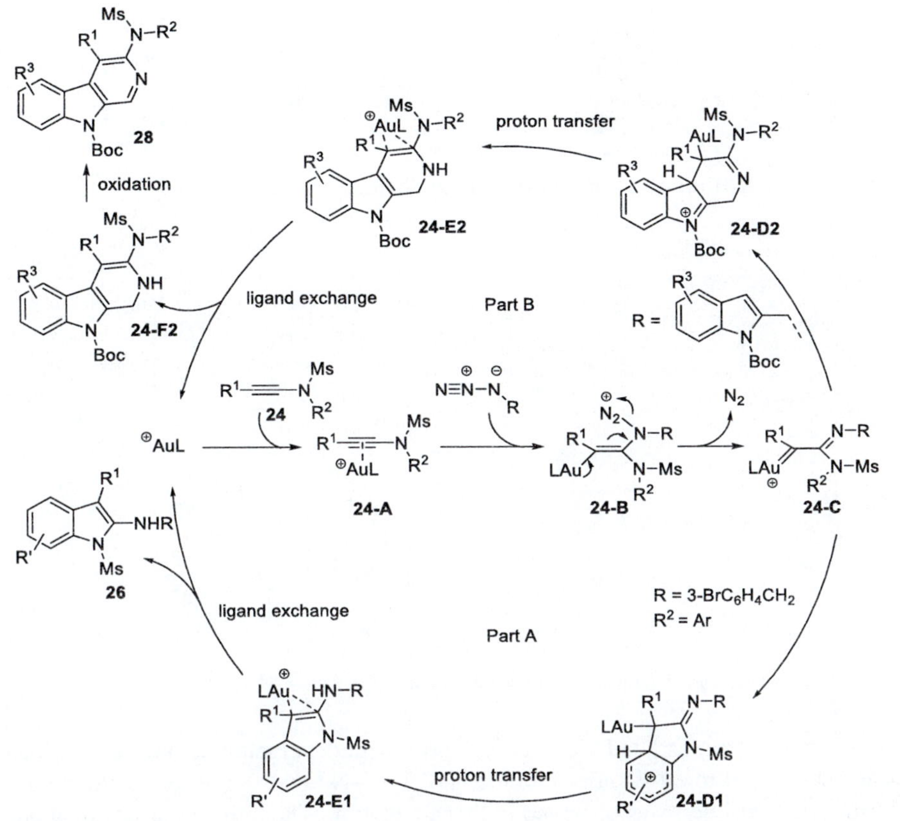

Scheme 18 Proposed mechanism for the synthesis of 2-aminoindoles **26** and 3-amino-*β*-carbolines **28**

transfer reactions. In 2011, the first example of intermolecular nitrogen transfer reaction of alkyne was developed by L. Zhang and co-workers, for the synthesis of *α*,*β*-unsaturated amidine derivatives [48]. In the same year, Davies and co-workers disclosed a regioselective gold(III)-catalyzed intermolecular formal [3 + 2]-annulation of ynamides **29** with iminopyridium ylides **30**, delivering a range of 1,3-oxazoles **31** in generally good to excellent yields (Scheme 19) [49]. Importantly, the iminopyridium ylide served as an *N*-nucleophilic 1,3-*N*,*O*-dipole equivalent in this reaction. Moreover, this strategy could also be applicable to the phenyl-substituted ynol ether under the same conditions, and produce the corresponding 4-ethoxyoxazole product in moderate yield.

As shown in Scheme 20, the reaction starts with nucleophilic attack of iminopyridium ylide **30** onto gold-activated ynamide **29-A/29-A′**, to generate the adduct **29-B**, followed by cyclization to form intermediate **29-E/29-E′**, which can be viewed as 4π-electrocyclization by the intermediate **29-D**. Finally, elimination of the gold-catalyst from **29-E** forms the desired 1,3-oxazole **31**. Of note, C–O bond

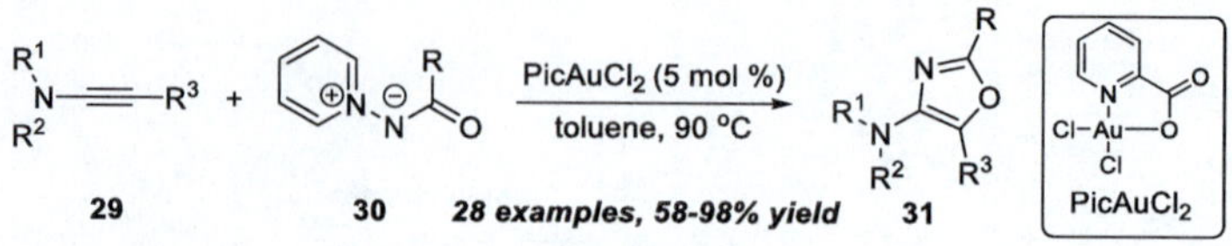

Scheme 19 Gold(III)-catalyzed intermolecular nitrene transfer of ynamides **29**

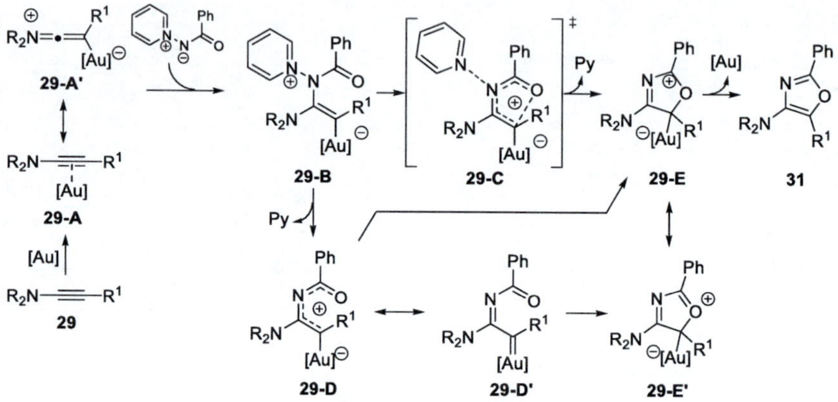

Scheme 20 Proposed mechanism for the synthesis of 1,3-oxazoles **31**

formation must be fast, and likely commences with developing carbocationic character (from **29-B** to **29-E**) prior to the complete N–N bond fission as no competing 1,2-insertion reaction was observed in this annulation. The nitrene transfer methodology has also been used for another formal [3 + 2] annulation by the same group, the synthesis of polyaromatic 2,4,5-trisubsituted oxazoles [50].

By using this strategy, Davies and co-workers further extended this nitrene transfer reagents from 1,3-*N*,*O*-dipole equivalents to 1,3-*N*,*N*-dipole equivalents, leading to facile synthesis of imidazodiazines, imidazopyridines [51] and *N*-(hetero)-aryl imidazole scaffolds [52] through gold-catalyzed formal [3 + 2] annulation. Of note, this kind of imidazole motif is found in a variety of bioactive molecules [53]. Thus, this protocol provides an access towards new types of heteroatom-substituted carbimidoyl nitrenoids.

3.2 Sulfur-Based Aza-Ylides

Sulfilimines have been developed as another type of nitrene transfer reagents by Hashmi and co-workers in 2019. According to their report, an Au(III)-catalyzed intermolecular annulation reaction between sulfilimines **32** and ynamides **33** via *α*-

imino gold carbenes has been established. The generated α-imino gold carbene intermediates were divergently trapped by different functional groups attached to either the sulfilimines or ynamides, leading to the efficient formation of four different types of aza-heterocycles, including 2-aminoindoles **34**, 3-azabicyclo[3.1.0]hexan-2-imines **35**, 4-acylquinolines **36,** and imidazoles **37** in mostly good to excellent yields (Scheme 21) [54].

The proposed mechanism for gold-catalyzed nitrene transfer is shown in Scheme 22. Regioselective nucleophilic attack of the sulfilimines **32** occurs at *N*-terminus of the gold-activated ynamides **33** to form the gold-substituted alkenes **33-A** or **33-C**, which are then converted into the α-imino gold carbenes **33-B** or **33-D** respectively through facile release of the thioether group by N–S cleavage of sulfilimines. Subsequent trapping of the α-imino gold carbenes by different functional groups attached to either the sulfilimines or ynamides affords the desired *aza*-heterocycles **34**–**37**.

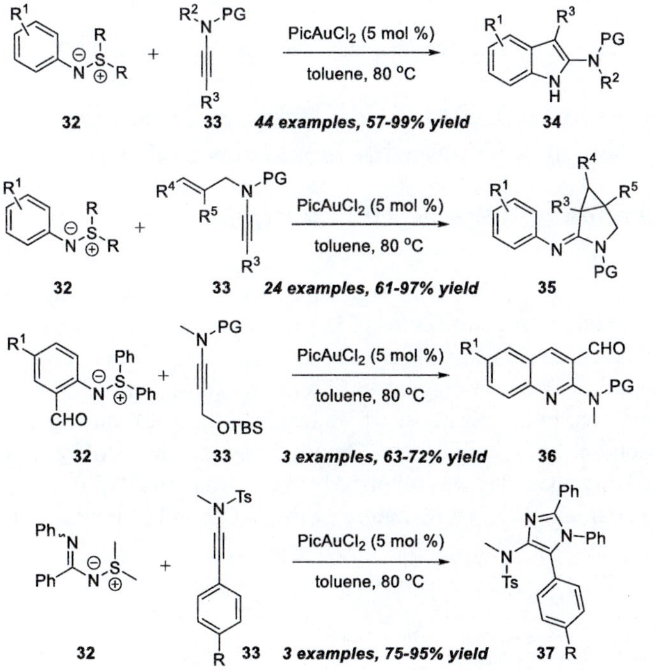

Scheme 21 Gold(III)-catalyzed aminative cyclization of ynamides with sulfilimines **32**

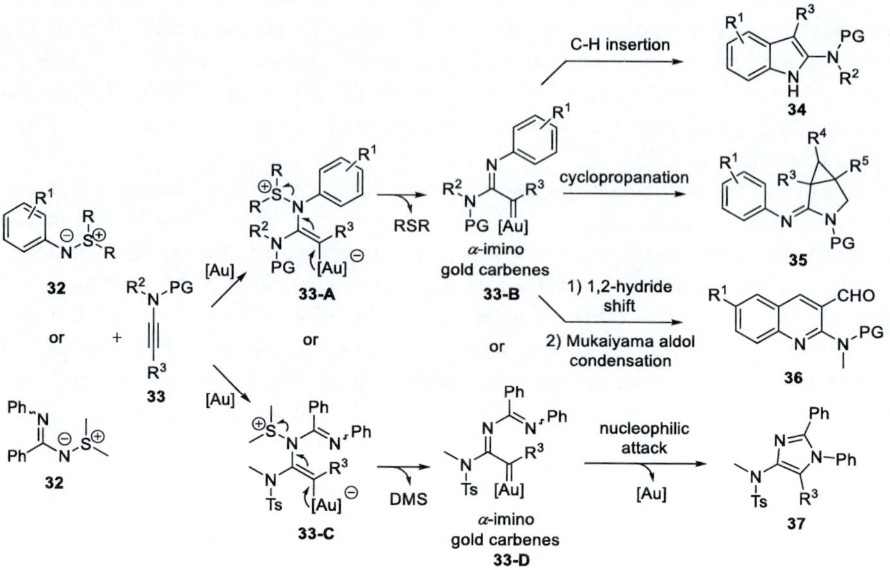

Scheme 22 Proposed mechanism for the synthesis of *aza*-heterocycles **34–37**

4 Formation of α-Imino Gold Carbenes Through Amination of Alkynes with Isoxazoles and Anthranils

4.1 Amination of Alkynes with Isoxazoles

As an important structural motif in organic chemistry, isoxazoles appear frequently in biologically active drug molecules [55, 56], however their use as a nucleophile in transition-metal catalysis was unknown for a long duration. The labile *N-O* bond of isoxazole makes it a perfect reagent in nitrene transfer reactions [57–59], but there were few reports after an initial use of isoxazoles in ring-expansion and *N-O* bond-insertion reactions by Manning and Davies [60, 61] until gold-catalyzed nitrene transfer based on isoxazoles was reported by Ye's group in 2015 [62]. As depicted in Scheme 23, Ye utilized the labile nature of the *N-O* bond of isoxazoles to construct

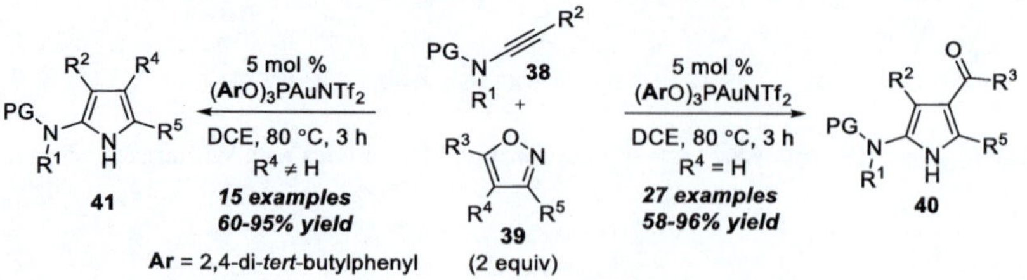

Scheme 23 Gold-catalyzed [3 + 2] annulation of ynamides **38**

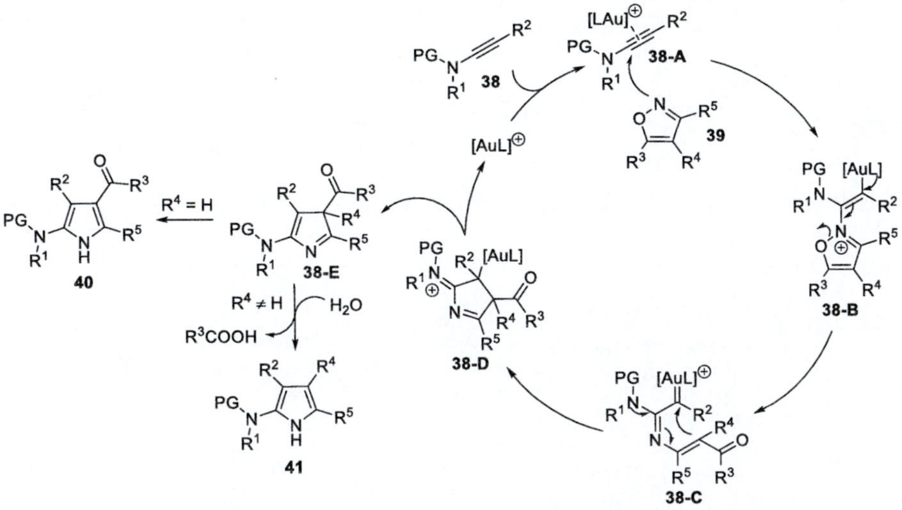

Scheme 24 Proposed mechanism for the preparation of 2-aminopyrroles **40** and **41**

polysubstituted 2-aminopyrroles **40–41** from ynamides **38** and isoxazoles **39**. Two distinct products were obtained depending on the substitution pattern of the isoxazoles; pyrroles **40** were a direct outcome of [3 + 2]-annulation for $R^4 = H$, whereas deacylated pyrroles **41** were obtained when the C4 position was substituted. This protocol served to inspire further development of isoxazoles as a nitrene-transfer reagent and uncovered several other *N-O*-containing reagents such as anthranils and 1,2-benzisoxazoles to become exploited in gold-catalyzed nitrene-transfer strategies.

The proposed mechanism of this gold-catalyzed [3 + 2]-annulation is shown in Scheme 24. The reaction begins with an activation of the ynamide triple bond by electrophilic gold towards nucleophilic *N*-attack of isoxazole **39** to generate *α*-imino gold carbene intermediate **38-C** through the ring opening of vinyl gold intermediate **38-B**. A subsequent 1,5-cyclization of intermediate **38-C** and protodeauration of intermediate **38-D** gives intermediate **38-E**. At this stage, depending on the substitution pattern at R^4, aromatization results in 2-aminopyrroles **40**; diacylation delivers 2-aminopyrroles **41**.

4.2 *Amination of Alkynes with Anthranils*

Hashmi and co-workers developed anthranil as another type of nitrene-transfer reagent in an elegant protocol to construct indoles from ynamides. In their report ynamide **42** was treated with anthranil **43** in the presence of $IPrAuCl/AgNTf_2$ (5 mol %) at −20°C to furnish 7-acyl indole **44** (Scheme 25) [63]. The reaction involved an *α*-imino gold carbene obtained by an intermolecular nitrene transfer from anthranils

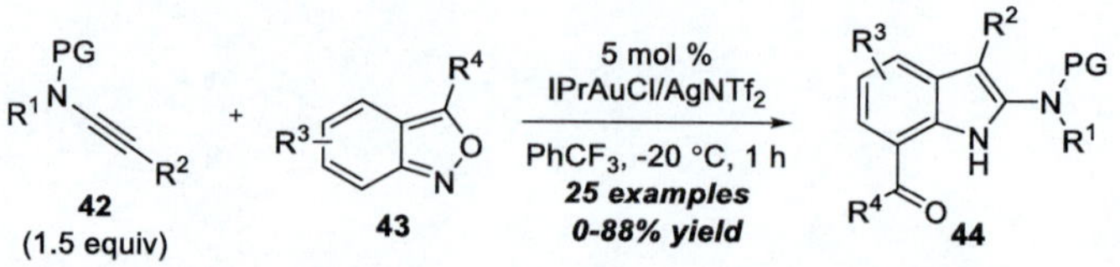

Scheme 25 Gold-catalyzed annulations of ynamides **42** with anthranils for the assembly of 7-acyl indoles **44**

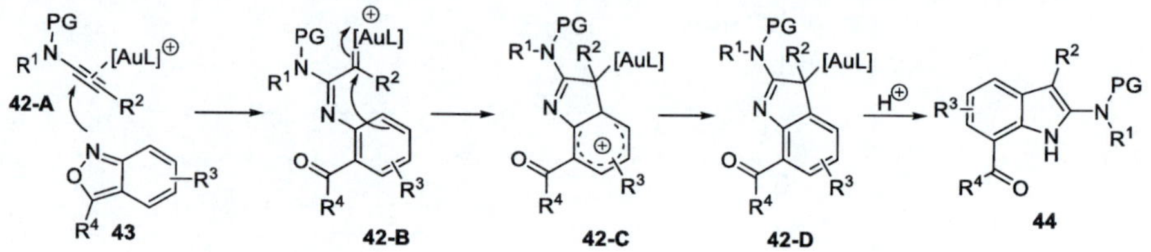

Scheme 26 Proposed mechanism for the synthesis of 7-acyl indoles **44**

and subsequent ortho C–H functionalization to deliver the desired indoles. The protocol was extendable to non-polarized internal and terminal alkynes, albeit at higher temperature and extended reaction times. This protocol paved a path for the use of anthranils in several other gold-catalyzed nitrene-transfer or annulation strategies.

The mechanism of this annulation is depicted in Scheme 26. Initially, a nucleophilic attack of anthranil **43** nitrogen on gold-tethered ynamide **42-A** occurs to provide α-imino gold-carbene intermediate **42-B** as a result of an anthranil ring opening. This carbene, being highly electrophilic, undergoes ortho C–H activation to furnish intermediate **42-C**, which then rearomatizes. After protodeauration **42-C** produces indole product **44**. Moreover, Hashmi and co-workers reported another gold-catalyzed cascade [4 + 2]-annulation between polarized propargyl silyl ethers and anthranils to produce quinoline rings through sequential ring-opening/1,2-*H* shift/protodeauration/Mukaiyama aldol condensation [64]. Two years later the same group developed a strategy employing non-polarized *o*-ethynylbiaryls to react with anthranils to construct dibenzo[*a*,*c*]acridines [65].

5 Formation of α-Oxo Gold Carbenes Through Oxidation of Alkynes with N–O Containing Oxidants

5.1 Nitro Compounds as Oxidants

Nitro compounds were first employed as oxygen atom transfer reagents by Asao and Yamamoto in 2003, leading to isatogens **46**, or anthranils **47** from *o*-(alkynyl)-nitrobenzenes **45**, through two distinct regioselective pathways (Scheme 27)

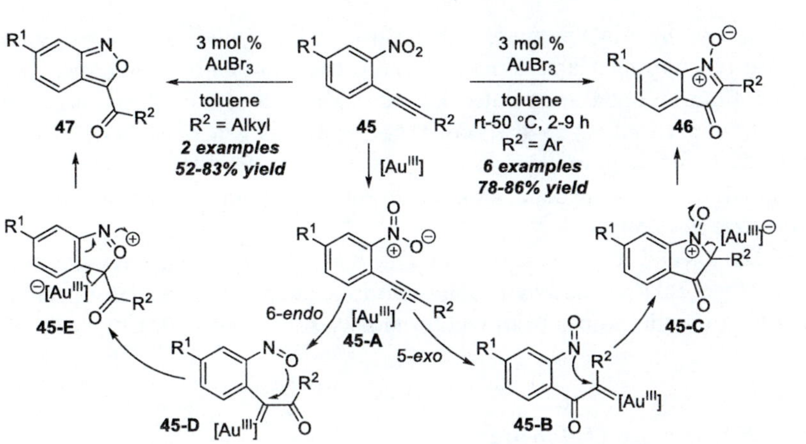

Scheme 27 Gold(III)-catalyzed cyclization of *o*-(alkynyl)nitrobenzenes **45** for the construction of isatogens **46** and anthranils **47**

Scheme 28 Gold-catalyzed formal [2 + 2 + 1] cycloaddition of nitroalkynes **48** with alkenes **49** for the assembly of azacycles **50**

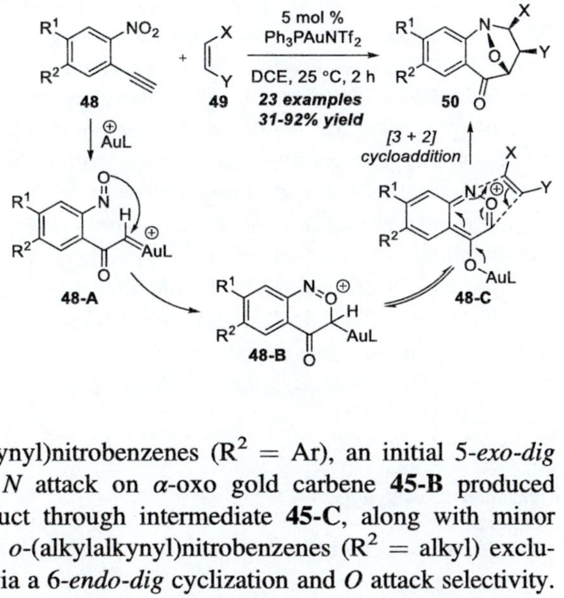

[66]. In the case of *o*-(arylalkynyl)nitrobenzenes (R^2 = Ar), an initial 5-*exo-dig* cyclization and a subsequent *N* attack on α-oxo gold carbene **45-B** produced isatogens **46** as a major product through intermediate **45-C**, along with minor amount of anthranils. Whereas *o*-(alkylalkynyl)nitrobenzenes (R^2 = alkyl) exclusively produced anthranils **47** via a 6-*endo-dig* cyclization and *O* attack selectivity. The reaction demonstrates good tolerance for several alkyl and aryl substituted *o*-(alkynyl)nitrobenzenes.

By using nitro compounds as oxygen transfer reagents, Liu and co-workers reported a gold-catalyzed intermolecular redox/cycloaddition cascade between nitroalkynes **48** and alkenes **49** to synthesize azacyclic compounds **50** (Scheme 28) [67]. The core structure is envisaged to be an outcome of a formal [2 + 2 + 1]-cycloaddition between an α-oxo gold carbene, a nitroso group, and an external alkene. The scope of this cycloaddition cascade was demonstrated by employing several electron-rich alkenes, such as enol ethers, cyclic enol ethers, and vinyl thioethers as compatible partners with electron-rich and -deficient nitroalkynes. The proposed reaction mechanism involved an α-oxo gold carbene intermediate **48-A** obtained from an initial intramolecular redox reaction of nitroalkyne **48**. A further intramolecular *O*-attack on the α-oxo gold carbene generated ketonyl oxonium species **48-B** that tautomerized to intermediate **48-C**. Finally, a concerted [3 + 2]-cycloaddition of intermediate **48-C** with enol ethers took place to generate the desired azacycles **50**.

In 2015, Jia and co-workers exploited the reactivity of isatogen towards indole by developing a gold(III) and chiral phosphoric acid-catalyzed enantioselective construction of indolin-3-ones bearing a quaternary stereocenter [68].

5.2 *Nitrones as Oxidants*

5.2.1 Intramolecular Annulation/Cyclization

Nitrones have also been developed as oxygen transfer reagents with alkynes. In 2009, Shin and co-workers developed a gold-catalyzed redox cascade cyclization of *o*-alkynyl tethered nitrones **51** to produce isoindoles **52** with a 7-*endo-dig* chemoselectivity (Scheme 29) [69]. The substitution pattern of the parent *o*-alkynyl nitrones governed the selection of suitable gold-catalyst out of IPrAuOTf and JohnPhosAuOTf to catalyze this transformation. Various substitutions on the aryl ring, alkynyl terminus, and nitrogen atom of the nitrone were compatible, as were *o*-alkynyl pyridinyl nitrones and a cycloalkenyl tether between the nitrone and alkyne functionalities. The mechanistic cycle is believed to involve a selective gold-assisted 7-*endo-dig* cyclization and a following intramolecular oxygen transfer to generate α-oxo gold intermediate **51-B** that underwent an intramolecular imine nitrogen attack on the gold carbene to generate azomethine ylide **51-C**. Finally, isoindoles **52** are generated after tautomerization and deauration.

Scheme 29 Gold-catalyzed cyclization of *o*-alkynylaryl nitrones **51** for the generation of isoindoles **52**

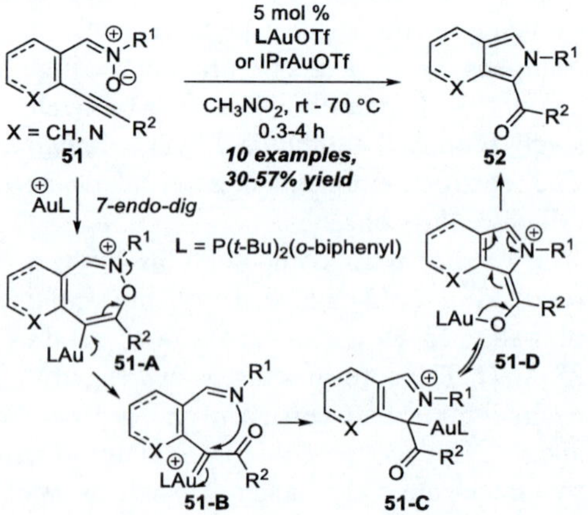

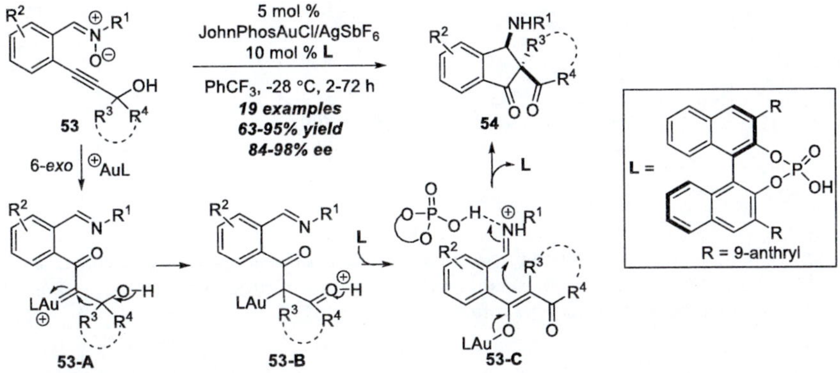

Scheme 30 Asymmetric cyclization of propargyl alcohol tethered nitrones **53** for the construction of spirocyclic diketones **54**

Notably, the enantioselective Redox-Pinacol-Mannich cascade of propargyl tertiary alcohol tethered nitrones **53** for the construction of spirocyclic diketones **54** was disclosed by J. Zhang and co-workers (Scheme 30) [70]. Inspired by a previous racemic version of a similar transformation [71], they introduced a chiral Brønsted acid for enantiocontrol. Electron-withdrawing and sterically demanding 3-trifluoromethyl phenyl (R^1) substitution on the nitrone nitrogen was crucial for higher yields and enantioselectivity; replacement with other substituents seriously affected enantioselectivity. In the proposed reaction mechanism, an initial 6-*exo* cyclization and internal redox generated the α-oxo gold carbene **53-A** that underwent an alkyl shift (pinacol rearrangement) to give intermediate **53-C** through **53-B**. Intermediate **53-C** then underwent Brønsted acid directed Mannich addition to produce chiral spirocyclic diketone **54**. Chiral spirocyclic diketones including 5- to 9-membered rings could be achieved using this strategy. Substrates containing a heteroatom in their cyclic skeleton, electron-donating and -withdrawing substituents on the aryl ring, and geminal dialkyl groups were tolerated in the transformation.

5.2.2 Intermolecular Annulation/Cyclization

Based on the special reactivity of nitrones and their previous study on 1,2-difunctionalization of aryl ynamides with nitrones [72], Liu and co-workers reported a gold-catalyzed oxidative cyclization/Mannich reaction between nitrones **55**, homopropargyl alcohols **56** and *o*-alkynyl phenols **60** to deliver 2-(α-aminobenzyl)dihydrofuran-3(2*H*)-ones **58** (**59**) and 2-(α-aminobenzyl)dihydrobenzofuran-3(2*H*)-ones **61** (**62**) with *syn*-selectivity through α-alkoxy enolates **57** (Scheme 31) [73]. Unlike classical gold-catalyzed *N*-oxide assisted oxidative cyclization, where the liberated *N*-heterocycles are not used at the end of the catalytic cycle, the new protocol employed the liberated imines for cooperative catalysis through a distinct O-H--N hydrogen bonding that facilitated a Mannich reaction

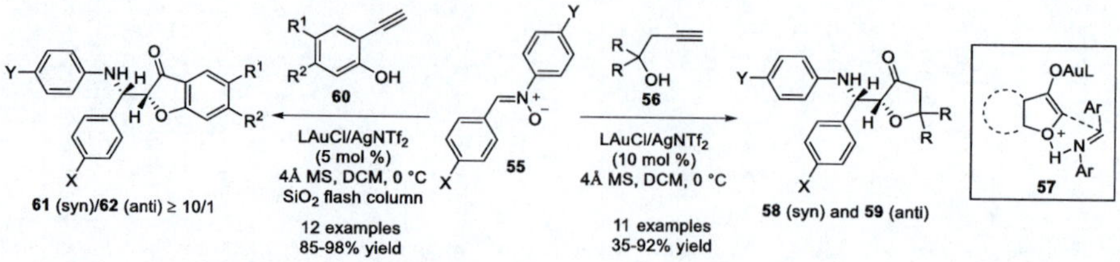

Scheme 31 Gold-catalyzed oxidative cyclization cascade of homopropargyl alcohols and *o*-alkynyl phenols with nitrones for the synthesis of 2-(α-aminobenzyl)dihydrofuran-3(2*H*)-ones **58** (**59**) and 2-(α-aminobenzyl)dihydro-benzofuran-3(2*H*)-ones **61** (**62**)

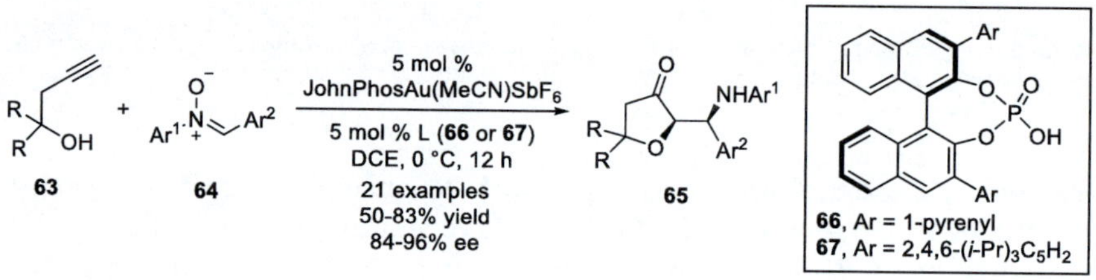

Scheme 32 Gold/CPA cooperative catalyzed enantioselective oxidative cyclization cascade of 3-butynols **63** with nitrones **64**

with α-alkoxy enolates **57**. Electron-rich and -deficient nitrones could be treated with simple or 1,1-disusbtituted homopropargyl alcohols **56** in the presence of 10 mol% of P(*t*-Bu)$_2$(*o*-biphenyl)goldtriflimide to generate the corresponding *syn*- and *anti*-dihydrofuran-3-(2*H*)-one Mannich products **58** (**59**) that could be separated on a silica column. On the other hand, the *syn*-selective dihydrobenzofuran-3-(2*H*)-one **61** (**62**) could be obtained with lower catalyst loading following flash chromatography on silica.

An enantioselective version of the gold-catalyzed oxidative cyclization/Mannich reaction was developed by Xu and co-workers in 2018. The reaction between homopropargyl alcohols **63** and nitrones **64** successfully constructed dihydrofuran-3-ones **65** with excellent enantioselectivity (Scheme 32) [74]. Chiral phosphoric acids were employed in this transformation to assist the asymmetric Mannich addition through hydrogen bonding, after initial gold-catalyzed alkyne oxidation and O–H insertion. This reaction of electron-rich and -deficient nitrones with 3-butynols resulted in various enantioenriched dihydrofuran-3-ones efficiently.

The proposed mechanism of the gold/chiral phosphoric acid-catalyzed oxidative cyclization/Mannich reaction is shown in Scheme 33. The catalytic cycle begins with the gold-activated alkyne oxidation with the help of nitrone **64** to generate α-oxo gold carbene **63-B** by liberating **64-A**. An intramolecular attack of the tethered hydroxyl group on gold-carbene **63-B** furnishes gold-enolate **63-D** through intermediate **63-C**. Subsequently, enantioenriched dihydrofuran-3-one **65** is formed via chiral phosphoric acid assisted Mannich addition of this gold-enolate **63-D** on imine **63-A**.

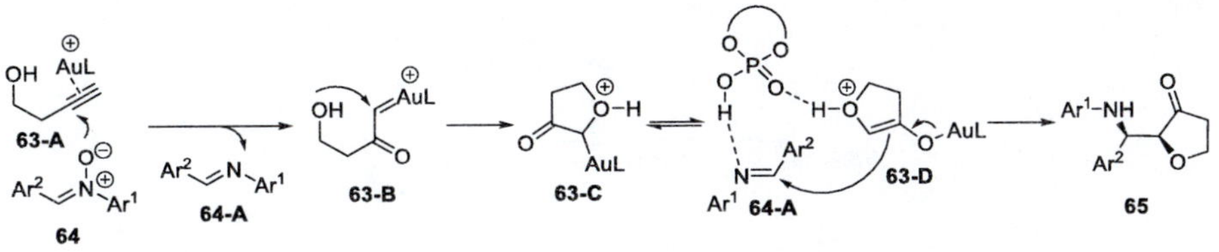

Scheme 33 Proposed mechanism for the synthesis of dihydrofuran-3-one **65**

6 Formation of *α*-Oxo Gold Carbenes Through Oxidation of Alkynes with Pyridine *N*-Oxides

6.1 X–H Insertion

6.1.1 O–H Insertion

While the gold-catalyzed oxidation of alkynes has been extensively explored by using various N-O containing oxidants such as nitro compounds, nitrones and sulfoxides, these protocols have been greatly limited to intramolecular reactions. In 2010, L. Zhang and co-workers developed *N*-oxides as the first intermolecular oxygen atom transfer reagent for carbene generation from alkynes. As shown in Scheme 34, they reported their pioneering work on the generation of *α*-oxo gold carbenes via gold-catalyzed intermolecular alkyne oxidation by *N*-oxides, thus making alkynes surrogates of hazardous *α*-diazoketones and enabling a non-diazo approach with gold carbene chemistry [75]. In the presence of 5 mol% of $Ph_3PAuNTf_2$ and 1.2 equiv. of MsOH, the reaction of homopropargylic alcohols **66** with 3,5-dichloropyridine *N*-oxide or 2-bromopyridine *N*-oxide allowed efficient formation of dihydrofuran-3-ones **67** in mostly good to excellent yields. The use of MsOH was believed to prevent the basic pyridine byproduct from deactivating the gold catalyst. In addition, a variety of functional groups including acid sensitive ones, such as OMOM and NBoc, were tolerated in this catalytic system. Presumably, this oxidative cyclization involves the generation of *α*-oxo gold carbenes **66-B** by

Scheme 34 Gold-catalyzed oxidative cyclization of homopropargylic alcohols **66** for the assembly of dihydrofuran-3-ones **67**

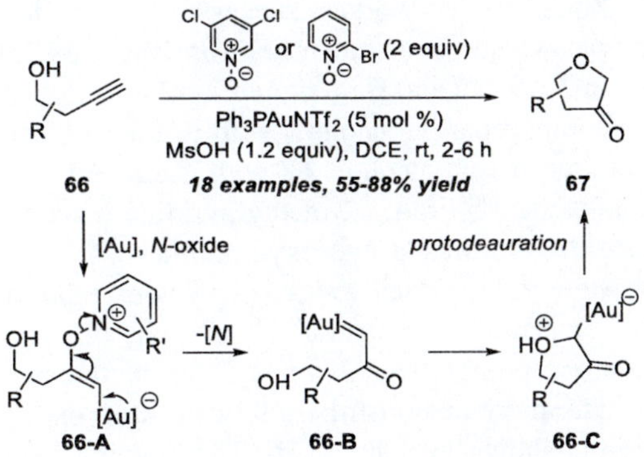

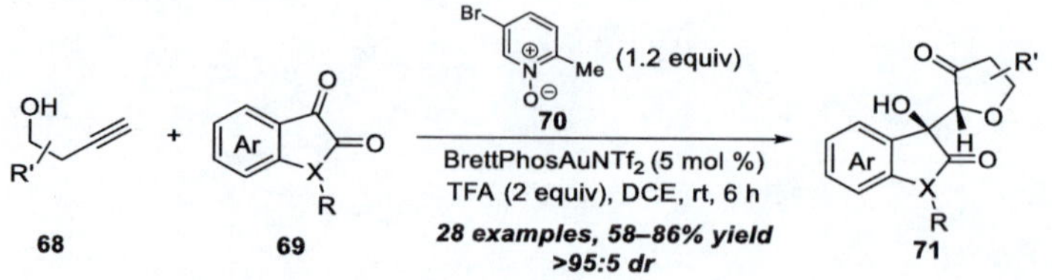

Scheme 35 Gold-catalyzed oxidative cyclization cascade of homopropargyl alcohols **68**

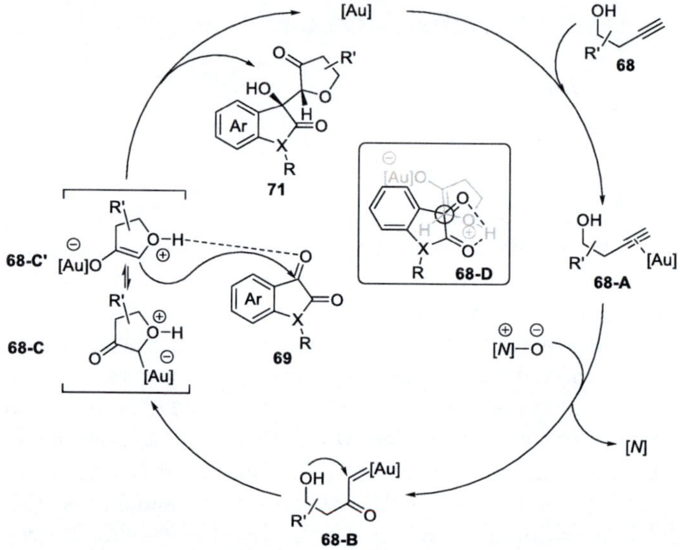

Scheme 36 Proposed mechanism for the synthesis of 3-hydroxyoxindoles **71**

oxidation of a terminal C–C triple bond, which was further trapped intramolecularly by the hydroxyl group. This protocol was successfully applied to the synthesis of chiral dihydrofuran-3-ones by Antilla and co-workers [76]. In the same year, L. Zhang and co-workers successfully extended the scope of the reaction to simple propargylic alcohol substrates, furnishing oxetan-3-ones efficiently [77].

In 2019, Xu and Hu reported a gold-catalyzed oxidative cyclization/aldol addition of homopropargyl alcohols with isatins. As shown in Scheme 35, the treatment of homopropargyl alcohols **68** with isatins **69** in the presence of BrettPhosAuNTf$_2$ as catalyst and 5-bromo-2-methylpyridine *N*-oxide as oxidant under room temperature afforded various 3-hydroxyoxindoles **71** in good to excellent yields with high diastereoselectivities (>95:5 dr) under mild reaction conditions [78]. Moreover, the reaction demonstrated full chirality transfer when an enantioenriched homopropargyl alcohol was used in the transformation.

The mechanism of the gold-catalyzed oxidative cyclization/aldol addition reaction is shown in Scheme 36. The reaction begins with attack of pyridine *N*-oxide onto gold-activated terminal alkynes **68-A** to produce the α-oxo gold carbene intermediates **68-B**. Then, cyclization of **68-B** affords the oxonium ylides **68-C**, which undergoes tautomerization to give **68-C′**. Subsequent intermolecular aldol addition delivers the final products **71**. The stereoselectivity in the enantio-determining aldol-type addition step originates from hydrogen bonding between the ylide intermediates and the carbonyl group of isatins.

6.1.2 N–H Insertion

The related N–H insertion of α-oxo gold carbenes was also feasible using *N*-oxides as oxygen transfer reagents. In 2011 L. Zhang and co-workers developed an oxidative N–H insertion enabled efficient construction of azetidin-3-ones **73** from *N*-propargylic amides **72** [79]. As shown in Scheme 37, these highly enantioselective amides, readily accessible via the *tert*-butylsulfinimide chemistry [80], cyclized smoothly to produce the strained *N*-heterocycles **73** in mostly good yields. Pre-oxidation was needed before the reaction, as the use of sulfoxide substrates failed to lead to the desired oxidative N–H insertion. In addition, the Bus group (tBuSO$_2$) could be easily removed from the azetidine ring under acidic conditions. The combination of the bulky BrettPhos ligand and the more reactive 2,6-dibromopyridine *N*-oxide is the key combination for the transformation.

6.1.3 C–H Insertion

The generated α-oxo gold carbenes have also been trapped by aryl groups to furnish the tandem formal C–H insertion reaction. In 2012 L. Zhang and co-workers reported a gold-catalyzed tandem alkyne oxidation/C–H insertion for the synthesis of various chroman-3-ones from readily accessible propargyl aryl ethers [81]. As described in Scheme 38, the oxidative cyclization of propargyl aryl ethers **74** in the presence of 5 mol% of Me$_4$(t-Bu)XPhosAuNTf$_2$ as a new bulky gold catalyst afforded the corresponding chroman-3-ones **75** in generally moderate to good yields. It should be mentioned that bulky electron-deficient *N*-oxides greatly improved the efficiency of the oxidative cyclization sequence. Similar C(sp^2)–H functionalization was developed by Gagosz and co-workers by employing a new class of

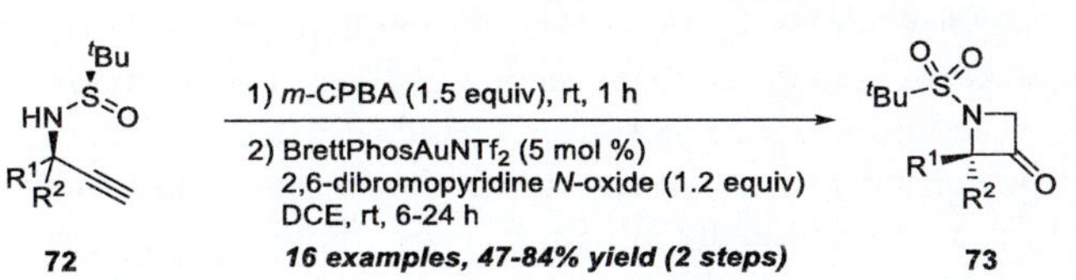

Scheme 37 Gold-catalyzed oxidation/intramolecular N–H insertion of *N*-propargylic amides **72** for the construction of azetidin-3-ones **73**

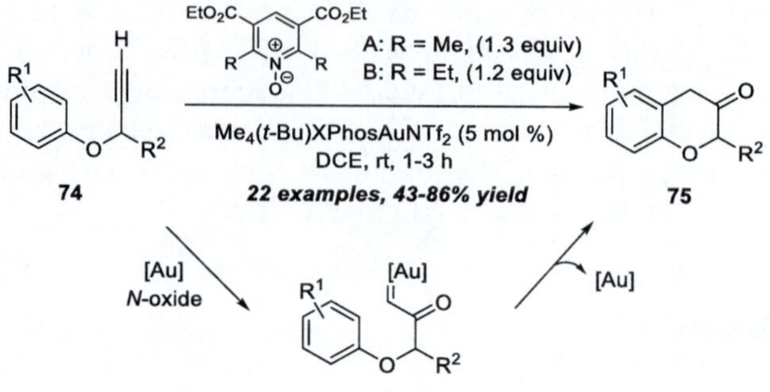

Scheme 38 Gold-catalyzed oxidative cyclization of propargyl aryl ethers **74** for the synthesis of chroman-3-ones **75**

Scheme 39 Gold-catalyzed oxidative cyclization of *N*-arylpropiolamides **76** for the synthesis of 3-acyloxindoles **77**

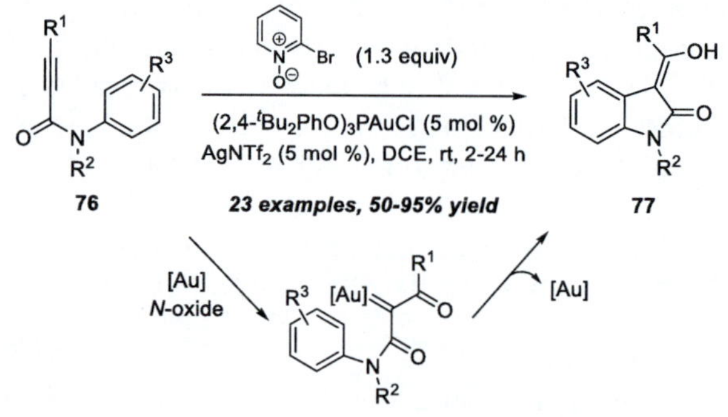

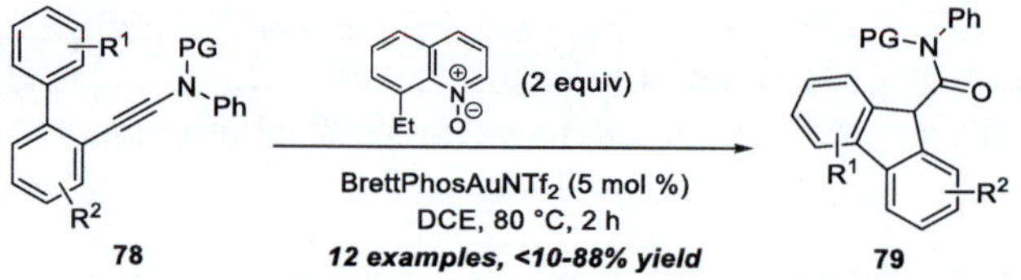

Scheme 40 Gold-catalyzed oxidative cyclization of *o*-alkynylbiaryls **78** for the preparation of functionalized fluorenes **79**

biarylphosphonite gold(I) complex as catalyst, leading to various synthetically useful indan-2-ones through S_N2' mechanism [82].

Furthermore, J. Zhang and co-workers also reported an elegant protocol for the synthesis of 3-acyloxindoles **77** from *N*-arylpropiolamides **76** under mild reaction conditions with wide substrate scope and mostly excellent yields (Scheme 39) [83]. However, according to the report from Li and co-workers in 2013, the oxidative cyclization of *N*-arylynamides delivered oxindoles in mostly low yields [84].

As an extension of terminal alkynes and electron-deficient alkynes, oxidative C–H functionalization is also applicable to ynamides. In 2014, Ye and co-workers disclosed the oxidative cyclization of *o*-alkynylbiaryls **78** for the preparation of functionalized fluorenes **79** in mostly good yields (Scheme 40) [85]. The use of bulky BrettPhosAuNTf$_2$ as catalyst and quinoline *N*-oxide as oxidant minimized the formation of diketone byproducts. Similar oxidative ynone and *N*-aryl ynamide cyclizations of via C(sp^2)–H functionalization were also developed for the construction of salicyl ketones [86] and fused γ-lactams [87]. Oxidative C–H insertion of alkynes has also been extended to related C(sp^3)–H bonds by Liu [88] and L. Zhang [89].

6.2 1,2-Migration

6.2.1 1,2-Alkynyl Migration

Due to the reactivity of α-oxo gold carbene intermediate, 1,2-*H* migration is another potential reaction pathway for the formation of unsaturated carbonyl compounds [90–94]. Hashmi and co-workers reported in 2014 an efficient gold-catalyzed cascade cyclization involving the generation of α-oxo gold carbenes followed by 1,2-alkynyl migration, which represents the first example of a 1,2-alkynyl migration onto gold carbenes [95]. The hydride migration outcompetes the alkyl migration reported in their previous work on the gold-catalyzed oxidative rearrangement of propargyl alcohols for the synthesis of 1,3-diketones [96]. As shown in Scheme 41, the oxidative cyclization of symmetric and unsymmetric 1,4-diyn-3-ols **80** could lead to a wide variety of 3-formylfurans **82** or **83** in good to excellent yields. After a series of mechanistic investigations involving isotope-labeling experiments and DFT calculations, the authors proposed an oxidation/1,2-alkynyl migration reaction pathway.

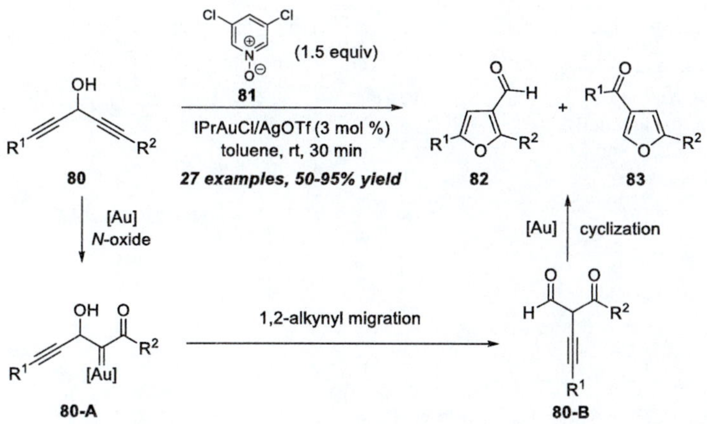

Scheme 41 Gold-catalyzed oxidation/1,2-alkynyl migration of 1,4-diyn-3-ols **80** for the synthesis of 3-formylfurans **82** and **83**

6.2.2 1,2-Enynyl Migration

In 2018 Liu and co-workers developed an efficient cascade cyclization of diynes involving gold-catalyzed highly regioselective oxidation, followed by 1,2-migration of an enynyl group, enabling the concise and regioselective synthesis of highly functionalized 1*H*-isochromene or 2*H*-pyran compounds **85** in 23–93% yields. (Scheme 42) [97]. This protocol proved to be quite general, and both aryl and alkyl-substituted alkynes proceeded smoothly. Isotopic labeling experiments revealed that an unprecedented 1,2-enynyl migration on an α-carbonyl gold carbenoid intermediate was presumably involved in this cascade cyclization.

6.3 Ring Expansion

Benefiting from the high reactivity of α-oxo gold carbene-like intermediates, ring expansion has also been achieved. In 2015, Liu and Li demonstrated an elegant protocol for gold-catalyzed oxidative ring expansion reaction (Scheme 43) [98]. In the presence of $Ph_3PAuNTf_2$ (2 mol%) as catalyst and pyridine *N*-oxide (2 equiv) as the oxidant, the reaction proceeded smoothly with a variety of 2-alkynyl-1,2-dihydro-pyridines or -quinolines **86** to produce the desired 1*H*-azepine or benzazepine derivatives **87** in 50–98% yields with broad substrate scope and good functional group tolerance. Of note, a non-carbene pathway was proposed in this case based on DFT calculations. The reaction most likely proceeds by a direct intramolecular S_N2' attack, followed by subsequent ring opening of cyclopropyl gold intermediate.

In the same year an efficient gold-catalyzed oxidative ring-expansion reaction involving the 1,2-heteroatom migration was developed by Liu and co-workers [99]. By employing gold as catalyst and *N*-oxide as oxidant, alkynes **88** could be

Scheme 42 Gold-catalyzed oxidative cyclization of diynes **84** for the assembly of 1*H*-isochromene and 2*H*-pyran compounds **85**

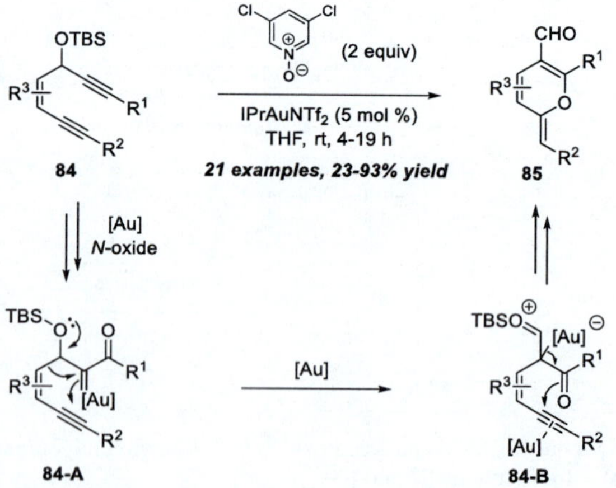

Scheme 43 Gold-catalyzed oxidative ring expansion of 2-alkynyl-1,2-dihydropyridines **86** for the formation of 1*H*-azepine or benzazepine derivatives **87**

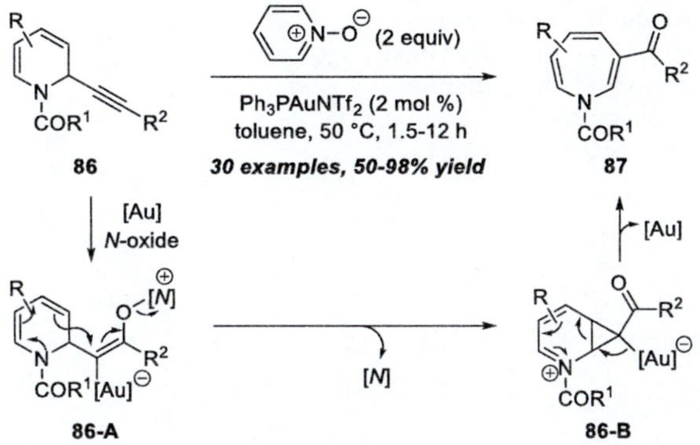

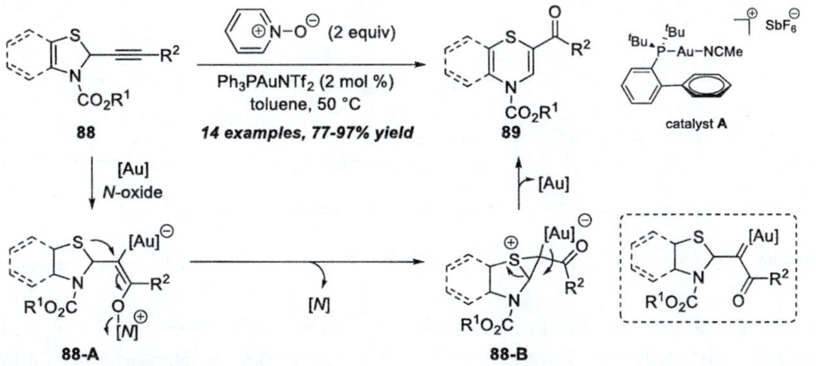

Scheme 44 Gold-catalyzed oxidative ring-expansion of alkynyl heterocycles **88** for the construction of benzothiazines **89**

readily transformed into the corresponding benzothiazines **89** in moderate to good yields by the oxidative ring-expansion reaction via 1,2-*S* migration, as shown in Scheme 44. Different *N*-protected and aryl-substituted alkynes bearing both electron-deficient and electron-rich groups were tolerated. Notably, this oxidative ring-expansion presumably proceeded via a direct S_N2' pathway rather than α-oxo gold carbene pathway. Additionally, the oxidative ring-expansion is also applicable to 1,2-*N* migration and 1,2-*O* migration, delivering 1,4-dihydroquinoxalines and 4*H*-benzo[*b*][1,4]oxazines efficiently.

6.4 Formal Annulation

The intermolecular oxidative formal [3 + 2]-annulation of alkynes with nitriles was also achieved via α-oxo gold carbene intermediates [100]. In 2011, L. Zhang and co-workers reported that various 2,5-disubstituted oxazoles **91** could be formed efficiently from commercially available terminal alkynes **90** and nitriles, which

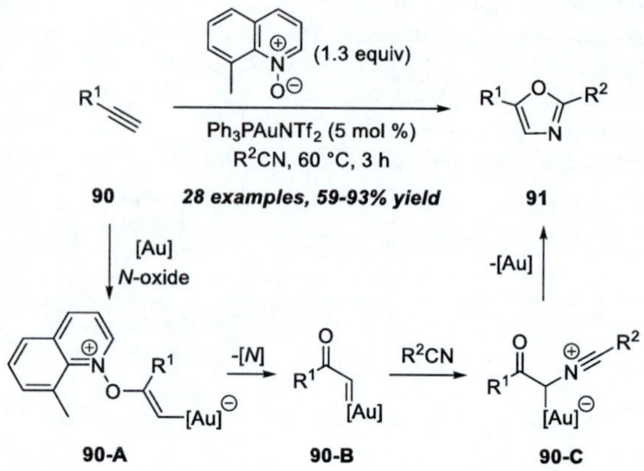

Scheme 45 Gold-catalyzed oxidation/[3 + 2]-annulation of alkynes with nitriles for the synthesis of 2,5-disubstituted oxazoles **91**

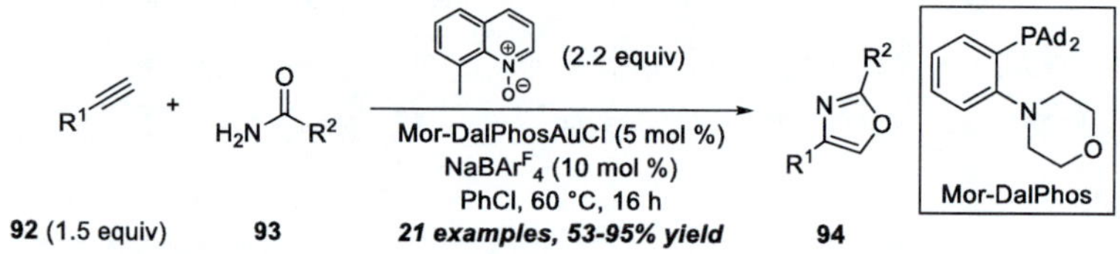

Scheme 46 Gold-catalyzed formal [3 + 2] annulation of terminal alkynes **92** with carboxamides **93**

were also the solvent, in the presence of 8-methylquinoline *N*-oxide as the oxidant, thus constituting a formal convergent [2 + 2 + 1]-annulation (Scheme 45). When a scarce and/or expensive nitrile was used, 3 equiv. of the nitrile was sufficient to deliver the desired product in a serviceable yield. Furthermore, Kukushkin and Rassadin reported in 2015 a related heterocyclization of terminal alkynes with cyanamides, which led to diverse 2-amino-1,3-oxazoles [101].

By using bidentate *P,N* ligand L. Zhang and co-workers reported a direct gold-catalyzed formal [3 + 2]-annulation of readily available terminal alkynes **92** and aromatic/alkenic carboxamides **93** for the synthesis of 2,4-oxazoles **94** in moderate to high yields under mild conditions (Scheme 46) [102]. The authors proposed that the gold complex, which leads to the formation of tricoordinated gold carbene intermediates via coordination of the non-phosphorus heteroatom, is the key for this annulation reaction.

The proposed mechanism of this formal [3 + 2]-annulation is shown in Scheme 47. The generatated α-oxo gold carbene intermediates **92-C** are selectively trapped by the oxygen of carboxamides and, following protodeauration, imidates **92-E** are formed. Subsequent cyclization and dehydration produce the final 2,4-oxazoles **94**. This gold-catalyzed oxidative methodology has also been extended to the [4 + 1]- and [4 + 2]-annulations by Liu and co-workers [103, 104].

Scheme 47 Proposed mechanism for the formation of 2,4-oxazoles **94**

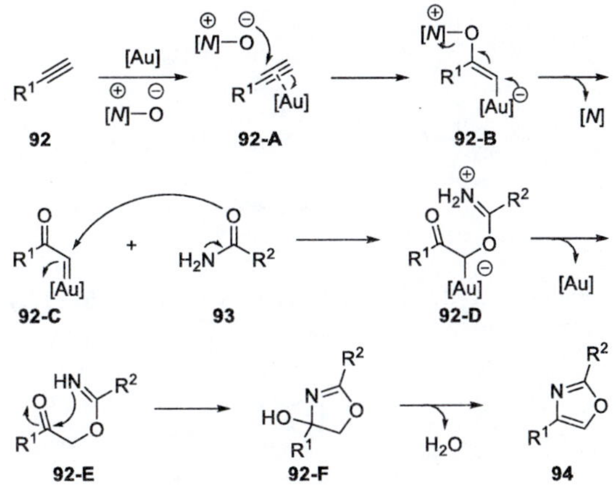

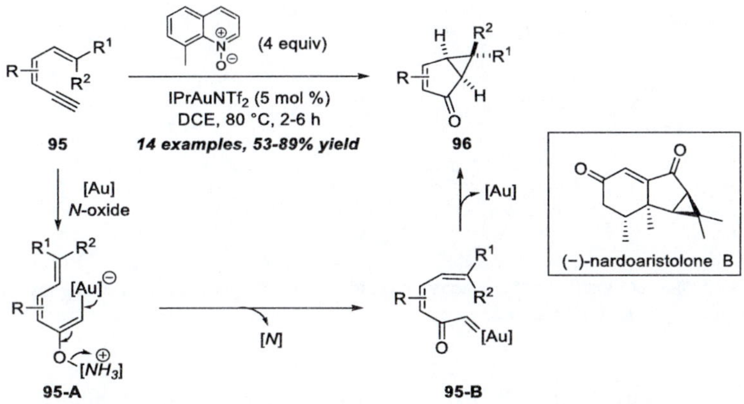

Scheme 48 Gold-catalyzed oxidative cyclopropanation of 1,5-enynes **95** for the formation of indanones or cyclopentenones **96**

6.5 *Cyclopropanation*

Liu and co-workers disclosed in 2011 the first gold-catalyzed oxidative cyclopropanation using *N*-oxides [105]. As shown in Scheme 48, treatment of 1,5-enynes **95** bearing terminal alkyne moieties with quinoline *N*-oxide in the presence of $IPrAuNTf_2$ (5 mol%) as catalyst allowed the synthesis of indanones or cyclopentenones **96** in 53–89% yields with excellent diastereoselectivities. Of note, in the case of 1,5-enynes with an aminoalkynyl substituent, gold-catalyzed 5-*exo-dig* oxidative cyclization occurs instead, delivering the corresponding 3-carbonyl-1*H*-indenes in moderate to excellent yields. On the basis of control experiments, both oxidative cyclizations presumably proceeded via α-oxo gold carbenes followed by intramolecular carbocyclizations. The reaction proved to be useful when

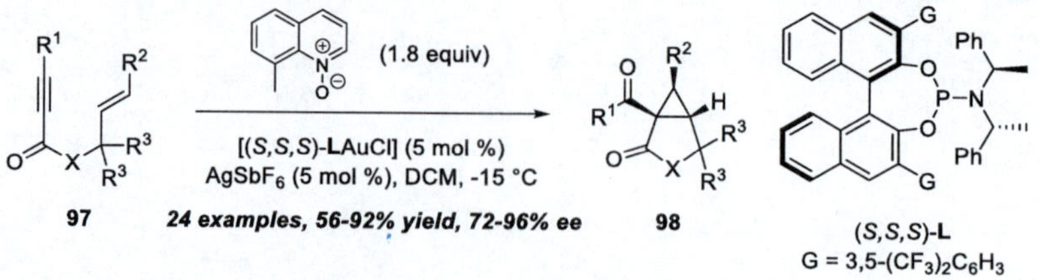

Scheme 49 Gold-catalyzed enantioselective oxidation/cyclopropanation of 1,6-enynes **97** for the generation of bicyclo[3.1.0]hexanes **98**

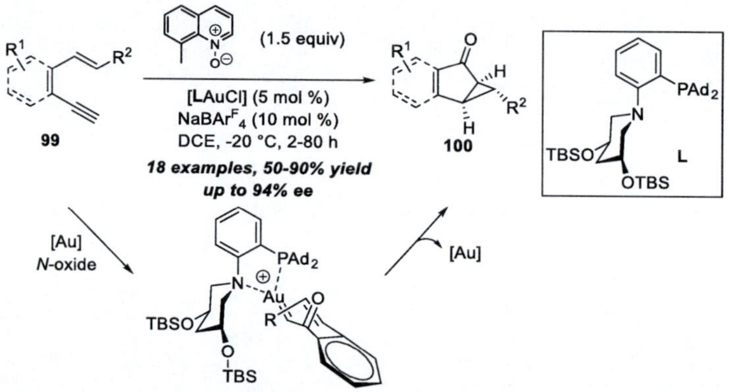

Scheme 50 Gold-catalyzed enantioselective oxidation/cyclopropanation of 1,5-enynes **99** for the formation of cyclopropyl ketones **100**

Echavarren's group accomplished the enantioselective total synthesis of (−)-nardoaristolone B employing it [106].

The first gold-catalyzed asymmetric oxidative cyclopropanation of an enyne was realized by J. Zhang and co-workers in 2014 [107]. By employing Au(I) and chiral phosphoramidite complexes, the oxidative cyclopropanation of 1,6-enynes **97** proceeded smoothly to produce the corresponding functionalized bicyclo[3.1.0] hexanes **98** bearing three contiguous quaternary and tertiary stereogenic centers with generally excellent enantioselectivities (Scheme 49).

After developing a novel chiral *P,N*-bidentate ligand, L. Zhang and co-workers reported in 2015 a gold-catalyzed enantioselective oxidative cyclopropanation of 1,5-enynes **99**, delivering versatile cyclopropyl ketones **100** with high efficiency and mostly excellent enantioselectivities (Scheme 50) [108]. They employed the *P,N*-bidentate ligand bearing a C_2-symmetric piperidine ring as the nitrogen-containing moiety, which led to a well-organized triscoordinated gold center. Different from J. Zhang's above-described protocol, α-oxo gold carbenes were proposed and cyclopropanation occurred at the *Si* face of α-oxo gold carbenes, leading to the formation of the desired products with high enantioselectivity.

In addition, the gold-catalyzed oxidative cyclization of alkynyl sulfoxides [109] and conjugated enynamides [110] were reported by Davies and Huang respectively, providing a range of cyclopropane-fused thiolane *S*-oxides and tetracyclic spiroindolines.

6.6 Rearrangement

6.6.1 Ylide Rearrangement

In addition to the above-mentioned oxidative cascades, oxidative rearrangement reactions enabled by *N*-oxides have also been discovered. Li and co-workers reported in 2012 an elegant gold-catalyzed oxidative rearrangement of homopropargylic ethers **101** via oxonium ylide intermediates (Scheme 51) [111]. With pyridine *N*-oxide as the oxidant, the reaction proceeded smoothly with various substituents, including alkyl as well as electron-withdrawing and electron-donating groups on the aromatic ring, leading to a range of α,β-unsaturated carbonyls **102** in moderate to good yields. Interestingly, substrates with electron-rich groups on the aryl ring underwent further cyclization to produce cyclobutanones **103**.

As shown in Scheme 52, the reaction begins with the formation of α-oxo gold carbenes **101-A** via nucleophilic attack of pyridine *N*-oxide onto gold-activated

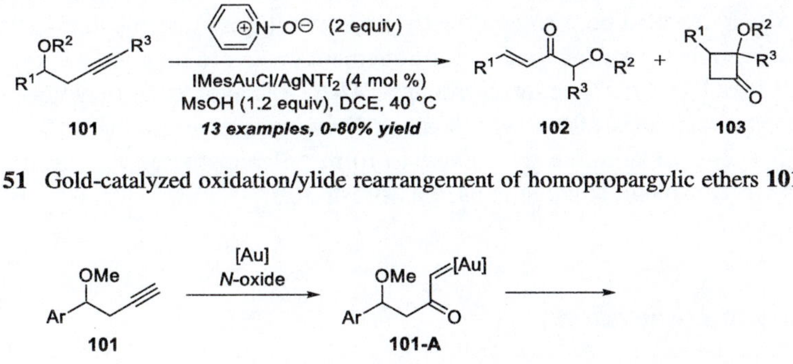

Scheme 51 Gold-catalyzed oxidation/ylide rearrangement of homopropargylic ethers **101**

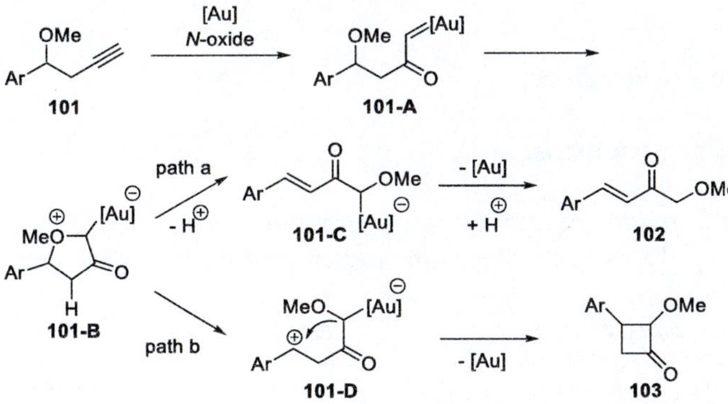

Scheme 52 Proposed mechanism for the synthesis of α,β-unsaturated carbonyls **102** and cyclobutanones **103**

alkynes **101** and subsequent expulsion of pyridine. In path a, gold carbenes **101-A** are trapped by the oxygen atom, leading to oxonium ylides **101-B**, which undergo subsequent C–O bond cleavage and protodeauration to eventually afford the final products **102** and regenerate the gold catalyst. For substrates bearing electron-rich groups on the aryl ring, intramolecular cyclization takes place after C–O bond cleavage, delivering cyclobutanone products **103** selectively. Similarly, a gold-catalyzed oxidative ylide rearrangement of homopropargylic ethers was reported by Li and co-workers for the selective synthesis of *cis*-cyclobutanones [112].

6.6.2 Desulfonylative Rearrangement

Another interesting gold-catalyzed desulfonylative rearrangement reaction was discovered by L. Zhang and co-workers in 2015. The use of alkynesulfonates **104** and styrenes **105** as substrates in the presence of 5 mol% of $IPrAuNTf_2$ led to formation of cyclopropyl ketones **106** in generally good to excellent yields by in situ expulsion of sulfur dioxide (Scheme 53) [113]. This protocol offers regiospecific access to donor-substituted acyl gold carbenes, which are seldom generated in traditional gold-catalyzed *N*-oxide oxidation of internal alkynes. Alternatively, a formal [3 + 2]-cycloaddition took place to give polysubstituted dihydrofuran products when α-methylstyrene was used as the substrate.

As proposed in Scheme 54, the sulfonyl acyl gold carbene intermediates **104-C**, generated from gold catalyzed alkyne oxidation of alkynesulfonates **104** by *N*-oxide, is trapped by intramolecular π-bonds to produce episulfone intermediates **104-D** via path a, which can also be formed directly from vinyl gold intermediates **104-B** via path b. Subsequent desulfonylative rearrangement occurs to produce the key donor-substituted acyl gold carbene intermediates **104-E**. Intermolecular cyclopropanation of α-oxo gold carbenes **104-E** with styrene derivatives give products **106**, and the formal [3 + 2]-cycloaddition took place to form polysubstituted dihydrofuran **107** products when α-methylstyrene is used as substrate.

6.7 Diyne Cyclization

6.7.1 Carbene Metathesis

The first gold-catalyzed oxidative diyne cyclization was developed by Hashmi and co-workers in 2013 (Scheme 55). In the presence of $Ph_3PAuNTf_2$ the treatment of

Scheme 53 Gold-catalyzed oxidation/desulfonylative rearrangement of alkynesulfonates **104**

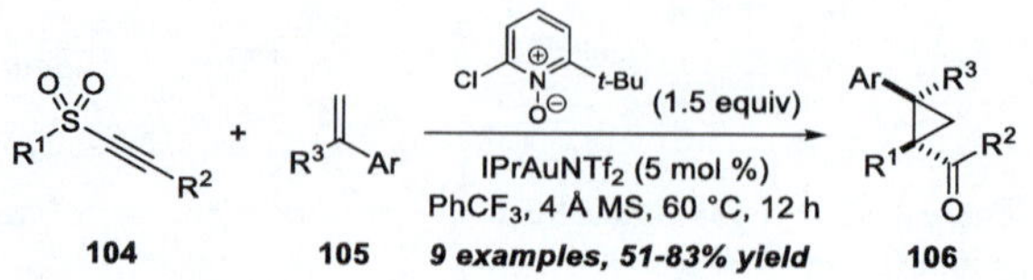

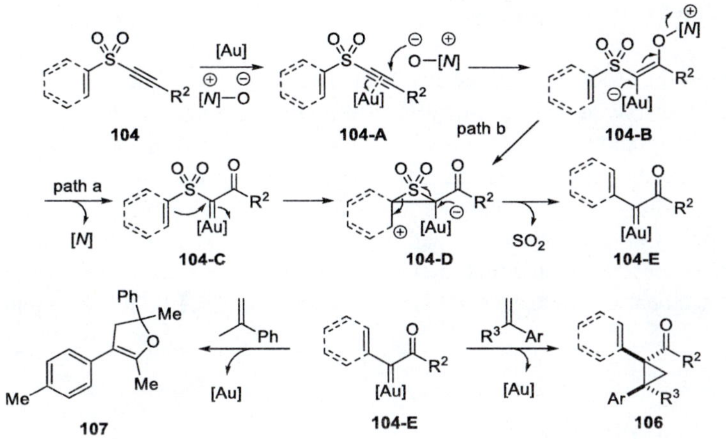

Scheme 54 Proposed mechanism for the synthesis of cyclopropyl ketones **106** and polysubstituted dihydrofuran **107**

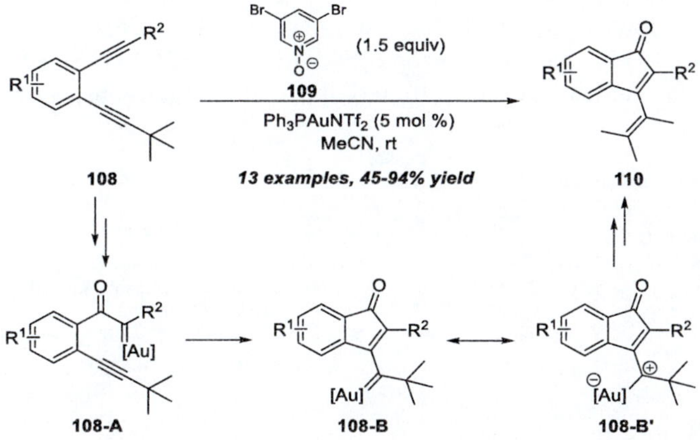

Scheme 55 Gold-catalyzed oxidative cyclization of diynes **108** for the preparation of substituted indenones **110**

haloalkynes **108** with 3,5-dibromopyridine *N*-oxide **109** allows the facile synthesis of highly substituted indenones **110** in 45–94% yields [114]. The reaction presumably proceeds through carbene metathesis by 1,6-carbene transfer followed by 1,2-alkyl migration. By replacing one of the *tert*-butyl alkynes by a phenyl acetylene, the less hindered aryl alkyne was attacked selectively. Furthermore, Ye and Lu reported an oxidative cyclization of terminal diynes via formal 1,5-carbene transfer, leading to diverse quinone derivatives [115].

6.7.2 Vinyl Cation Pathway

By combining $Me_4{}^tBuXPhosAuCl$ and $NaBAr^F{}_4$ as catalyst, L. Zhang and co-workers discovered a gold-catalyzed oxidative cyclization of diynes **111**, furnishing the pyranone-fused bridged [3.2.1] skeletons **112** in 40–70% yields with good diasteroselectivities (Scheme 56) [116]. Based on the control experiments and previous studies on the related donor–/acceptor-substituted gold carbenes, the authors proposed a vinyl cation pathway for the oxidative cyclization. The study of substrate scope indicated that the C(sp^3)–H functionalization proceeded smoothly with diynes bearing methine C–H bonds, methylene C–H bonds, and methyl C–H bonds.

6.7.3 1,2-Alkynyl Migration

In addition to the cyclization of diynes, 1,2-alkynyl migration of diynes were also developed. Ohno and co-workers reported in 2018 an efficient gold-catalyzed oxidative cascade cyclization of 1,4-diyn-3-ones **113**, where four new bonds were formed in one step. This protocol led to a variety of benzo[6,7]cyclohepta[1,2-*b*] furan skeletons **114** in moderate to good yields (Scheme 57) [117]. In addition, these products were further transformed into the corresponding benzotropone-fused

Scheme 56 Gold-catalyzed oxidative cyclization of cyclohexanone-derived diynes **111** for the synthesis of pyranone-fused bridged [3.2.1] skeletons **112**

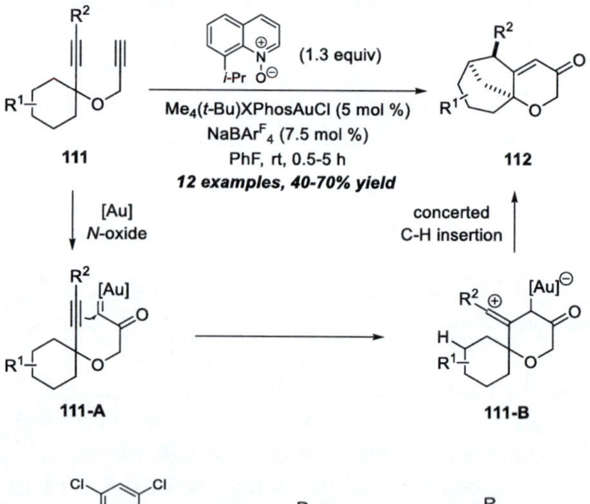

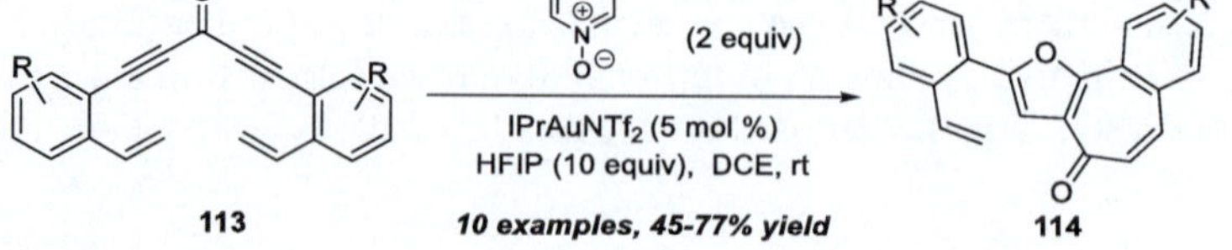

Scheme 57 Gold-catalyzed oxidative cyclization of 1,4-diyn-3-ones **113**

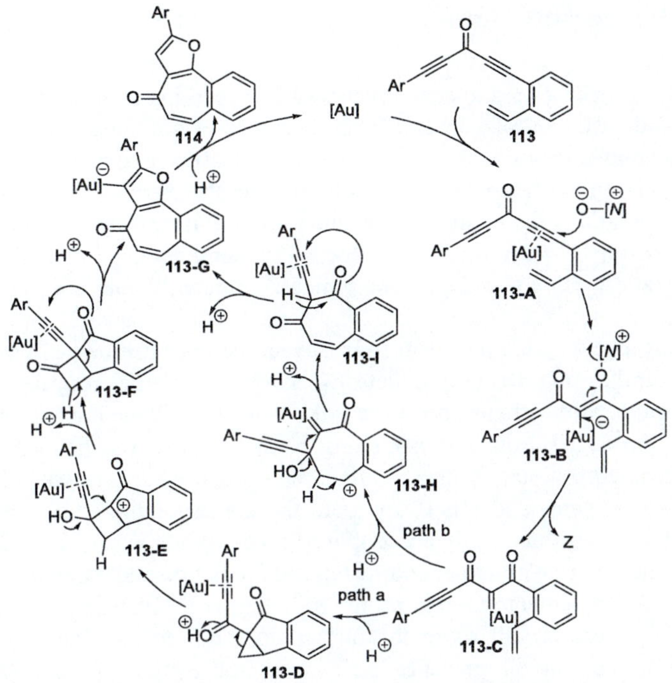

Scheme 58 Proposed mechanism for the generation of benzo[6,7]cyclohepta[1,2-*b*]furans **114**

naphtho[1,2-*b*]furans, through epoxidation and Lewis acid-catalyzed intramolecular C – C bond formation.

A mechanism was proposed based on the experimental observations (Scheme 58). The reaction starts with the generation of β-diketone-α-gold carbenes **113-C** that are trapped by the alkenyl group to afford cyclopropanation intermediates **113-D.** Subsequent ring expansion takes place to form the cationic cyclobutane intermediates **113-E.** A 1,2-alkynyl shift occurs to form intermediates **113-F,** which are transformed into the final products **114** by gold-catalyzed intramolecular 5-*endo-dig* cyclization with ring expansion and protodeauration. Alternatively, the nucleophilic addition of the styrene moiety on the carbonyl group forms benzyl cation intermediates **113-H** that undergo a further 1,2-alkynyl shift, delivering diketone intermediates **113-I**. Finally, products **114** forms after 5-*endo-dig* cyclization and protodeauration of **113-I**.

7 Conclusion and Outlook

During the past decades significant advances have been made in the gold-catalyzed amination and oxidation of alkynes. In the presence of gold catalysts, treatment of readily available alkynes with amination and oxidation reagents resulted in the efficient formation of versatile α-imino gold carbene species and α-oxo gold carbene species. These gold carbenes are able to undergo various transformations including X–H insertion, 1,2-migration, formal annulation, cyclopropanation, electrocyclization, ylide rearrangement, which significantly enriching gold carbene chemistry.

These protocols provide powerful and efficient methods for the rapid assembly of valuable natural or non-natural products, especially diverse heterocycles. It is notable that α-imino gold carbene species in gold-catalyzed alkyne amination reactions were rarely produced from corresponding α-imino diazo compounds by metal-catalyzed dediazotizations of diazo compounds. Meanwhile, α-oxo gold carbene species could be generated efficiently from the corresponding α-oxo diazo compounds, but the use of not easily accessible diazo compounds has limited its further synthetic applications. Therefore, gold-catalyzed amination and oxidation of alkynes provide more convenient operations in gold carbene chemistry. Despite these achievements, there is still room for further exploration: (a) There is no direct structural evidence for the presence of α-imino and α-oxo gold carbene species because they have not been isolated or spectroscopically detected and characterized. (b) It is sometimes difficult to discern if gold carbenes are actually intermediates since non-carbene pathways are also reasonable in some cases. (c) Catalytic asymmetric versions based on gold-catalyzed amination and oxidation of alkyne have been rarely demonstrated, and they are always challenging in gold catalysis. (d) Low atom-economy still exists, especially in gold-catalyzed atom transfer reactions, so the development of more atom-economic and environmentally friendly amination and oxidation reagents remains highly desirable. We anticipate that studies of gold-catalyzed amination and oxidation of alkynes will attract more interest from the chemistry community as well as from other areas to promote the development in both the methodologies and their applications.

References

1. Zi W, Toste FD (2016) Recent advances in enantioselective gold catalysis. Chem Soc Rev 45: 4567–4589
2. Harris RJ, Widenhoefer RA (2016) Gold carbenes, gold-stabilized carbocations, and cationic intermediates relevant to gold-catalysed enyne cycloaddition. Chem Soc Rev 45:4533–4551
3. Asiri AM, Hashmi ASK (2016) Gold-catalysed reactions of diynes. Chem Soc Rev 45:4471–4503
4. Liu L, Zhang J (2016) Gold-catalyzed transformations of α-diazocarbonyl compounds: selectivity and diversity. Chem Soc Rev 45:506–516

5. Jia M, Ma S (2016) New approaches to the synthesis of metal carbenes. Angew Chem Int Ed 55:9134–9166
6. Dorel R, Echavarren AM (2015) Gold(I)-catalyzed activation of alkynes for the construction of molecular complexity. Chem Rev 115:9028–9072
7. Wang Y, Muratore ME, Echavarren AM (2015) Gold carbene or carbenoid: is there a difference? Chem Eur J 21:7332–7339
8. Qian D, Zhang J (2015) Gold-catalyzed cyclopropanation reactions using a carbenoid precursor toolbox. Chem Soc Rev 44:677–698
9. Wei F, Song C, Ma Y, Zhou L, Tung C-H, Xu Z (2015) Gold carbene chemistry from diazo compounds. Sci Bull 60:1479–1492
10. Fensterbank L, Malacria M (2014) Molecular complexity from polyunsaturated substrates: the gold catalysis approach. Acc Chem Res 47:953–965
11. Obradors C, Echavarren AM (2014) Gold-catalyzed rearrangements and beyond. Acc Chem Res 47:902–912
12. Wang Y-M, Lackner AD, Toste FD (2014) Development of catalysts and ligands for enantioselective gold catalysis. Acc Chem Res 47:889–901
13. Hashmi ASK (2014) Dual gold catalysis. Acc Chem Res 47:864–876
14. Pflästerer D, Hashmi ASK (2016) Gold catalysis in total synthesis – recent achievements. Chem Soc Rev 45:1331–1367
15. Rudolph M, Hashmi ASK (2012) Gold catalysis in total synthesis – an update. Chem Soc Rev 41:2448–2462
16. Hashmi ASK, Rudolph M (2008) Gold catalysis in total synthesis. Chem Soc Rev 37:1766–1775
17. Darses B, Rodrigues R, Neuville L, Mazurais M, Dauban P (2017) Transition metal-catalyzed iodine(III)-mediated nitrene transfer reactions: efficient tools for challenging syntheses. Chem Commun 53:493–508
18. Xiong T, Zhang Q (2016) New amination strategies based on nitrogen-centered radical chemistry. Chem Soc Rev 45:3069–3087
19. Bizet V, Hendriks CMM, Bolm C (2015) Sulfur imidations: access to sulfimides and sulfoximines. Chem Soc Rev 44:3378–3390
20. Uchida T, Katsuki T (2014) Asymmetric nitrene transfer reactions: sulfimidation, aziridination and C–H amination using azide compounds as nitrene precursors. Chem Rec 14:117–129
21. Degennaro L, Trinchera P, Luisi R (2014) Recent advances in the stereoselective synthesis of aziridines. Chem Rev 114:7881–7929
22. Sahani RL, Ye L-W, Liu R-S (2020) Chapter 3 – Synthesis of nitrogen-containing molecules via transition metal-catalyzed reactions on isoxazoles, anthranils and benzoisoxazoles. Adv Organomet Chem 73:195–251
23. Aguilar E, Santamaría J (2019) Gold-catalyzed heterocyclic syntheses through α-imino gold carbene complexes as intermediates. Org Chem Front 6:1513–1540
24. Davies PW, Garzón M (2015) Nucleophilic nitrenoids through π-acid catalysis: providing a common basis for rapid access into diverse nitrogen heterocycles. Asian J Org Chem 4:694–708
25. Fabian WMF, Bakulev VA, Kappe CO (1998) Pericyclic versus pseudopericyclic 1,5-electrocyclization of iminodiazomethanes. An ab initio and density functional theory study. J Org Chem 63:5801–5805
26. Harmon RE, Stanley F, Gupta SK, Johnson J (1970) N,N-Dialkylamino-1,2,3-triazole-.alpha.-diazoamidine tautomers from substituted benzenesulfonyl azides and ynamines. J Org Chem 35:3444–3448
27. Zheng Z, Wang Z, Wang Y, Zhang L (2016) Au-catalysed oxidative cyclisation. Chem Soc Rev 45:4448–4458
28. Yeom H-S, Shin S (2014) Catalytic access to α-oxo gold carbenes by N–O bond oxidants. Acc Chem Res 47:966–977

29. Zhang L (2014) A non-diazo approach to α-oxo gold carbenes *via* gold-catalyzed alkyne oxidation. Acc Chem Res 47:877–888
30. Xiao J, Li X (2011) Gold α-oxo carbenoids in catalysis: catalytic oxygen-atom transfer to alkynes. Angew Chem Int Ed 50:7226–7236
31. Gorin DJ, Davis NR, Toste FD (2005) Gold(I)-catalyzed intramolecular acetylenic Schmidt reaction. J Am Chem Soc 127:11260–11261
32. Gronnier C, Boissonnat G, Gagosz F (2013) Au-catalyzed formation of functionalized quinolines from 2-alkynyl arylazide derivatives. Org Lett 15:4234–4237
33. Lonca GH, Tejo C, Chan HL, Chiba S, Gagosz F (2017) Gold(I)-catalyzed 6-*endo-dig* azide–yne cyclization: efficient access to 2*H*-1,3-oxazines. Chem Commun 53:736–739
34. Witham CA, Mauleón P, Shapiro ND, Sherry BD, Toste FD (2007) Gold(I)-catalyzed oxidative rearrangements. J Am Chem Soc 129:5838–5839
35. Wetzel A, Gagosz F (2011) Gold-catalyzed transformation of 2-alkynyl arylazides: efficient access to the valuable pseudoindoxyl and indolyl frameworks. Angew Chem Int Ed 50:7354–7358
36. Lu B, Luo Y, Liu L, Ye L, Wang Y, Zhang L (2011) Umpolung reactivity of indole through gold catalysis. Angew Chem Int Ed 50:8358–8362
37. Xiao Y, Zhang L (2012) Synthesis of bicyclic imidazoles via [2 + 3] cycloaddition between nitriles and regioselectively generated α-imino gold carbene intermediates. Org Lett 14:4662–4665
38. Li N, Wang T-Y, Gong L-Z, Zhang L (2015) Gold-catalyzed multiple cascade reaction of 2-alkynylphenylazides with propargyl alcohols. Chem Eur J 21:3585–3588
39. Sherry BD, Toste FD (2004) Gold(I)-catalyzed propargyl Claisen rearrangement. J Am Chem Soc 126:15978–15979
40. Su H, Bao M, Pei C, Hu W, Qiu L, Xu X (2019) Gold-catalyzed dual annulation of azide-tethered alkynes with nitriles: expeditious synthesis of oxazolo[4,5-*c*]quinolines. Org Chem Front 6:2404–2409
41. Matsuoka J, Matsuda Y, Kawada Y, Oishi S, Ohno H (2017) Total synthesis of dictyodendrins by the gold-catalyzed cascade cyclization of conjugated diynes with pyrroles. Angew Chem Int Ed 56:7444–7448
42. Tokimizu Y, Oishi S, Fujii N, Ohno H (2014) Gold-catalyzed cascade cyclization of (azido)-ynamides: an efficient strategy for the construction of indoloquinolines. Org Lett 16:3138–3141
43. Li N, Lian X-L, Li Y-H, Wang T-Y, Han Z-Y, Zhang L, Gong L-Z (2016) Gold-catalyzed direct assembly of aryl-annulated carbazoles from 2-alkynyl arylazides and alkynes. Org Lett 18:4178–4181
44. Yan Z-Y, Xiao Y, Zhang L (2012) Gold-catalyzed one-step construction of 2,3-dihydro-1*H*-pyrrolizines with an electron-withdrawing group in the 5-position: a formal synthesis of 7-methoxymitosene. Angew Chem Int Ed 51:8624–8627
45. Shen W-B, Sun Q, Li L, Liu X, Zhou B, Yan J-Z, Lu X, Ye L-W (2017) Divergent synthesis of *N*-heterocycles via controllable cyclization of azido-diynes catalyzed by copper and gold. Nat Commun 8:1748–1756
46. Shu C, Wang Y-H, Zhou B, Li X-L, Ping Y-F, Lu X, Ye L-W (2015) Generation of α-imino gold carbenes through gold-catalyzed intermolecular reaction of azides with ynamides. J Am Chem Soc 137:9567–9570
47. Zhou B, Zhang Y-Q, Liu X, Ye L-W (2017) Gold-catalyzed intermolecular reaction of ynamides with 3-indolyl azides via an unexpected 1,2-alkyl migration. Sci Bull 62:1201–1206
48. Li C, Zhang L (2011) Gold-catalyzed nitrene transfer to activated alkynes: formation of α,β-unsaturated amidines. Org Lett 13:1738–1741
49. Davies PW, Cremonesi A, Dumitrescu L (2011) Intermolecular and selective synthesis of 2,4,5-trisubstituted oxazoles by a gold-catalyzed formal [3+2] cycloaddition. Angew Chem Int Ed 50:8931–8935

50. Chatzopoulou E, Davies PW (2013) Highly regioselective synthesis of 2,4,5-(hetero)aryl substituted oxazoles by intermolecular [3+2]-cycloaddition of unsymmetrical internal alkynes. Chem Commun 49:8617–8619
51. Garzón M, Davies PW (2014) A direct route into fused imidazo-diazines and imidazo-pyridines using nucleophilic nitrenoids in a gold-catalyzed formal [3 + 2]-dipolar cycloaddition. Org Lett 16:4850–4853
52. Garzón M, Arce EM, Reddy RJ, Davies PW (2017) General entry into o-,o′-heteroatom-linked *N*-(hetero)aryl-imidazole motifs by gold-catalysed formal [3+2]-dipolar cycloaddition. Adv Synth Catal 359:1837–1843
53. Yousefi BH, Manook A, Drzezga A, Reutern B, Schwaiger M, Wester H-J, Henriksen G (2011) Synthesis and evaluation of ^{11}C-labeled imidazo[2,1-*b*]benzothiazoles (IBTs) as PET tracers for imaging β-amyloid plaques in Alzheimer's disease. J Med Chem 54:949–956
54. Tian X, Song L, Rudolph M, Rominger F, Oeser T, Hashmi ASK (2019) Sulfilimines as versatile nitrene transfer reagents: facile access to diverse aza-heterocycles. Angew Chem Int Ed 58:3589–3593
55. Wang J, Wu Y, Ma C, Fiorin G, Wang J, Pinto LH, Lamb RA, Klein ML, DeGrado WF (2013) Structure and inhibition of the drug-resistant S31N mutant of the M2 ion channel of influenza A virus. Proc Natl Acad Sci U S A 110:1315–1320
56. Jensen AA, Plath N, Pedersen MHF, Isberg V, Krall J, Wellendorph P, Stensbøl TB, Gloriam DE, Krogsgaard-Larsen P, Frølund B (2013) Design, synthesis, and pharmacological characterization of *N*- and *O*-substituted 5,6,7,8-tetrahydro-4*H*-isoxazolo[4,5-*d*]azepin-3-ol analogues: novel 5-HT_{2A}/5-HT_{2C} receptor agonists with pro-cognitive properties. J Med Chem 56:1211–1227
57. Hu F, Szostak M (2015) Recent developments in the synthesis and reactivity of isoxazoles: metal catalysis and beyond. Adv Synth Catal 357:2583–2614
58. Huple DB, Ghorpade S, Liu R-S (2016) Recent advances in gold-catalyzed *N*- and *O*-functionalizations of alkynes with nitrones, nitroso, nitro and nitroxy species. Adv Synth Catal 358:1348–1367
59. Li L, Tan T-D, Zhang Y-Q, Liu X, Ye L-W (2017) Recent advances in transition-metal-catalyzed reactions of alkynes with isoxazoles. Org Biomol Chem 15:8483–8492
60. Manning JR, Davies HML (2008) One-pot synthesis of highly functionalized pyridines via a rhodium carbenoid induced ring expansion of isoxazoles. J Am Chem Soc 130:8602–8603
61. Manning JR, Davies HML (2008) Efficient route to 2*H*-1,3-oxazines through ring expansion of isoxazoles by rhodium carbenoids. Tetrahedron 64:6901–6908
62. Zhou A-H, He Q, Shu C, Yu Y-F, Liu S, Zhao T, Zhang W, Lu X, Ye L-W (2015) Atom-economic generation of gold carbenes: gold-catalyzed formal [3+2] cycloaddition between ynamides and isoxazoles. Chem Sci 6:1265–1271
63. Jin H, Huang L, Xie J, Rudolph M, Rominger F, Hashmi ASK (2016) Gold-catalyzed C−H annulation of anthranils with alkynes: a facile, flexible, and atom-economical synthesis of unprotected 7-acylindoles. Angew Chem Int Ed 55:794–797
64. Jin H, Tian B, Song X, Xie J, Rudolph M, Rominger F, Hashmi ASK (2016) Gold-catalyzed synthesis of quinolines from propargyl silyl ethers and anthranils through the umpolung of a gold carbene carbon. Angew Chem Int Ed 55:12688–12692
65. Zeng Z, Jin H, Sekine K, Rudolph M, Rominger F, Hashmi ASK (2018) Gold-catalyzed regiospecific C−H annulation of *o*-ethynylbiaryls with anthranils: π-extension by ring-expansion en route to *N*-doped PAHs. Angew Chem Int Ed 57:6935–6939
66. Asao N, Sato K, Yamamoto Y (2003) $AuBr_3$-catalyzed cyclization of *o*-(alkynyl)-nitrobenzenes. Efficient synthesis of isatogens and anthranils. Tetrahedron Lett 44:5675–5677
67. Jadhav AM, Bhunia S, Liao H-Y, Liu R-S (2011) Gold-catalyzed stereoselective synthesis of azacyclic compounds through a redox/[2 + 2 + 1] cycloaddition cascade of nitroalkyne substrates. J Am Chem Soc 133:1769–1771

68. Liu R-R, Ye S-C, Lu C-J, Zhuang G-L, Gao J-R, Jia Y-X (2015) Dual catalysis for the redox annulation of nitroalkynes with indoles: enantioselective construction of indolin-3-ones bearing quaternary stereocenters. Angew Chem Int Ed 54:11205–11208
69. Yeom H-S, Lee Y, Lee J-E, Shin S (2009) Geometry-dependent divergence in the gold-catalyzed redox cascade cyclization of *o*-alkynylaryl ketoximes and nitrones leading to isoindoles. Org Biomol Chem 7:4744–4752
70. Qian D, Zhang J (2013) Gold/Brønsted acid relay catalysis for enantioselective construction of spirocyclic diketones. Chem Eur J 19:6984–6988
71. Yeom H-S, Lee Y, Jeong J, So E, Hwang S, Lee J-E, Lee SS, Shin S (2010) Stereoselective one-pot synthesis of 1-aminoindanes and 5,6-fused azacycles using a gold-catalyzed redox-pinacol-Mannich-Michael cascade. Angew Chem Int Ed 49:1611–1614
72. Mukherjee A, Dateer RB, Chaudhuri R, Bhunia S, Karad SN, Liu R-S (2011) Gold-catalyzed 1,2-difunctionalizations of aminoalkynes using only N- and O-containing oxidants. J Am Chem Soc 133:15372–15375
73. Sahani RL, Patil MD, Wagh SB, Liu R-S (2018) Catalytic transformations of alkynes into either α-alkoxy or α-aryl enolates: Mannich reactions by cooperative catalysis and evidence for nucleophile-directed chemoselectivity. Angew Chem Int Ed 57:14878–14882
74. Wei H, Bao M, Dong K, Qiu L, Wu B, Hu W, Xu X (2018) Enantioselective oxidative cyclization/Mannich addition enabled by gold(I)/chiral phosphoric acid cooperative catalysis. Angew Chem Int Ed 57:17200–17204
75. Ye L, Cui L, Zhang G, Zhang L (2010) Alkynes as equivalents of α-diazo ketones in generating α-oxo metal carbenes: a gold-catalyzed expedient synthesis of dihydrofuran-3-ones. J Am Chem Soc 132:3258–3259
76. Jain P, Wang H, Houk KN, Antilla JC (2012) Brønsted acid catalyzed asymmetric propargylation of aldehydes. Angew Chem Int Ed 51:1391–1394
77. Ye L, He W, Zhang L (2010) Gold-catalyzed one-step practical synthesis of oxetan-3-ones from readily available propargylic alcohols. J Am Chem Soc 132:8550–8551
78. Cai J, Wang X, Qian Y, Qiu L, Hu W, Xu X (2019) Gold-catalyzed oxidative cyclization/aldol addition of homopropargyl alcohols with isatins. Org Lett 21:369–372
79. Ye L, He W, Zhang L (2011) A flexible and stereoselective synthesis of azetidine-3-one via a key gold-catalyzed intermolecular alkyne oxidation. Angew Chem Int Ed 50:3236–3239
80. Chen B-L, Wang B, Lin G-Q (2010) Highly diastereoselective addition of alkynylmagnesium chlorides to *N-tert*-butanesulfinyl aldimines: a practical and general access to chiral α-branched amines. J Org Chem 75:941–944
81. Wang Y, Ji K, Lan S, Zhang L (2012) Rapid access to chroman-3-ones through gold-catalyzed oxidation of propargyl aryl ethers. Angew Chem Int Ed 51:1915–1918
82. Henrion G, Chavas TEJ, Goff XL, Gagosz F (2013) Biarylphosphonite gold(I) complexes as superior catalysts for oxidative cyclization of propynyl arenes into indan-2-ones. Angew Chem Int Ed 52:6277–6282
83. Qian D, Zhang J (2012) Catalytic oxidation/C–H functionalization of *N*-arylpropiolamides by means of gold carbenoids: concise route to 3-acyloxindoles. Chem Commun 48:7082–7084
84. Yang L-Q, Wang K-B, Li C-Y (2013) Synthesis of oxindoles through the gold-catalyzed oxidation of *N*-arylynamides. Eur J Org Chem 2013:2775–2779
85. Pan F, Liu S, Shu C, Lin R-K, Yu Y-F, Zhou J-M, Ye L-W (2014) Gold-catalyzed intermolecular oxidation of *o*-alkynylbiaryls: an easy and practical access to functionalized fluorenes. Chem Commun 50:10726–10729
86. Ji K, Yang F, Gao S, Tang J, Gao J (2016) Gold-catalyzed oxidation/C−H functionalization of ynones: efficient and rapid access to functionalized polycyclic salicyl ketones. Chem Eur J 22: 10225–10229
87. Sánchez-Cantalejo F, Priest JD, Davies PW (2018) A gold carbene manifold to prepare fused γ-lactams by oxidative cyclisation of ynamides. Chem Eur J 24:17215–17219
88. Bhunia S, Ghorpade S, Huple DB, Liu R-S (2012) Gold-catalyzed oxidative cyclizations of *cis*-3-en-1-ynes to form cyclopentenone derivatives. Angew Chem Int Ed 51:2939–2942

89. Wang Y, Zheng Z, Zhang L (2015) Intramolecular insertions into unactivated C(sp^3)–H bonds by oxidatively generated β-diketone-α-gold carbenes: synthesis of cyclopentanones. J Am Chem Soc 137:5316–5319
90. Lu B, Li C, Zhang L (2010) Gold-catalyzed highly regioselective oxidation of C−C triple bonds without acid additives: propargyl moieties as masked α,β-unsaturated carbonyls. J Am Chem Soc 132:14070–14072
91. Bian Z, Marvin CC, Martin SF (2013) Enantioselective total synthesis of (−)-citrinadin A and revision of its stereochemical structure. J Am Chem Soc 135:10886–10889
92. Kong K, JrJA E, McCallum ME, Smith GM, Matsumaru T, Menhaji-Klotz E, Wood JL (2013) An enantioselective total synthesis and stereochemical revision of (+)-citrinadin B. J Am Chem Soc 135:10890–10893
93. Bian Z, Marvin CC, Pettersson M, Martin SF (2014) Enantioselective total syntheses of citrinadins A and B. stereochemical revision of their assigned structures. J Am Chem Soc 136:14184–14192
94. Davies PW, Cremonesi A, Martin N (2011) Site-specific introduction of gold-carbenoids by intermolecular oxidation of ynamides or ynol ethers. Chem Commun 47:379–381
95. Wang T, Shi S, Hansmann MM, Rettenmeier E, Rudolph M, Hashmi ASK (2014) Synthesis of highly substituted 3-formylfurans by a gold(I)-catalyzed oxidation/1,2-alkynyl migration/cyclization cascade. Angew Chem Int Ed 53:3715–3719
96. Hashmi ASK, Wang T, Shi S, Rudolph M (2012) Regioselectivity switch: gold(I)-catalyzed oxidative rearrangement of propargyl alcohols to 1,3-diketones. J Org Chem 77:7761–7767
97. Zhao J, Xu W, Xie X, Sun N, Li X, Liu Y (2018) Gold-catalyzed oxidative cyclizations of {*o*-(alkynyl)phenyl propargyl} silyl ether derivatives involving 1,2-enynyl migration: synthesis of functionalized 1*H*-isochromenes and 2*H*-pyrans. Org Lett 20:5461–5465
98. Chen M, Chen Y, Sun N, Zhao J, Liu Y, Li Y (2015) Gold-catalyzed oxidative ring expansion of 2-alkynyl-1,2-dihydropyridines or -quinolines: highly efficient synthesis of functionalized azepine or benzazepine scaffolds. Angew Chem Int Ed 54:1200–1204
99. Chen M, Sun N, Xu W, Zhao J, Wang G, Liu Y (2015) Gold-catalyzed ring expansion of alkynyl heterocycles through 1,2-migration of an endocyclic carbon–heteroatom bond. Chem Eur J 21:18571–18575
100. He W, Li C, Zhang L (2011) An efficient [2 + 2 + 1] synthesis of 2,5-disubstituted oxazoles *via* gold-catalyzed intermolecular alkyne oxidation. J Am Chem Soc 133:8482–8485
101. Rassadin VA, Boyarskiy VP, Kukushkin VY (2015) Facile gold-catalyzed heterocyclization of terminal alkynes and cyanamides leading to substituted 2-amino-1,3-oxazoles. Org Lett 17: 3502–3505
102. Luo Y, Ji K, Li Y, Zhang L (2012) Tempering the reactivities of postulated α-oxo gold carbenes using bidentate ligands: implication of tricoordinated gold intermediates and the development of an expedient bimolecular assembly of 2,4-disubstituted oxazoles. J Am Chem Soc 134:17412–17415
103. Dateer RB, Pati K, Liu R-S (2012) Gold-catalyzed synthesis of substituted 2-aminofurans via formal [4+1]-cycloadditions on 3-en-1-ynamides. Chem Commun 48:7200–7202
104. Wang L, Xie X, Liu Y (2013) Gold-catalyzed oxidative rearrangement involving 1,2-acyl migration: efficient synthesis of functionalized dihydro-γ-carbolines from α-(2-indolyl) propargylic alcohols and imines. Angew Chem Int Ed 52:13302–13306
105. Vasu D, Hung H-H, Bhunia S, Gawade SA, Das A, Liu R-S (2011) Gold-catalyzed oxidative cyclization of 1,5-enynes using external oxidants. Angew Chem Int Ed 50:6911–6914
106. Homs A, Muratore ME, Echavarren AM (2015) Enantioselective total synthesis of (−)-nardoaristolone B *via* a gold(I)-catalyzed oxidative cyclization. Org Lett 17:461–463
107. Qian D, Hu H, Liu F, Tang B, Ye W, Wang Y, Zhang J (2014) Gold(I)-catalyzed highly diastereo- and enantioselective alkyne oxidation/cyclopropanation of 1,6-enynes. Angew Chem Int Ed 53:13751–13755
108. Ji K, Zheng Z, Wang Z, Zhang L (2015) Enantioselective oxidative gold catalysis enabled by a designed chiral *P,N*-bidentate ligand. Angew Chem Int Ed 54:1245–1249

109. Barrett MJ, Khan GF, Davies PW, Grainger RS (2017) Alkynyl sulfoxides as α-sulfinyl carbene equivalents: gold-catalysed oxidative cyclopropanation. Chem Commun 53:5733–5736
110. Lin M, Zhu L, Xia J, Yu Y, Chen J, Mao Z, Huang X (2018) Gold-catalyzed oxidative cyclization of tryptamine derived enynamides: a stereoselective approach to tetracyclic spiroindolines. Adv Synth Catal 360:2280–2284
111. Xu M, Ren T-T, Li C-Y (2012) Gold-catalyzed oxidative rearrangement of homopropargylic ether via oxonium ylide. Org Lett 14:4902–4905
112. Xu M, Ren T-T, Wang K-B, Li C-Y (2013) Synthesis of cyclobutanones *via* gold-catalyzed oxidative rearrangement of homopropargylic ethers. Adv Synth Catal 355:2488–2494
113. Chen H, Zhang L (2015) A desulfonylative approach in oxidative gold catalysis: regiospecific access to donor-substituted acyl gold carbenes. Angew Chem Int Ed 54:11775–11779
114. Nösel P, dos Santos Comprido LN, Lauterbach T, Rudolph M, Rominger F, ASK H (2013) 1,6-carbene transfer: gold-catalyzed oxidative diyne cyclizations. J Am Chem Soc 135:15662–15666
115. Shu C, Shi C-Y, Sun Q, Zhou B, Li T-Y, He Q, Lu X, Liu R-S, Ye L-W (2019) Generation of endocyclic vinyl carbene complexes *via* gold-catalyzed oxidative cyclization of terminal diynes: towards naphthoquinones and carbazolequinones. ACS Catal 9:1019–1025
116. Zheng Z, Zhang L (2015) C–H insertions in oxidative gold catalysis: synthesis of polycyclic 2*H*-pyran-3(6*H*)-ones *via* a relay strategy. Org Chem Front 2:1556–1560
117. Hamada N, Yamaguchi A, Inuki S, Oishi S, Ohno H (2018) Gold(I)-catalyzed oxidative cascade cyclization of 1,4-diyn-3-ones for the construction of tropone-fused furan scaffolds. Org Lett 20:4401–4405

Heterocycles from Donor and Donor/Donor Carbenes

Dong Zhu and Shifa Zhu

Contents

Abstract This chapter reviews the recent advances of donor and donor–donor carbenes for the synthesis of heterocycles. Since new sources of donor-type carbenes have been developed, heterocycles can be obtained through carbene involved reactions in a safe manner. This chapter is organized based on the approaches of generating donor-type carbenes, including carbene precursors, such as diazo-type carbene sources, alkyne-type carbene sources, cyclopropenes, aryl silyl ketones,

D. Zhu and S. Zhu (✉)
Key Laboratory of Functional Molecular Engineering of Guangdong Province, School of Chemistry and Chemical Engineering, South China University of Technology, Guangzhou, P. R. China
e-mail: zhusf@scut.edu.cn

simple aryl aldehydes and typical reaction mode, such as alkyne cycloisomerization, carbene/alkyne metathesis, and Brook rearrangement. With these carbene sources as reaction substrates, the synthetic examples of heterocycles, as well as the reaction mechanisms are discussed.

Keywords Alkyne-cycloisomerization · Alkyne-type carbene sources · Aryl aldehydes · Aryl silyl ketones · Brook rearrangement · Carbene/alkyne metathesis · Cyclopropenes · Diazo-type carbene sources · Donor-type carbenes · Heterocycles

1 Introduction

Heterocycles are a very important class of organic compounds that play a vital role in biological processes. Widely used in pharmaceuticals, food additives, pesticides, and materials [1–4], the development of new methods for their construction is always intriguing for synthetic chemists. Among the various synthetic strategies, metal carbene transfer reactions are one of the most efficient and direct approaches to synthesize heterocycles [5], especially with the rapid progress of donor-type metal carbenes in recent decades.

Based on the different electronic effects of substituents adjacent to the carbene center, metal carbenes can be divided into three major groups [6], which are known as acceptor-type metal carbenes (including acceptor and acceptor–acceptor metal carbenes), acceptor–donor metal carbenes, and donor-type metal carbenes (including donor and donor–donor metal carbenes, Scheme 1).

Significant achievements have been made during the past half century with acceptor-type metal carbenes and donor–acceptor metal carbenes [7–9]. These metal carbenes are safely prepared from readily available acceptor-type diazo compounds. However, the donor-type metal carbenes have not been developed to the same extent [10]. The most commonly used diazo compounds with electron-donating groups are unstable and can be explosive. Mass and Aggarwal et al. reported that diazo compounds without electron-withdrawing groups have caused some unexpected incidents of explosions [11, 12]. Thus, when preparing and handling these unstable donor-type diazo compounds, careful operation is always required, and low temperature and dilution are also necessary for their safe storage and transportation. More importantly, the homo-dimerization of diazo compounds, which are their major side reaction for carbene chemistry, usually happens because

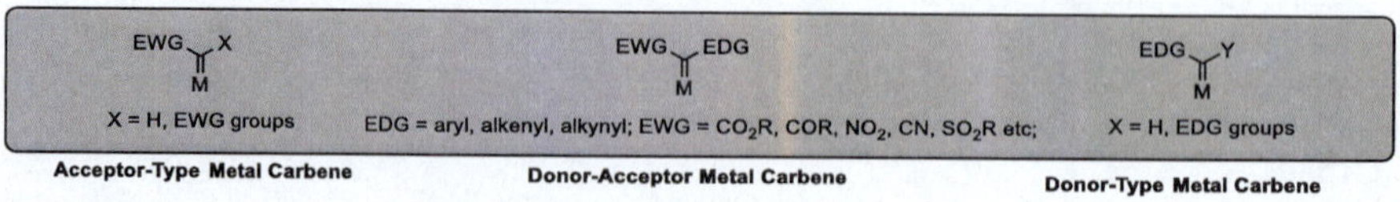

Scheme 1 Classification of metal carbenes

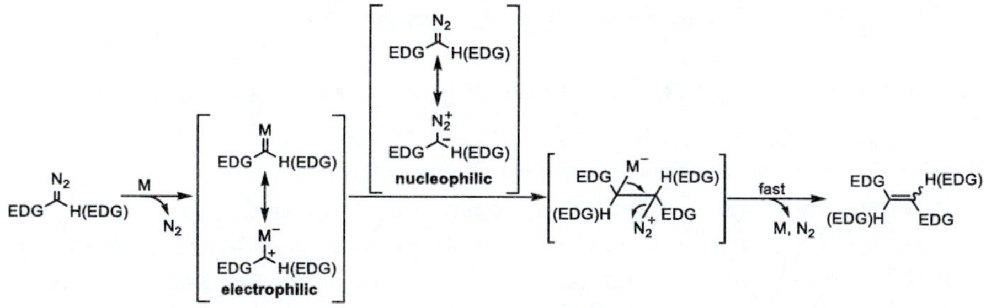

Scheme 2 Homo-dimerization of donor-type diazo compounds with umpolung of reactivity

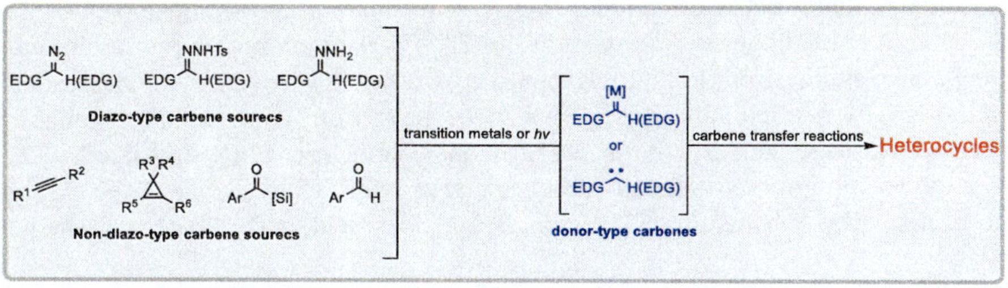

Scheme 3 Heterocycles from different types of donor-type carbene sources

the nucleophilic diazo compounds react fast with electrophilic metal carbene intermediates due to the umpolung of reactivity (Scheme 2). For these reasons, low concentration is especially needed for their application in reactions. These problems severely limited the progress of donor-type metal carbenes.

In order to overcome the above limitations, great efforts have been made to find safer alternative precursors for donor-type carbenes, and hydrazones have become the most widely used precursors. The most significant advantage of hydrazone-based systems is to avoid the direct handling of unstable diazo compounds. However, since this system generates relatively unstable diazo compounds, it is still potentially dangerous. Furthermore, fast homo-dimerization occurs in the hydrazone-based system and is difficult to avoid. As a consequence, other non-diazo-type carbene precursors for the generation of donor-type metal carbenes have attracted great attention. A variety of non-diazo reagents, including enynes, enynones, propargyl ethers, cyclopropenes, aryl silyl ketones and even simple aryl aldehydes, have been found to be efficient donor-type carbene precursors (Scheme 3). With diazo-type and non-diazo-type compounds as precursors, many carbene transfer reactions, such as C–H insertion, X–H (X = N, O, Si, B, S, P, etc.) insertion, cyclopropanation, cycloaddition, ylide formation, and migratory insertion involved cross-coupling reaction, can be conducted under simple and mild conditions [13]. These new methods open a door for the synthesis of valuable heterocycles.

2 Heterocycles from Diazo-Type Carbene Sources

2.1 *Heterocycles from Diazo Compounds as Carbene Sources*

Compared to aryl diazo compounds, diaryl diazo compounds are more stable and can be easily prepared under straightforward conditions and stored at low temperature for days. In 2015, Sun and co-workers demonstrated an unprecedented rhodium(III)-catalyzed C–H functionalization of aryl diazo compounds for the synthesis of isoquinolines **3** (Scheme 4) [14]. This novel transformation features a tandem process of diazo coupling, C–H bond activation, and N–N bond cleavage in the presence of a single rhodium(III) catalyst.

Although diaryl diazo compounds that serve as donor–donor carbene precursors can be kept stable under low temperature (−20 °C), they are normally used immediately once synthesized. In 2020, Kan and co-workers achieved a stereocontrolled total synthesis of sophoraflavanone H **9** in 14 steps from commercially available starting materials. This synthesis features an asymmetric Rh(II)-mediated C–H insertion reaction of donor–donor carbenes that provides ready access to the highly substituted dihydrobenzofuran skeleton **8** in 97% yield with 79% *ee* (Scheme 5) [15].

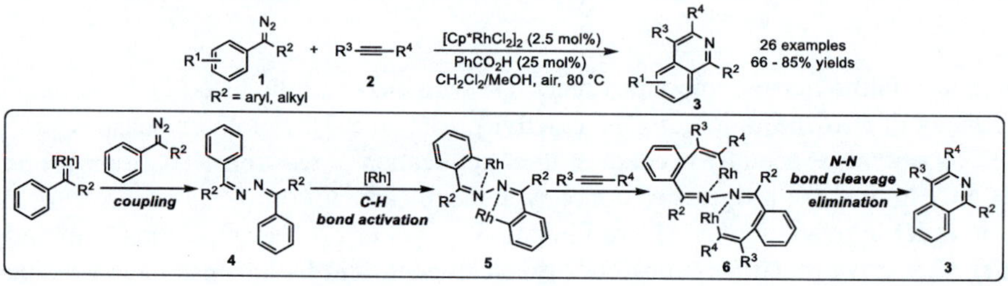

Scheme 4 Rhodium-catalyzed reaction of diazo compounds with alkynes for the synthesis of isoquinolines

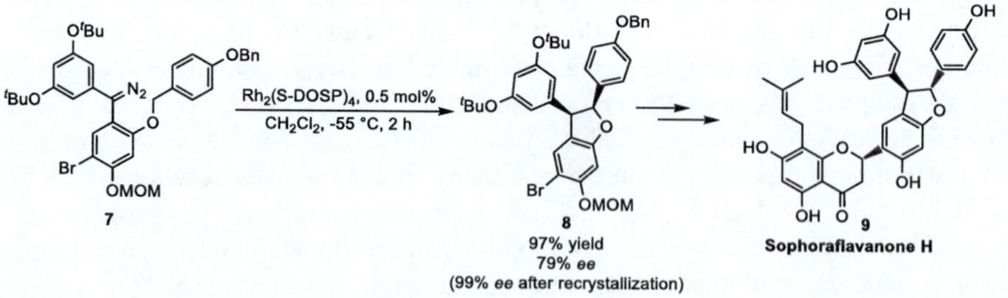

Scheme 5 Enantioselective synthesis of sophoraflavanone H via donor–donor carbenes

2.2 Heterocycles from Hydrazones as Carbene Sources

Hydrazones from hydrazine monohydrate and aldehydes or ketones can generate corresponding diazo compounds quickly by MnO_2 oxidation [16]. This method allows the one-pot synthesis of heterocycles through donor–donor carbene-involved intramolecular C–H insertion, X–H insertion, or intermolecular cycloaddition reactions. In 2014, Shaw and co-workers reported the first example of asymmetric C–H insertion reactions of donor–donor carbenes (Scheme 6) [17]. This method enabled the synthesis of chiral benzodihydrofurans **11** with high enantio- and diastereoselectivities. Furthermore, this protocol was applied for the enantioselective synthesis of the natural product *E*-δ-viniferin **13**, which was a *trans*-resveratrol dimer containing a dihydrobenzofuran ring (Scheme 6). Another similar asymmetric synthesis of *E*-δ-viniferin **13** was also reported by Hashimoto and co-workers with intramolecular enantioselective C–H insertion as a key step [18].

Shaw and co-workers also employed the above oxidative/Rh(II)-catalytic system to facilitate asymmetric syntheses of indolines **15A** and **B**, indanes **15C,** and benzodihydrothiophenes **15D** through enantioselective intramolecular 1,5-C–H insertion of donor–donor carbenes in 2018 (Scheme 7) [19]. This approach offered rapid access to chiral benzo five-membered heterocycles **15**, which are important skeletons in natural products and drug molecules. In 2020, asymmetric intramolecular 1,6-C–H insertions forming isochromans **17** were also achieved with good diastereo- and enantioselectivity, and no competitive Stevens rearrangement products were observed [20]. Based on this 1,6-C–H insertion approach, they disclosed a divergent asymmetric synthesis of Panowamycin A **18**, Panowamycin B **19**, TM-135 **20**, and Veramycin F **21** [21]. These syntheses were accomplished asymmetrically by a C–H insertion reaction using a donor–donor Rh(II)-carbene as the key intermediate.

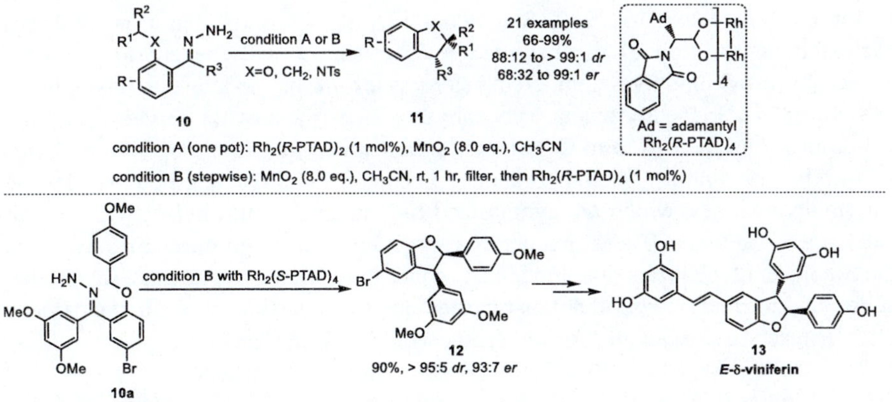

Scheme 6 Enantioselective C–H insertion for the synthesis of chiral benzodihydrofurans and (−)-*E*-δ-viniferin

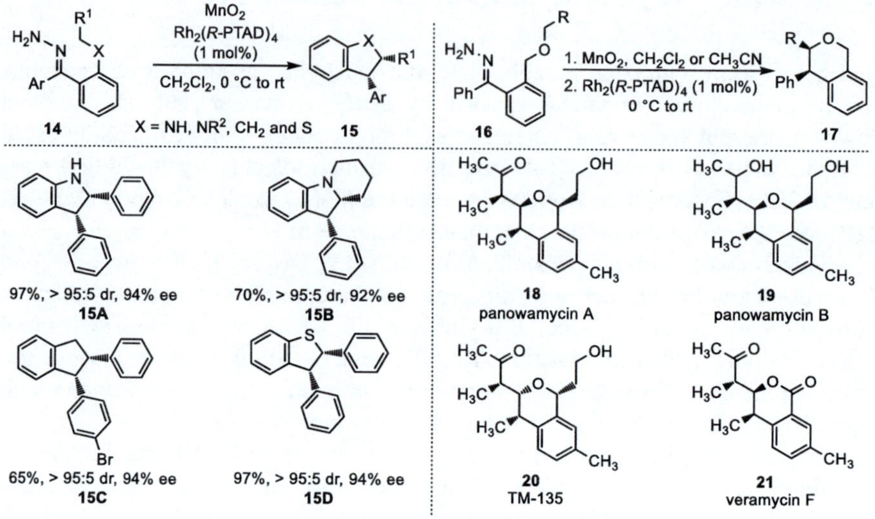

Scheme 7 Asymmetric C–H insertion for the synthesis of chiral 5- or 6-membered heterocycles

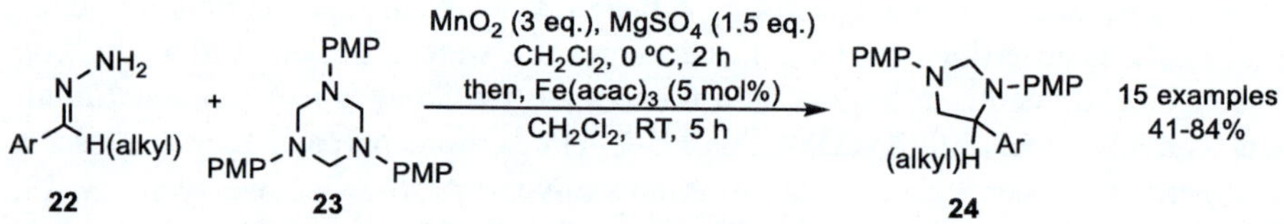

Scheme 8 Cycloaddition of hydrazones with triazines

In addition to intramolecular C–H insertion, the donor–donor metal carbenes derived from the oxidation pathway can also undergo intermolecular cycloaddition to form heterocycles. In 2017, Sun and co-workers reported an iron-catalyzed cycloaddition reaction of disubstituted hydrazones **22** with hexahydro-1,3,5-triazines **23**, providing five-membered heterocycles **24** in moderate to high yields (Scheme 8) [22]. In this reaction, hexahydro-1,3,5-triazines **23** served as formal 1,4-dipoles, and the C–C and C–N bonds were formed in a single operation.

Another commonly used hydrazone precursor to diazo compounds are the sulfonylhydrazones, which are synthesized by reaction of sulfonyl hydrazide with aldehydes or ketones. These precursors can generate diazo compounds slowly via the Bamford–Stevens reaction under basic conditions. Many reports have shown that sulfonylhydrazones are good donor-type carbene sources widely used in the carbene transfer reactions, especially for the synthesis of heterocycles.

In 2013, Ignacio Rodríguez-García and co-workers reported that 2,3-dihydrobenzo[*b*]furan neolignan acuminatin **27** could be prepared using sulfonylhydrazone as the starting material [23]. The key step was an intramolecular C–H insertion via a donor carbene, which was prepared through the decomposition

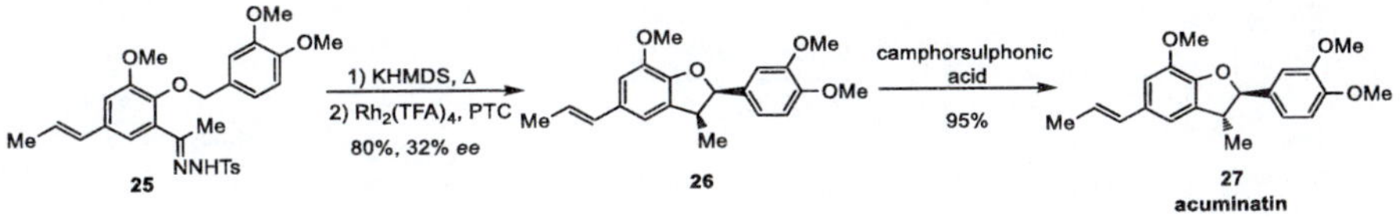

Scheme 9 Synthesis of acuminatin via C–H insertion with sulfonylhydrazones as donor carbene sources

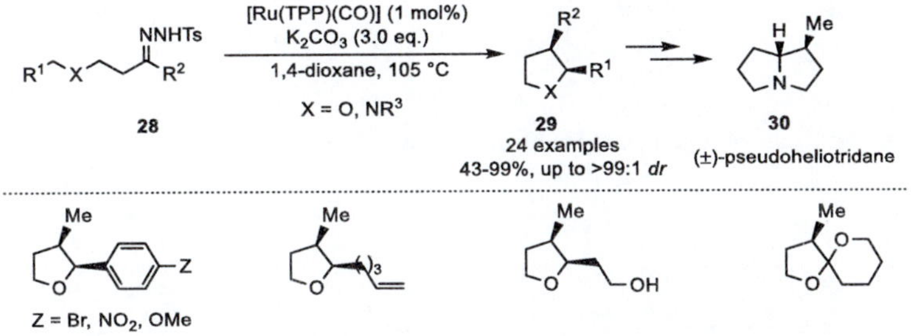

Scheme 10 Ruthenium porphyrin-catalyzed diastereoselective synthesis of tetrahydrofurans and pyrrolidines

of tosylhydrazone **25** in the presence of the quaternary ammonium salt cinchonidine, which is a phase-transfer catalyst (Scheme 9). This C–H insertion reaction was also conducted with chiral rhodium(II) catalysts to achieve chirality induction, albeit with low levels of enantioselectivity (<32% *ee*).

In 2014, Che and co-workers developed a ruthenium porphyrin-catalyzed diastereoselective intramolecular alkyl carbene C–H insertion reaction [24]. Using [Ru(TTP)(CO)] as the catalyst, *N*-tosylhydrazones **28** were converted to tetrahydrofurans and pyrrolidines **29** in good yields and remarkable *cis* selectivity (up to >99% dr). The reaction displays good tolerance of many functionalities, such as halo, nitro, methoxy, alkene, hydroxy, and acetal groups; and, notably, the procedure is quite simple and has no need for slow addition with a syringe pump. Moreover, this reaction was successfully applied in a concise synthesis of (±)-pseudoheliotridane **30** (Scheme 10).

Co(II) complex is a distinctive catalyst in the transformations of carbene precursors. This Co-based catalyst can promote traditional carbene transfer reactions efficiently with unique cobalt(III)-carbene radicals as the key intermediates. For example, in 2018, de Bruin reported a cobalt(II) porphyrin-catalyzed intramolecular formal C–H insertion reactions with *o*-aminobenzylidine *N*-tosylhydrazones **31** for the synthesis of indolines **33** through Co(III) carbene radical intermediate **32** and followed 1,5-*H* transfer process (Scheme 11) [25]. This reaction shows excellent yields and requires simple conditions; more importantly, the catalyst is inexpensive and avoids the use of more expensive metals such as Rh, Ru, etc.

A similar enantioselective formal C(sp3)–H insertion reaction for the synthesis of chiral indolines was achieved by Zhang and co-workers in the same year. As shown

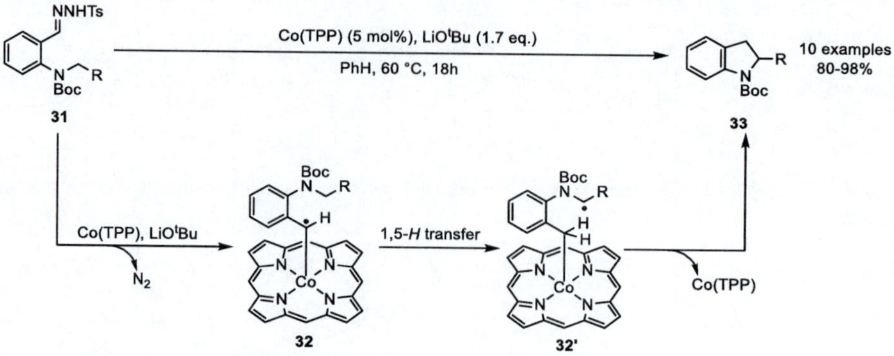

Scheme 11 Synthesis of indolines by cobalt(III)-carbene radicals

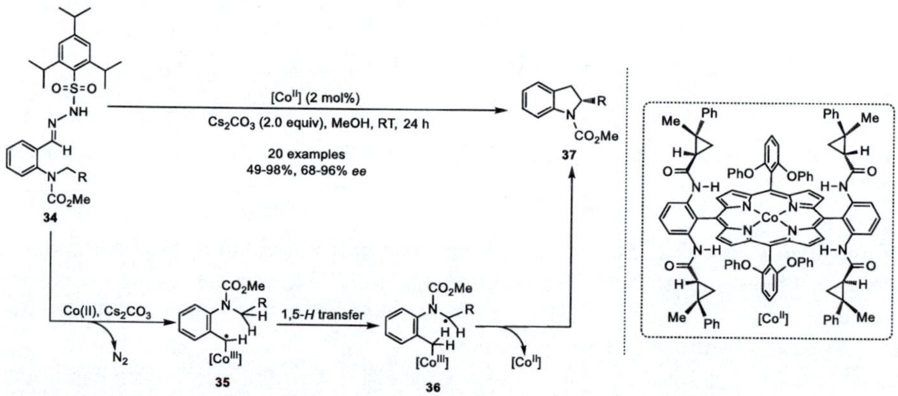

Scheme 12 Formal C–H insertion of Co(III) carbene radicals for enantioselective synthesis of chiral indolines

in Scheme 12, with chiral Co(II) porphyrin complex as catalyst, 2,4,6-triisopropylbenzenesulfonyl hydrazones **34** were successfully converted to indolines **37** with moderate to excellent enantioselectivities under mild conditions [26]. In addition to regioselectivity and chemoselectivity, this Co(II)-catalyzed formal C–H insertion featured functional group tolerance as well as compatibility with heteroaryl units.

In addition to formal C–H insertion reactions, the Co(II) porphyrin complex is also a capable catalyst for cyclopropanation. For example, in 2016, Che et al disclosed a intramolecular cyclopropanation of pyrroles and indoles with imbedded *N*-tosylhydrazones **38** as substrates under Co(II) porphyrin-catalyzed conditions [27]. A range of *N*-containing fused polycyclic compounds **39** were constructed in up to excellent yields (Scheme 13).

In 2014, De Bruin et al developed a one-pot protocol for the synthesis of 2*H*-chromenes. These reactions proceed via radical-induced addition–cyclization of terminal alkynes with carbene radicals [28]. Salicyl *N*-tosylhydrazones **40** were

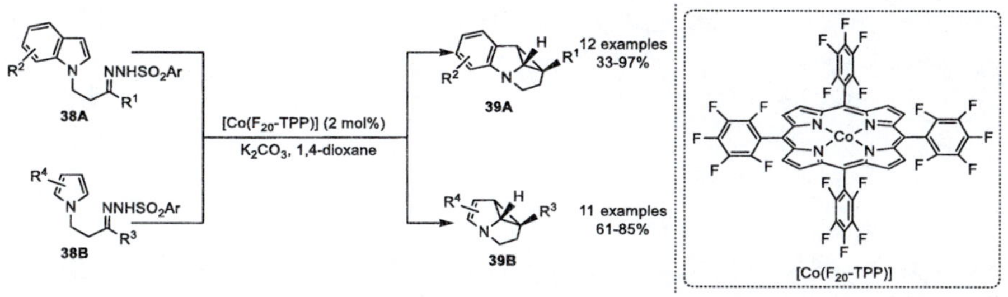

Scheme 13 Efficient synthesis of fused polycyclic *N*-heterocycles via Co(II) catalyzed cyclopropanation

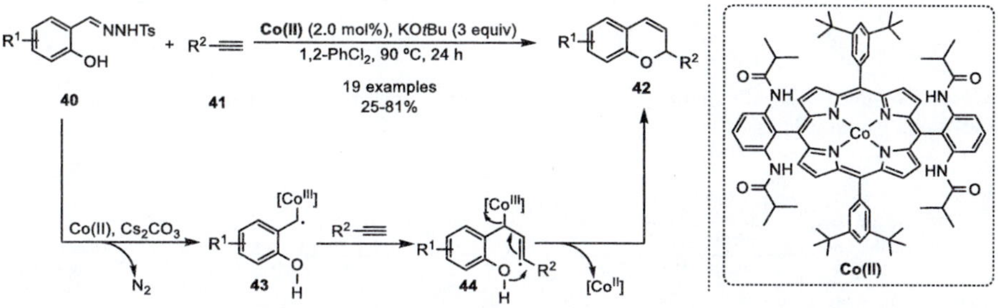

Scheme 14 Co(II)-catalyzed synthesis of 2*H*-chromenes via a carbene radical process

used as carbene radical precursors by treatment with base and the Co(II) porphyrin in these reactions, and Co(III)-carbene radicals **43** were formed. The vinyl radicals **44** were subsequently formed after Co(III)-carbene radicals **43** add to alkynes **41**. Then hydrogen atom transfer process from the hydroxyl moiety in salicyl *N*-tosylhydrazones **40** to the vinyl radicals took place and generated an *o*-quinone methide species, which dissociated from the cobalt center and underwent ring-closing reaction to produce the final 2*H*-chromene product **42** (Scheme 14).

In addition to traditional carbene transfer reactions, sulfonylhydrazones have been found to undergo a new reaction process, known as carbene migratory insertion. After this new reaction pathway was discovered, much new research in this area has been reported. This finding provides abundant approaches to construct C–X bonds and form desired heterocycles. In this carbene migratory insertion process, transition metals such as palladium, copper, and rhodium are effective catalysts for cross-coupling of sulfonylhydrazones with different coupling partners. The coupling partners are numerous, including aryl halides, vinyl halides, alkyl halides, boronic acids, silanes, and alkynes. In 2012, Van Vranken and co-workers demonstrated a Pd (0)-catalyzed reaction of vinyl iodides **46** with *N*-tosylhydrazones **45** that assembles η^3-allyl ligands through carbene migratory insertion. This method easily generates pyrrolidine and piperidine ring systems **47** that are common building blocks to

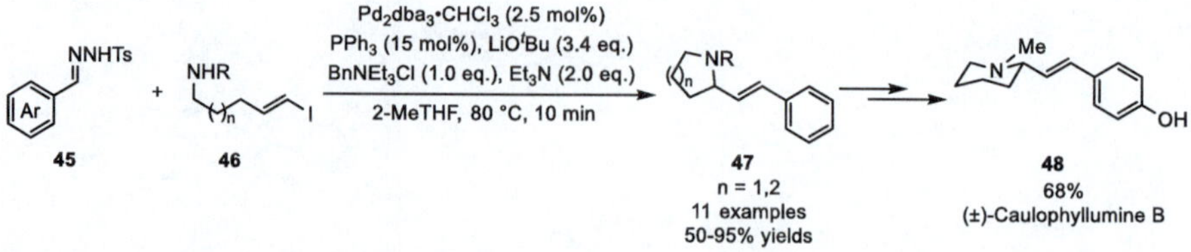

Scheme 15 Pd-catalyzed reaction of *N*-tosylhydrazones with vinyl iodides for the synthesis of pyrrolidines and piperidines

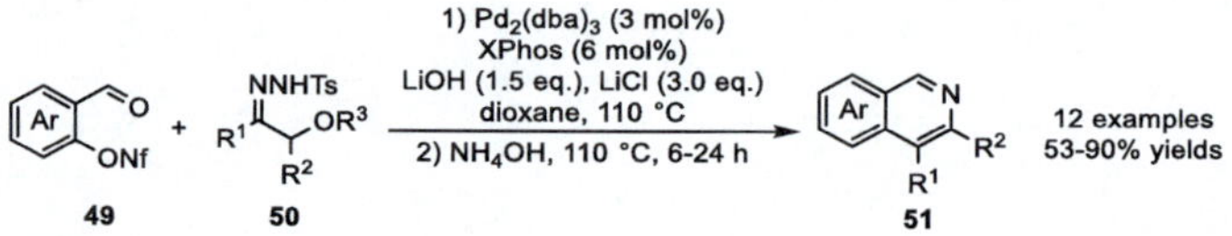

Scheme 16 Pd-catalyzed reaction of *α*-alkoxytosylhydrazones with nonaflates for the synthesis of substituted isoquinolines

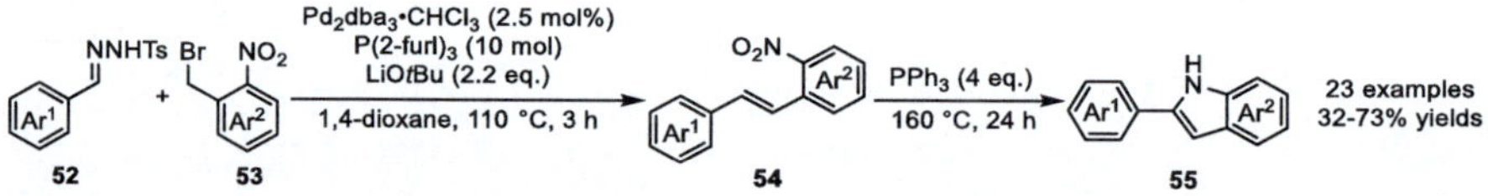

Scheme 17 Pd-catalyzed reaction of *N*-tosylhydrazones with nitro-benzyl bromide for the synthesis of indoles

alkaloid natural products, such as (±)-Caulophyllumine B **48** with 68% yields in one step (Scheme 15) [29].

In addition to halides, nonaflates ($CF_3CF_2CF_2CF_2SO_2^- = Nf$) are also good coupling partners in the carbene migratory insertion process. In 2012, Carlos Valdés and co-workers reported a Pd(0)-catalyzed cross-coupling reaction of α-alkoxytosylhydrazones **50** with nonaflates **49** derived from salicylaldehydes, giving substituted isoquinolines **51** by simple treatment of the coupling adducts with ammonium hydroxide (Scheme 16) [30].

In 2019, Alami and Hamze reported a one-pot coupling between *N*-tosylhydrazones **52** and *o*-nitrobenzyl bromide **53**, followed by deoxygenation of *ortho*-nitrostyrenes **54** and subsequent cyclization, leading to the synthesis of C2-aryl substituted indoles **55** in moderate to high yields (Scheme 17) [31]. This method may find use in medicinal chemistry programs as it allows the synthesis of NH-free indole libraries for direct biological testing. The indoles can be subsequently alkylated in order to increase the molecular diversity of the methodology and enable drug discovery.

In 2021, Huang and co-workers demonstrated a palladium-catalyzed formal [4 + 3]-annulation for the construction of tricyclic lactams **58** [32]. The key step of

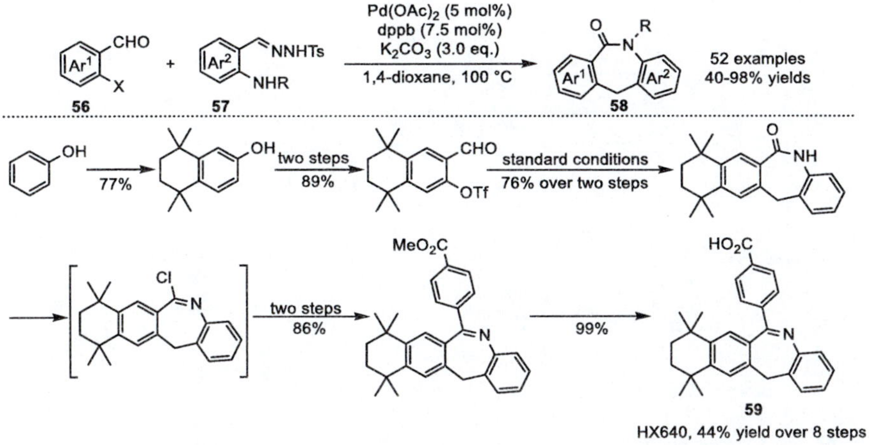

Scheme 18 Pd-catalyzed formal [4 + 3] annulation for the synthesis of seven-membered lactams

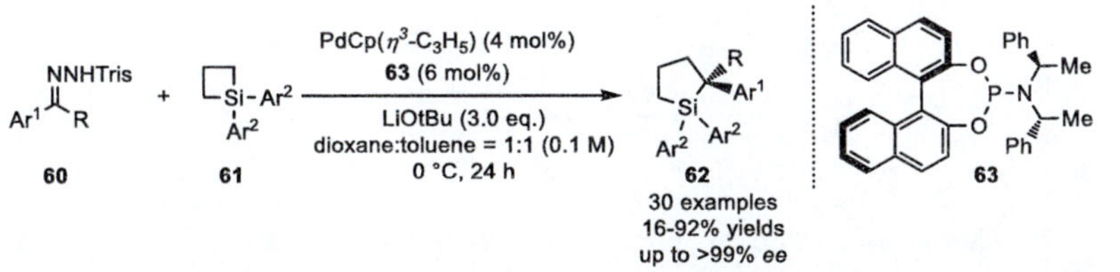

Scheme 19 Pd-catalyzed formal Si–C bonds insertions for the synthesis of silacyclopentanes

the reaction mechanism involves a Pd-carbene migratory insertion enabled 1,4-palladium shift. This reaction features easily accessible starting materials, broad substrate scope, and excellent functional group compatibility. Notably, this synthetic method can be used to prepare the retinoid synergist HX640 **59** in 44% yield over eight linear steps from phenol (Scheme 18).

Silanes also can be used as efficient partners in the palladium-catalyzed carbene migratory insertion reactions. For example, using silacyclobutanes **61** and trisylhydrazones **60** as starting materials, Wang and co-workers reported highly efficient palladium(II)-catalyzed formal carbene Si–C bond insertions into strained silacyclobutanes with excellent enantioselectivity, which provides a rapid method to asymmetrically access silacyclopentanes **62** (Scheme 19) [33].

Recently, the Wang group also expanded this method to synthesize benzosilacyclobutanes **65** and **67**. The corresponding benzo[*b*]siloles **66** and **68** were obtained in good yields, high enantioselectivity, and site-selectivity [34]. When **63** was used as the ligand, the intermediate palladium carbene preferred insertion into the C(sp^3)–Si bond of benzosilacyclobutanes **65** and gave the 2,3-dihydro-1*H*-benzo[*b*]siloles **66** with good results. Whereas, when the ligand was changed to **69**, the carbene intermediate site-selectively inserted into the C

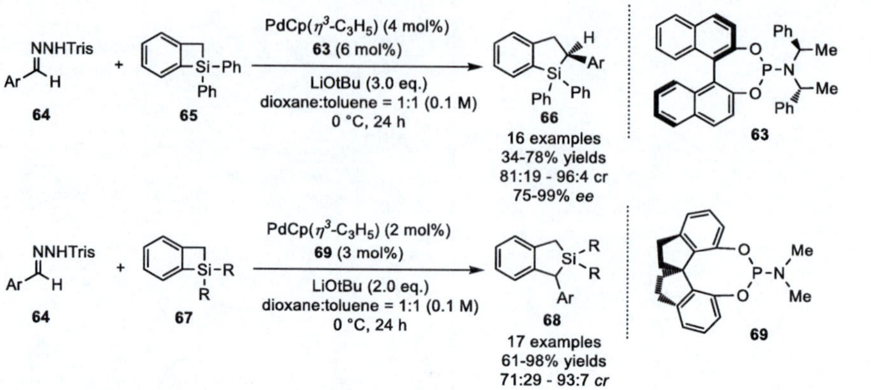

Scheme 20 Ligand-controlled site- and enantioselective carbene insertion into carbon–silicon bonds of benzosilacyclobutanese

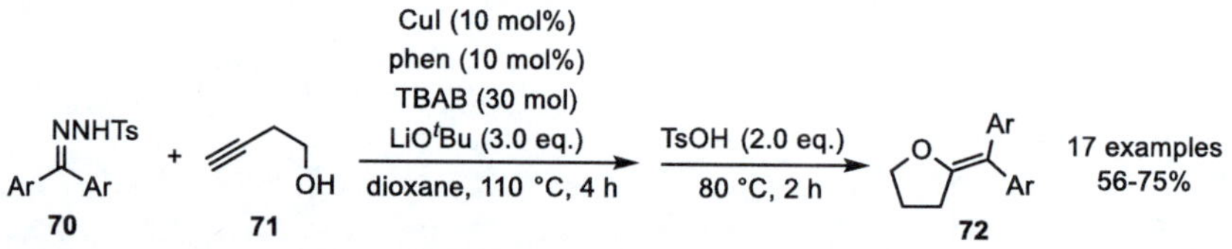

Scheme 21 Cu-catalyzed cascade reaction of *N*-tosylhydrazones for the synthesis of tetrahydrofurans

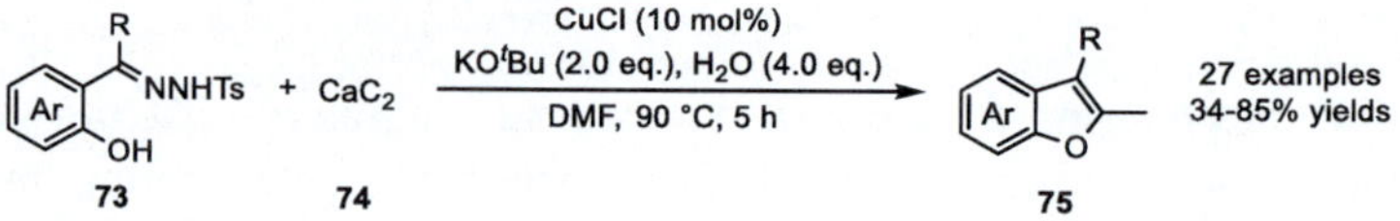

Scheme 22 Cu-catalyzed coupling reaction for the synthesis of 2-methylbenzofurans

(sp^2)–Si bond of benzosilacyclobutanes **67** to give the corresponding 2,3-dihydro-1*H*-benzo[*c*]siloles **68** in high to excellent yields with good selectivity (Scheme 20).

In addition to palladium-catalyzed carbene migratory insertions, copper-catalyzed reactions of *N*-tosylhydrazones via migratory insertion can also be used to synthesize heterocycles. For example, Wang and co-workers reported a Cu(I)-catalyzed cascade coupling/cyclization reaction of *N*-tosylhydrazones **70** with 3-butyn-1-ol **71** for the synthesis of tetrahydrofuran derivatives (Scheme 21) [35]. This reaction proceeds via the formation of a carbene intermediate, followed by sequential cyclization and isomerization, representing a straightforward approach for the synthesis of 2-(diarylmethylene)tetrahydrofurans **72** in good to high yields.

Alkynes can also be the compatible partners in carbene migratory insertions. In 2018, Li demonstrated a direct synthesis of 2-methylbenzofurans **75** from calcium carbide **74** and salicylaldehyde *N*-tosylhydrazones **73** (Scheme 22) [36]. In this

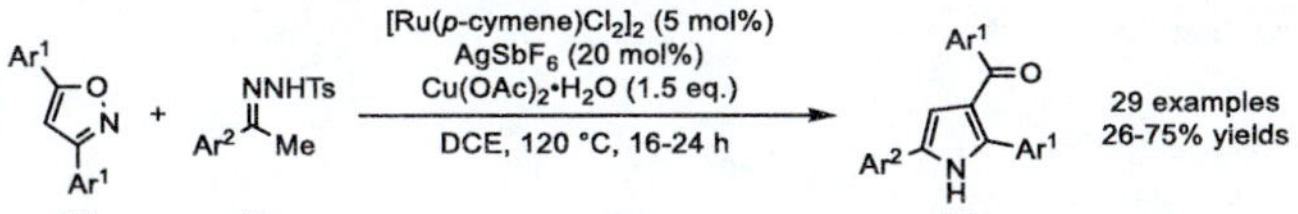

Scheme 23 Ru(II), Cu(II)-mediated carbene migratory insertion in the synthesis of trisubstituted pyrroles

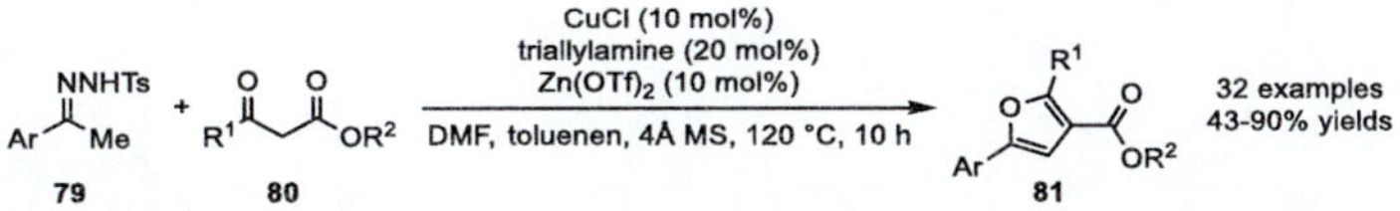

Scheme 24 Copper-catalyzed [3 + 2] oxidative cyclization reaction for the synthesis of trisubstituted furans

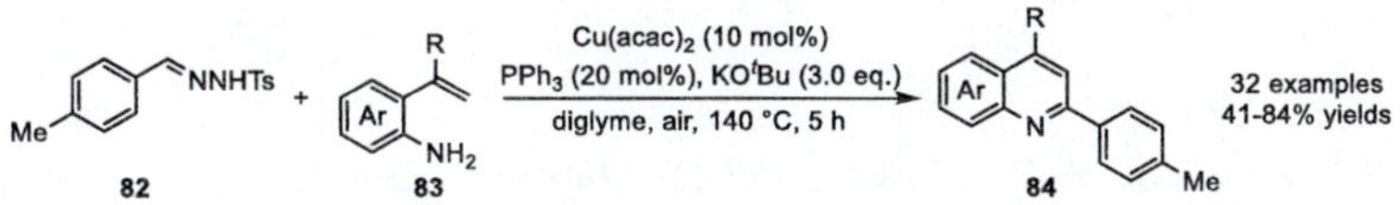

Scheme 25 Copper-catalyzed oxidative cyclization reaction for the synthesis of quinoline derivatives

transformation, acetylene was generated from calcium carbide **74** with water, followed by alkynyl copper carbene migratory insertion, generating the desired 2-methylbenzofurans **75** in synthetically useful yields.

Another example is the synthesis of trisubstituted pyrroles from isoxazoles. In 2021, Kapur and co-workers developed a simple and convenient "one-pot" synthesis of trisubstituted pyrroles **78** via a Ru(II)-catalyzed, Cu(II)-mediated reaction of substituted isoxazoles **76** with sulfonylhydrazones **77** under mild reaction conditions. This protocol resulted in a synergistic formation of C–C and C–N bonds, which provides a simple and efficient method for the synthesis of highly functionalized pyrroles (Scheme 23) [37].

Besides their uses for carbene migratory insertion, sulfonylhydrazones can also participate in other carbene transfer reactions to form heterocycles. In 2016, Jiang and Wu reported a copper-catalyzed [3 + 2] oxidative cyclization reaction to afford 2,3,5-trisubstituted furans **81** from *N*-tosylhydrazones **79** and *β*-ketoesters **80** (Scheme 24) [38]. This transformation provides a novel application of hydrazone compounds in cyclization reactions and cross-couplings.

In 2018, Anbarasan and co-workers also reported a Cu(II)-catalyzed oxidative coupling of *ortho*-vinylanilines **83** with *N*-tosylhydrazones **82**. Various substituted quinoline derivatives **84** of biological importance were produced in good to excellent yield (Scheme 25) [39]. The important features are the high functional group tolerance, scale-up to gram scale synthesis, and possible one-pot syntheses of quinolines from corresponding carboxaldehydes.

Scheme 26 The reaction for the synthesis of trifluoromethylated oxiranes

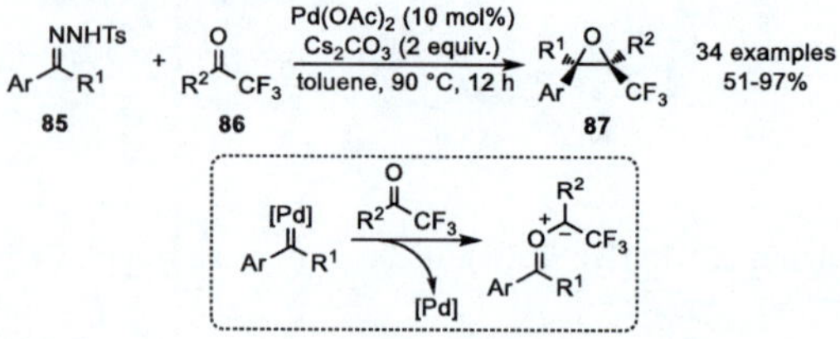

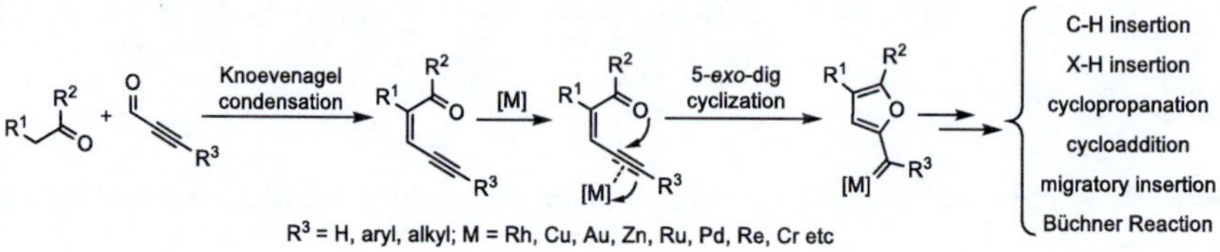

Scheme 27 Formation of donor-type 2-furyl metal carbenes with enynones used as non-diazo carbene precursors

In 2017, Jiang and Wu reported a Pd(II)-catalyzed example for the synthesis of trifluoromethylated oxiranes with *N*-tosylhydrazones **85** and trifluoromethyl ketones **86** [40]. In this transformation, various trifluoromethylated oxiranes **87** were constructed in high yields and diastereoselectivities and a remarkable fluorine effect on the reactivity was observed. A carbonyl ylide from donor-type Pd(II)-carbene was proposed as the key intermediate. (Scheme 26).

3 Heterocycles from Alkyne-Type Carbene Sources

3.1 Heterocycles from Enynones as Carbene Sources

Enynones can be easily synthesized with ynals and active methylene ketones through Knoevenagel condensation reactions. As one of the safest and practical non-diazo carbene precursors, enynones have attracted increasing attention. These compounds can generate donor-type 2-furyl metal carbenes via 5-*exo*-dig cyclization under the catalysis of various transition metals, such as Rh, Cu, Au, Pd, Zn, Ru, etc. (Scheme 27). These active species also have been characterized with a series of Cr, W, and Re solid-state metal carbene complexes [41]. Under this catalytic mode, enynones can be used to synthesize a series of valuable heterocycles by traditional carbene transfer reactions, such as C–H and X–H insertion, cyclopropanation, cycloaddition, carbene migratory insertion, and Büchner reactions [42].

In 2012, López and Vicente developed an efficient approach for cyclopropanation and Si–H insertion reactions using enynones as the 2-furyl Zn(II)-carbene precursors in the catalytic of cheap $ZnCl_2$ [43]. The cyclopropanation of alkenes with Zn(II)-carbenes **89** proceeded smoothly in inter- and intramolecular manner and gave rise to

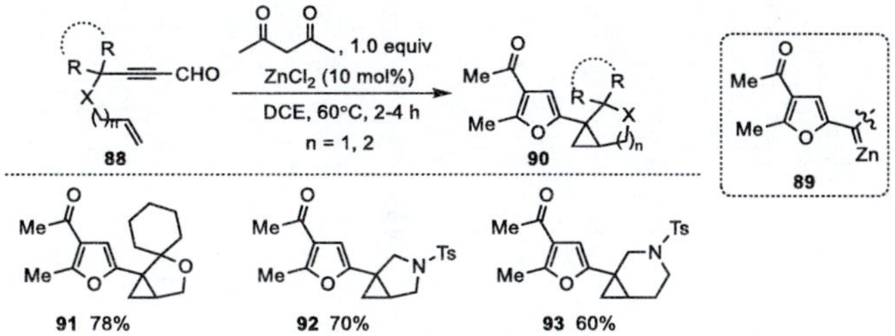

Scheme 28 One-pot synthesis of tetrahydrofuran, pyrrolidine, and piperidine derivatives

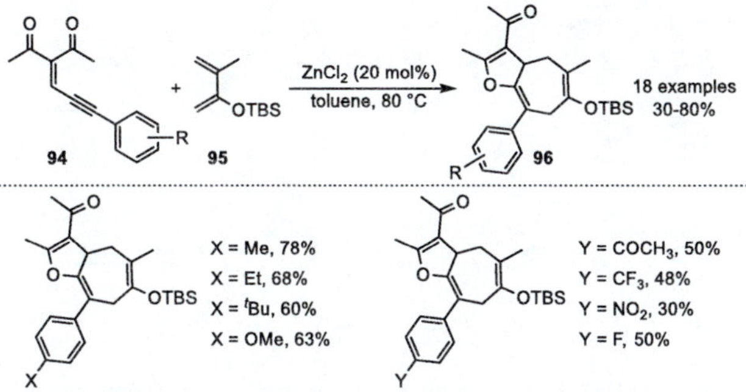

Scheme 29 Zn(II)-catalyzed [4 + 3] cycloaddition of enynones with diene

the products **90** with high yields and broad reaction scope. Notably, the Zn (II) catalytic conditions can be compatible well with Knoevenagel condensation and realize a one-pot synthesis of heterocycles through condensation/intramolecular cyclopropanation tandem reaction. As shown in Scheme 28, ynals **88** underwent Knoevenagel condensation with an equimolecular amount of acetylacetone and then produced the tetrahydrofuran **91**, pyrrolidine **92,** and piperidine **93** through intramolecular cyclopropanation in good yields.

Cycloaddition of metal carbenes with dienes or dipoles is an important approach to assemble heterocycles in organic synthesis. In 2014, Liang and co-workers reported a Zn(II)-catalyzed [4 + 3]-cycloaddition of enynones **94** with diene **95** and various cyclohepta[*b*]furan rings **96** were obtained in moderate to good yields [44]. In this reaction, electronic effects are observed to influence the reaction to some extent, for instance, substrates with electron-donating groups (EDG) on the phenyl of enynones **94** give the desired products in good yields, while substrates with electron-withdrawing groups (EWG) on the aromatic rings lead to the corresponding products in lower yields (Scheme 29).

In 2018, an efficient Zn(II)-promoted cyclopropanation of indoles **97** with enynones **98** through directing group strategy was reported by Xu and co-workers [45]. The directing groups, including *N,N*-dimethylacetyl, 2-pyridyl and 2-pyrimidyl, played a vital role in the cyclization process and acted as a bridge between the Zn(II) carbene intermediates and indoles. The reaction offered the

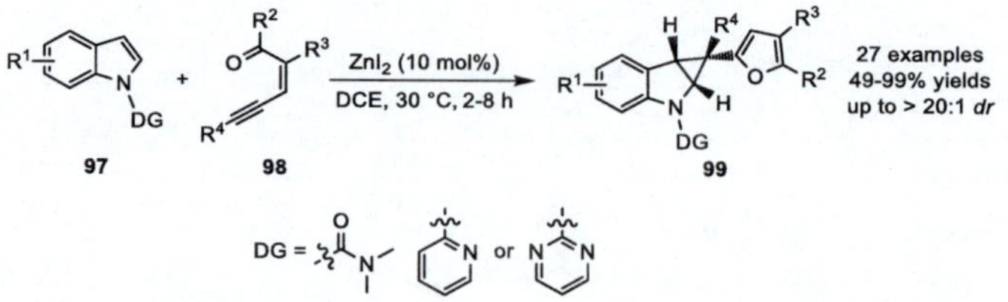

Scheme 30 Zn(II)-catalyzed cyclopropanation of enynones with indoles through directing group strategy

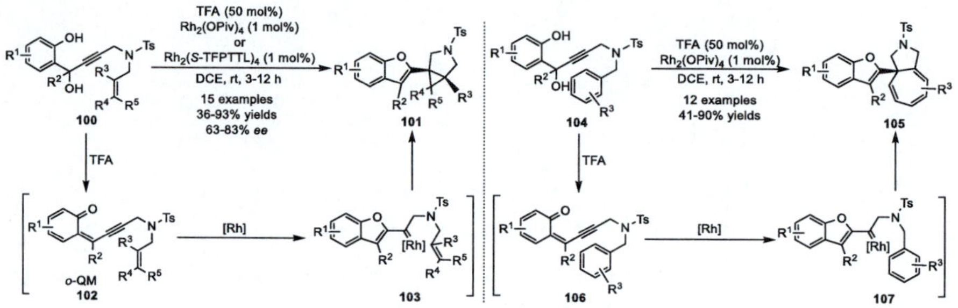

Scheme 31 Brønsted acid/Rh(II) catalyzed cyclopropanation and Büchner reaction

resulting polycyclic compounds **99** with up to quantitative yields and excellent diastereoselectivities (up to >20:1 *dr*) in simple and mild condition (Scheme 30).

In 2016, an efficient protonic acid/metal-catalyzed cascade benzofuran annulation/carbene transfer reaction for the synthesis of various benzofuryl-substituted pyrrolidines **101** was developed by Zhu and co-workers [46]. Mechanistic studies indicated that the reaction was initiated through the dehydration of *o*-hydroxylbenzyl alcohol **100** via TFA-mediated dehydration process, forming the key intermediate *o*-QM **102**. Then benzofuryl Rh(II)-carbene **103** was generated through intramolecular cyclization, followed by an enantioselective cyclopropanation with chiral Rh(II) carboxylate (Scheme 31). When the allyl group was replaced by a benzyl group, the Büchner reaction took place smoothly after the formation of the benzofuryl-metallocarbene **107**, liberating the cycloheptatriene product **105** in up to 90% yield.

With *N*-allyl tethered enynones as substrates, Zhu and co-workers realized a highly asymmetric intramolecular cyclopropanation reaction using enynones **108** as carbene precursors in the presence of a chiral rhodium(II) carboxylate catalyst [47, 48]. The reactions were conducted in a one-pot manner and provided the desired tetrahydropyrrole derivatives **109** with up to 100% atom efficiency and 97% *ee* (Scheme 32).

Using benzyl and alkyl groups instead of allyl tethered enynones **110** as starting materials, the same group reported an efficient Rh(II)-catalyzed enantioselective intramolecular C–H insertion reactions of donor–donor carbenes in 2016 [47]. The reactions were conducted in one-pot with 100% atom economy, and provided the

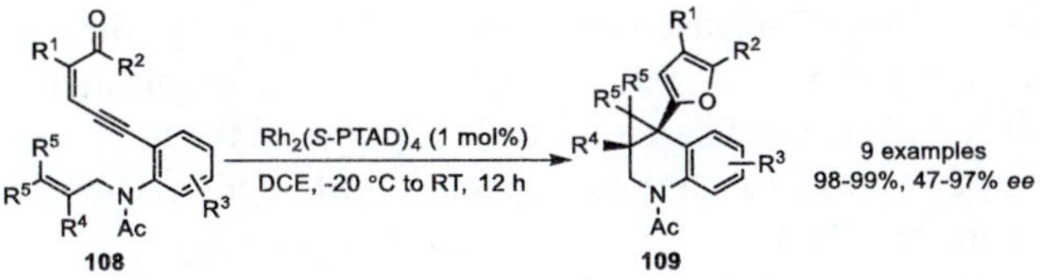

Scheme 32 Rh(II)-catalyzed asymmetric cyclopropanation for the synthesis of chiral tetrahydropyrrole derivatives

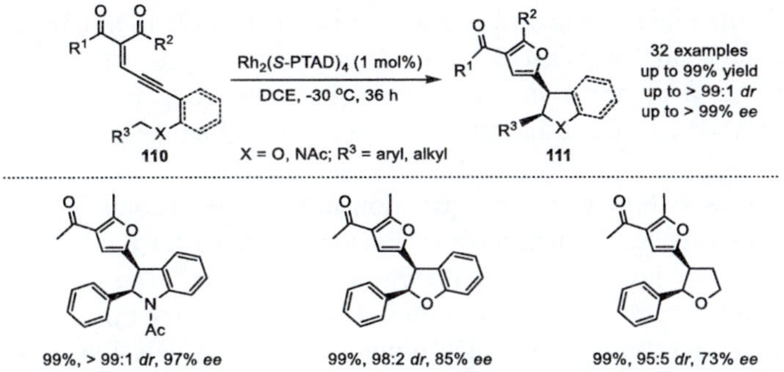

Scheme 33 Rh(II)-catalyzed enantioselective intramolecular C–H insertion of enynones

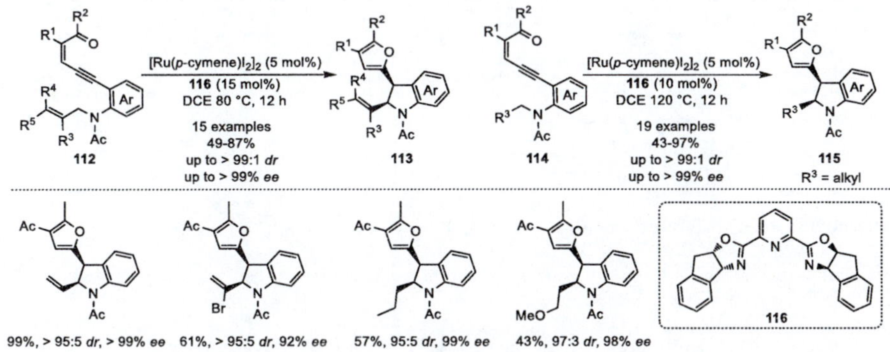

Scheme 34 Ru(II)-catalyzed enantioselective allyl and alkyl C–H insertion

desired dihydroindole, dihydrobenzofuran, and tetrahydrofuran derivatives **111** with excellent yields, high diastereoselectivities, and up to >99% *ee* (Scheme 33).

Interestingly, with *N*-allyl substituted enynones **112** as carbene sources, the Zhu group realized a catalyst-controlled highly stereo- and chemoselective intramolecular allylic C–H insertion of an in situ formed donor–donor carbene in 2018 (Scheme 34) [48]. In this transformation the Ru(II)/Pybox **116** complex selectively catalyzed the intramolecular allylic C–H insertion. A series of vinyl-substituted dihydroindoles **113** were produced with up to >99% *ee* and > 95:5 chemoselectivity. In addition, this Ru(II)/Pybox **116** catalytic system was also applied to the high challenging alkyl C–H insertion of *N*-alkyl substituted enynones **114**, which afforded the desired products **115** with excellent diastereo- and enantioselectivity.

Apart from the Rh(II) and Ru(II) catalytic systems that lead to the *cis* selective C–H insertion products (**111**, **113** and **115**), Pd(II)/Pybox **122** system also performed well in the enantioselective C–H insertion of *N*-benzyl tethered enynones [49]. Highly *cis* selective furyl substituted dihydroindoles **118** were obtained from enynones with good yields and excellent enantioselectivities. Impressively, the *cis* selectivity is inhibited and the *trans* selective products **119** are dominant under the Pd(II)-catalytic conditions when diaryl diazo compounds with *N*-Ts protected amino groups were used as donor–donor carbene precursors (Scheme 35). Mechanistic studies suggest that a palladium carbene is the key intermediate in the C–H insertion reaction.

In addition to traditional cyclopropanation and C–H insertion reactions, Zhu and co-workers also reported other carbene transfer reactions for the synthesis of heterocycles using enynones as donor–donor carbene sources. These carbene transfer reactions include aziridination with imines **124** [50] and enantioselective intramolecular Büchner reactions [51], which provide a series of valuable polysubstituted aziridines **125** and chiral polycyclic heterocycles **127** (Scheme 36).

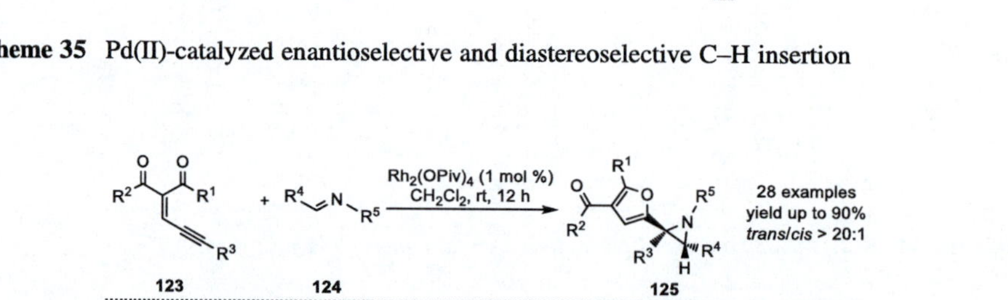

Scheme 35 Pd(II)-catalyzed enantioselective and diastereoselective C–H insertion

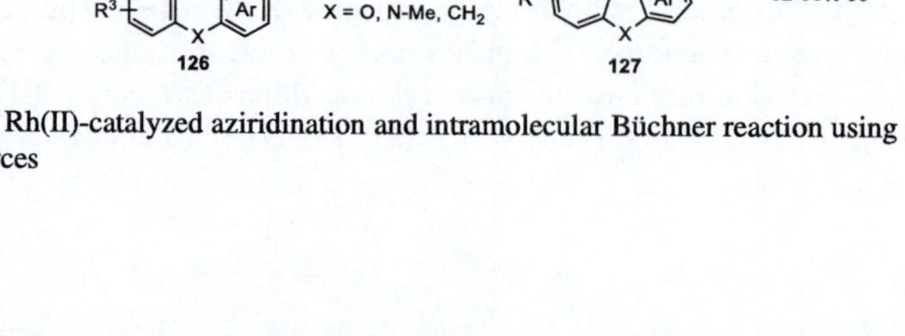

Scheme 36 Rh(II)-catalyzed aziridination and intramolecular Büchner reaction using enynones as carbene sources

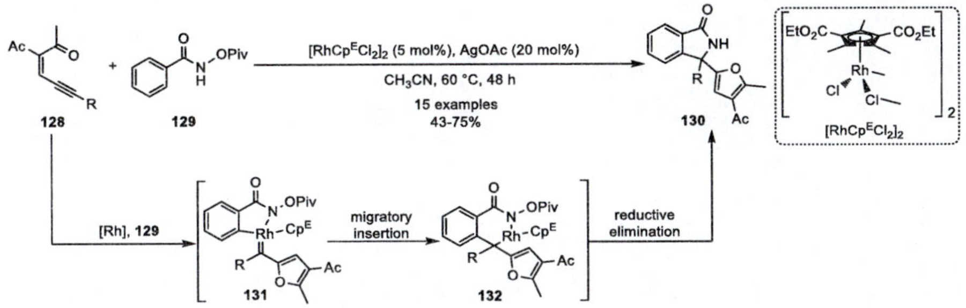

Scheme 37 Rh(III)-catalyzed formal [4 + 1] annulation of enynones with *N*-pivaloyloxy benzamides

Chang and co-workers disclosed a Rh(I)-catalyzed formal [4 + 1]-cycloaddition from *N*-pivaloyloxy benzamides **128** and enynones **129**, affording the cyclization products **130** (Scheme 37) [52]. The rhodium(I) center bearing Cp^E ligands can readily initiate a 5-*exo*-dig cyclization to afford the key 2-furyl Rh(I) carbene intermediate **131**, which then underwent migratory insertion to form the rhodacycle intermediate **132**. Subsequent C–N bond formation through reductive elimination and N–O bond cleavage gave the formal [4 + 1] heterocycle product **130**.

Combined C–H activation and carbene insertion can quickly construct structurally complicated molecules. In 2021, Yang and co-workers developed two types of Pd(II)-catalyzed intermolecular unactivated Csp^3–H bond olefination reactions mediated by using enynones as donor/donor carbene precursors, allowing for the construction of diverse furan and dihydrofuran substituted alkenes **135** [53]. Furthermore, alkenyl substituted furans serve as key building blocks for the successful assembly via a short and modular biomimetic strategy of macrolactams **136**, which showed significant anti-inflammatory activity against TNF-α, IL-6, and IL-1β and the cytotoxicity is comparable to Dexamethasone (Scheme 38).

Apart from enynones, enynals can also serve as donor–donor carbene precursors. Zhu and co-workers developed a bottom-up modular construction of chemically and structurally well-defined oligo(arylfuran)s by de novo synthesis of α,β′-bifuran monomers **142** via a carbene-involved semi-pinacol rearrangement [54]. By late-stage bromination, stannylation, and subsequent coupling reactions of the bifuran monomers, tetrafuran **143**, sexifuran **144**, and decafuran **145** were synthesized successfully (Scheme 39). These twisted oligo-(arylfuran)s exhibit excellent solubility in common organic solvents (such as toluene, chloroform, DCM, and THF), and they have high stability toward heat, oxygen, and moisture. More importantly, their photophysical properties, such as tunable and polarity sensitive fluorescence emission and high quantum yields, could also be well-modulated by installing different aryl groups.

This strategy of carbene-involved semi-pinacol rearrangement from enynals is not only used for the synthesis of bifurans, but also applied for the construction of α-furyl ketones **147** [55] and [2,5]-furanophanes **149** [56] in moderate to good yields under $PtCl_2$ catalytic conditions. As shown in Scheme 40, different ring types and

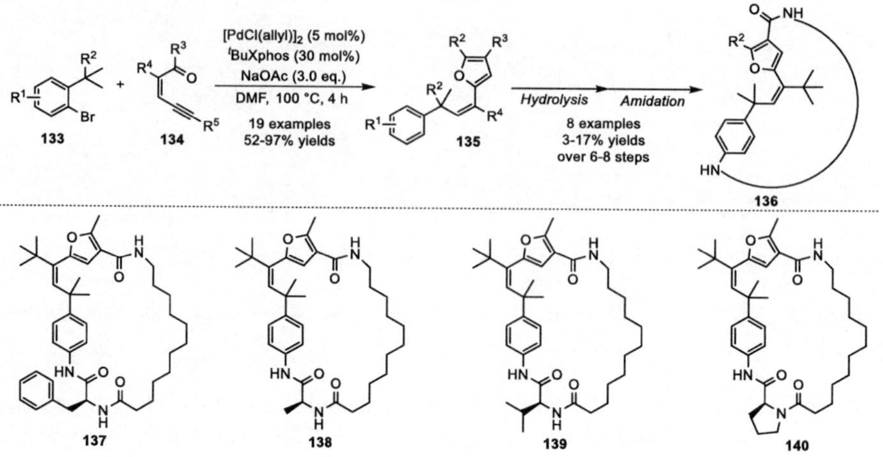

Scheme 38 Synthesis of macrocycles via Pd-catalyzed unactivated C(sp^3)-H olefination using enynones as carbene sources

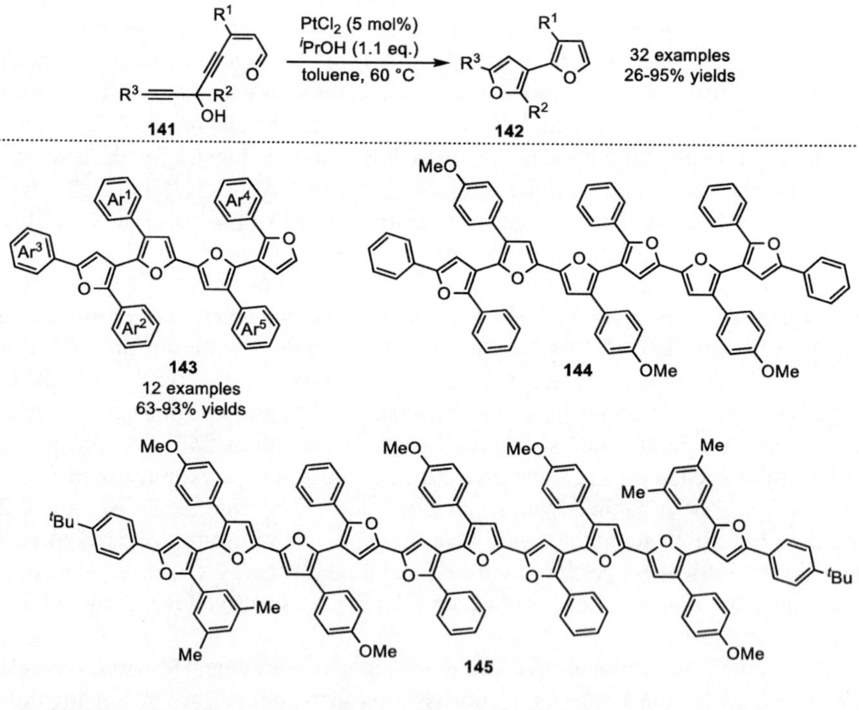

Scheme 39 Pd-catalyzed bottom-up modular synthesis of well-defined oligo(arylfuran)s

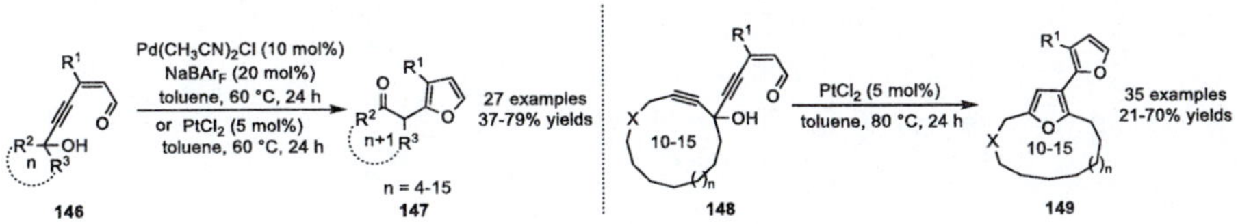

Scheme 40 Carbene-involved semi-pinacol rearrangement of enynals for the synthesis of heterocycles

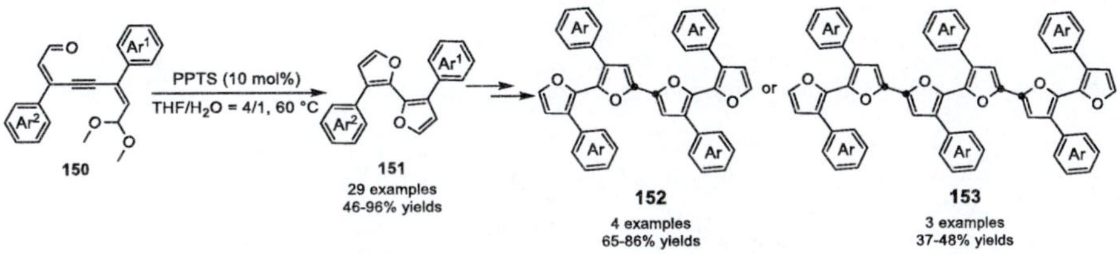

Scheme 41 Metal free approach for the synthesis of α-bi(arylfuran)s

ring sizes have been obtained under mild conditions. The two practical protocols feature broad substrate scope, easy operation, high atom economy, and ease of further transformations.

A metal free approach to α-bi(arylfuran)s has also been developed by Zhu and co-workers very recently [57]. With hemi-protected dienynals **150** as starting materials, they formed α-bi(arylfuran)s **151** under simple pyridinium *p*-toluenesulfonate (PPTS) catalytic conditions (Scheme 41). Most of the products were obtained in good to excellent yields (up to 96%), and the reactions could be easily scaled up to 5-g without a drastic decrease in yield. Moreover, these α-bi(arylfuran)s **151** can be transformed to α-tetra(arylfuran)s **152** and α-hexa(arylfuran)s **153** smoothly through cross-coupling reactions. X-ray crystallographic analysis of α-tetra(arylfuran)s **152** showed that the furan backbone of these molecules exhibits almost coplanar configurations. These α-oligo(arylfuran)s also showed high thermal and electrochemical stabilities and exhibit good hole mobility, which enables them to function well as a hole-transporting layer in OLEDs.

3.2 *Heterocycles from Propargyl Ethers as Carbene Sources*

Propargyl ethers can also serve as donor and donor–donor carbene precursors. They generate vinyl carbenes under transition metal catalysis, and the alkoxy ether moiety such as a methoxy group usually acts as a leaving group in the nucleophilic intramolecular reaction. In 2017, Zhu and co-workers reported an efficient approach to generate aryl gold carbenes from 1,6-diyne ether **154** [58]. The reaction was

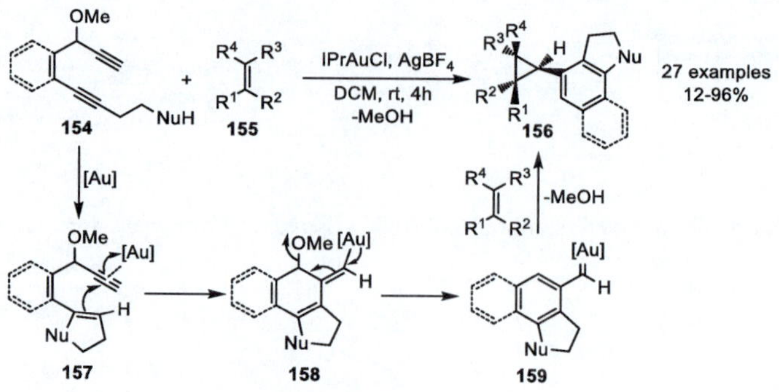

Scheme 42 Intermolecular cyclopropanation of propargylic ethers

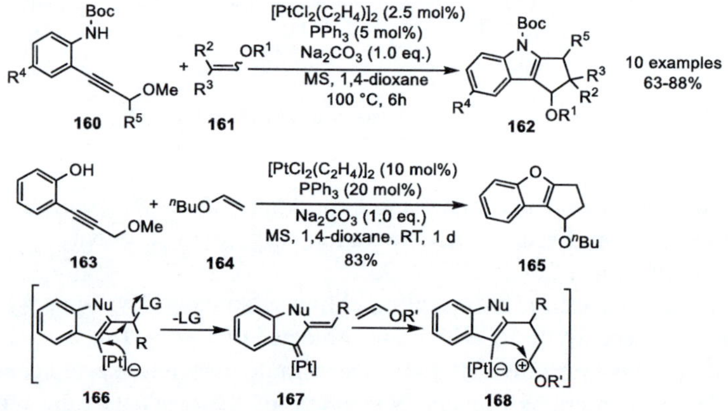

Scheme 43 Pt(II)-catalyzed [3 + 2]-cycloaddition of propargylic ethers

proposed to proceed through cyclization of propargyl ether **154** to form the intermediate **157** then, followed by a second cyclization and aromatization, to the key aryl carbene intermediate **159**, which was trapped through a typical carbene transfer reaction, such as cyclopropanation, with good to high yields (Scheme 42).

In 2011, Iwasawa and co-workers described the Pt(II)-catalyzed generation of a donor carbene complex from various propargyl ether derivatives **160** and **163** [59]. The in situ generated carbene complexes **167** underwent a stepwise [3 + 2]-cycloaddition reaction with various vinyl ethers, leading to efficient formation of indoles **162** and benzofuran **165** through addition to electronic rich alkenes in high yields (Scheme 43).

Ferreira and co-workers reported that a number of diversely substituted furans **171** and **172** could be synthesized via a cycloisomerization process with Pt(II)-vinyl carbene species **170** as the key intermediate, which was derived from propargylic ethers **169** in the presence of $PtCl_2$ [60]. The reactivity of the carbene intermediate can be modulated by different reaction conditions, resulting in highly selective

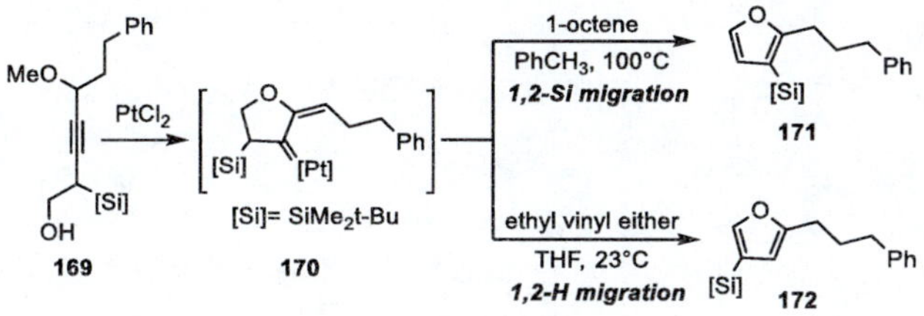

Scheme 44 Pt(II)-catalyzed 1,2-shift of propargylic ethers

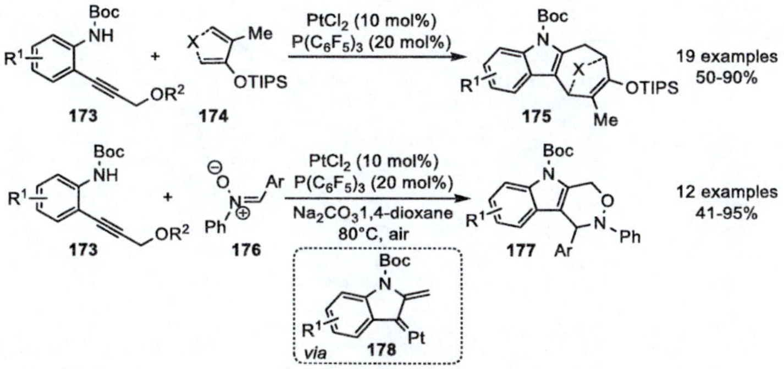

Scheme 45 Pt(II)-catalyzed [n + 3] cycloaddition of propargylic ethers

1,2-Si or 1,2-H migration to afford 1,2- or 1,4-disubstituted furan derivatives (Scheme 44).

Tang and co-workers reported an interesting indole annulation/[4 + 3]-cycloaddition sequence for the synthesis of various substituted cyclohepta[b]indoles **175** in 2013 [61]. Both cyclic dienes **174** and acyclic ones successfully participated in this tandem reaction, and high regioselectivity was observed in most cases (Scheme 45). In the same year, Hashmi and co-workers reported a platinum-catalyzed reaction of Boc-protected 2-(3-methoxy-1-propynyl)anilines **173** with nitrones **176** to deliver [1,2]oxazino-[5,4-*b*]indoles **177** [62]. With regard to the mechanism, the reactions probably involve an initial intramolecular cyclization/elimination to vinylcarbenoid species **178**, followed by a stepwise intermolecular [3 + 3]-cycloaddition.

3.3 Heterocycles from N-Propargyl Ynamides as Carbene Sources

In 2019, Ye and co-workers reported a new kind of donor–donor copper carbene through copper-catalyzed cascade cyclization of *N*-propargyl ynamides [63]. Mechanistic studies revealed that pyrrol-3-yl copper carbene **183** was generated from *N*-

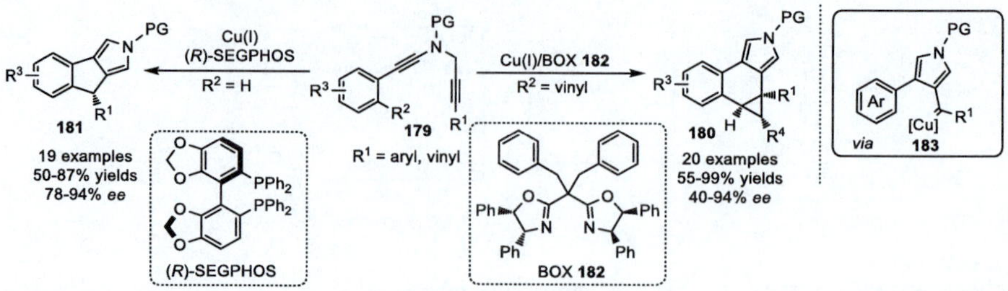

Scheme 46 Asymmetric cyclopropanation and aromatic substitution of donor–donor pyrrol-3-yl copper carbenes

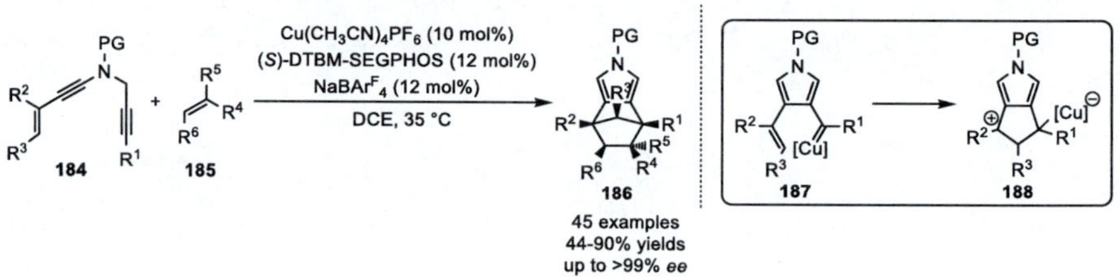

Scheme 47 Enantioselective formal [3 + 2] cycloaddition of styrenes with alkenyl *N*-propargyl ynamides

propargyl ynamides **179** under copper-catalyzed conditions. When R^2 was a vinyl group, enantioselective intramolecular cyclopropanation was achieved with the chiral Cu(I)/BOX **182** catalytic system. When R^2 was a hydrogen atom, then asymmetric aromatic substitution reaction occurred under Cu(I)/(R)-SEGPHOS catalytic conditions (Scheme 46). Both of these transformations showed broad substrate scope, and the chiral polycyclic pyrroles **180** and **181** were obtained in generally good to excellent yields with excellent enantioselectivities (up to 94% *ee*).

With the novel methodology of pyrrol-3-yl copper carbene from *N*-propargyl ynamides that has been developed, several classic enantioselective carbene transfer reactions have been demonstrated by the Ye group. These transformations include intramolecular [3 + 2]-cycloaddition, [1,2]-Stevens-type rearrangement, formal vinylic C(sp^2)–H insertion, intermolecular B–H bond insertion, and formal [2 + 1] and [4 + 1] annulations. In 2020, they reported a copper-catalyzed enantioselective formal [3 + 2]-cycloaddition of styrenes **185** with alkenyl *N*-propargyl ynamides **184** as donor–donor copper carbene precursors (Scheme 47) [64]. This tandem cyclization–cycloaddition reaction allows practical and economic assembly of diverse chiral pyrrole-fused bridged [2.2.1]-heterocycles **186** in moderate to good yields with generally excellent enantioselectivities (up to >99% *ee*). Mechanistic studies revealed that Cu-containing all-carbon 1,3-dipoles **188** from pyrrol-3-yl copper carbenes **187** were the key ingredient of this reaction.

When the linear vinyl group was replaced by a cyclic alkenyl group of the *N*-propargyl ynamides, formal intramolecular vinylic C(sp^2)–H insertion occurred

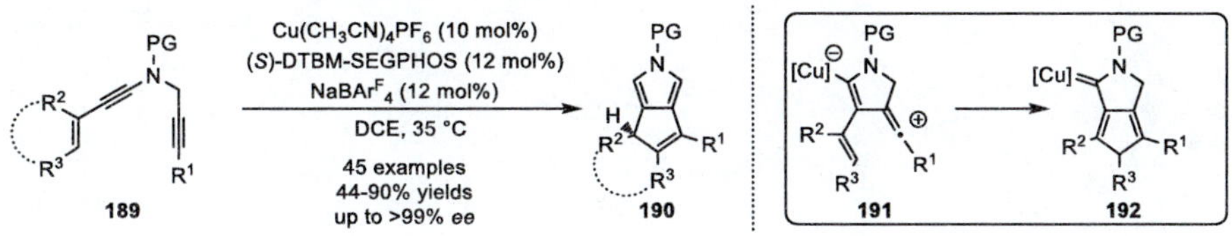

Scheme 48 Enantioselective formal vinyl C(sp^2)–H insertion of alkenyl *N*-propargyl ynamide

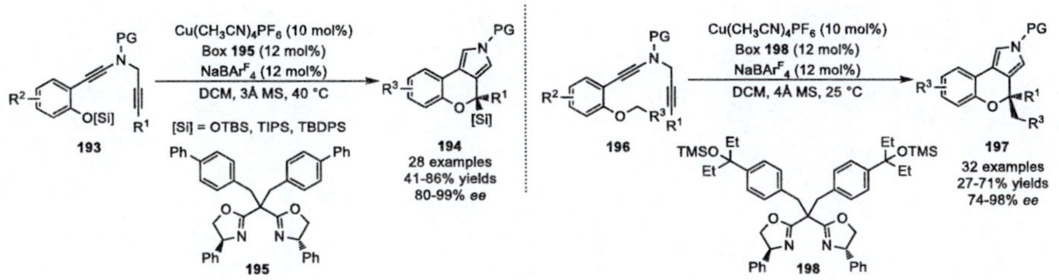

Scheme 49 Enantioselective cyclization via [1,2]-Stevens-type rearrangement for the synthesis of chiral chromeno[3,4-*c*]pyrroles

(Scheme 48) [65]. This method enables the practical and atom-economical construction of a diverse array of valuable chiral dicyclic- and polycyclic-pyrroles **190** in moderate to excellent yields with wide substrate compatibility and high enantioselectivities (up to 99% *ee*) under mild conditions. Interestingly, mechanistic studies indicated that a new endocyclic donor–donor copper carbene species **192** from vinyl cation intermediates **191,** rather than a pyrrol-3-yl copper carbene, is the key step in this reaction.

Based on the discovery of endocyclic donor–donor copper carbenes from *N*-propargyl ynamide, Ye et al. developed a copper-catalyzed enantioselective cascade cyclization/[1,2]-Stevens-type rearrangement reaction, leading to the practical and atom-economic assembly of various valuable chiral polycyclic pyrroles bearing a quaternary carbon stereocenter. This transformation occurred in generally moderate to good yields with wide substrate scope and excellent enantioselectivities (up to 99% *ee*) [66]. As shown in Scheme 49, *N*-propargyl ynamide **193** bearing a siloxy group gave chromeno[3,4-*c*]pyrrole products **194** with moderate to good yields and excellent enantioselectivities under Cu(I)/BOX **195** catalytic conditions. When the siloxy group was replaced by other groups, such as benzyl, allyl, propargyl, and alkyl, the related [1,2]-Stevens rearrangements were conducted smoothly only through changing the ligand to BOX **198**.

Having investigated the intramolecular carbene transfer reactions, the Ye group also explored the intermolecular reactions of these new kind of donor–donor copper carbenes. In 2021, they reported an efficient copper-catalyzed formal B–H bond insertion via cyclization of *N*-propargyl ynamides with borane adducts [67]. This reaction provided valuable pyrrolyl contained organoboron compounds **200** with

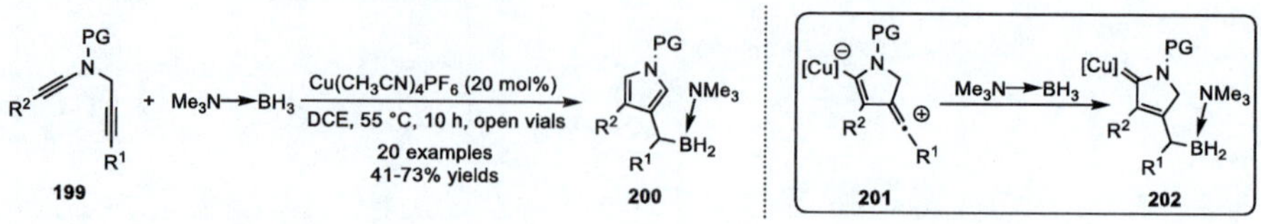

Scheme 50 Formal B–H bond insertion via cyclization of *N*-propargyl ynamides

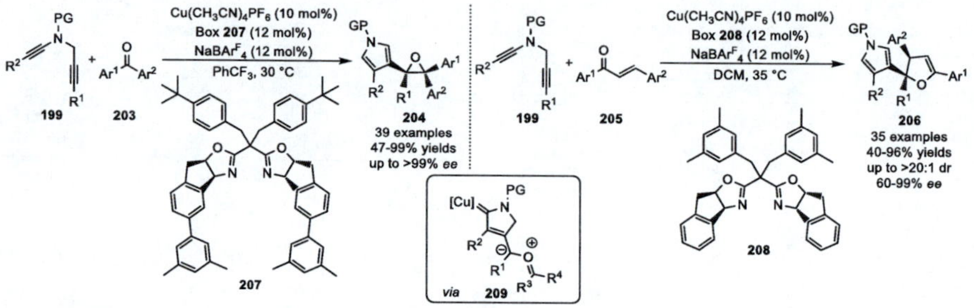

Scheme 51 Synthesis of chiral oxiranes and dihydrofurans from formal [2 + 1] and [4 + 1] annulations of *N*-propargyl ynamides

good yields and a wide substrate scope and good functional group tolerance under mild reaction conditions (Scheme 50). The key intermediate was the endocyclic donor–donor copper carbene **202** from vinyl cations **201**.

Recently, the Ye group achieved the copper-catalyzed asymmetric formal [2 + 1]- and [4 + 1]-annulations of *N*-propargyl ynamides with ketones via carbonyl ylides [68]. As shown in Scheme 51, when diaryl ketones **203** were used, epoxidation occurred under Cu(I)/BOX **207** catalytic conditions and the reaction gave 3-pyrrolyl oxiranes **204** in moderate to excellent yields (47–99%) and up to >99% enantioselectivities with broad substrate scope. Interestingly, when chalcones **205** were used instead of diaryl ketones, a series of pyrrolyl-substituted chiral dihydrofurans **206** were obtained under Cu(I)/BOX **208** conditions. Both of the transformations provided a large number of structurally rich pyrrolyl contained heterocycles.

3.4 Heterocycles from Alkyne-Cycloisomerization as Carbene Sources

Alkyne-cycloisomerization is one of the most representative strategies for the construction of cyclic frameworks, especially heterocycles. From the point of view of donor-type metal carbene intermediates involved reactions, serval examples have successfully built heterocyclic skeletons in the past decade. In 2012, Iwasawa and

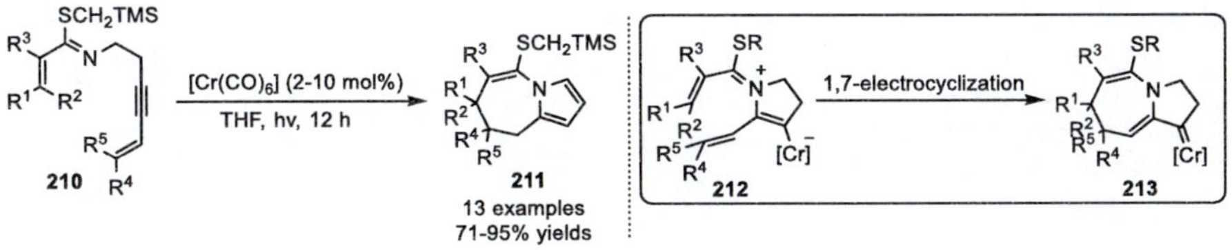

Scheme 52 Synthesis of 1-azabicyclo[5.3.0]decane derivatives via alkyne-cycloisomerization

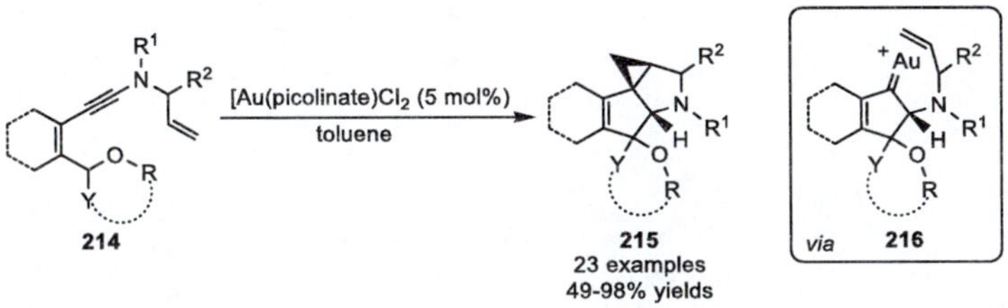

Scheme 53 Gold-catalyzed cycloisomerization for the synthesis of nitrogen-heterocycle scaffolds

co-workers developed a chromium(0)-catalyzed/photoirradiation tandem cyclization of α,β-unsaturated thioimidates **210**, which bear an enyne moiety [69]. This reaction provides a useful method for the synthesis of the 1-azabicyclo[5.3.0]decane derivatives **211** in good yields and diastereoselectivities (Scheme 52). The reaction mechanism was proposed to proceed via carbene intermediate **213** through 1,7-electrocyclization of zwitterionic intermediate **212**.

Pyrrolidine-contained polycycles **215** were formed through the cascade polycyclization of *N*-allyl ynamides **214** by Davies and co-workers in 2015 [70]. Under gold-catalyzed conditions, the fused nitrogen-heterocycle scaffolds were obtained in moderate to excellent yields with superb selectivity through cyclopropanation of donor gold carbene intermediates **216** (Scheme 53).

Compared to the typically good catalysts for the activation of alkynes, including Au(I) and Pt(II), dirhodium(II) catalysts are considered less efficient, especially in the generation of metal carbenes via alkyne-cycloisomerization. However, in 2021, Zhu and co-workers developed an enantioselective Rh(II)-catalyzed desymmetric cycloisomerization of diynes **217.** This methodology provides efficient access to highly functionalized and enantiomerically enriched alkynyl-substituted furan-fused dihydropiperidine derivatives **218** with wide substrate scope [71]. Two heterocycles and an optically active alkynyl-substituted aza-quaternary carbon stereocenter are constructed in one step (Scheme 54). This method not only represents a new cycloisomerization of diynes but also constitutes the first Rh(II)-catalyzed asymmetric intramolecular cycloisomerization of 1,6-diynes. Calculational studies and control experiments indicate that the reaction undergoes a concerted [3 + 2]-cycloaddition/[1,2]-*H* shift of the donor-type rhodium carbene intermediate **219.**

Scheme 54 Rh(II)-catalyzed enantioselective cycloisomerization of triynals for the synthesis of furan-fused dihydropiperidine derivative

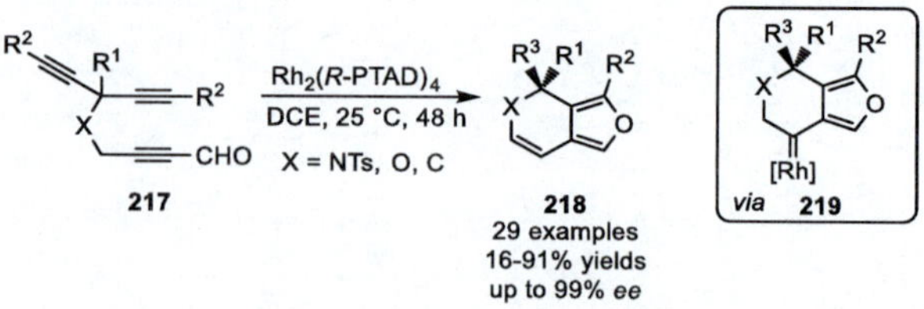

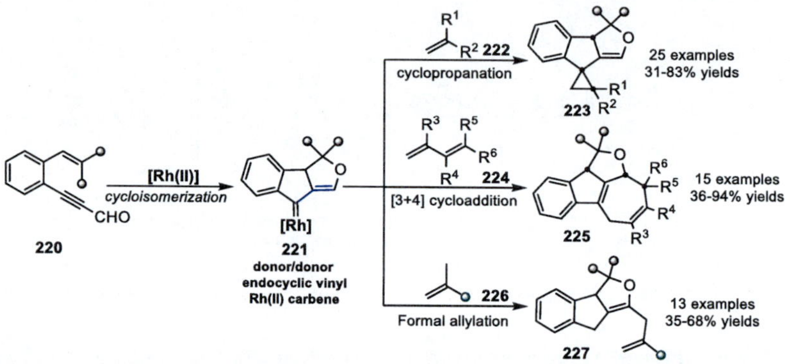

Scheme 55 Rh(II)-catalyzed cycloisomerization of phenyl enynals for the synthesis of polycyclic heterocycles.

When the aliphatic diynals are replaced by phenyl enynals **220**, a new kind of endocyclic donor–donor phenyl vinyl carbene **221** is generated under dirhodium(II)-catalyzed cycloisomerization [72]. As shown in Scheme 55, these vinyl Rh (II) carbenes are trapped by different alkenes, such as 1,1-disubstituted alkenes **222**, dienes **224**, and α-methyl styrenes **226**, and undergo cyclopropanation, [4 + 3] cycloaddition, and formal allylation, respectively. All of these reactions provide divergent spiro and fused polycyclic heterocycles **223**, **225**, and **227** with wide substrate scope and good yields, as well as excellent selectivity.

Recently, a dirhodium(II)-catalyzed asymmetric cycloisomerization reaction of azaenyne **228** through a cap-tether synergistic modulation strategy was developed by Zhu and co-workers [73]. Benefiting from this approach, a diverse array of centrally chiral isoindazole derivatives **230** could be constructed in up to 99% yield, 99:1 *dr*, and 98% *ee* in an efficient and practical manner (Scheme 56). Compared with the enynone system, this work represents a significant progress in the enantioselective cyclization of conjugated enyne motifs. The introduction of the nitrogen tether atom and capping group provided strategic advantages in both further applications of the products and reactivity regulation of reactants. This cycloisomerization of azaenynes features an additional cap in the formed heterocycle, which enables access to various enantiomerically enriched atropisomers **233** bearing five-five-membered heteroaryls through an oxidative central-to-axial chirality transfer strategy.

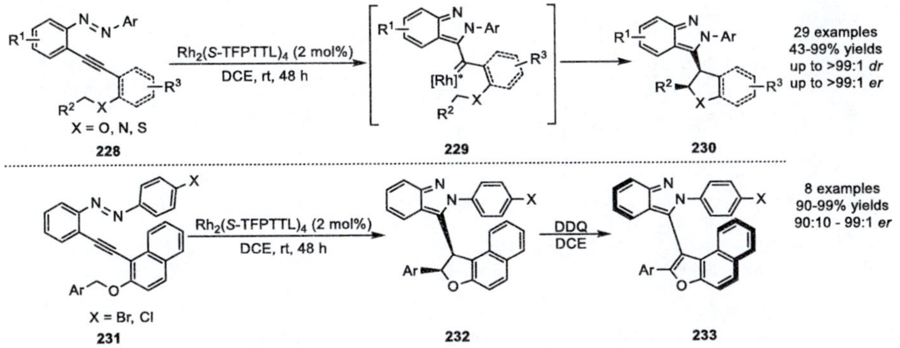

Scheme 56 Rh(II)-catalyzed enantioselective cycloisomerization of azaenynes for the synthesis of centrally chiral isoindazole derivatives

4 Heterocycles from Cyclopropenes as Carbene Sources

Cyclopropenes can serve as vinyl metal carbene precursors under the catalysis of transition metals. The donor-type vinyl metal carbenes are generated by ring opening reaction due to intrinsic ring strain of cyclopropenes. Using cyclopropenes as starting materials, heterocycles can be obtained through intramolecular cyclopropanation or C–H insertion reactions. In 2010, Cossy and Meyer first developed a gold-catalyzed intramolecular cyclopropanation of allyl tethered cyclopropenes **234** under mild condition. The corresponding six-membered heterocyclic products including 5-isopropylidene-3-oxa and 3-azabicyclo[4.1.0]heptanes **236** were formed in excellent yields and high diastereoselectivities (Scheme 57) [74, 75]. Vinyl gold carbene **235** was probably the key intermediate.

In 2011, the same group developed a Rh(II)-catalyzed intramolecular cyclopropanation of cyclopropenes **237** for the synthesis of eight-membered oxygen heterocycles **238** (Scheme 58) [76]. The reaction was conducted smoothly in mild condition with the loading of $Rh_2(OAc)_4$ as low as 0.5 mol%. Notably, the key vinyl Rh(II) carbene intermediates reacted selectively with C=C bonds of **237** and no possible Stevens rearrangement products were detected when O- or N-tethered allyl substrates were used in this reaction.

C–H insertion reactions can also be realized by using cyclopropenes as donor-type vinyl metal carbene precursors, and the resulting carbene intermediates can be used for the synthesis of heterocycles through intramolecular reactions. Cossy et al reported that vinyl Rh(II) carbenes, derived from the ring opening of cyclopropenes **239** and **241** under rhodium(II) catalytic conditions, displayed high efficiency and reactivity in intramolecular insertion of C–H bonds (Scheme 59) [77]. Heterocyclic products **240** and **242** were obtained in good yields with excellent diastereoselectivities via the C–H insertion reactions.

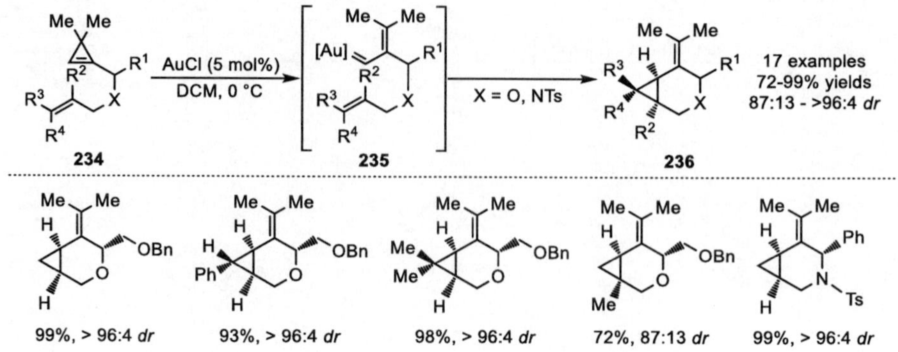

Scheme 57 Au(I)-catalyzed intramolecular cyclopropanation of cyclopropenes

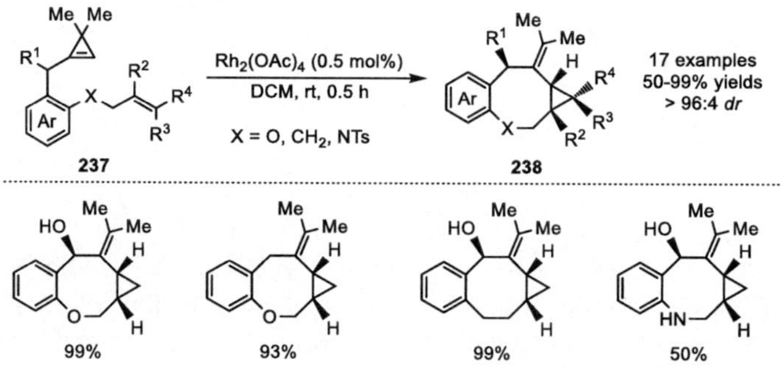

Scheme 58 Rh(II)-catalyzed cyclopropanation of cyclopropenes for the synthesis of polycyclic heterocycles

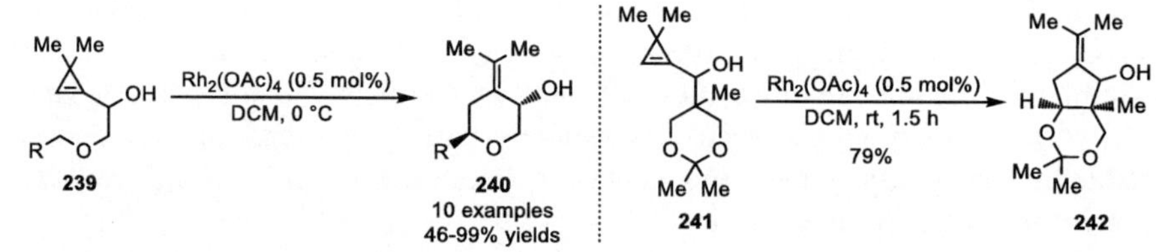

Scheme 59 Rh(II)-catalyzed C–H insertion of cyclopropenes for the synthesis of heterocycles

5 Heterocycles from Donor-Type Carbenes via Carbene/Alkyne Metathesis

Carbene/alkyne metathesis, known as CAM, is a tandem process in which metal carbene intermediates are attacked by alkynes in an intra- or intermolecular manner and a new kind of vinyl metal carbene species are generated via a formal metathesis

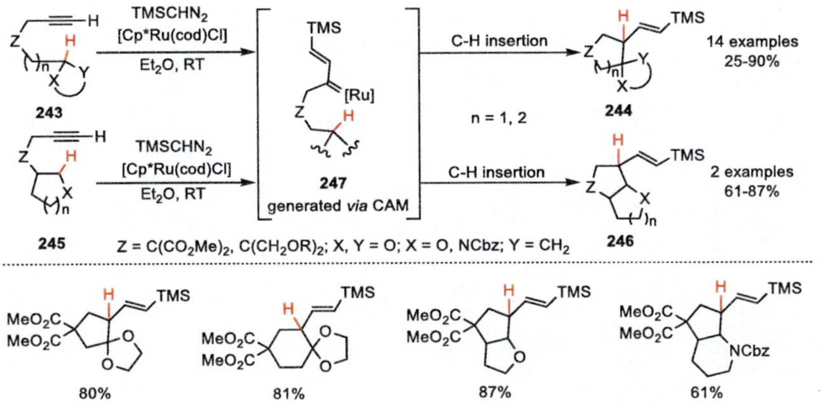

Scheme 60 C–H insertion of vinyl Ru(II)-carbene via CAM for the synthesis of spirobicycles and fused bicycles

Scheme 61 Cyclization of vinyl carbene via CAM process for the synthesis of dihydropyrans and dihydrooxazines

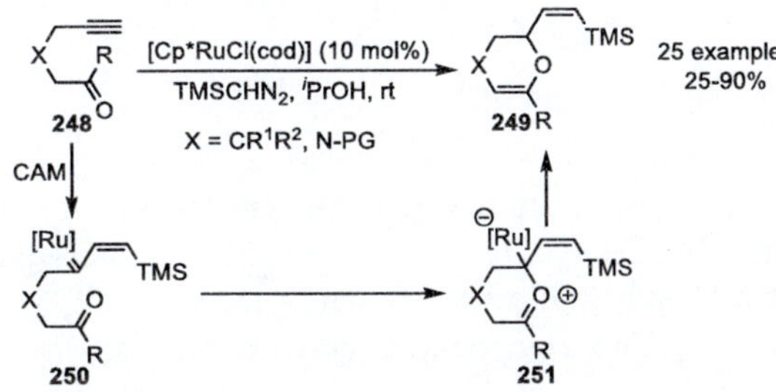

process. This reaction mode has also been used in the formation of donor-type carbenes. In 2012, Saa et al reported that the easily available linear alkynyl ethers, acetals, and amines **243** and **245** can be converted into heterocyclic products **244** and **246** through a Ru(II)-catalyzed intramolecular C(sp^3)–H insertion reactions with CAM process [78]. Mechanism studies supported the vinyl Ru(II)-carbene **247** as key intermediate, which initially underwent 1,5- or 1,6-hydride shift and followed cyclization. These cyclization reactions enabled the efficient conversion of both secondary and tertiary C(sp^3)–H bonds into new C–C bonds and finally gave the heterocyclic products under mild reaction conditions (Scheme 60).

A novel synthesis of 2-vinyl dihydropyrans and dihydrooxazines **249** from readily available alkynals and alkynones **248** through Ru(II)-catalyzed cyclization has been developed by Saa and co-workers [79]. Vinyl ruthenium carbenes **250** derived from alkynes **248** and (trimethylsilyl)diazomethane ($TMSCHN_2$) via a CAM process are proposed as the key intermediates in the cyclization processes. Dihydropyrans and dihydrooxazines **249** were obtained in moderate to high yields under mild reaction conditions with good functional group tolerance (Scheme 61).

In 2015, Saa and Rodríguez disclosed an approach to 1,3-benzoxazines **253** by Ru(II)-catalyzed annulation reaction of *ortho*-(alkynyloxy)benzylamines **252**

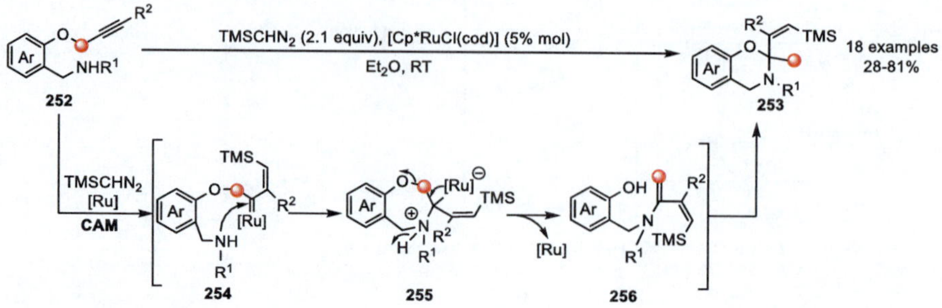

Scheme 62 Cyclization of Ru(II) vinyl carbene via a CAM process for the synthesis of 1,3-benzoxazines

Scheme 63 Intramolecular CAM process for the synthesis of heterocycles

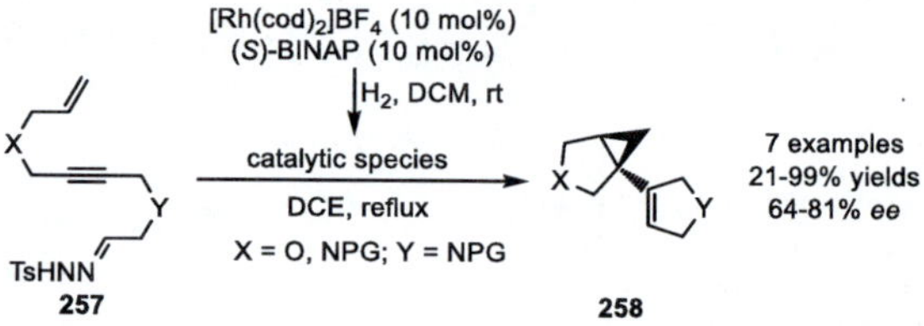

[80]. The cyclization began with Ru(II)-carbenes in the presence of [Cp*RuCl(cod)] and $TMSCHN_2$. This carbene species was subsequently attacked by the triple bonds of **252** and simultaneously generated new vinyl Ru(II)-carbenes **254** via the CAM process. The electrophilic vinyl Ru(II)-carbenes **254** could induce nucleophilic attack by the amine group to afford the zwitterionic intermediates **255**. With the phenoxide moiety as the living group, the cyclic intermediates **255** facilitated ring opening to afford enamine intermediates **256** due to ring strain. Finally, the enamine intermediates **256** were trapped with the phenoxide to give the desired 1,3-benzoxazines **253,** and the ruthenium complex was released to re-enter the catalytic cycle (Scheme 62).

The CAM process can also be conducted in an intramolecular manner to provide the desired heterocycles. For example, Pla-Quintana and co-workers demonstrated that a chiral cationic rhodium(I) catalytic system enables the enantioselective synthesis of heterocycles **258** through base-free rhodium carbene formation and a carbene/alkyne metathesis-cyclopropanation sequence (Scheme 63) [81]. Thus, non-stabilized rhodium carbenes are capable of mediating this efficient formation of bicyclic cyclopropane-containing scaffolds.

A new chiral dirhodium(II) paddlewheel complex $Rh_2(R\text{-KC4N})_4$ was synthesized by May and co-workers, and the resulting complex was used in enantioselective carbene/alkyne cascade reactions [82]. A structurally intricate bridged polycyclic molecule **260** was synthesized via the CAM process with good yield and excellent enantioselectivity compared to the known dirhodium(II) catalyst Rh(*S*-PTAD) (Scheme 64).

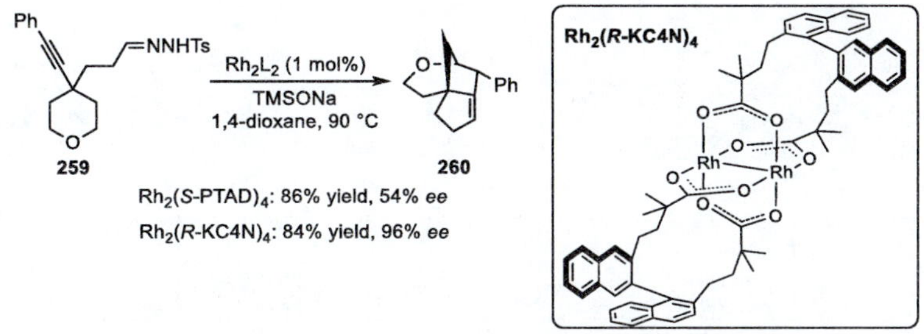

Scheme 64 Synthesis of bridged polycyclic heterocycle via CAM process with new chiral dirhodium(II) catalyst

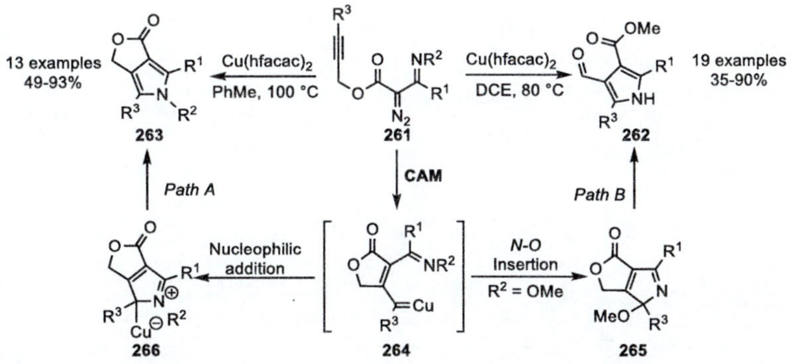

Scheme 65 Divergent reaction of Cu(I)-catalyzed CAM process for the synthesis of pyrroles

Xu and co-workers realized a cascade reaction of alkynyl-tethered α-iminodiazoacetates **261**, which provided general access to both multi-substituted and fused pyrroles **262** and **263** in high yields (Scheme 65) [83]. The γ-imino carbene **264** via a CAM process was proposed as the key intermediate in this divergent reaction. When the substituent on the imino group was a OMe group, the reaction proceeded through nucleophilic addition to afford fused pyrroles **263**. For comparison, instead of nucleophilic addition, the N–O insertion to the Cu (II) carbene intermediate took place when the R_2 substituent on the imino group was OMe, leading to the multi-substituted pyrroles **263** through a stepwise aromatization process.

Xu and Doyle reported a general access to chiral dihydroindole derivatives **268** and **269** in high yields and high site-/enantioselectivity via C–H functionalization of propargyl diazoacetates **267** [84]. The high site-selectivity was realized by catalyst control. Sterically demanding dirhodium carboxylates, with Rh_2(*S*-BTPCP)$_4$ being optimal, favored C–H insertion into 1° C–H bonds with regioselectivities and enantioselectivities up to 95:5 (1° > 2° benzylic) and > 90% *ee*, respectively (Scheme 66). The preferential 2° and 3° C–H bond insertion occurred under catalysis

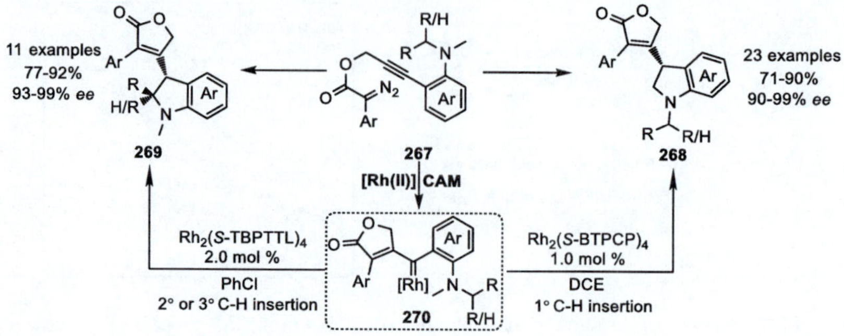

Scheme 66 Rh(II)-catalyzed asymmetric C–H insertion via CAM process for the synthesis of chiral dihydroindoles

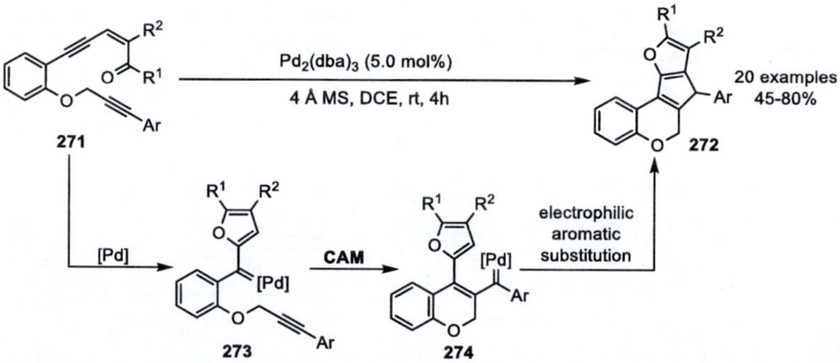

Scheme 67 Pd(0)-catalyzed reaction of alkyne-tethered enynones via CAM for the synthesis of polyheterocycles

by Rh_2(*S*-TBPTTL)$_4$ due to the configuration of catalyst and electronic effects. The chiral dirhodium catalysts played dual roles, which not only promoted the CAM reaction to generate the donor–donor carbene **270**, but also induced high enantioselectivity in the C–H bond insertion reaction.

Enynones can also be applied as good precursors for the CAM process. In 2018, Xu and co-workers developed a Pd(0)-catalyzed CAM tandem reaction of alkyne-tethered enynones **271** for the synthesis of various fused polyheterocycles **272** with good yields (Scheme 67) [85]. In the presence of $Pd_2(dba)_3$, enynones **271** generated donor–donor 2-furyl Pd(0)-carbenes **273** via 5-*exo*-dig cyclization. New donor-type aryl-vinyl Pd(0)-carbene intermediates **274** then formed through CAM process with Pd(0)-carbenes **273** attacked by the propargyl moiety. Finally, intramolecular electrophilic aromatic substitution of **274** took place to give fused polyheterocycles **272**.

6 Heterocycles from Donor-Type Carbenes via Brook Rearrangement

In recent years, a new donor-type aryl siloxycarbenes was generated via a 1,2-Brook rearrangement from aryl silyl ketones under microwave irradiation or photo-induced conditions. This kind of carbene intermediate can undergo traditional carbene transfer reactions, such as intramolecular C–H insertion and cyclopropanation, to construct heterocycles. In 2009, Dong and co-workers reported a formal C–H insertion to form 2,3-dihydrobenzofuran and dihydrobenzothiophene derivatives **276** from benzyl ether tethered aryl silyl ketones **275** under microwave irradiation [86]. This C–H bond functionalization strategy involves a thermally induced Brook rearrangement to form a putative siloxycarbene intermediate **277** (Scheme 68).

In 2015, Bolm and co-workers developed a highly efficient intramolecular C–H insertion of *N*-functionalized aryl silyl ketones to give dihydroquinolinones **279** in up to quantitative yields under photochemically or thermally induced conditions (Scheme 69) [87]. Mechanistically, the transformations involve in situ generated siloxycarbenes that react intramolecularly with excellent selectivity.

Recently, Priebbenow and co-workers discovered that visible light-induced donor-type carbene intermediates via Brook rearrangement underwent intramolecular cyclopropanation with tethered olefins, generating valuable bicyclo[3.1.0]hexane and bicyclo[4.1.0]heptane scaffolds **281** with simultaneous formation of two new ring systems [88]. Advantageously, this highly stereospecific reaction requires only visible light irradiation, which avoids the requirement for exogenous additives, sensitizers, or (photo)catalysts, affording a unique class of silicon-derived heterocycles (Scheme 70).

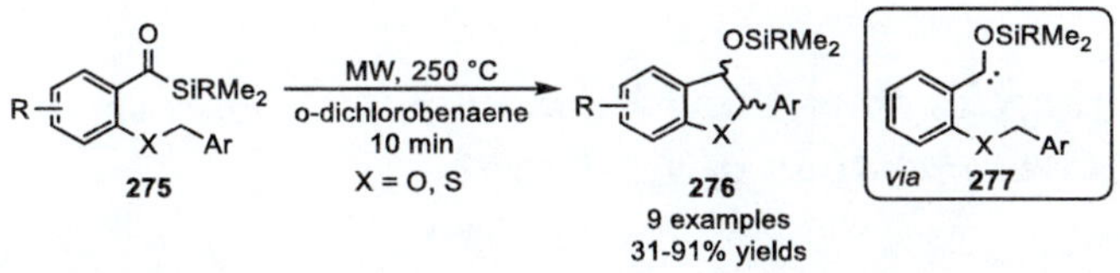

Scheme 68 Synthesis of dihydrobenzofuran and dihydrobenzothiophene derivatives involving Brook rearrangement

Scheme 69 Synthesis of dihydroquinolinones involving Brook rearrangement

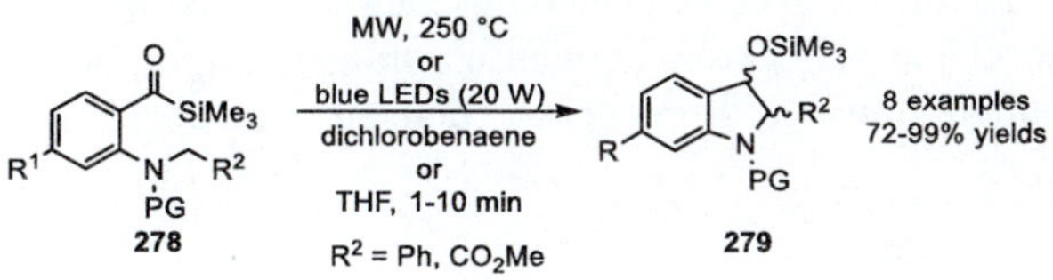

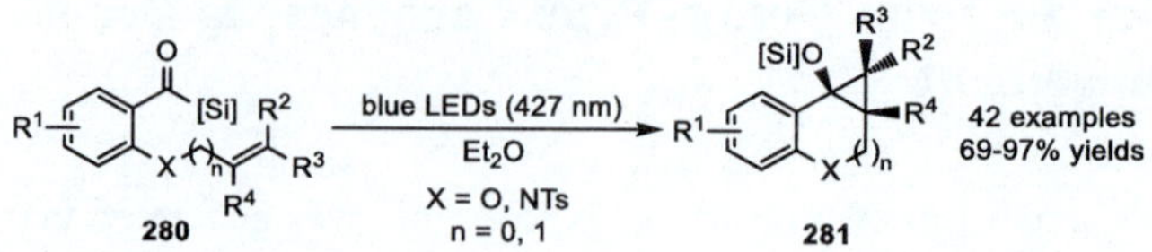

Scheme 70 Synthesis of silicon-derived heterocycles involving Brook rearrangement

7 Heterocycles from Other Sources of Donor-Type Carbenes

7.1 *Heterocycles from Donor-Type Fischer Carbenes*

In 2016, Sampedro and Santamaría reported a formal [3 + 3]-carbocyclization between phenyl alkynylcarbene complexes **283** and furfural imine **284** to give the corresponding benzofurans **285** (Scheme 71) [89]. These phenyl alkynylcarbene complexes, which are non-heteroatom-stabilized Fischer carbenes, represent a new type donor–donor carbenes, providing benzofurans in good yields.

Wang and co-workers have explored vinyl chromium(0) Fischer carbene complexes **287** as precursors of palladium(II) carbenes through transmetallation in 2017 [90]. The reaction represents novel access to π-allylic palladium(II) species and provides a series of flavonones products **288** from 2-iodophenols/2-iodoanilines **286** and carbene complexes **287** in moderate to good yields (Scheme 72). Mechanistically, this cascade transformation involves a transmetallation, palladium carbene migratory insertion, and intramolecular nucleophilic addition of a π-allylic palladium(II) intermediate **291**.

7.2 *Heterocycles from Vinyl Ruthenium Carbenes via gem-Hydrogenation of 1,3-Enynes*

Donor-type of ruthenium carbene complexes **294** derived from 1,3-enynes **292** via *gem*-hydrogenation with $[Cp^*RuCl]_4$ as the catalyst has been recently developed by Fürstner and co-workers [91]. Under this catalytic mode, intramolecular C–H insertion was successfully realized, and tetrahydrofuran derivatives **293** were obtained in moderate to excellent yields (Scheme 73). Notably, this reaction provides ready access to spirocyclic as well as bridged ring systems of immediate relevance as building blocks for medicinal chemistry and chemical biology.

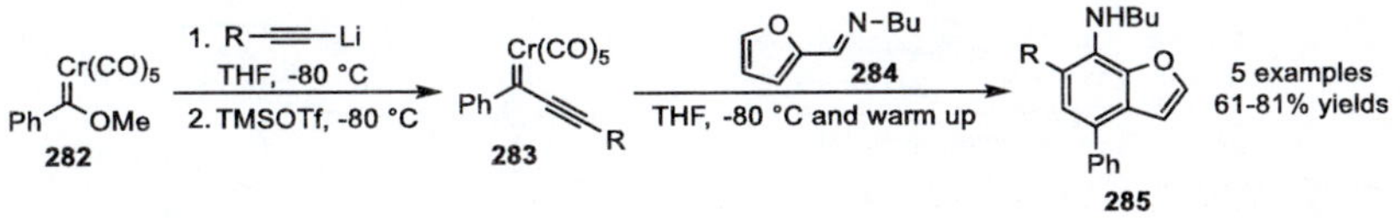

Scheme 71 Synthesis of benzofurans from in situ synthesized phenyl alkynylcarbene complexes

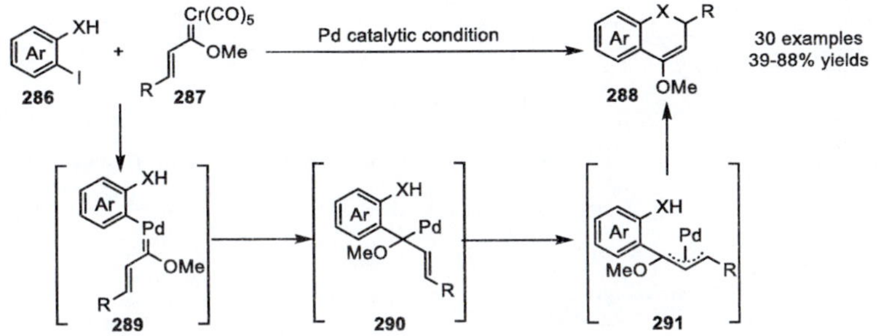

Scheme 72 Synthesis of flavonones and quinoline derivatives via vinyl Cr(0) carbene complexes

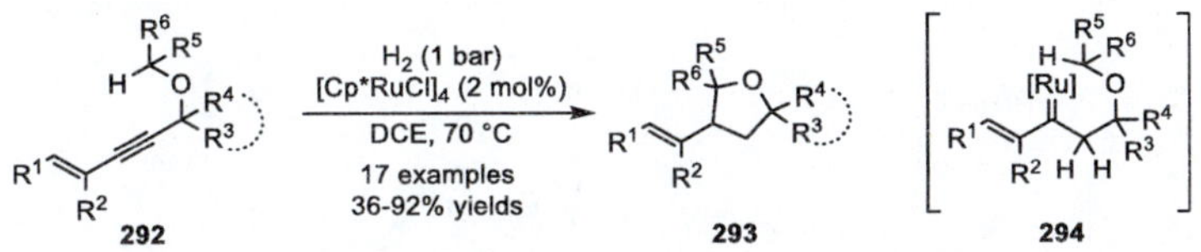

Scheme 73 Synthesis of tetrahydrofuran derivatives from vinyl ruthenium carbenes via *gem*-hydrogenation

7.3 *Heterocycles from Cyclopropyl/Carbonyl Compounds as Donor-Type Carbenes Sources*

In 2018, Asako and Takai demonstrated that aryl cyclopropanes can be used as donor and donor–donor carbenes sources by releasing ethylene with a simple molybdenum/quinone **297** catalyst [92]. The strategic use of a pyridyl group with the catalyst enabled the efficient retro-cyclopropanation of cyclopropanes **295** to generate Mo(0) carbenes **299** without extra strain. This reaction displayed a convenient synthesis of pyridoisoindole derivatives **296** without potential explosive and toxic disadvantages compared to traditional donor–donor diazo compounds as carbene precursors (Scheme 74).

Similarly, aryl carbonyl compounds **300** could also serve as carbene equivalents through deoxygenation. Asako and Takai reported in 2019 that the $Mo(CO)_6$/quinone **303** reagent can be easily prepared in situ using commercially available compounds without purification or specially designed ligands. This catalytic system effects deoxygenative insertion of a carbonyl carbon by the carbene intermediate **305**, followed by an intramolecular C(sp^3)–H bond insertion to give the indole

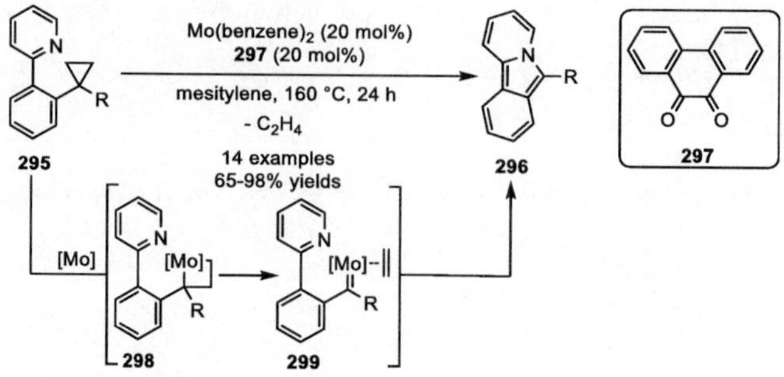

Scheme 74 Synthesis of pyridoisoindole derivatives with Mo(0) carbenes from retro-cyclopropanation

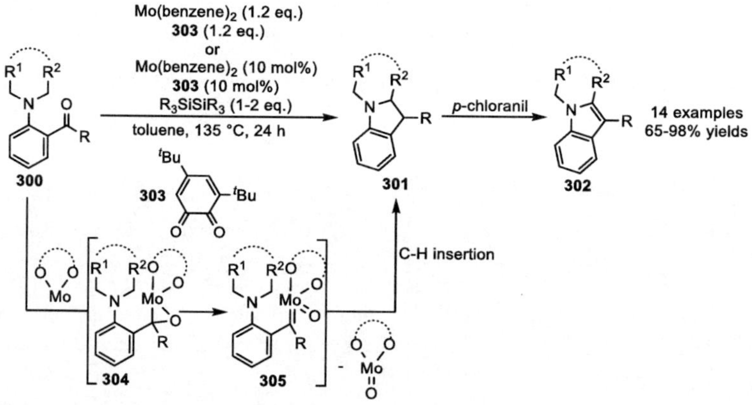

Scheme 75 Synthesis of indole derivatives via formal C–H insertion of Mo(0) carbenes decried from deoxygenation process

products **302** in generally high yields (Scheme 75) [93]. The deoxygenative cyclization also proceeded with a catalytic amount of a Mo(0)/quinone complex using R_3SiSiR_3 as an oxygen atom acceptor. This new protocol allows the direct use of carbonyl compounds as carbene equivalents upon deoxygenation, and omits the transformation of carbonyl groups to hydrazones or related diazo compounds, streamlining the overall process.

Remarkably, an efficient method for the generation of donor-type metal carbenes from common aldehydes **306** has been reported recently by Nagib and co-workers [94]. As shown in Scheme 76, a range of traditional carbene transfer reactions, such as dimerization, cyclopropanation, [2 + 1]-cycloaddition with unsaturated carbon–heteroatom bonds, and X–H (X = N, P, Si, B, S) insertions, have been explored with good yields and high efficiency. Importantly, this strategy enables safe reactions of non-stabilized formal carbenes from simple alkyl, aryl, and formyl aldehydes via

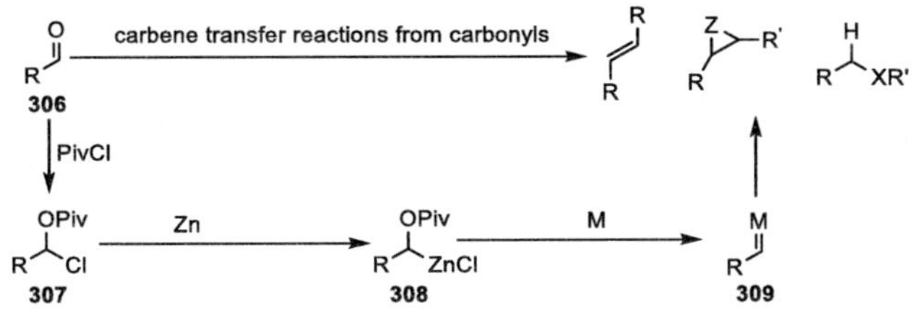

Scheme 76 Synthesis of heterocycles from common aldehydes as donor-type carbene sources

metal carbenoids with earth-abundant metal salts ($FeCl_2$, $CoCl_2$, CuCl) and has great potential for the construction of structural complicated heterocycles.

8 Conclusions

In this chapter, the synthesis of heterocyclic compounds from donor and donor–donor carbene intermediates is summarized. These synthetic methods include classic carbene transfer reactions, such as cyclopropanation, C–H insertion, X–H insertion, cycloaddition, and new reaction modes, including carbene migratory insertion, developed during the past two decades, Among the numerous catalytic reactions, the typical donor-type carbene sources are also discussed. Beyond traditional diazo compounds, surrogate precursors such as hydrazones have been developed. In addition, more and more attention has been paid to the non-diazo carbene sources in recent years. The functional alkynes, cyclopropenes, and even the common carbonyl compounds could be used as precursors of donor and donor–donor carbenes. With the rapid progress on the discovery of new and safe donor-type carbene sources, and with the development of compatible catalytic methods, more practical synthetic methods will be disclosed for the construction of valuable heterocycles with even greater structural diversity in the future.

Acknowledgments The authors acknowledge funding of their work by the National Natural Science Foundation of China (22071062, 21871096, 22201078), Ministry of Science and Technology of the People's Republic of China (2016YFA0602900), 111 Project (B20003), Guangdong Science and Technology Department (2018B030308007, 2021A1515012331), China Postdoctoral Science Foundation (2021M701244).

References

1. Rotella DP (2021) Heterocycles in drug discovery: properties and preparation. In: Advances in heterocyclic chemistry. Academic Press, Cambridge

2. Krafft C, Brennecke S, Ott F, Backes M, Salzer R, Grunenberg J, Ley JP, Krammer GE, Weber B (2008) High-impact sulfur compounds: constitutional and configurational assignment of sulfur-containing heterocycles. Chem Biodivers 5:1204–1212
3. Beltran-Hortelano I, Alcolea V, Font M, Pérez-Silanes S (2020) The role of imidazole and benzimidazole heterocycles in Chagas disease: a review. Eur J Med Chem 206:112692–112705
4. Huang J, Yu G (2021) Structural engineering in polymer semiconductors with aromatic *N*-heterocycles. Chem Mater 33:1513–1539
5. Soam P, Kamboj P, Tyagi V (2022) Rhodium-catalyzed cascade reactions using diazo compounds as a carbene precursor to construct diverse heterocycles. Asian J Org Chem 11: e202100570
6. Davies HML, Beckwith REJ (2003) Catalytic enantioselective C-H activation by means of metal-carbenoid-induced C-H insertion. Chem Rev 103:2861–2904
7. Doyle MP, Duffy R, Ratnikov M, Zhou L (2010) Catalytic carbene insertion into C-H bonds. Chem Rev 110:704–724
8. Davies HML, Morton D (2011) Guiding principles for site selective and stereoselective intermolecular C-H functionalization by donor/acceptor rhodium carbenes. Chem Soc Rev 40:1857–1869
9. Cheng Q-Q, Deng Y, Lankelma M, Doyle MP (2017) Cycloaddition reactions of enoldiazo compounds. Chem Soc Rev 46:5425–5443
10. Denney DB, Klemchuk PP (1958) Deuterium isotope effects in some intramolecular aromatic substitutions. J Am Chem Soc 80:3289–3290
11. Maas G (2009) New syntheses of diazo compounds. Angew Chem Int Ed 48:8186–8195
12. Fulton JR, Aggarwal VK, de Vicent J (2005) The use of tosylhydrazone salts as a safe alternative for handling diazo compounds and their applications in organic synthesis. Eur J Org Chem 2005:1479–1492
13. Zhu D, Chen L, Fan H, Yao Q, Zhu S (2020) Recent progress on donor and donor-donor carbenes. Chem Soc Rev 49:908–950
14. Qiu L, Huang D, Xu G, Dai Z, Sun J (2015) Realized C-H functionalization of aryldiazo compounds *via* rhodium relay catalysis. Org Lett 17:1810–1813
15. Murakami H, Asakawa T, Muramatsu Y, Ishikawa R, Hiza A, Tsukaguchi Y, Tokumaru Y, Egi M, Inai M, Ouchi H, Yoshimura F, Taniguchi T, Ishikawa Y, Kondo M, Kan T (2020) Total synthesis of sophoraflavanone H and confirmation of its absolute configuration. Org Lett 22: 3820–3824
16. Javed MI, Brewer M (2007) Diazo preparation *via* dehydrogenation of hydrazones with "activated" DMSO. Org Lett 9:1789–1792
17. Soldi C, Lamb KN, Squitieri RA, Gonzalez-López M, Di MMJ, Shaw JT (2014) Enantioselective intramolecular C-H insertion reactions of donor-donor metal carbenoids. J Am Chem Soc 136:15142–15145
18. Natori Y, Ito M, Anada M, Nambu H, Hashimoto S (2015) Catalytic asymmetric synthesis of (−)-*E*-δ-viniferin *via* an intramolecular C-H insertion of diaryldiazomethane using Rh_2(*S*-TFPTTL)$_4$. Tetrahedron Lett 56:4324–4327
19. Souza LW, Squitieri RA, Dimirjian CA, Hodur BM, Nickerson LA, Penrod CN, Cordova J, Fettinger JC, Shaw JT (2018) Enantioselective synthesis of indolines, benzodihydrothiophenes, and indanes by C-H insertion of donor/donor carbenes. Angew Chem Int Ed 57:15213–15216
20. Nickerson LA, Bergstrom BD, Gao M, Shiue Y-S, Laconsay CJ, Culberson MR, Knauss WA, Fettinger JC, Tantillo DJ, Shaw JT (2020) Enantioselective synthesis of isochromans and tetrahydroisoquinolines by C-H insertion of donor/donor carbenes. Chem Sci 11:494–498
21. Bergstrom BD, Merrill AT, Fettinger JC, Tantillo DJ, Shaw JT (2022) Divergent asymmetric synthesis of Panowamycins, TM-135, and Veramycin F using C-H insertion with donor/donor carbenes. Angew Chem Int Ed 61:e202203072
22. Liu P, Zhu C, Xu G, Sun J (2017) Iron-catalyzed intermolecular cycloaddition of diazo surrogates with hexahydro-1,3,5-triazines. Org Biomol Chem 15:7743–7746

23. López-Sánchez C, Álvarez-Corral M, Jiménez-González L, Muñoz-Dorado M, Rodríguez-García I (2013) Rh(II)-catalyzed enantioselective synthesis of acuminatin through a C-H insertion reaction of a non-stabilized carbenoid. Tetrahedron 69:5511–5516
24. Reddy AR, Zhou C-Y, Guo Z, Wei J, Che C-M (2014) Ruthenium–porphyrin-catalyzed diastereoselective intramolecular alkyl carbene insertion into C-H bonds of alkyl diazomethanes generated *in situ* from *N*-tosylhydrazones. Angew Chem Int Ed 53:14175–14180
25. Karns AS, Goswami M, de Bruin B (2018) Catalytic synthesis of indolines by hydrogen atom transfer to cobalt(III)–carbene radicals. Chem A Eur J 24:5253–5258
26. Wen X, Wang Y, Zhang XP (2018) Enantioselective radical process for synthesis of chiral indolines by metalloradical alkylation of diverse C(sp^3)-H bonds. Chem Sci 9:5082–5086
27. Reddy AR, Hao F, Wu K, Zhou C-Y, Che C-M (2016) Cobalt(II) porphyrin-catalyzed intramolecular cyclopropanation of *N*-alkyl indoles/pyrroles with alkylcarbene: efficient synthesis of polycyclic *N*-heterocycles. Angew Chem Int Ed 55:1810–1815
28. Paul ND, Mandal S, Otte M, Cui X, Zhang XP, de Bruin B (2014) Metalloradical approach to 2*H*-chromenes. J Am Chem Soc 136:1090–1096
29. Khanna A, Maung C, Johnson KR, Luong TT, Vranken DLV (2012) Carbenylative amination with *N*-tosylhydrazones. Org Lett 14:3233–3235
30. Florentino L, Aznar F, Valdés C (2012) Synthesis of polysubstituted isoquinolines through cross-coupling reactions with α-alkoxytosylhydrazones. Org Lett 14:2323–2325
31. Bzeih T, Zhang K, Khalaf A, Hachem A, Alami M, Hamze A (2019) One-pot reaction between *N*-tosylhydrazones and 2-nitrobenzyl bromide: route to NH-free C2-arylindoles. J Org Chem 84:228–238
32. Yu Y, Ma L, Xia J, Xin L, Zhu L, Huang X (2020) A modular approach to dibenzo-fused ε-lactams: palladium-catalyzed bridging-C-H activation. Angew Chem Int Ed 59:18261–18266
33. Huo J, Zhong K, Xue Y, Lyu M, Ping Y, Liu Z, Lan Y, Wang J (2021) Palladium-catalyzed enantioselective carbene insertion into carbon-silicon bonds of silacyclobutanes. J Am Chem Soc 143:12968–12973
34. Huo J, Zhong K, Xue Y, Lyu M, Ping Y, Ouyang W, Liu Z, Lan Y, Wang J (2022) Ligand-controlled site- and enantioselective carbene insertion into carbon-silicon bonds of benzosilacyclobutanes. Chem A Eur J 28:e202200191
35. Hossain ML, Wang K, Ye F, Zhang Y, Wang J (2017) Cu(I)-catalyzed cascade reaction of *N*-tosylhydrazones with 3-butyn-1-ol: a new synthesis of tetrahydrofurans. Chin J Catal 38:115–122
36. Fu R, Li Z (2018) Direct synthesis of 2-methylbenzofurans from calcium carbide and salicylaldehyde *p*-tosylhydrazones. Org Lett 20:2342–2345
37. Kumar P, Keshri SK, Kapur M (2021) Ru(II)-catalyzed, Cu(II)-mediated carbene migratory insertion in the synthesis of trisubstituted pyrroles from isoxazoles. Org Biomol Chem 19: 3428–3433
38. Huang Y, Li X, Yu Y, Zhu C, Wu W, Jiang H (2016) Copper-mediated [3+2] oxidative cyclization reaction of *N*-tosylhydrazones and β-ketoesters: synthesis of 2,3,5-trisubstituted furans. J Org Chem 81:5014–5020
39. Reddy ACS, Anbarasan P (2018) Copper catalyzed oxidative coupling of *ortho*-vinylanilines with *N*-tosylhydrazones: efficient synthesis of polysubstituted quinoline derivatives. J Catal 363:102–108
40. Zhu C, Zhu R, Chen P, Chen F, Wu W, Jiang H (2017) Carbonyl ylides derived from palladium carbenes: the impressive fluorine effect. Adv Synth Catal 359:3154–3159
41. Iwasawa N, Shido M, Kusama H (2001) Generation and reaction of metal-containing carbonyl ylides: tandem [3+2]-cycloaddition-carbene insertion leading to novel polycyclic compounds. J Am Chem Soc 123:5814–5815
42. Chen L, Chen K, Zhu S (2018) Transition-metal-catalyzed intramolecular nucleophilic addition of carbonyl groups to alkynes. Chem 4:1208–1262

43. Vicente R, González J, Riesgo L, González J, López LA (2012) Catalytic generation of zinc carbenes from alkynes: zinc-catalyzed cyclopropanation and Si-H bond insertion reactions. Angew Chem Int Ed 51:8063–8067
44. Song B, Li L-H, Song X-R, Qiu Y-F, Zhong M-J, Zhou P-X, Liang Y-M (2014) Zinc-catalyzed [4+3] cycloaddition with concomitant furan annulation: formation of cyclohepta[*b*]furans. Chem A Eur J 20:5910–5913
45. Chen W, Ji D-S, Luo Y-C, Wang Z-Y, Xu P-F (2018) Directing group assisted ZnI_2-catalyzed cyclopropanation of indoles via 2-furylcarbenoids. Org Chem Front 5:1768–1771
46. Ma J, Chen K, Fu H, Zhang L, Wu W, Jiang H, Zhu S (2016) Dual catalysis: proton/metal-catalyzed tandem benzofuran annulation/carbene transfer reaction. Org Lett 18:1322–1325
47. Zhu D, Ma J, Luo K, Fu H, Zhang L, Zhu S (2016) Enantioselective intramolecular C-H insertion of donor and donor/donor carbenes by a nondiazo approach. Angew Chem Int Ed 55: 8452–8456
48. Zhu D, Chen L, Zhang H, Ma Z, Jiang H, Zhu S (2018) Highly chemo- and stereoselective catalyst-controlled allylic C-H insertion and cyclopropanation using donor/donor carbenes. Angew Chem Int Ed 57:12405–12409
49. Li W, Zhang H, Chen K, Jiang H, Sun J, Zhu S (2022) Palladium-catalyzed intramolecular enantioselective $C(sp^3)$-H insertion of donor/donor carbenes. Chem Sci 13:12396–12402
50. Luo H, Chen K, Jiang H, Zhu S (2016) A route to polysubstituted aziridines from carbenes and imines through a nondiazo approach. Org Lett 18:5208–5211
51. Zhu D, Cao T, Chen K, Zhu S (2022) $Rh_2(II)$-catalyzed enantioselective intramolecular Büchner reaction and aromatic substitution of donor-donor carbenes. Chem Sci 13:1992–2000
52. Hong SY, Jeong J, Chang S (2017) [4+2] or [4+1] annulation: changing the reaction pathway of a rhodium-catalyzed process by tuning the Cp ligand. Angew Chem Int Ed 56:2408–2412
53. Hao J, Guo X, He S, Xu Z, Chen L, Li Z, Song B, Zuo J, Lin Z, Yang W (2021) Marine furanocembranoids-inspired macrocycles enabled by Pd-catalyzed unactivated $C(sp^3)$-H olefination mediated by donor/donor carbenes. Nat Commun 12:1304–1314
54. Chen Y, Shen P, Cao T, Chen H, Zhao Z, Zhu S (2021) Bottom-up modular synthesis of well-defined oligo(arylfuran)s. Nat Commun 12:6165–6177
55. Chen Y, Zhu S (2022) Modular construction of α-furanyl ketones via semi-pinacol rearrangement-mediated ring expansion. Org Chem Front 9:6194–6199
56. Shi Q, Chen Y, Cao T, Zhu S (2022) Construction of [2,5]-furanophanes by carbene-mediated alkynyl migration cyclization. Org Lett 24:8142–8146
57. Wei Z, Chen Y, Wang J, Yang T, Zhao Z, Zhu S (2023) *De novo* synthesis of α-oligo (arylfuran)s and its application in OLED as hole-transporting material. Chem A Eur J 29: e202203444
58. Cao T, Chen K, Zhu S (2017) An efficient approach to generate aryl carbenes: g non-stabilized sequential activation of 1,6-diynes. Org Chem Front 4:450–454
59. Saito K, Sogou H, Suga T, Kusama H, Iwasawa N (2011) Platinum(II)-catalyzed generation and [3+2] cycloaddition reaction of α,β-unsaturated carbene complex intermediates for the preparation of polycyclic compounds. J Am Chem Soc 133:689–691
60. Allegretti PA, Ferreira EM (2011) Generation of α,β-unsaturated platinum carbenes from homopropargylic alcohols: rearrangements to polysubstituted furans. Org Lett 13:5924–5927
61. Shu D, Song W, Li X, Tang W (2013) Rhodium- and platinum-catalyzed [4+3] cycloaddition with concomitant indole annulation: synthesis of cyclohepta[*b*]indoles. Angew Chem Int Ed 52: 3237–3240
62. Yang W, Wang T, Yu Y, Shi S, Zhang T, Hashmi ASK (2013) Nitrones as trapping reagents of α,β-unsaturated carbene intermediates–[1,2]oxazino[5,4-*b*]indoles by a platinum-catalyzed intermolecular [3+3] cycloaddition. Adv Synth Catal 355:1523–1528
63. Hong F-L, Wang Z-S, Wei D-D, Zhai T-Y, Deng G-C, Lu X, Liu R-S, Ye L-W (2019) Generation of donor/donor copper carbenes through copper-catalyzed diyne cyclization: enantioselective and divergent synthesis of chiral polycyclic pyrroles. J Am Chem Soc 141: 16961–16970

64. Hong F-L, Chen Y-B, Ye S-H, Zhu G-Y, Zhu X-Q, Lu X, Liu R-S, Ye L-W (2020) Copper-catalyzed asymmetric reaction of alkenyl diynes with styrenes by formal [3+2] cycloaddition *via* Cu-containing all-carbon 1,3-dipoles: access to chiral pyrrole-fused bridged [2.2.1] skeletons. J Am Chem Soc 142:7618–7626
65. Zhu X-Q, Hong P, Zheng Y-X, Zhen Y-Y, Hong F-L, Lu X, Ye L-W (2021) Copper-catalyzed asymmetric cyclization of alkenyl diynes: method development and new mechanistic insights. Chem Sci 12:9466–9474
66. Hong F-L, Shi C-Y, Hong P, Zhai T-Y, Zhu X-Q, Lu X, Ye L-W (2022) Copper-catalyzed asymmetric diyne cyclization via [1,2]-Stevens-type rearrangement for the synthesis of chiral chromeno[3,4-*c*]pyrroles. Angew Chem Int Ed 61:e202115554
67. Zhu G-Y, Zhai T-Y, Li X, Shi C-Y, Zhu X-Q, Ye L-W (2021) Copper-catalyzed cyclization of *N*-propargyl ynamides with borane adducts through B-H bond insertion. Org Lett 23:8067–8071
68. Qi L-J, Li C-T, Huang Z-Q, Jiang J-T, Zhu X-Q, Lu X, Ye L-W (2022) Enantioselective copper-catalyzed formal [2+1] and [4+1] annulations of diynes with ketones via carbonyl ylides. Angew Chem Int Ed 61:e202210637
69. Karibe Y, Kusama H, Iwasawa N (2012) Chromium(0)-catalyzed tandem cyclization of α,-β-unsaturated thioimidates containing an enyne moiety. Angew Chem Int Ed 51:6214–6218
70. Adcock HV, Chatzopoulou E, Davies PW (2015) Divergent C-H insertion-cyclization cascades of *N*-allyl ynamides. Angew Chem Int Ed 54:15525–15529
71. Wu R, Lu J, Cao T, Ma J, Chen K, Zhu S (2021) Enantioselective Rh(II)-catalyzed desymmetric cycloisomerization of diynes: constructing furan-fused dihydropiperidines with an alkyne-substituted aza-quaternary stereocenter. J Am Chem Soc 143:14916–14925
72. Wu R, Chen Y, Zhu S (2023) Rh(II)-catalyzed enynal cycloisomerization for generation of vinyl carbene: divergent access to polycyclic heterocycles. ACS Catal 13:132–140
73. Qiu S, Gao X, Zhu S (2021) Dirhodium(II)-catalysed cycloisomerization of azaenyne: rapid assembly of centrally and axially chiral isoindazole frameworks. Chem Sci 12:13730–13736
74. Miege F, Meyer C, Cossy J (2010) Synthesis of 3-oxa- and 3-azabicyclo[4.1.0]heptanes by gold-catalyzed cycloisomerization of cyclopropenes. Org Lett 12:4144–4147
75. Miege F, Meyer C, Cossy J (2012) Gold(I)-catalysed cycloisomerisation of 1,6-cyclopropene-enes. Chem A Eur J 18:7810–7822
76. Miege F, Meyer C, Cossy J (2011) Rhodium-catalyzed cycloisomerization involving cyclopropenes: efficient stereoselective synthesis of medium-sized heterocyclic scaffolds. Angew Chem Int Ed 50:5932–5937
77. Archambeau A, Miege F, Meyer C, Cossy J (2012) Highly efficient stereoselective catalytic C (sp^3)-H insertions with donor rhodium carbenoids generated from cyclopropenes. Angew Chem Int Ed 51:11540–11544
78. Cambeiro F, López S, Varela JA, Saá C (2012) Cyclization by catalytic ruthenium carbene insertion into Csp^3-H bonds. Angew Chem Int Ed 51:723–727
79. Cambeiro F, López S, Varela JA, Saá C (2014) Vinyl dihydropyrans and dihydrooxazines: Cyclizations of catalytic ruthenium carbenes derived from alkynals and alkynones. Angew Chem Int Ed 53:5959–5963
80. González-Rodríguez C, Suárez JR, Varela JA, Saá C (2015) Nucleophilic addition of amines to ruthenium carbenes: *ortho*-(alkynyloxy)benzylamine cyclizations towards 1,3-benzoxazines. Angew Chem Int Ed 54:2724–2728
81. Torres Ò, Roglans A, Pla-Quintana A (2016) An enantioselective cascade cyclopropanation reaction catalyzed by rhodium(I): asymmetric synthesis of vinylcyclopropanes. Adv Synth Catal 358:3512–3516
82. Chen P-A, Setthakarn K, May JA (2017) A binaphthyl-based scaffold for a chiral dirhodium (II) biscarboxylate ligand with α-quaternary carbon centers. ACS Catal 7:6155–6161
83. Zhang C, Chang S, Qiu L, Xu X (2016) Chemodivergent synthesis of multi-substituted/fused pyrroles *via* copper-catalyzed carbene cascade reaction of propargyl α-iminodiazoacetates. Chem Commun 52:12470–12473

84. Dong K, Pei C, Zeng Q, Wei H, Doyle MP, Xu X (2018) Selective C(sp^3)-H bond insertion in carbene/alkyne metathesis reactions. Enantioselective construction of dihydroindoles. ACS Catal 8:9543–9549
85. Zheng Y, Bao M, Yao R, Qiu L, Xu X (2018) Palladium-catalyzed carbene/alkyne metathesis with enynones as carbene precursors: synthesis of fused polyheterocycles. Chem Commun 54: 350–353
86. Shen Z, Dong VM (2009) Benzofurans prepared by C-H bond functionalization with acylsilanes. Angew Chem Int Ed 48:784–786
87. Becker P, Pirwerdjan R, Bolm C (2015) Acylsilanes in iridium-catalyzed directed amidation reactions and formation of heterocycles via siloxycarbenes. Angew Chem Int Ed 54:15493–15496
88. Bunyamin A, Hua C, Polyzos A, Priebbenow DL (2022) Intramolecular photochemical [2+1]-cycloadditions of nucleophilic siloxy carbenes. Chem Sci 13:3273–3280
89. Funes-Ardoiz I, González J, Santamaría J, Sampedro D (2016) Understanding the mechanism of the divergent reactivity of non-heteroatom-stabilized chromium carbene complexes with furfural imines: formation of benzofurans and azetines. J Org Chem 81:1565–1570
90. Wang K, Ping Y, Chang T, Wang J (2017) Palladium-catalyzed [3+3] annulation of vinyl chromium(0) carbene complexes through carbene migratory insertion/Tsuji–Trost reaction. Angew Chem Int Ed 56:13140–13144
91. Peil S, González AG, Leutzsch M, Fürstner A (2022) C-H insertion *via* ruthenium catalyzed *gem*-hydrogenation of 1,3-enynes. J Am Chem Soc 144:4158–4167
92. Asako S, Kobashi T, Takai K (2018) Use of cyclopropane as C1 synthetic unit by directed retro-cyclopropanation with ethylene release. J Am Chem Soc 140:15425–15429
93. Asako S, Ishihara S, Hirata K, Takai K (2019) Deoxygenative insertion of carbonyl carbon into a C(sp^3)-H bond: synthesis of indolines and indoles. J Am Chem Soc 141:9832–9836
94. Zhang L, Demuynck BM, Paneque AN, Rutherford JE, Nagib DA (2022) Carbene reactivity from alkyl and aryl aldehydes. Science 377:649–654

An Overview of *N*-Heterocycle Syntheses Involving Nitrene Transfer Reactions

Ken Lee, Kyeongdeok Seo, Mahzad Dehghany, Yun Hu, Anh Trinh, and Jennifer M. Schomaker

Contents

K. Lee, K. Seo, M. Dehghany, Y. Hu, A. Trinh, and J. M. Schomaker (✉)
Department of Chemistry, University of Wisconsin, Madison, WI, USA
e-mail: schomakerj@chem.wisc.edu

Abstract The addition of either free or transition metal-supported nitrenes to unsaturated C–C bonds and heteroatom lone pairs, as well as intramolecular insertions of these reactive intermediates into $C(sp^2)$–H and $C(sp^3)$–H bonds, are efficient strategies for the formation of nitrogen-containing heterocycles. This chapter highlights early examples of thermally-induced formation and reactions to form *N*-heterocycles, as well as modern examples of transition metal-catalyzed nitrene transfer.

Keywords Amination · Amines · Asymmetric · Aziridine · Catalysis · Heterocycles · Nitrene transfer · Silver

1 Introduction

Nitrene transfer (NT) reactions are a popular strategy for the preparation of diverse heterocycles, including pyrroles, pyridines, indoles, pyrrolidines, piperidines, and rings containing multiple heteroatoms [1–16]. Many of these structural motifs are prevalent in agrochemicals, pharmaceuticals, and other useful building blocks. The goal of this chapter is to present diverse examples of heterocycle syntheses that involve nitrenes as key intermediates. In a broad sense, reactions of nitrenes leading to heterocycles can be subdivided into five general strategies (Scheme 1) [17, 18]: (1) a nitrene adds to a π-bond to furnish an aziridine or other 3-membered heterocycle; (2) the nitrene inserts into a X–H or C–H bond to form a heterocycle; (3) a NT reaction in tandem with a subsequent transformation that results in the ring expansion of an initial strained intermediate; (4) addition of the nitrene to lone pairs on various heteroatoms; and (5) cases where acyl nitrenes and related species serve as dipoles in cycloaddition reactions. This chapter presents curated and informative examples that engage both metal-free and metal-catalyzed NT reactions to furnish diverse heterocyclic scaffolds. Representative reactions are first divided according to the five categories in Scheme 1, then further classified by type of nitrene precursor. This chapter is not intended to be comprehensive, but rather highlights a selection of

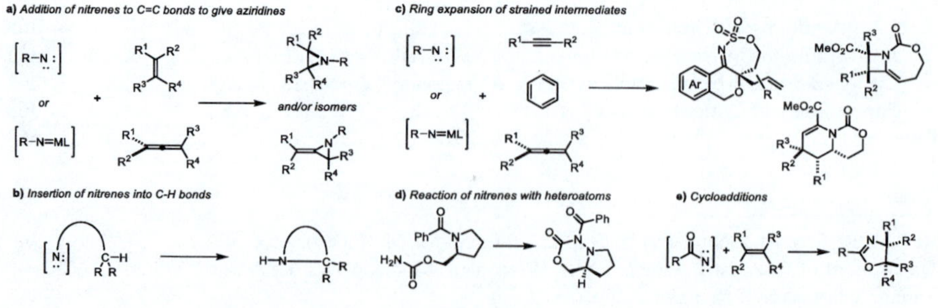

Scheme 1 Nitrene transfer reactions that furnish heterocycles covered in this chapter

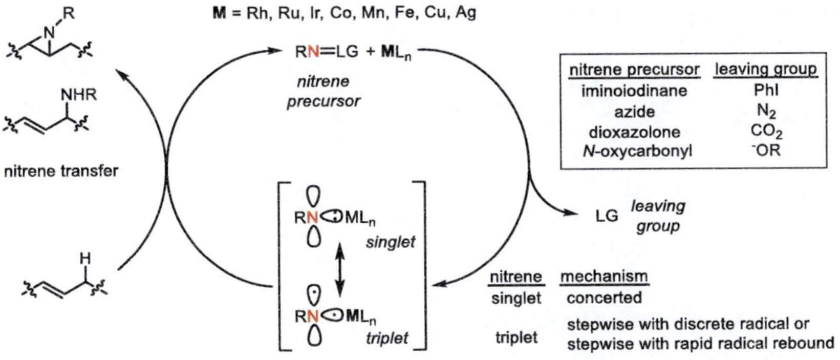

Scheme 2 General overview of the mechanism for metal-catalyzed nitrene transfer reactions

widely employed and synthetically useful transformations. Only NT promoted by chemocatalysts are included here; the reader is referred elsewhere for reviews on enzyme-mediated processes [19–24]. NT reactions proposed to be catalyzed by gold are also not covered in this chapter, as there are many extensive reviews on this topic [25]. The chapter concludes with a brief discussion of areas where opportunities and challenges still remain in the field.

General features of nitrenes and metal-catalyzed nitrene transfer (NT) reactions. Nitrenes are reactive species that contain a monovalent nitrogen atom and six valence electrons. These highly electrophilic, neutral intermediates can engage with a variety of π bonds, C–H bonds, and lone pairs on heteroatoms to form new C–N and X–N bonds. There is an extensive early literature describing reactions believed to involve free nitrenes, but modern synthetic methods for NT use transition metals to stabilize and exert control over the ultimate fate of the nitrene. The first suggestion of a metallonitrene intermediate was reported in 1967 by Kwart and Khan in the Cu(0)-catalyzed decomposition of benzosulfonyl azides en route to the formation of aziridines and vinyl amines [26]. Since this early example, NT catalyzed by diverse transition metals have been studied, with Rh, Ru, Ir, Cu, Fe, Co, Mn, and Ag among the most popular choices [1–16]. The general mechanism for metal-catalyzed NT (Scheme 2) involves oxidation of an appropriate nitrene precursor (or elimination of nitrogen gas from an azide) to give an oxidized nitrogen species that is transferred to a metal to furnish the metallonitrene. Depending on the electronic configuration of the nitrene (singlet or triplet), it engages the substrate in either a concerted or stepwise NT pathway.

Since Kwart and Khan's initial report of Cu-catalyzed NT, a number of transition metal complexes have been developed for similar transformations; some of the more common examples are illustrated in Fig. 1. Rhodium and ruthenium catalysts supported by tetracarboxylates and tetracarboxamidates are among the most commonly used, while inexpensive catalysts based on first-row transition metals are attractive and include cobalt, manganese, and iron supported by porphyrin- and phthalocyanine-type ligands are also effectively employed. Ruthenium and iridium-based catalysts capable of asymmetric NT are supported by a variety of ligands that

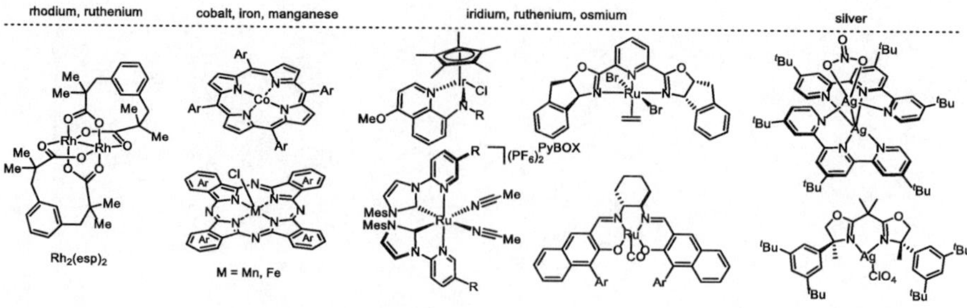

Fig. 1 Selected transition metal catalysts employed for nitrene transfer reactions

include diamines, PyBOX, and salen ligands. More recently, the diverse coordination geometries of silver(I) complexes supported by *N*-dentate ligands have been shown to offer predictable tunability in chemo-, site-, and enantioselective NT reactions.

2 Aziridines and Related Heterocycles via Nitrene Transfer

2.1 Background

The utility of aziridines in synthetic transformations has spurred the development of many strategies for the addition of free or metal-supported nitrenes to unsaturated carbon–carbon bonds present in alkenes and allenes. This section highlights some representative examples organized according to the type of nitrene precursor. Several comprehensive overviews of metal-catalyzed aziridination reactions via NT can be found in a host of review articles [3, 27–29].

2.2 Unactivated Azides as Nitrene Precursors

Azides are convenient precursors for NT, as only innocuous nitrogen gas is released to generate the reactive nitrene. Formation of a free or metal-stabilized nitrene can be promoted using heat, light, or transition metals, with metal-catalyzed methods providing the most control over the ultimate outcome of the reaction. Alkyl azides are discussed in Sect. 2.2, while activated azides that bear electron-withdrawing groups, exemplified by sulfonylazides, phosphoryl [30, 31], sulfonyl, and aryl azides [32], are discussed in Sect. 2.3.

The thermal extrusion of nitrogen from alkyl azides that subsequently engage in aziridination can be challenging to achieve, as multiple other pathways are often competitive, especially when the reaction is intermolecular. Hudlicky and coworkers demonstrated a synthetically useful intramolecular aziridination in the synthesis of

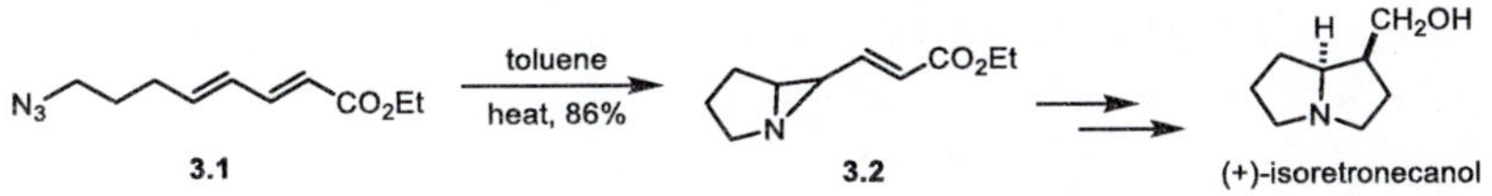

Scheme 3 Thermally-induced aziridination via nitrene transfer

Scheme 4 Selected scope of intermolecular aziridination using Betley's Fe-dipyrrinato catalyst

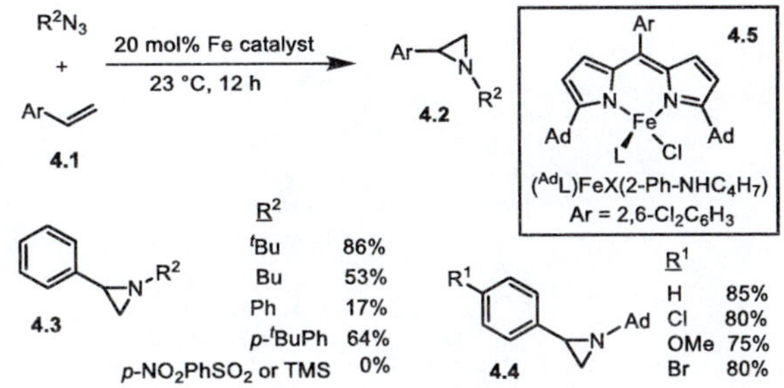

several pyrrolizidine alkaloids [33]. The use of the conjugated dienoate **3.1** was required to achieve high yield in the reaction to the vinyl aziridine **3.2** (Scheme 3).

Transition metal-catalyzed aziridinations using alkyl azide precursors typically employ 3d, first-row transition metals (Fe, Co, Cu) supported by *N*-dentate ligands, with metal-porphyrinates and metal-phthalocyanines among the most frequently used. The Betley group developed a new ligand class of "half-porphyrin" ligands for iron and found that exposure of terminal styrenes **4.1** to diverse alkyl and aryl azides in the presence of iron dipyrrinato catalysts of the form **4.5** could furnish aziridines **4.2**–**4.4** in moderate-to-good yields (Scheme 4) [34].

Iron catalysts supported by porphyrin ligands have also been reported to furnish aziridines using tosyl azide with a variety of terminal styrenes and alkyl-substituted alkenes [35, 36]. Despite the low cost and abundance of iron, one of the current major drawbacks using these types of catalysts is their limited substrate scope. The Che group demonstrated that an electron-withdrawing porphyrin-supported Fe catalyst can be used to perform aziridinations with electron-rich substituted styrenes and sulfonyl azides with good-to-excellent yields (Scheme 5) [37]. Aliphatic alkenes and aryl azides were also shown to perform aziridinations with good yields.

The Jenkins group designed a unique iron catalyst supported by an *N,B*-tetradentate ligand to promote the intermolecular aziridination of terminal and cyclic disubstituted alkenes using either alkyl or aryl azides as nitrene precursors (Table 1) [38, 39]. The aziridine products were generally obtained in good yields, provided the alkene substrate was employed in large excess.

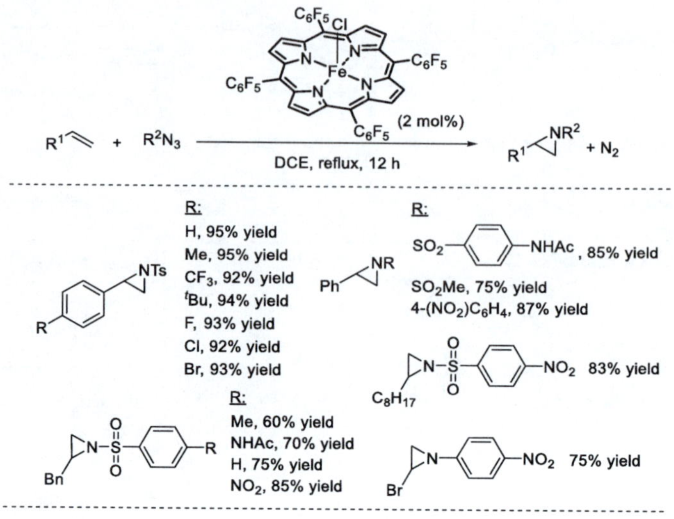

Scheme 5 Selected scope of intermolecular iron-catalyzed aziridination using a porphyrin-supported Fe catalyst

2.3 Activated Azides as Nitrene Precursors

Activated azides, including carbamoyl and sulfamoyl azides, have been used to great effect in enantioselective aziridination via NT. The Zhang group has reported some of the most powerful examples of chemo- and enantioselective aziridinations using activated azides as nitrene precursors [30–32]. A key design feature of their work is the use of open-shell Co complexes that facilitate and control reactivity through the transfer of radical character. These metalloradical catalysts (MRC) are based on well-defined Co(II) complexes **6.1–6.4** supported by D_2-symmetric chiral amidoporphyrins [Co(D_2-Por*)] (Scheme 6, top) and deliver aziridines in high yields and *ee* in most cases (Scheme 6, bottom). Phosphoryl [30, 31], sulfonyl, and aryl azides [32] can all be employed as precursors to furnish Co(II)-supported nitrogen-centered radicals that engage in the desired asymmetric aziridination. Computational and experimental studies suggest Co(III)–nitrene radicals react with alkenes in a stepwise manner, involving a radical addition–substitution pathway and aziridines in high *ee* [40, 41].

Ruthenium-based catalysts supported by bulky salen ligands have also been employed for asymmetric aziridination using sulfamoyl azides as precursors (Scheme 7) [42]. The transformation was most effective using terminal alkenes, which gave the best balance of yield and *ee*. *Cis*-substituted β-styrenes were also effective, giving stereoselective formation of the aziridines in excellent *ee*. However, even a minor increase in the steric bulk of the β-alkyl group on the styrene significantly decreased the yield and *ee* and promoted competing C–H insertion.

Table 1 Selected scope of Jenkin's Fe-catalyzed intermolecular aziridination using alkyl and aryl azide precursors

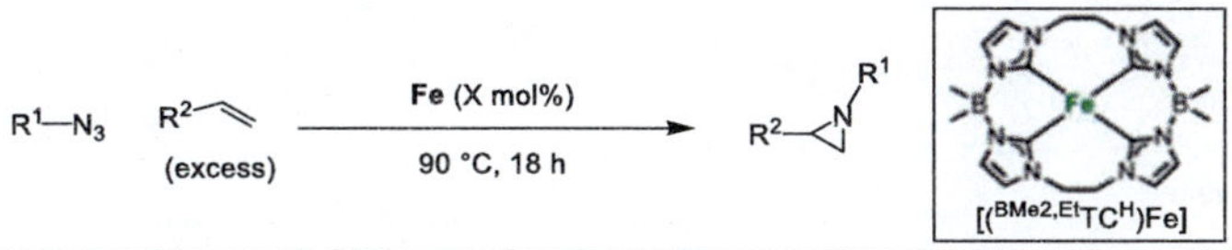

Entry	Alkene	Azide	Aziridine	Catalyst (%)	Yield[a] [%]
1	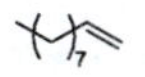	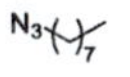		1	82
2	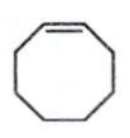		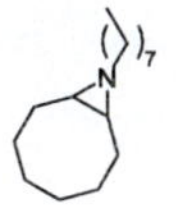	1	92
3	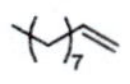	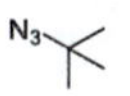	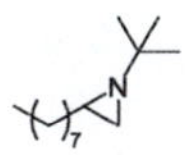	1	49
4	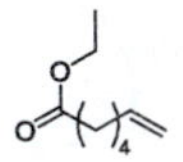		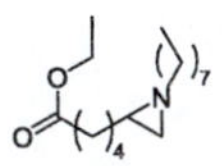	10	36
5	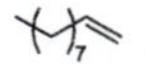	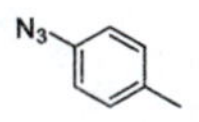	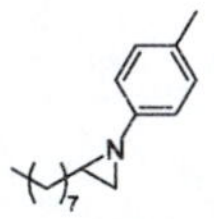	1	95
6			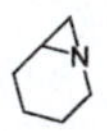	0.1	83

2.4 Carbamates as Nitrene Precursors

Carbamates serve as effective nitrene precursors in the presence of hypervalent iodine reagents as external oxidants. The earliest examples were catalyzed by dinuclear Rh(II) complexes supported by bridging tetracarboxylate ligands [43, 44], but a major challenge is the lack of predictable chemoselectivity, as C–H insertion is a competing pathway [45, 46]. Schomaker and coworkers showed that the diversity of coordination geometries available to Ag(I) catalysts using simple sp^2 *N*-dentate ligands could promote tunable, chemodivergent C–H amination or aziridination of allene **T2.1** by simple modifications to the ligand-Ag ratio [47]. Ratios of $<1{:}2$ of AgOTf:phen selected for the aziridine product **T2.2**, while ratios $>1{:}2$ reversed the chemoselectivity to favor the C−H insertion product **T2.3** (Table 2). A broad range of homoallenic and homoallylic carbamates (not shown)

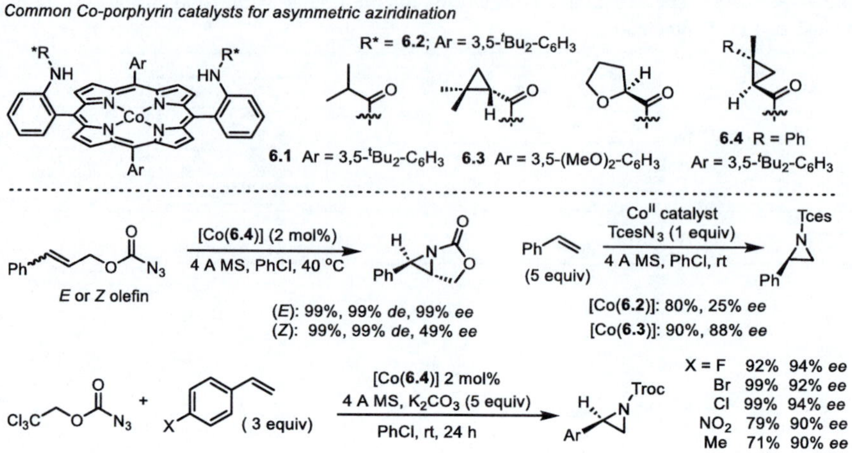

Scheme 6 Asymmetric aziridinations promoted by D_2-symmetric chiral amidoporphyrins [Co(D_2-Por*)]

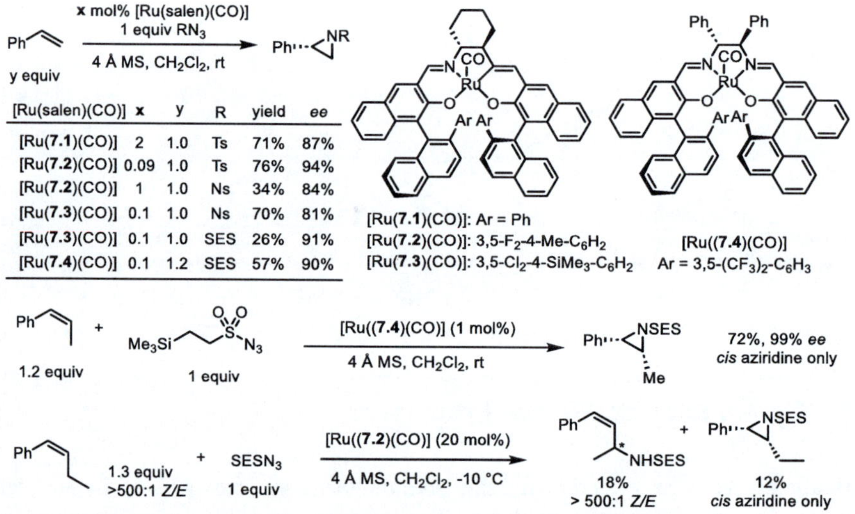

[Ru(salen)(CO)]	x	y	R	yield	ee
[Ru(7.1)(CO)]	2	1.0	Ts	71%	87%
[Ru(7.2)(CO)]	0.09	1.0	Ts	76%	94%
[Ru(7.2)(CO)]	1	1.0	Ns	34%	84%
[Ru(7.3)(CO)]	0.1	1.0	Ns	70%	81%
[Ru(7.3)(CO)]	0.1	1.0	SES	26%	91%
[Ru(7.4)(CO)]	0.1	1.2	SES	57%	90%

Scheme 7 Asymmetric aziridinations promoted by salen-supported Ru catalysts

successfully underwent chemoselective aziridination to furnish methyleneaziridines (MA) and bicyclic aziridines, respectively.

Schomaker and coworkers have described examples of intramolecular asymmetric transition metal-catalyzed reactions of homoallylic and homoallenic carbamates **T3.1** (Table 3) to give bicyclic aziridines of the form **T3.2** [48]. A silver salt supported by an enantioenriched (*S*,*S*)-^{t}BuBOX ligand provided good chemo- and enantioselectivity for the aziridination of diverse 1,2-substituted and

Table 2 Effect of the metal:ligand ratio on the chemoselectivity of intramolecular nitrene transfer with carbamates

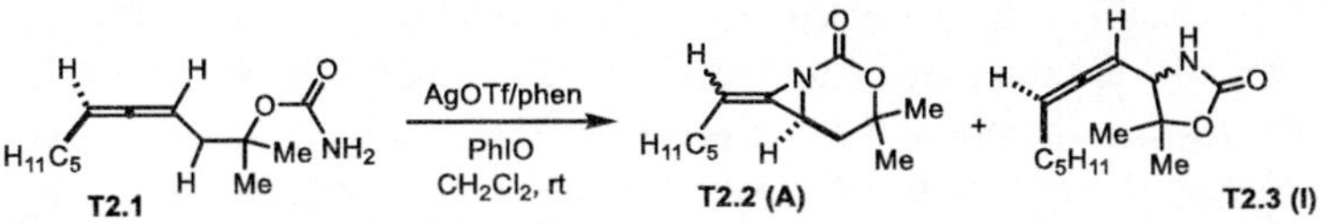

entry	Ag : phen	aziridination	C–H insertion	A : I
1	1 : 0.5	60%	12%	5 : 1
2	1 : 1	75%	13%	5.8 : 1
3	1 : 1.25	80%	13%	6.2 : 1
4	1 : 1.5	70%	12%	5.8 : 1
5	1 : 2	18%	72%	1 : 4
6	1 : 3	2%	76%	1 : 38

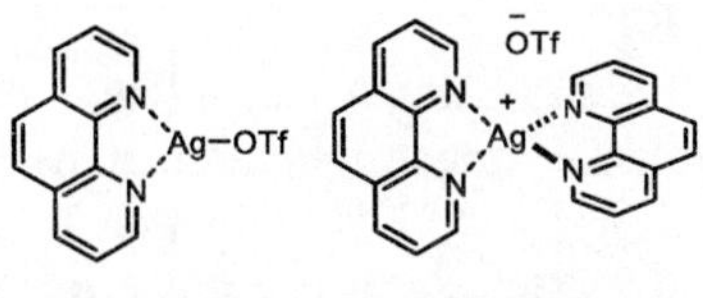

aziridination catalyst *C–H insertion catalyst*

Table 3 Selected scope of silver-catalyzed asymmetric aziridination

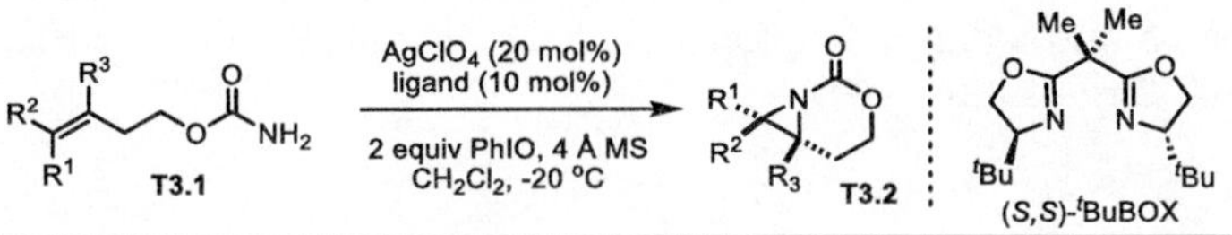

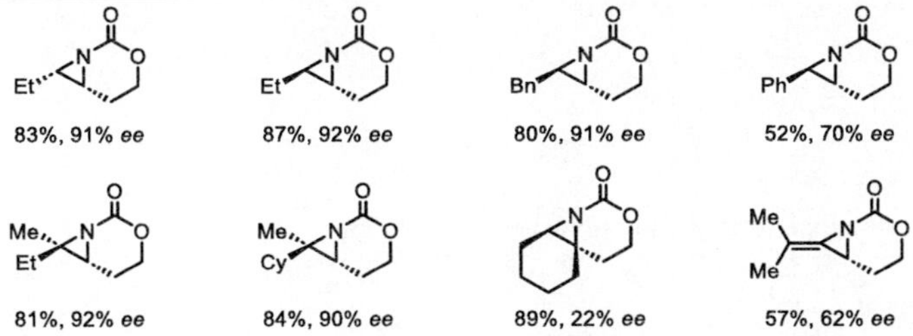

1,1′,2-substituted homoallylic carbamates (Table 3). Substrates that did not perform well with this chemistry include styrenes and 1,2,2′-substituted alkenes. Interestingly, an achiral allene could be transformed into an enantioenriched bicyclic methyleneaziridine in moderate *ee*.

2.5 Sulfamates as Nitrene Precursors

Sulfamates are arguably the most widely utilized nitrene precursors for alkene aziridination. Due to the number of reviews that extensively cover this chemistry [5, 6, 9–14], only a few chemo- and enantioselective examples are highlighted here. Dauban and coworkers showed that Cu salts supported by an (*S,S*)-^{t}BuBOX ligand successfully carried out asymmetric intramolecular aziridination of sulfamates (Table 4) [49].

Somewhat surprisingly, there have been only a few reported examples of intermolecular asymmetric aziridinations that utilize sulfamates, partly due to the propensity for allylic C–H amination to compete with the desired pathway. The Dauban and Darses groups recently reported [50] a highly efficient asymmetric aziridination of alkenes and aromatic sulfamates catalyzed by a chiral C4–symmetric dirhodium(II) tetracarboxylate (Table 5). Highlights of this method include excellent chemoselectivity for aziridination and *ee* of up to 99%, as well as low catalyst loadings (0.1–1 mol%), scalability to multi-gram quantities and broad scope (mono-, di-, and trisubstituted alkenes). Computational studies suggested a two-spin-state mechanism that invokes the formation of an enantiodiscriminating triplet transition state to rationalize the high *ee*. Key interactions that dictate *ee* occur in the hydrophobic pocket of the chiral rhodium complex and engage the phthaloyl rings, the sulfamoyl nitrene, and the alkene.

Table 4 Selected scope of copper-catalyzed asymmetric aziridination

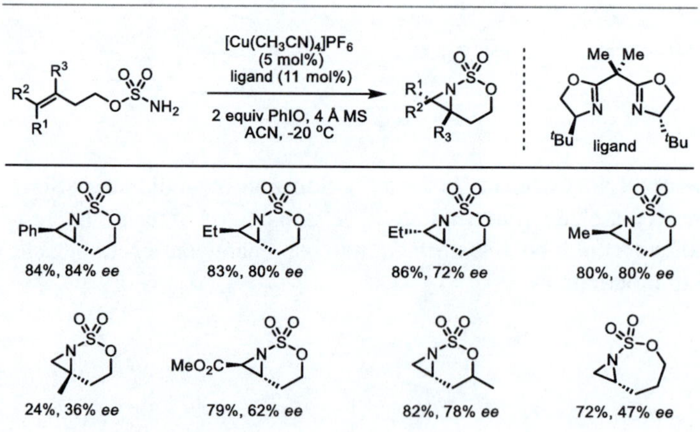

Table 5 Selected scope of Rh-catalyzed intermolecular asymmetric aziridination of sulfamates

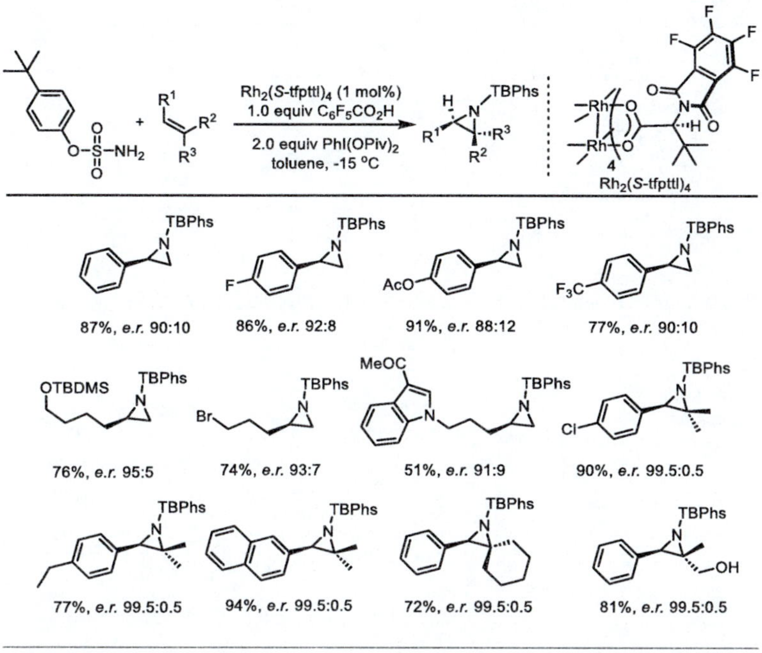

2.6 *Pre-Oxidized Nitrene Precursors with N–O Bonds*

Aziridinations that utilize carbamates and sulfamates as nitrene precursors require an external oxidant to activate the nitrogen source, with hypervalent iodine reagents as the most common oxidants. The use of an external oxidant is not required if the nitrene precursor is "pre-oxidized." Typically, this is achieved by the conversion of one of the N–H bonds of the carbamate or sulfamate to a N–O bond. For example, the Kürti, Falck, and Ess groups reported a convenient aziridination method that furnishes the free N-H aziridines as the end-products, eliminating the need for subsequent deprotection (Scheme 8) [51]. The initial amine source was *O*-(2,4-dinitrophenyl)hydroxylamine (DHP) to convert a range of mono-, di-, tri-, and tetrasubstituted olefins to NH and *N*-methyl aziridines.

The Kürti lab later addressed limitations in the initial method that included the need for stoichiometric DHP, significant by-product formation, and difficulty in purifying the products. DHP was replaced with hydroxylamine-*O*-sulfonic acid (HOSA), although it reacted sluggishly under the initial reaction conditions. Further optimization furnished substituted N–H aziridines in good-to-excellent yields (Scheme 9a) [52]. The reaction is stereospecific, yielding single enantiomers in many cases. The mechanism (Scheme 9b) is proposed to involve formation of a

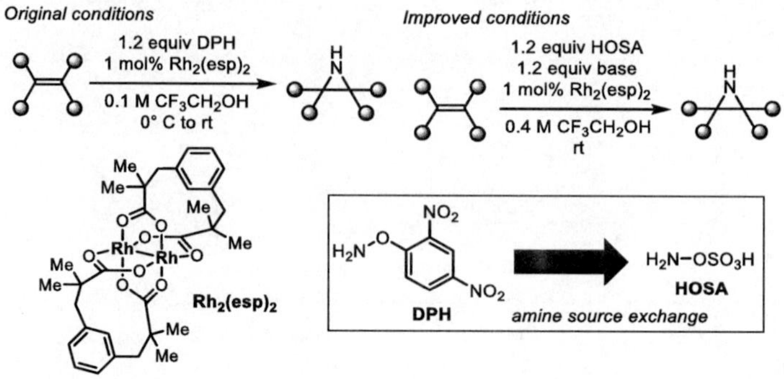

Scheme 8 Initial and improved conditions for rhodium-catalyzed synthesis of N-H aziridines

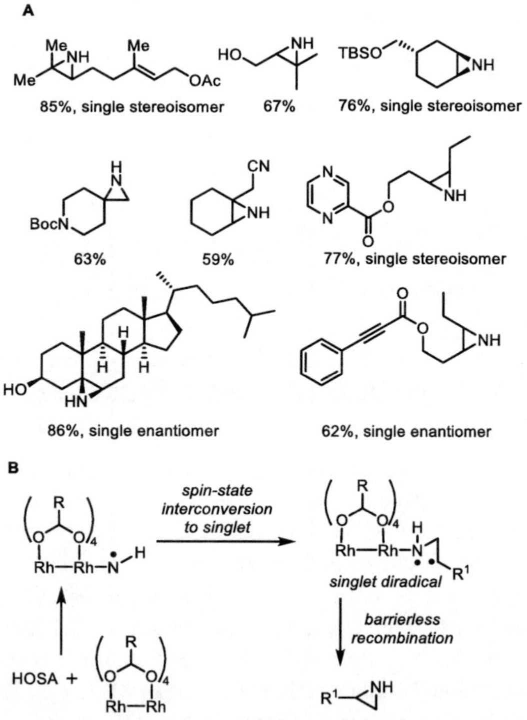

Scheme 9 (**a**) Selected examples of reaction scope. (**b**) Proposed mechanism

Rh-nitrenoid via amino group coordination and loss of sulfate anion, prior to olefin capture. DFT (Density Functional Theory) calculations suggest a triplet transition state that then furnishes to a singlet diradical intermediate, followed by barrierless recombination to yield the aziridine products.

Another valuable approach that employs nitrene precursors containing an N–O bond to avoid the need for an external oxidant is exemplified by Lebel's use of *N*-tosyloxycarbamates in either the presence of a Rh or Cu-based catalyst [53–

Table 6 Selected scope of aziridines prepared from allylic *N*-tosyloxycarbamates

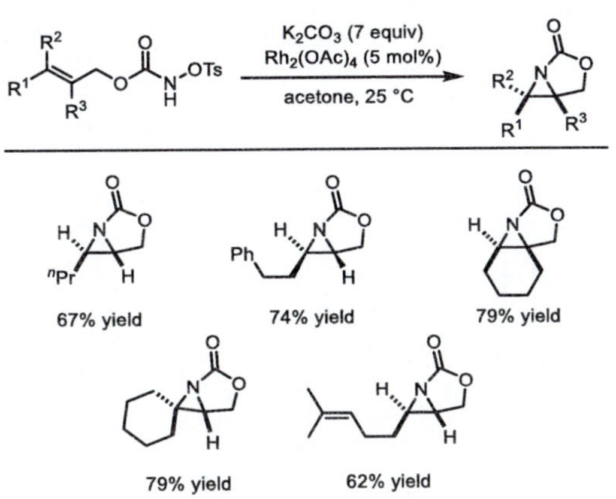

56]. Table 6 highlights a selection of substrates that successfully underwent Rh-catalyzed intramolecular aziridination to give bicyclic aziridines. The aziridination proved to be stereospecific, although the chemoselectivity was decreased if the tether between the nitrogen and the alkene was extended due to competing C–H insertion.

2.7 Unactivated Amines as Nitrene Precursors

Ideally, simple amines could serve as convenient nitrogen sources for aziridination via NT. In this respect, Driver and coworkers have recently demonstrated a Rh_2(II)-catalyzed intermolecular *N*-aryl aziridination of olefins using non-activated N-atom precursors [57]. The chemistry utilizes anilines as nitrene precursors in the presence of a hypervalent iodine as the stoichiometric oxidant. The iminoiodinane intermediate transfers the *N*-aryl nitrene fragment to the Rh catalyst, followed by aziridination. Di-, tri-, and tetrasubstituted cyclic and acyclic alkenes are all well tolerated in the reaction, which is highly stereospecific, as well as chemo- and diastereoselective (Table 7).

Table 7 Selected scope for Rh_2(II)-catalyzed intermolecular *N*-aryl aziridination of olefins with anilines

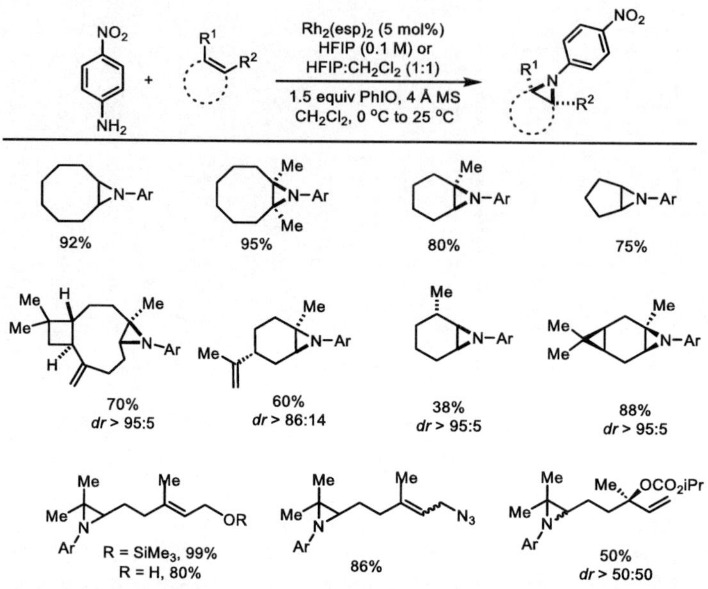

3 Heterocycles from Intramolecular C–H Amination via Nitrene Transfer

3.1 Background

In addition to the addition reactions across multiple C–X bonds, another powerful reaction of nitrenes involves their insertions into diverse C(sp^2)–H and C(sp^3) C–H bonds to form *N*-heterocycles. C–H bond functionalization is an efficient way to elaborate simple, readily available precursors into synthetically useful building blocks. In particular, there has been significant recent interest in the development of highly chemo-, site-, and enantioselective methods of nitrogen insertion into C–H bonds for the synthesis of amines that are prevalent in natural products and pharmaceuticals. A number of nitrene precursors have been employed, including azides, carbamates, sulfamates, and ureas, to prepare an array of mainly five- to seven-member *N*-heterocycles. A selection of these examples is described in this section.

3.2 Unactivated Azides as Nitrene Precursors

In addition to engaging in aziridinations of unsaturated carbon–carbon bonds, free and metal-supported nitrenes formed from azides readily insert into C(sp^2)–H and C(sp^3)–H bonds to form a variety of *N*-heterocycles, including indoles, fused indoles, carbazoles, pyrrolidines, piperidines and sultams, piperidines, pyrrolidines, cyclic sulfamides, cyclic sulfamates, cyclic carbamates, and cyclic ureas. Selected examples are highlighted in this section.

Indoles and related heterocycles from azides. Indoles and larger heterocycles that contain an indole subunit can be formed from azides invoking either free or metal-supported nitrene intermediates. For example, Smith and Brown reported in 1951 [58] that heating a 2-azidobiphenyl **10.1** (Scheme 10a) furnished the carbazole **10.4** in 76% yield, presumably through loss of N_2 to form the nitrene **10.2**, followed by cyclization. In the presence of an amine nucleophile, an alternate pathway was invoked that proceeds through intermediate formation of an azirine **10.5** from **10.2**, which is followed by ring-opening to **10.6**. Similar examples of proposed nitrene insertions into the C(sp^2)–H bonds of alkenes have been reported; *ortho*-vinyl substituted aryl azides of the form **10.7** furnish products of the form **10.8** in good yields (Scheme 10b) [59–62].

The Hemetsberger–Knittel synthesis is a popular method to access indoles from vinyl azides. Moody and Rees reported that heating a 3-aryl substituted 2-azidopropenoic acid ester furnished an indole-2-carboxylic acid ester; while the mechanism has not been determined, azirines have been isolated in reactions run at lower temperatures (Scheme 11A) [63, 64]. A large number of indoles have been prepared using this strategy. For example, Smolinsky and Pryde reported the formation of 3-phenylindoles via the pyrolysis of 2,2-diphenyl-1-azidoethenes in boiling xylene (Scheme 11b) [65]. The use of ethanol as the solvent decreased the yield of the 3-phenylindole and formed a dimerization by-product, 2,2,5,5-tetraphenyldihydropyrazine. Isomura and coworkers demonstrated that either the *E*- or *Z*-isomer of an azide precursor successfully formed the indole products (Scheme 11c), an observation that pointed to a rapid equilibrium between a vinyl nitrene species and an azirine intermediate [66]. Formation of the indole is proposed to occur via insertion of the nitrene into the aromatic ring to deliver the 7a*H*-indole, which undergoes a final [1,5]-sigmatropic rearrangement.

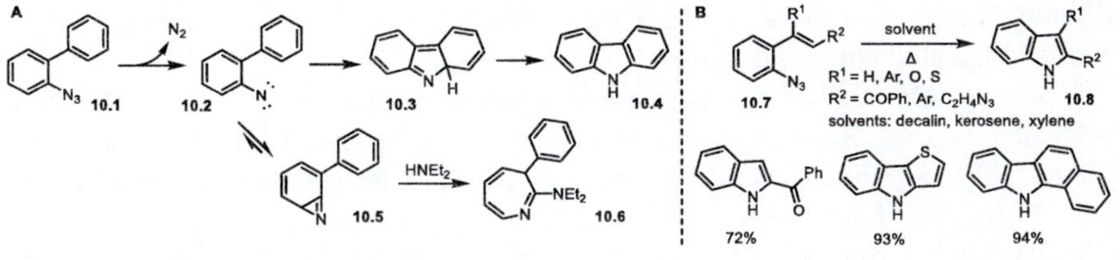

Scheme 10 Synthesis of carbazoles, indoles, and related *N*-heterocycles via thermal decomposition of aryl azides to free nitrenes

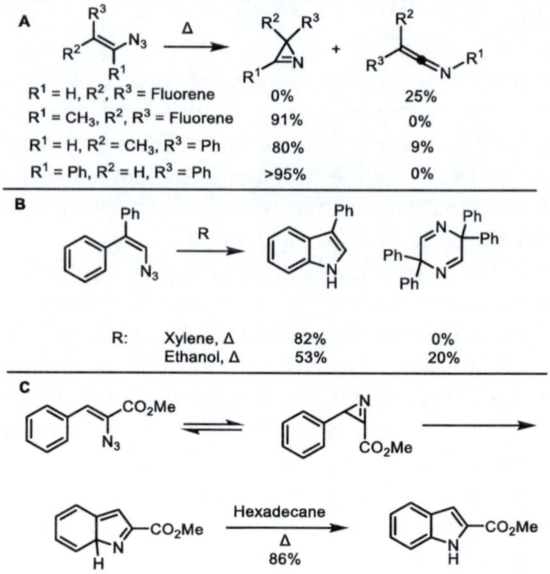

Scheme 11 Hemetsberger–Knittel reaction to prepare indoles

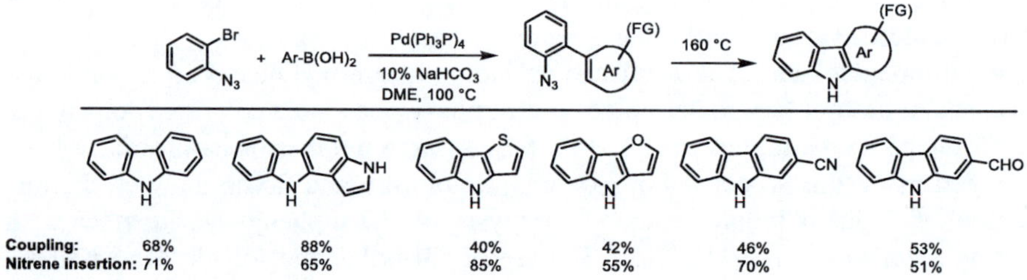

Scheme 12 Synthesis of fused indole heterocycles via a Pd-catalyzed cross-coupling/nitrene insertion

A preparation of fused indole-containing heterocycles was reported using a Pd-catalyzed cross-coupling between azido-2-bromobenzene and diverse arylboronic acids to generate an azide-bearing biphenyl precursor. Heating to 160 °C initiated loss of nitrogen gas to generate the free nitrene, which then inserts into an aryl C–H bond (Scheme 12) to deliver the products [67]. Other heterocycles, including thiophene and furan motifs, were also tolerated, as were functional groups that included aldehydes and nitriles.

More recently, Driver and coworkers reported the preparation of indoles and carbazoles via a Rh-catalyzed decomposition of aryl azides (Scheme 13a) [68, 69]. The addition of olefins or arenes proceeds with excellent yields of up to 99% employing highly electrophilic rhodium(II) perfluorobutyrate or rhodium (II) octanoate as the catalyst. The chemistry is compatible with either electron-donating or electron-withdrawing substituents on either aryl ring of the precursor. The mechanism (Scheme 13b) is proposed to proceed through the reaction of the

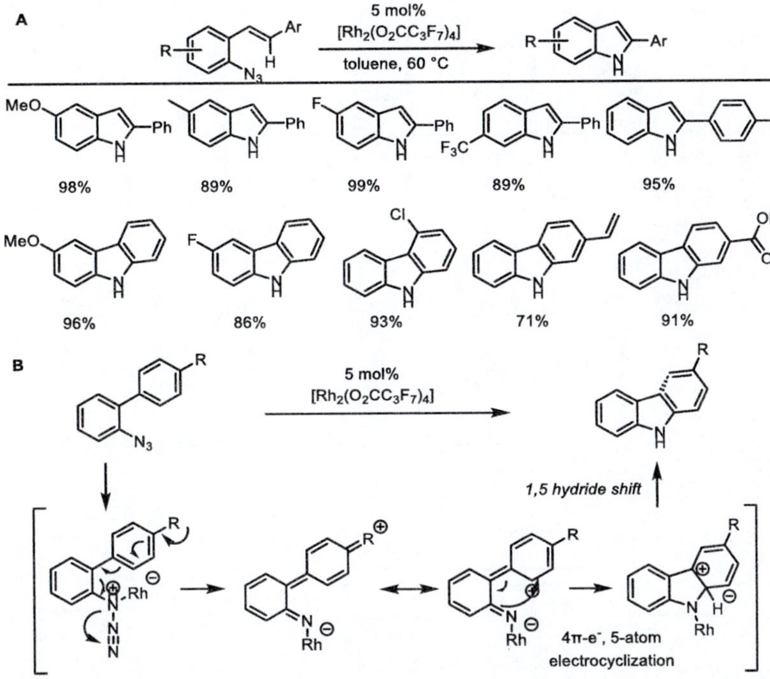

Scheme 13 Driver's synthesis of indoles via Rh-catalyzed decomposition and insertion into vinyl C–H bonds

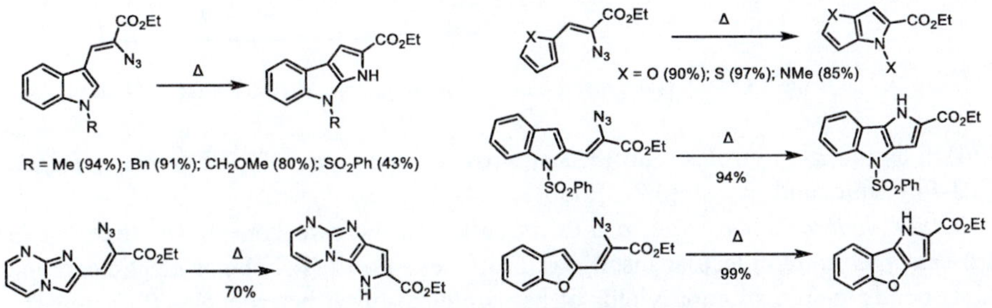

Scheme 14 Synthesis of fused *N*-heterocycles via a Hemetsberger–Knittel-type approach

azide with the Rh catalyst to give an intermediate arylnitrenium ion, which then undergoes a 4π-electron electrocyclization to form a new C–N bond. A subsequent 1,5-hydride shift furnishes the final heterocyclic product.

Annulation of pyrroles, indoles, and other heterocycles with azides. The sp^2 C–H bonds of pyrroles, indoles, and other *N*-heterocycles have been reported to undergo annulation with azides under thermal conditions. Scheme 14 highlights some of the diverse polyfused aromatic systems that can be obtained in this fashion [70], including pyrrolo[2,3-*b*]indoles [71], pyrrolo[3,2-*d*]thiazole [72], furo[2,3-*b*]pyrrole

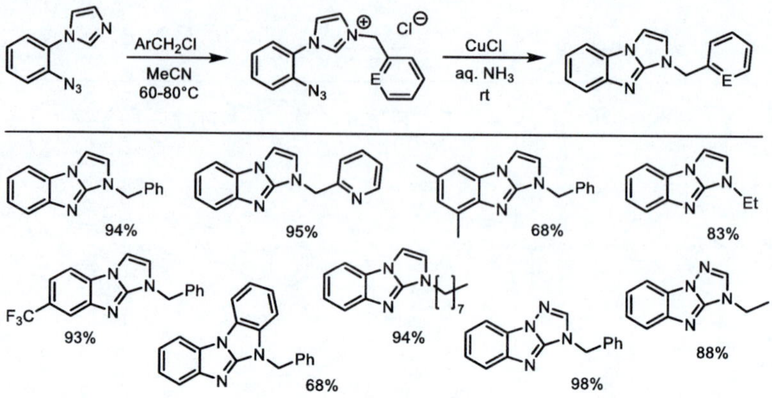

Scheme 15 Catalytic intramolecular nitrene insertion into Cu(I)–NHC bond to give fused nitrogen heterocycles

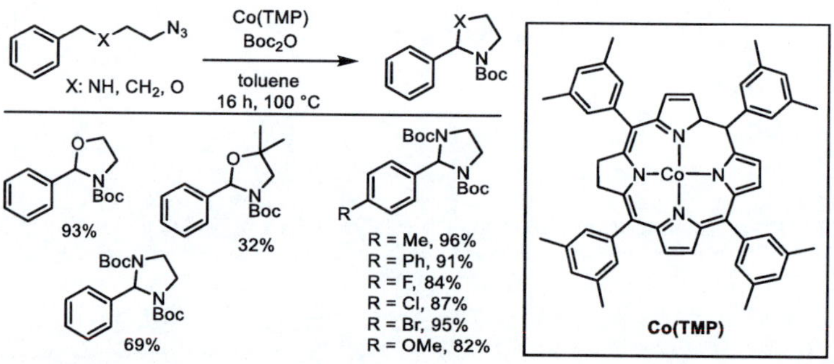

Scheme 16 de Bruin's Co-catalyzed insertion of alkyl azides into activated C(sp^3)–H bonds

[73] thieno[3,2-*b*]pyrroles, furo[3,2-*b*]pyrroles, and pyrrolo[3,2-*b*]pyrrole, thieno [2,3-*b*]pyrrole, and pyrrolo[3,2-*b*]indoles [74–78].

Other heterocycles from azides. Gautier and coworkers have reported a Cu-catalyzed intramolecular insertion of nitrenes into Cu(I)-N-heterocyclic carbene C–H bonds to furnish good yields of benzimidazo-fused heterocycles (Scheme 15) [79, 80]. The precursor *N*-(2-azidophenyl)azolium salts are readily prepared and undergo an initial reaction with copper(I) to form a Cu–NHC complex. The Cu is then proposed to coordinate to the internal nitrogen of the azide, followed by loss of nitrogen to yield the copper carbene, which then inserts into the imidazole C(sp^2)–H bond to form the final fused ring in good-to-excellent yields.

Saturated nitrogenated heterocycles from azides. Transition metal-supported nitrenes formed from azides can also insert into C(sp^3)–H bonds to form saturated nitrogen heterocycles. For example, the de Bruin group reported that a cobalt catalyst supported by a porphyrin ligand promoted a ring-closing C–H amination of aliphatic azides to give saturated heterocycles (Scheme 16) [81]. The reaction gave the best results when the C–H bonds were activated and had low bond

Table 8 Selected scope of intramolecular iron-catalyzed C–H bond amination from alkyl azides

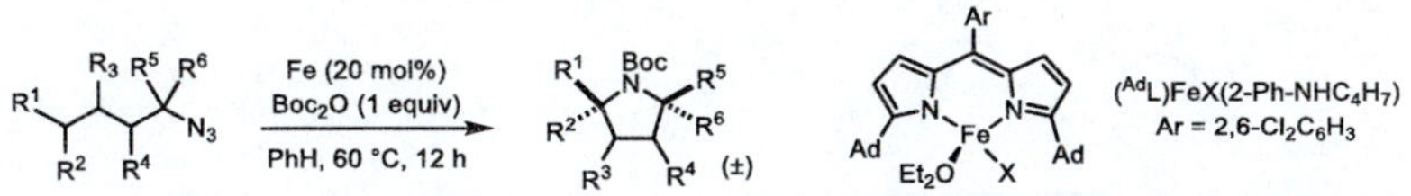

entry	substrate	product	yield[a] (%)	*dr*
1	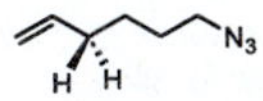	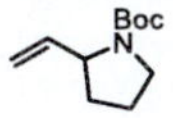	72[c]	–
2	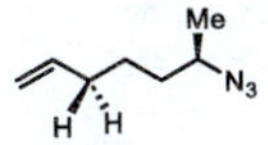	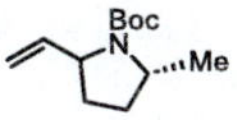	84[c]	1.1:1[c]
3	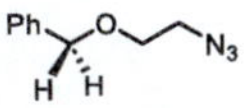	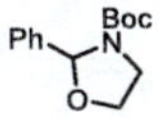	47[c]	–
4	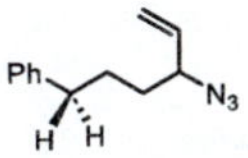	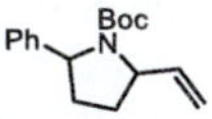	72[c] 72[d]	3.9:1[c] >20:1[d]
5	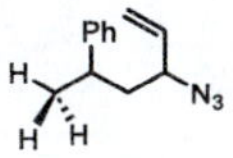	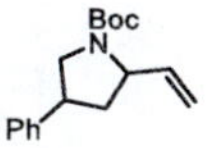	66[c] 72[d]	1.5:1[c] 2:1[d]
6	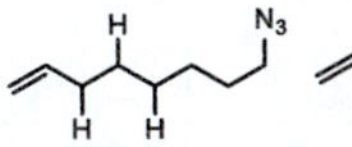	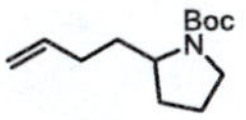	82[c]	–
7	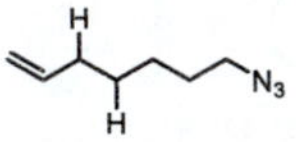	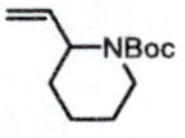	45[c]	–

dissociation energies. The product amine required in situ protection with a Boc group in order to promote efficient catalyst turnover.

The Betley group has reported intramolecular C(sp^3)–H aminations of alkyl azides using iron complexes supported by dipyrrinato ligands that are modeled on a half-porphyrin structure (Table 8). Pyrrolidines, piperidines, and other *N*-heterocycles are formed in moderate-to-good yields (Table 8). Five-membered rings are typically favored (entries 1–6), unless there is an option to insert the metal nitrene into an activated (allylic, benzylic) C–H bond to form a six-membered ring (entry 7) [82, 83].

While metal-supported nitrenes formed from azides typically engage in alkene aziridination or insertion into a C–H bond, some unusual modes of reactivity to form heterocycles have also been reported. One such example forms bicyclic lactams of the form **17.2** from symmetric 1,3-diketones **17.1** (Scheme 17). The mechanism is shown with a six-membered ring 1,3-diketone and is proposed to occur via initial coordination of both the azide and one of the carbonyl groups to the Cu catalyst. Loss of nitrogen forms a Cu-supported nitrene, which then engages the carbonyl to form an intermediate oxaziridine. A 1,2-alkyl shift completes the sequence to form enantioenriched bicyclic lactams [84].

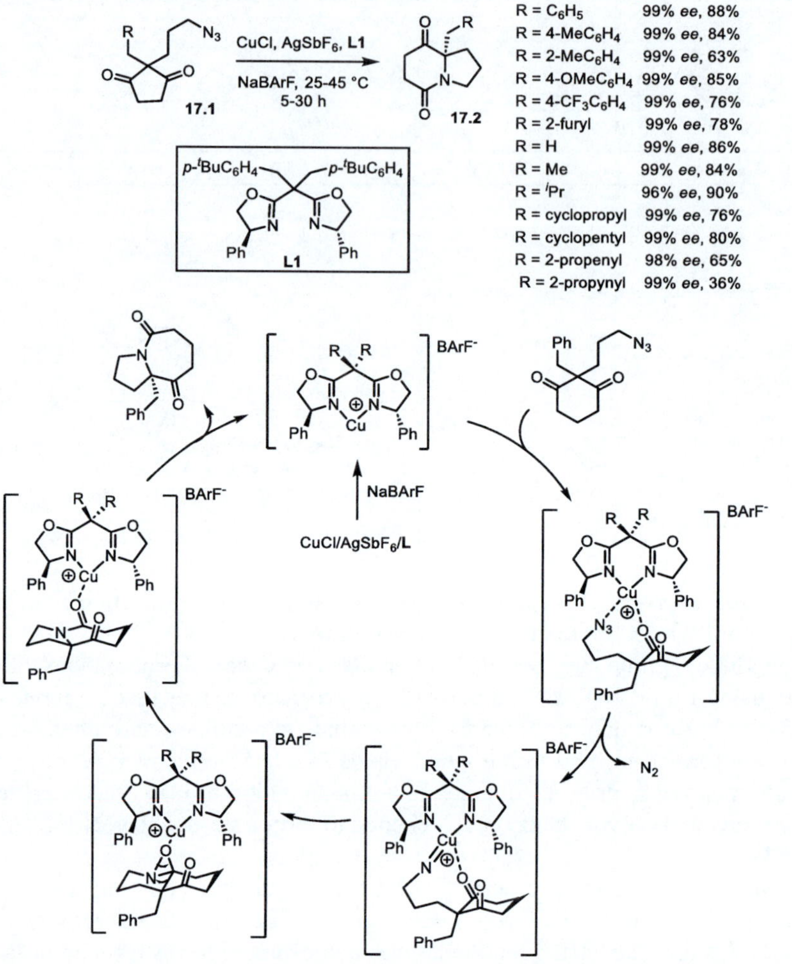

Scheme 17 Lactams from reaction of azides with carbonyl groups

3.3 *Sulfonyl Azides and Other Activated Azides as Nitrene Precursors*

Zhang and coworkers have reported a class of D_2-symmetric chiral cobalt–porphyrin complexes (selected examples in Fig. 2) that can control both the site- and enantioselectivity of carbene and nitrene transfer reactions [85–88]. In NT the radical nature of Co(II) effectively forms metal–nitrene intermediates from organic azides via homolytic activation and elimination of dinitrogen. The Co(II)-based metalloradical complex catalyzes a stepwise C–H amination via an initial H-atom abstraction (HAA) by the Co-supported nitrene radical, followed by substitution between the resultant alkyl radical and the Co(III)–amido complex to form the new

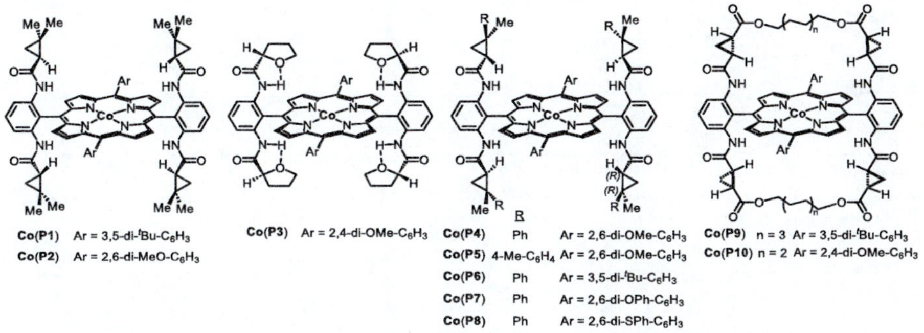

Fig. 2 Chiral cobalt–porphyrin catalysts for asymmetric C–H amination via nitrene transfer

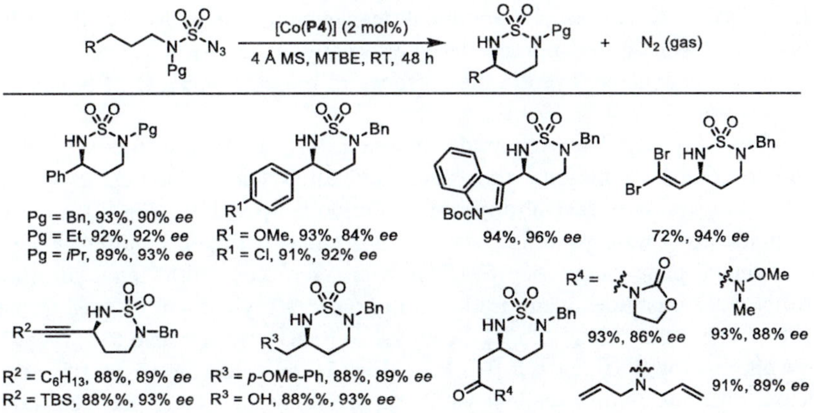

Scheme 18 Enantioselective intramolecular C–H amination with [Co(Por*)]

C–N bond. These systems are very flexible in the mode of asymmetric induction, depending on which elementary step proves to be the enantio-determining one. Another advantage is the high tunability of the steric and electronic properties of the chiral ligands without interfering with the primary coordination sphere of the porphyrin ligand.

Cobalt-catalyzed enantioselective C–H aminations of sulfamoyl azides (Scheme 18) [85] proceeds via intramolecular, radical-based amidation to furnish 6-membered sulfamides via a 1,6-C–H activation process. Both the chiral amide arms and the achiral aryl substituents of the D_2-symmetric porphyrin ligand exert a substantial influence on reactivity and enantioselectivity. Bulky substituents on the cyclopropane units of the optimal catalyst [Co(**P4**)] (Fig. 2) furnished the products in high yield and good *ee* under mild conditions (Scheme 18). Both electron-rich and electron-poor aryl groups were tolerated. In addition to amination of benzylic C–H bonds, intramolecular functionalizations of allylic and propargylic C–H bonds were also achieved in high chemo- and enantioselectivity. The Co-based catalysts were also highly selective for the amination of non-activated and electron-deficient C–H

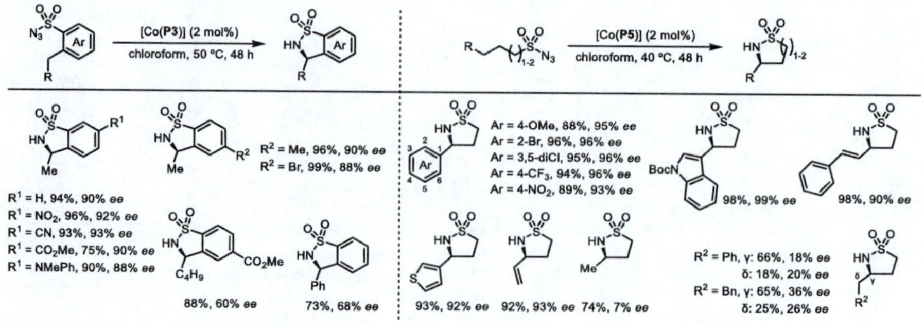

Scheme 19 Enantioselective intramolecular C–H amination with [Co(Por*)]

bonds adjacent to carbonyl functionalities, which are typically difficult to functionalize with other nitrene transfer catalysts. A strong preference for 1,6-amination was noted, even in the presence of more activated C–H bonds at the 5- or 7-positions.

The Wojtas group later reported asymmetric 1,5-C–H aminations to furnish 5-member sultams employing aryl- and alkylsulfonyl azide precursors (Scheme 19) [89]. Reaction with arylsulfonyl azides employed [Co(**P3**)] (Fig. 2), where the chiral amide arms were rigidified via intramolecular hydrogen bonding, when the initial cyclopropane rings of [Co(**P1**)] were replaced with tetrahydrofurans. Benzofused sulfonamides were produced in up to 99% yield and 93% *ee* (Scheme 19, left). Neither the parent [Co(**P1**)] catalyst or [Co(**P3**)] were able to successfully engage alkylsulfonyl azides, but [Co(**P5**)] (Fig. 2) delivered the desired 5-member cyclic sulfonamides with excellent yield and *ee*. When the linker between the aryl and azide groups was extended, the 5-member rings were the major regioisomers, despite the potential for competitive 1,6- and 1,7-pathways, although the yields and *ee* were decreased (Scheme 19, right).

One of the challenges in asymmetric nitrene transfer is the stereoselective amination of chiral, racemic tertiary C(sp^3)–H bonds, since the original stereochemical information must be ablated and reset by the catalyst. Most NTs proceed through either a concerted mechanism that retains stereochemistry, or occur in a stepwise fashion via HAA followed by a barrierless radical recombination [90]. Zhang and coworkers demonstrated that their [Co(Por)] scaffolds could promote a temperature-dependent racemization and enantioconvergence of a Co(III)–alkyl radical intermediate. HAA of the tertiary C–H of the starting material results in the formation of a long-lived, planar carbon radical (Scheme 20) [91]. In the case of racemic, chiral alkyl sulfamoyl azide **20.1**, treatment with [Co(**P4**)] favored formation of one enantiomer of **20.2** in up to 86% ee. The authors probed the nature of asymmetric induction by using chirality-matching KIE (kinetic isotope effect) experiments. They found a slight preference for H-atom abstraction of one substrate configuration over the other for most of the screened chiral catalysts.

Another example of asymmetric nitrene transfer using an activated azide was reported by the Meggers group in 2020 (Table 9). Sulfonyl azides [92] and

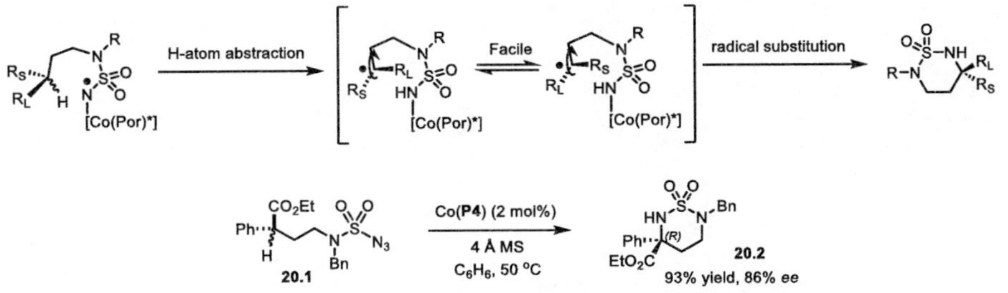

Scheme 20 Enantioconvergent, intramolecular C–H amination with [Co(Por*)]

Table 9 Selected scope for Ru-catalyzed enantioselective C–H amination of sulfonyl and sulfamoyl azides

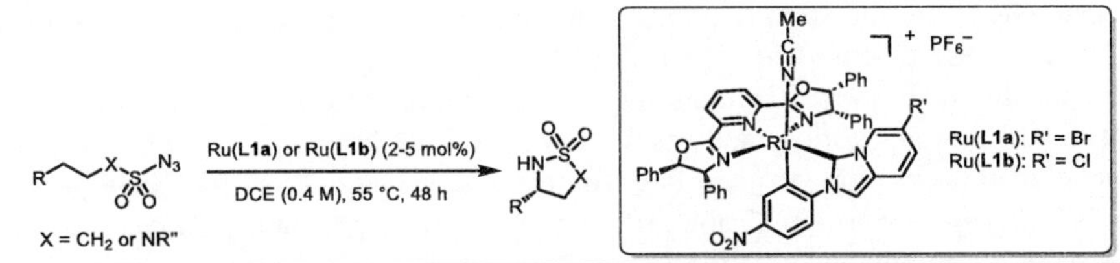

Entry	Product	Catalyst	Yield (%)	*ee* (%)	Entry	Product	Catalyst	Yield (%)	*ee* (%)
1	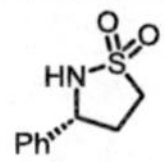	Ru(**L1a**)	99	97	5	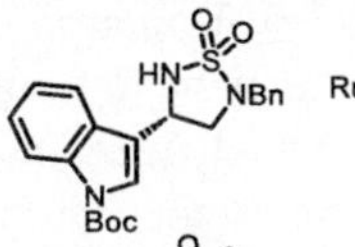	Ru(**L1b**)	99	95
2	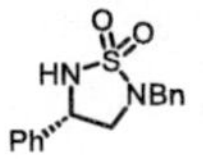	Ru(**L1b**)	93	95	6	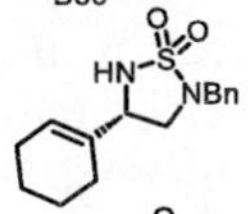	Ru(**L1b**)	76	77
3	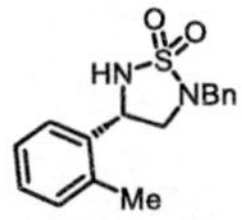	Ru(**L1b**)	82	95	7	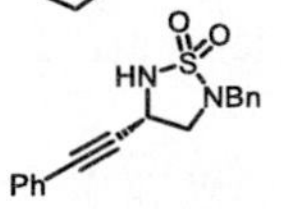	Ru(**L1b**)	98	68
4	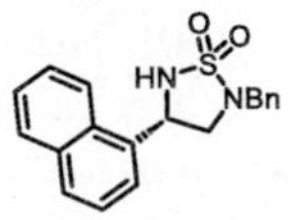	Ru(**L1b**)	76	90	8	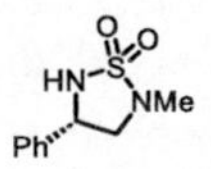	Ru(**L1b**)	87	94

sulfamoyl azides could be employed as nitrene precursors in the presence of chiral Ru catalysts Ru(**L1a**) and Ru(**L1b**) to give the heterocyclic products in excellent yields and *ee* [93]. The "arms" of the PyBOX ligand could be altered to control the reactivity, while the primary role of the NHC-type ligand was to control the electronics at the Ru center.

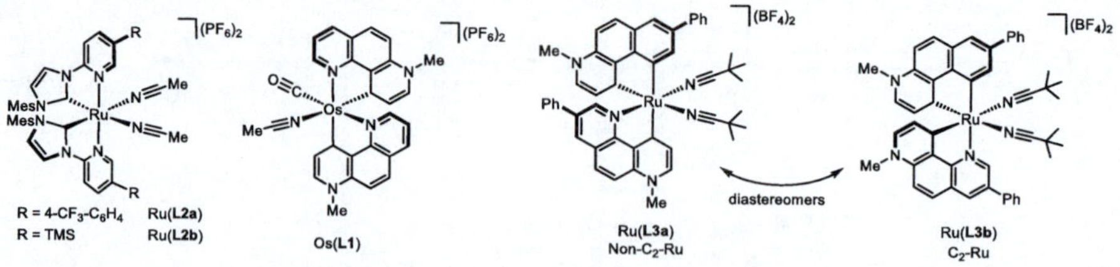

Fig. 3 Chiral-at-metal ruthenium and osmium metal complexes

Meggers and coworkers have also employed "chiral-at-metal" Ru and Os catalysts (Fig. 3) in asymmetric NT reactions to great effect. Even though the *N*-heterocyclic carbene (NHC) and pyridine-derived ligands themselves are achiral, the combination results in an absolute Λ or Δ stereogenic center [94–97]. Asymmetric benzylic C–H amination methods using this suite of catalysts have successfully delivered enantioenriched pyrrolidines [96], 2-imidazolidinones [95], 4-imidazolidinones [94], γ-sultams and 2-oxazolidinones [97] in moderate-to-excellent *ee*. The first-generation catalysts Ru(**L2a**) and Ru(**L2b**) varied the R group at the 3-position of the pyridine to identify the optimal steric features that furnish high *ee* in the NT of aliphatic azides (Table 10). Interestingly, a phosphine co-catalyst was necessary to promote the reaction of azides to substituted pyrrolidines, which were obtained in moderate yields and with excellent *ee* in most cases [96]. The mechanism is proposed to first involve formation of an iminophosphorane by loss of nitrogen, followed by transfer to the metal to furnish a Ru-supported nitrene and regenerate the phosphine co-catalyst. A concerted insertion of the nitrene into the C–H bond and amine protection completes the catalytic cycle (not shown). A related Ru(**L2b**) catalyst was reported to convert alkyl azide precursors to 4-imidazolidinones in moderate yields and in good *ee*. An osmium catalyst (Os(**L1**)) proved superior to the ruthenium version when sulfonyl azides and azidoformates were used as the nitrene precursors to furnish γ-sultams and 2-oxazolidinones as the products in good yields and moderate *ee* [97].

In addition to Ru–salen catalysts for asymmetric intermolecular C–H aminations [98, 99], Katsuki and coworkers recently reported Ir-based versions [100]. The Ir^{III}–salen complex Ir(**S1**) promoted the formation of benzosultams in a regio- and enantioselective C–H amination of benzylic C–H bonds (selected examples in Table 11). When 2-ethylbenzenesulfonyl azides were used, amination at the benzylic C–H bond gave the corresponding five-member sultams in high *ee*. Extending the length of the pendant alkyl chain switched the site-selectivity to furnish six-member sultams. Katsuki hypothesized that in addition to bond dissociation energies, selectivity can depend on the ability of the substrate to adopt a conformation that yields the required orbital interaction between the C–H and iridium–nitrenoid bonds for nitrene insertion.

Table 10 Chiral-at-metal Ru and Os catalysts for enantioselective C–H amination with azide derivatives

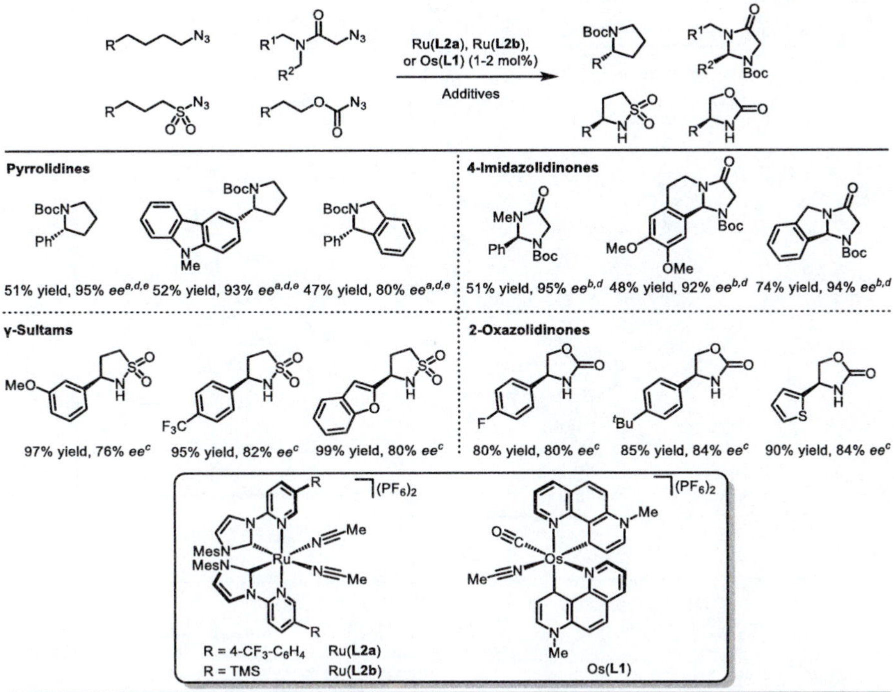

[a]Ru(**L2a**) as a catalyst. [b]Ru(**L2b**) as a catalyst. [c]Os(**L1**) as a catalyst. [d]Boc_2O as an additive. [e]P(4-F-Ph)$_3$ as an additive.

3.4 Carbamates as Nitrene Precursors

Carbamates have been used as nitrene precursors for C–H amination, notably in efforts by the Padwa and Du Bois groups using Rh catalysts [43, 44]. A strong preference for formation of a five-membered ring was noted when the C–H bond undergoing reaction was activated. However, the lower reactivity of carbamates as compared to sulfamates has made them traditionally less appealing as nitrene precursors.

Recent work in silver-catalyzed NT has shown that the nature of the ligand on silver can be used to control whether a five- or six-member ring is produced from the carbamate precursor [101, 102]. A dmBOX ligand, which is proposed to generate a monomeric silver complex, favors formation of a six-membered heterocycle, as opposed to $[(Py_5Me_2)Ag(ClO_4)]_2$, a sterically congested dimer in solution, which gave excellent selectivity for reaction at the δ-site to form a 5-member ring (Scheme 21). The scope displayed good-to-excellent tunability, with $[(Py_5Me_2)Ag(ClO_4)]_2$

Table 11 Enantioselective Ir-catalyzed C–H amination

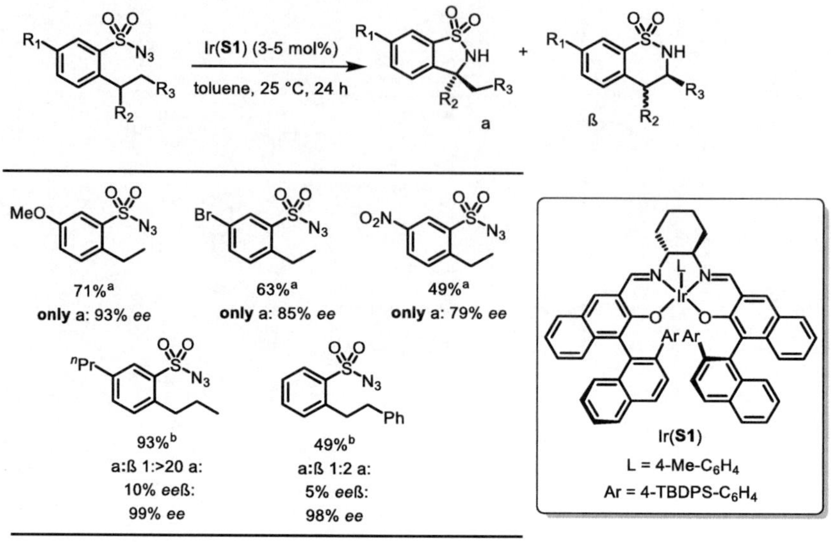

a) 5 mol%. b) 3 mol%.

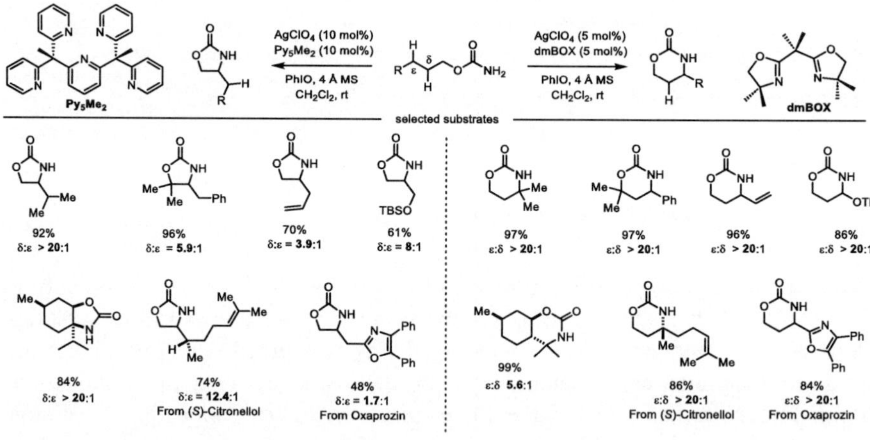

Scheme 21 Representative examples of tunable, silver-catalyzed heterocycle formation

showing moderate-to-good selectivity for unactivated C–H bonds, even in the presence of weaker benzylic and allylic C–H bonds. The versatility of the chemistry was showcased in late-stage C–H aminations of selected biologically relevant compounds. DFT studies of a monomeric (dmBOX)$AgClO_4$ complex showed a near-linear N···H···C geometry favoring a 7-membered transition state, while the

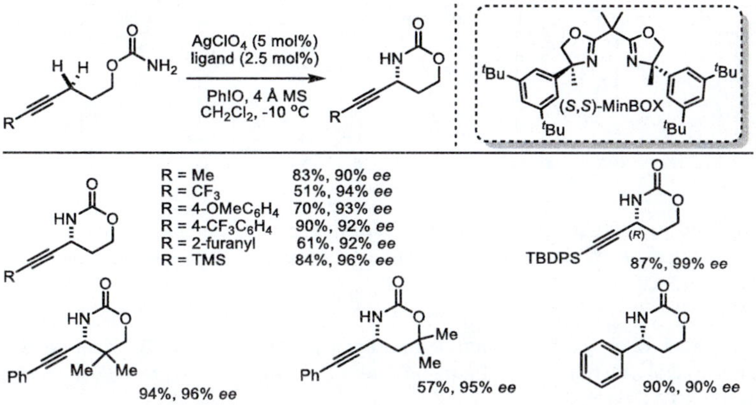

Scheme 22 Selected scope for asymmetric amination of propargylic and benzylic C–H bonds

dimeric silver complex using Py_5Me_2 as the ligand presented a congested binding pocket that prefers a 6-member ring transition state.

Enantioselective NT involving carbamate precursors are scarce. Schomaker and coworkers reported the first examples of silver-catalyzed asymmetric C–H aminations of activated propargylic C–H bonds to furnish enantioenriched γ-alkynyl γ-aminoalcohols [48]. Aryl-substituted bis(oxazoline) (BOX) ligands gave excellent chemoselectivity and promising *ee*; increasing the bulk in the aryl groups and installing an additional methyl group at the chiral center (Min-BOX) significantly improved the *ee*. A selected scope in Scheme 22 highlights several functional groups that are tolerated, including alkyl, aryl, heteroaryl, and silyl substitutions at the distal alkyne carbon, with little impact of the steric and electronic modifications on *ee*.

3.5 *Sulfamates as Nitrene Precursors*

Sulfamates are arguably the most widely used nitrene precursors for the formation of heterocycles via NT processes. Du Bois and coworkers, in particular, were instrumental in resurrecting interest in Rh-catalyzed nitrene transfer through their development of the Rh_2esp_2 catalyst (Fig. 4a) that shows improved oxidative stability compared to $Rh_2(OAc)_4$. Trends in the reactivity of this catalyst with sulfamates, including chemo- and site-selectivity, have been covered in numerous reviews and book chapters; the reader is referred to these resources for further details [9, 10, 29, 103].

The development of general dinuclear Rh catalysts for asymmetric C–H amination of sulfamates has been challenging. Recent advances have been spurred by reports of chiral tetracarboxylate and tetracarboxamidate ligands that bear a bulky chiral center adjacent to the carboxylate or carboxamidate (Fig. 4b, c).

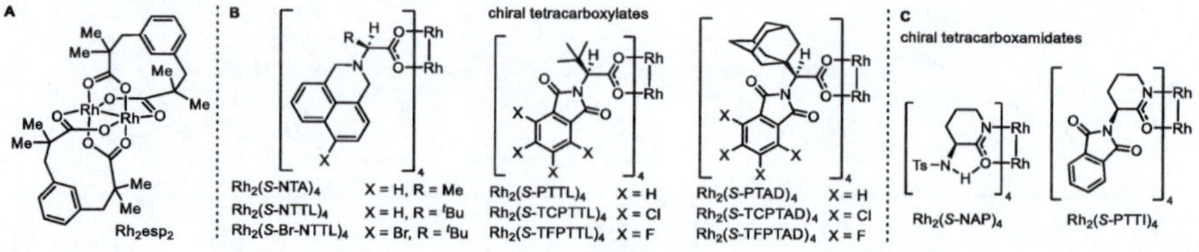

Fig. 4 (**a**) Rh_2esp_2. (**b**) Chiral rhodium tetracarboxylate catalysts employed in asymmetric C–H amination. (**c**) Chiral rhodium tetracarboxamidate catalysts employed in asymmetric C–H amination

Table 12 (A) NT catalyzed by rhodium tetracarboxamidates (Fig. 4c). (B) Chemoselectivity of carboxamidate catalysts. (C) Scope of intramolecular C–H amination with *N*-tosyloxycarbamates with Rh_2(*S*-TCPTAD)$_4$ (Fig. 4b)

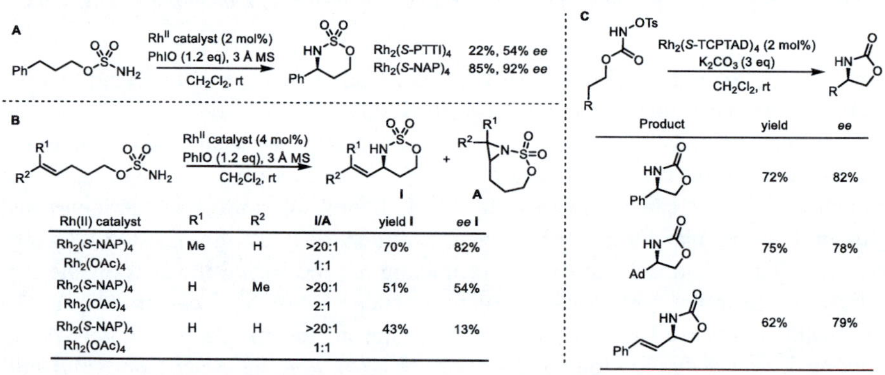

Rh(II) catalyst	R[1]	R[2]	I/A	yield I	ee I
Rh_2(*S*-NAP)$_4$	Me	H	>20:1	70%	82%
$Rh_2(OAc)_4$			1:1		
Rh_2(*S*-NAP)$_4$	H	Me	>20:1	51%	54%
$Rh_2(OAc)_4$			2:1		
Rh_2(*S*-NAP)$_4$	H	H	>20:1	43%	13%
$Rh_2(OAc)_4$			1:1		

Product	yield	ee
Ph (benzylic)	72%	82%
Ad	75%	78%
Ph (allylic)	62%	79%

Du Bois reported that Rh_2(*S*-NAP)$_4$ (Fig. 4c) [104], which bears an electron-donating carboxamide to better stabilize the intermediate nitrene as compared to a carboxylate, furnishes benzylic amines in *ee* up to 92% (Table 12A), with good chemoselectivity for insertion over aziridination in homoallyl sulfamates (not shown). The *ee* for insertion into allylic C–H bonds are modest, with much higher *ee* observed for *cis* alkenes vs. *trans*- or mono-substituted substrates (Table 12B). Davies and coworkers reported that Rh_2(*S*-TCPTAD)$_4$ (Fig. 4b) [105] shows moderate *ee* for benzylic, allylic, and unactivated C–H bonds using *N*-tosyloxycarbamates as nitrene precursors (Table 12C).

Ruthenium, iridium, and osmium complexes are attractive catalysts for NT, as they are resistant to oxidative degradation and are much less expensive than Rh and Ir. The Che group has reported Ru catalysts, such as Ru(**L1**), supported by chiral porphyrin-based ligands give asymmetric C–H bond amidation (Scheme 23) in *ee*'s of up to 82% [106].

Achiral tetracarboxylate- and tetraamidate-supported diruthenium(II,III) catalysts also promote NT due to their high oxidation potentials [107]. The Matsunaga group

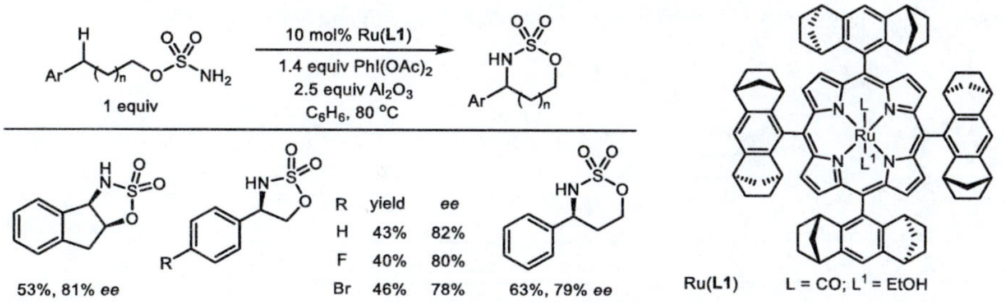

Scheme 23 Porphyrin-supported ruthenium catalysts for enantioselective C–H amination

Table 13 Enantioselective C–H amination with diruthenium paddle-wheel complexes

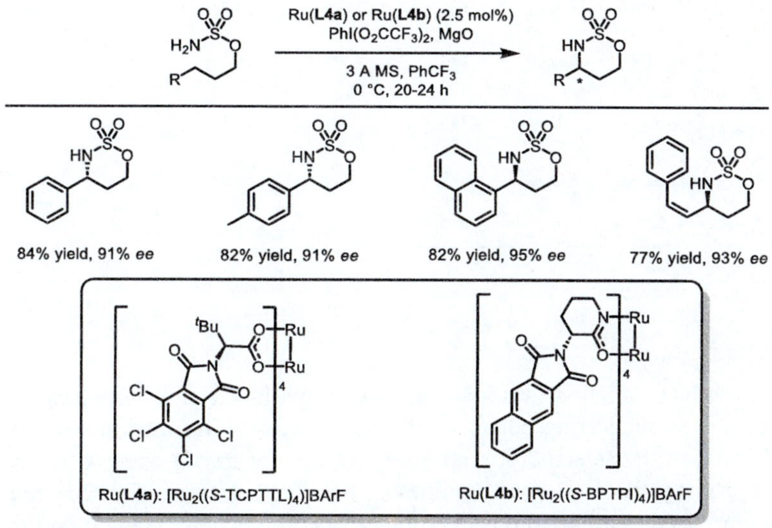

developed diruthenium paddle-wheel complexes that promoted asymmetric NT of sulfamates at benzylic C–H bonds. The yields ranged from 72–84% with high *ee* of 91–95% (Table 13) [108].

An interesting feature of silver-catalyzed C–H amination of sulfamates is that the identity of the ligand can be used to tune between reaction at a 3 ° C(sp^3)–H (**T**) bond and diverse C–H bonds with low BDEs (bond dissociation energy) (**A**) (Table 14) [109]. A silver salt supported by a ^{t}Bubipy ligand favors reaction at **T**, while a tris(2-pyridylmethyl)amine (tpa) ligand gives selective amination at **A**. Steric effects did not affect the preference for reaction at the **T** C–H bond to any great extent (Table 14, entries 1, 3), but **A**-functionalization did benefit from increased

Table 14 Regioselective C−H amination between tertiary (T) and activated (A) C−H bonds

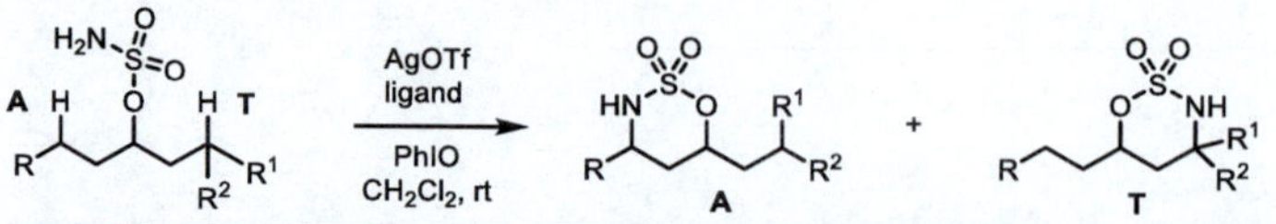

entry	R, R^1, R^2	ligand	activated C–H	tertiary C–H	A : T
1	Ph, Me, Me	^{t}Bubipy	20%	55%	1 : 2.8
2	Ph, Me, Me	tpa	59%	25%	2.4 : 1
3	Ph, ^{i}Pr, ^{i}Pr	^{t}Bubipy	21%	54%	1 : 2.6
4	Ph, ^{i}Pr, ^{i}Pr	tpa	63%	-	>20 : 1
5	4-$CF_3C_6H_4$, Me, Me	^{t}Bubipy	11%	69%	1 : 6.3
6	4-$CF_3C_6H_4$, Me, Me	tpa	42%	54%	1 : 1.3
7	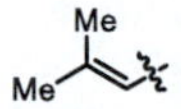 , Me, Me	^{t}Bubipy	28%	30%	1 : 1.1
8	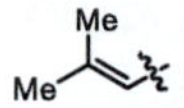 , Me, Me	tpa	74%	8%	9.3 : 1
9	$H_{11}C_5C\equiv C$–, Me, Me	^{t}Bubipy	21%	51%	1 : 2.4
10	$H_{11}C_5C\equiv C$–, Me, Me	tpa	37%	26%	1.4 : 1

steric hindrance at the competing sites (entries 2, 4). Similarly, electron-withdrawing substituents on the aromatic ring enhanced the preference for **T**-functionalization (entry 5) and reversed the selectivity trend (entry 6). Similar trends were observed for allylic and propargylic C−H bonds (entries 7–10). Site-selectivity for activated C–H bonds is proposed to arise from NCIs (non-covalent interactions) between the ligand and the substrate, including aromatic π•••π and Ag•••π interactions [110–112].

While examples in Table 14 provided indirect evidence for the role of NCIs in silver-catalyzed NT, further support was provided by noting that an increase in electron density of the pyridines on the tpa ligand through installation of a *para* dimethylamino group (Table 15, entry 5) showed improved selectivity for an electron-poor C–H bond as compared to the parent tpa ligand. Computations also supported the presence of directing, attractive π•••π NCIs (Table 15, right, **A**), where the lowest energy active species can be formulated as an Ag(II)–nitrene radical anion. The calculated attractive π•••π interactions occur between the aryl moiety of the substrate and the pyridine arm of the tpa ligand **14**, supporting the hypothesis that NCIs are at least partially responsible for the preference for amination of benzylic C–H bonds using tpa ligands [47–49]. Computational studies also show a Ag•••π interaction in the transition state of allylic amination (Table 15B) to drive the high

Table 15 Experimental evidence suggesting NCIs play a role in selective NT into benzylic C–H bonds

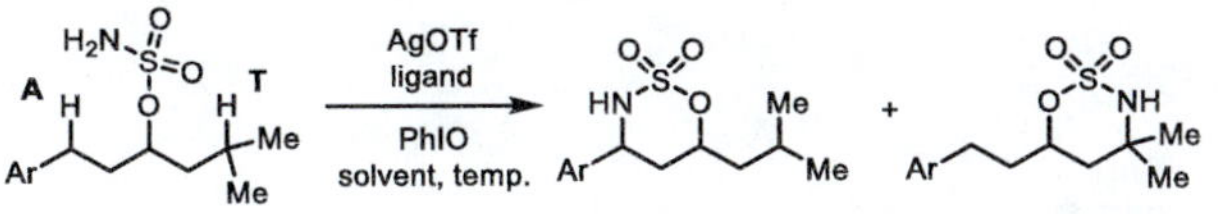

entry	Ar group	ligand	activated C–H	tertiary C–H	A : T
1	Ph	tpa[a]	56%	20%	**2.8** : 1
2		tpa[b]	73%	17%	**4.3** : 1
3	4-BrC$_6$H$_4$	tpa[a]	57%	31%	**1.8** : 1
4		tpa[b]	65%	18%	**3.6** : 1
5		(p-Me$_2$N)$_3$tpa[c]	71%	10%	**7.1** : 1
6	4-OMeC$_6$H$_4$	tpa[a]	70%	15%	**4.7** : 1
7		tpa[b]	84%	–	**>20** : 1

[a] CH$_2$Cl$_2$, rt; [b] CHCl$_3$, -20 °C; [c] CH$_2$Cl$_2$, -20 °C.

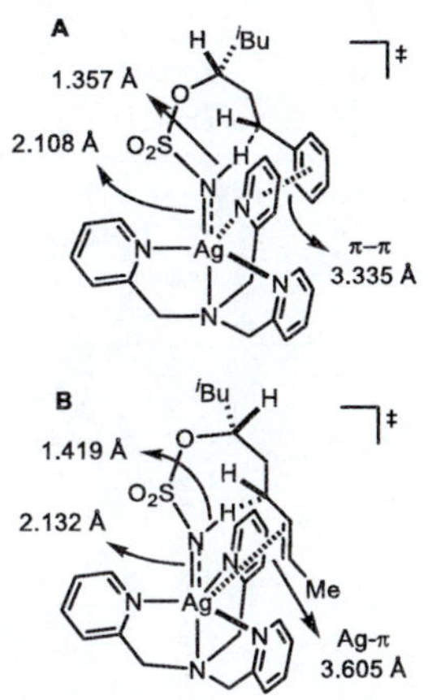

chemo- and site-selectivity over competing aziridination or insertion into the tertiary alkyl C(sp^3)–H bond (>19:1 site-selectivity).

Ruthenium and iron salts supported by 2,6-bis(oxazolin-2-yl)pyridine (PyBOX) ligands have been shown to promote asymmetric C–H aminations of sulfamates. Blakey and coworkers reported a cationic Ru complex supported by an enantioenriched PyBOX ligand that promotes asymmetric intramolecular C–H amination of sulfamates at either benzylic or *trans*-allylic C–H bonds [113]. While the yields were moderate, the enantioselectivities were all near 90% (Table 16).

While there are numerous examples of Fe complexes supported by porphyrin or phthalocyanine ligands [8, 114, 115], iron catalysts supported by non-heme *N*-dentate ligands have been reported to promote intramolecular C–H amination of sulfamates (Table 17). White and coworkers reported intramolecular reactions of sulfamates into activated C–H bonds using an iron salt supported by **L1**, while **L2** and **L3** also proved effective catalysts.

3.6 *Pre-Oxidized Nitrene Precursors with N–O Bonds*

The Lebel group reported that various *N*-(acyloxy) or *N*-(sulfonyloxy)carbamates, particularly tosyloxycarbamates, undergo Rh- and Cu-catalyzed NT with no need for an external oxidant (Table 18 for selected examples) [53–56]. Not surprisingly, higher yields were noted for substrates that contain either weaker or more electron-rich C–H bonds.

Table 16 Blakey's Ru-catalyzed asymmetric C–H amination

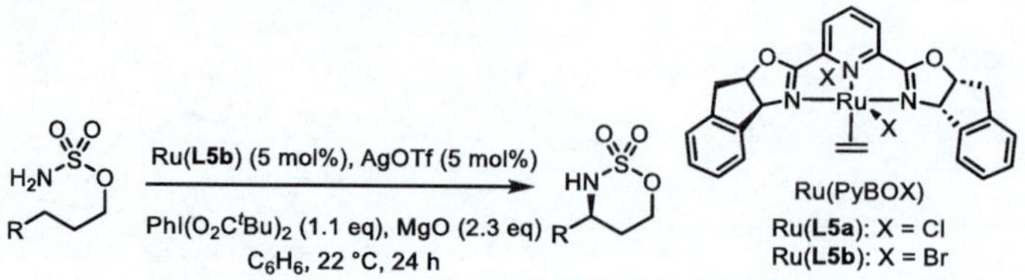

Entry	Substrate	Product	Yield (%)	*ee* (%)
1	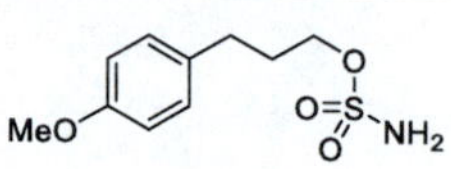	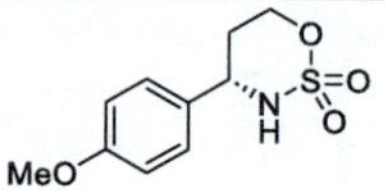	68	90
2	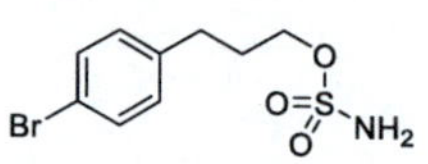	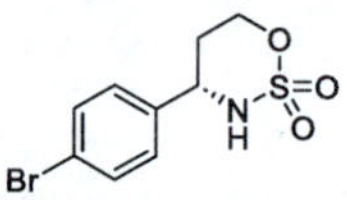	60	88
3	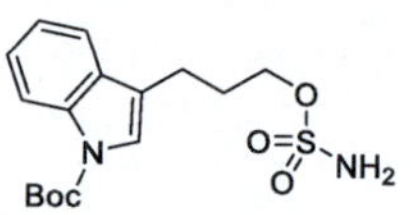	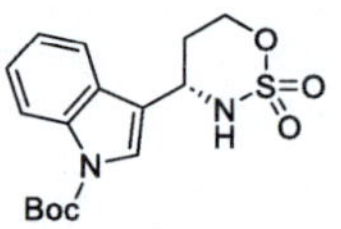	58	92
4	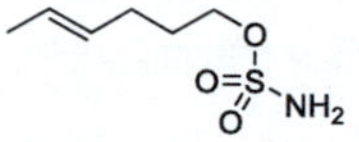	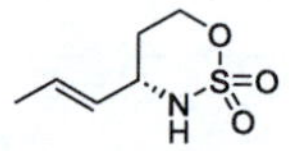	60	89

O-Benzoylhydroxylamines also serve as alkyl nitrene precursors and have been reported to furnish saturated *N*-heterocycles from primary amines (Table 19A) when catalyzed by Rh_2esp_2 [116]. The addition of Boc_2O is necessary to obtain high yields of the pyrrolidine products, presumably due to the catalyst binding strongly to the free amine. In addition to insertion into C(sp^3)–H bonds, *O*-benzoylhydroxylamines can insert into aryl C(sp^2)–H bonds to give fused heterocycles (Table 19B).

The Meggers group reported the first catalytic asymmetric formation of 2-imidazolidinones from pre-oxidized ureas using a chiral-at-ruthenium complex (Table 20) [95]. The scope and *ee* were excellent for amination at benzylic C(sp^3)–H benzylic bonds, giving high yields and *ee*, irrespective of the electronic features of the aryl group. The chemistry performed well for amination of a propargylic C–H bond. Changing the internal urea protecting group from a methyl to a benzyl group decreased the yield, although the *ee* was still high. Finally, the chemistry was able to tolerate ureas where the internal nitrogen was unprotected.

An unusual transformation involving a proposed nitrene intermediate furnishes cyclic carbonates in good yield and *ee*, again using a chiral-at-ruthenium catalyst (Table 21) [117]. The mechanism is proposed to proceed via formation of the Ru-supported nitrene, which then abstracts a hydrogen atom to generate a

Table 17 Intramolecular NT reactions catalyzed by Fe complexes and non-heme ligands

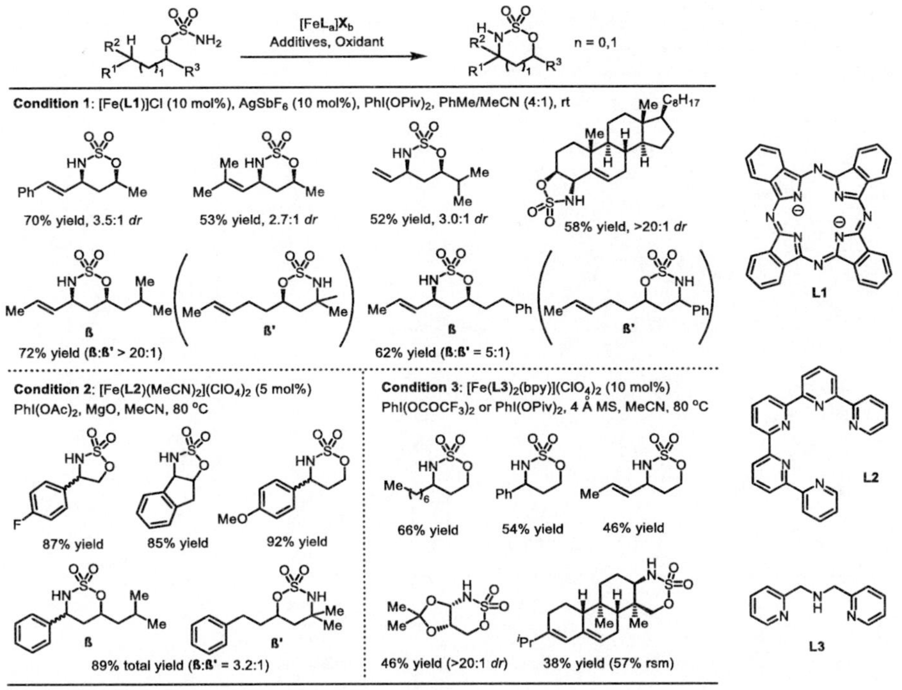

Table 18 *N*-Tosyloxycarbamates as nitrene precursors for Rh-catalyzed C–H insertion

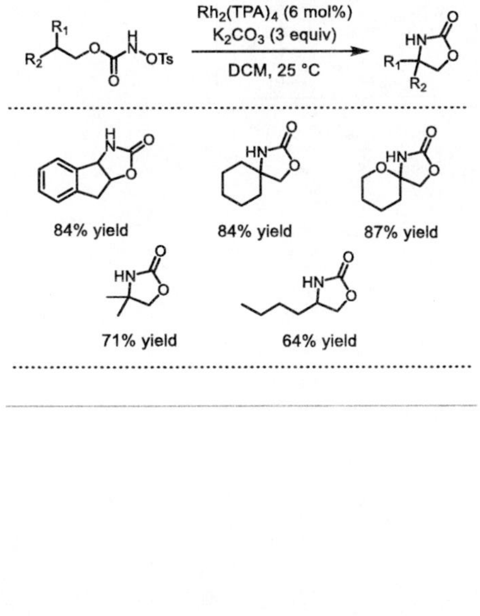

Table 19 (A) Selected scope of O-benzoylhydroxylamines as nitrene precursors for Rh-catalyzed C(sp^3)–H insertion. (B) Rh-catalyzed C(sp^2)–H insertion

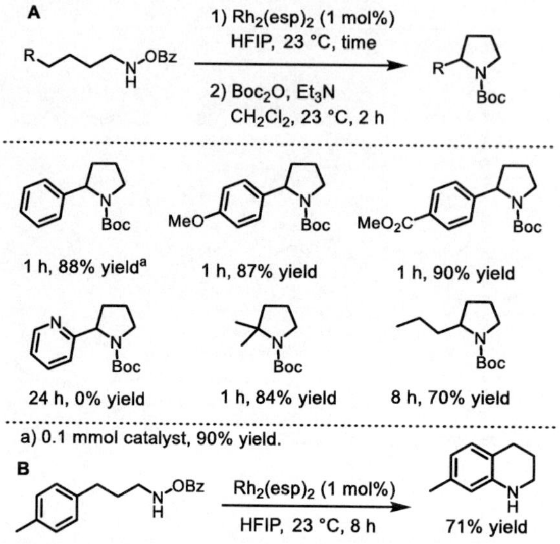

Table 20 Ruthenium-catalyzed enantioselective ring-closing C–H amination of urea derivatives

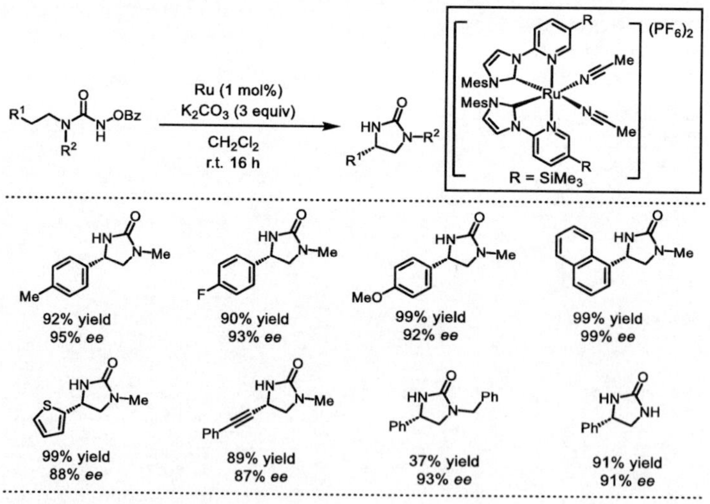

Table 21 Ruthenium-catalyzed enantioselective C–H oxygenation of cyclic organic carbonates and mechanism

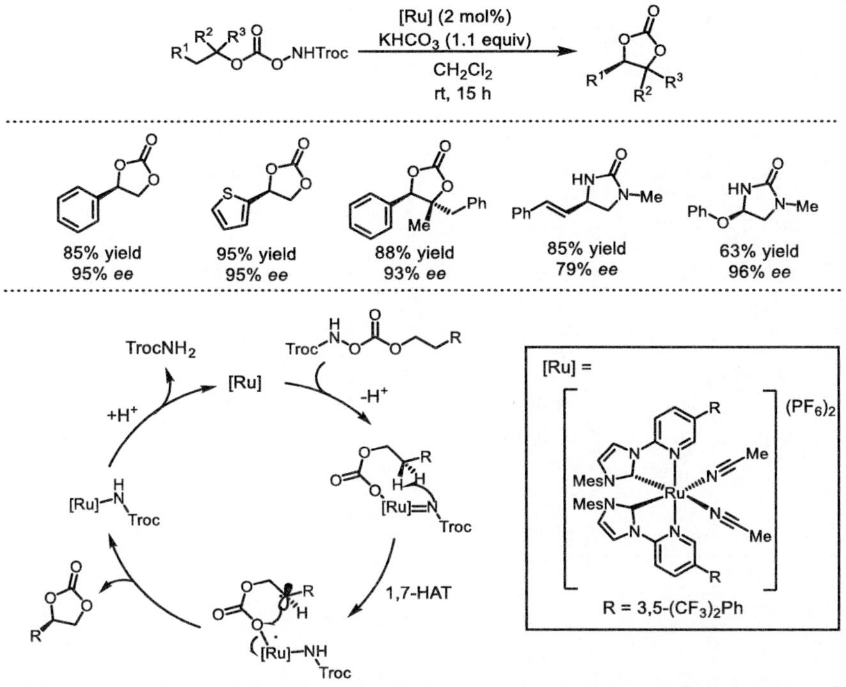

carbon-centered radical through a 1,7-hydrogen atom transfer (HAT) process. The radical engages the metal-bound oxygen to form the cyclic carbonate and an amide as the nitrogen-containing by-product.

3.7 Ureas, Guanidines, and Sulfonamides as Nitrene Precursors

The Schomaker group showed that sterically-differentiated Ag complexes control the approach of the nitrene to the C–H bond to form either 5- or 6-membered heterocycles from simple sulfonamide precursors bearing both benzylic and homobenzylic C–H bonds (Table 22) [118]. Silver-catalyzed amination using $[(Py_5Me_2)AgOTf]_2$ gave preferred insertion at the benzylic position to form the five-member sultam, while $[(^tBu_3tpy)AgOTf]_2$ favored formation of the six-member sultam. More benzylic amination was observed with Py_5Me_2 when the steric bulk

Table 22 Regioselective C–H amination forming 5- and 6-membered heterocycles

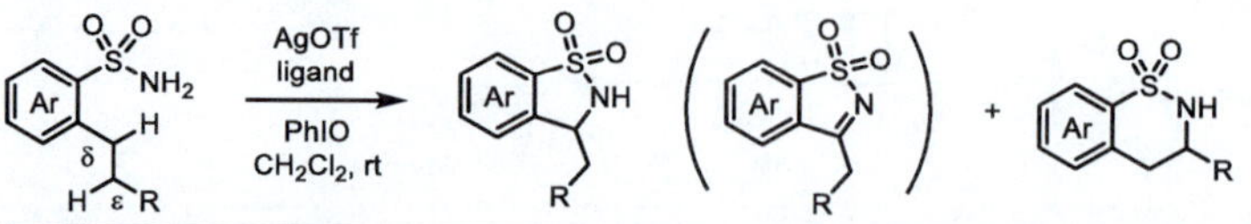

Entry	Substrate	Ligand	δ-C–H	ε-C–H	δ : ε
1	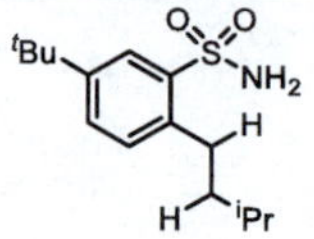	Py_5Me_2	49%	30%	1.6 : 1
2		$^tBu_3tpy$	11%	65%	1 : 5.9
3	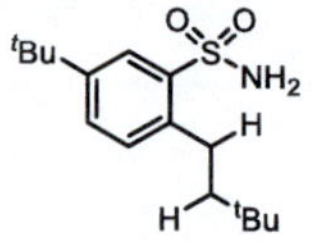	Py_5Me_2	76%	12%	6.3 : 1
4		$^tBu_3tpy$	18%	68%	1 : 3.8
5	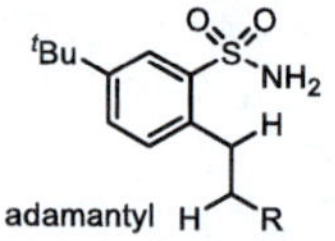	Py_5Me_2	46%	10%	4.6 : 1
6	R = adamantyl	$^tBu_3tpy$	12%	72%	1 : 5.9

was increased at the ζ-carbon, suggesting the sterics of the catalyst are likely the key factor in determining the selectivity.

In 2006, Du Bois and coworkers demonstrated one of the earliest examples of nitrene transfers with ureas by performing a Rh-catalyzed amination of benzylic and alkyl C–H bonds to generate cyclic ureas (Scheme 24) [119]. Both secondary and tertiary C–H bonds underwent amination with good-to-excellent yields, with tertiary C–H bonds resulting in higher yields. The Meggers group also reported an alternative Fe-catalyzed method that expanded the substrate scope with comparable results as the work done by the Du Bois group [120]. Later, the Meggers group reported an enantioselective version of this reaction catalyzed by Ru(**L2a**) with excellent yields and *ee* [95]. This result was consistent with heterocycles, propargylic C–H bonds, and substituents α to the urea. Along with cyclic urea formation, Du Bois and coworkers also reported a method of Rh-catalyzed nitrene transfers with guanidines to generate cyclic guanidines in poor to good yields with tertiary C–H bonds being generally higher yielding. Furthermore, the Dauban group expanded the scope of this amination to include carbamimidates via Rh-catalyzed nitrene transfer to generate cyclic carbamimidates [121]. This amination proceeded with good-to-excellent yields with secondary or tertiary C–H bonds and heterocycles. Propargylic C–H bonds were aminated, albeit in reduced yields.

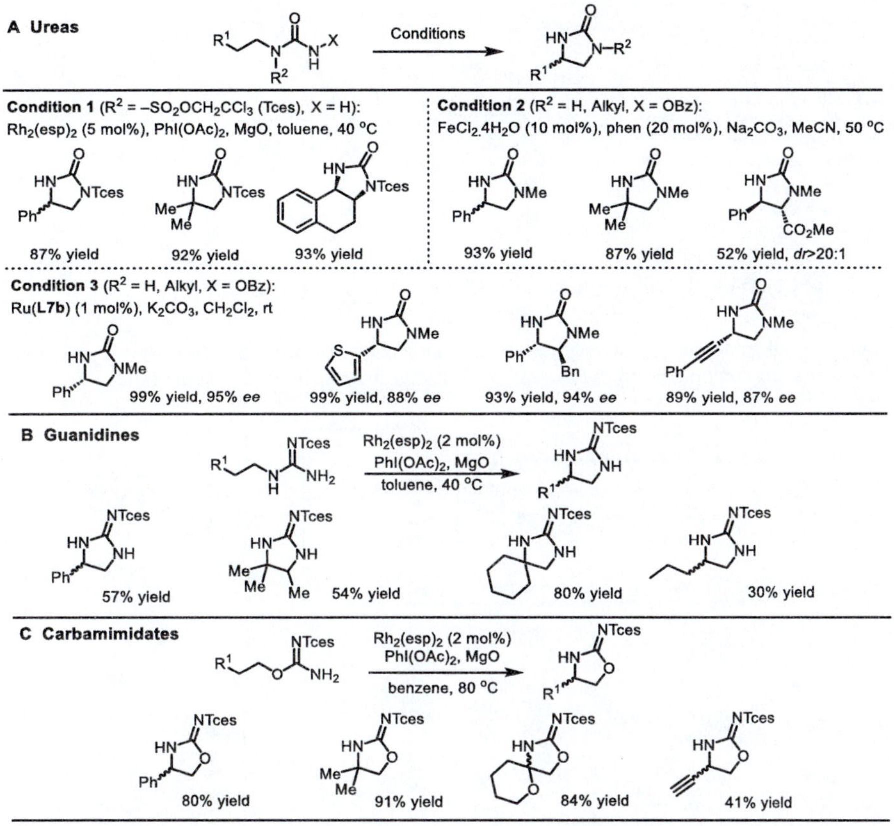

Scheme 24 Heterocycles via intramolecular nitrene transfer with ureas, guanidines, and carbamimidates

3.8 *Dioxazolones as Nitrene Precursors to Form Lactams*

Traditionally, amides have not been utilized as nitrene precursors, as exposing them to oxidants triggers the competing Curtius rearrangement. However, 1,4,2-dioxazol-5-ones can be used as effective nitrene precursors to furnish lactam products, particularly when the reactions are catalyzed by metal half-sandwich complex. The Chang group has pioneered this chemistry and has developed several different methods utilizing dioxazolones as nitrene precursors [122] that employ Ir catalysts supported by highly electron-donating bidentate ligands (Fig. 5).

Chang and coworkers also reported enantioenriched Ir catalysts for asymmetric C–H bond amidation. The chiral diamine scaffold of Ir(**S3**) (Fig. 5) engages in hydrogen bonding with the carbonyl group of the substrate [123, 124], which is proposed to account for the observed high *ee* (Table 23A). A second-generation catalyst Ir(**S4**) also engages in diverse NCIs to form an enzyme-like chiral

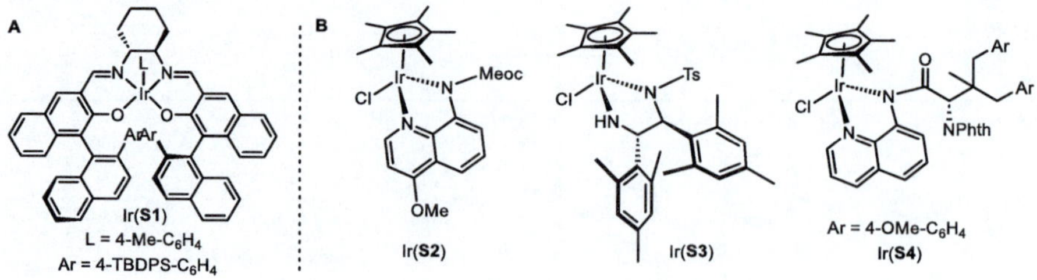

Fig. 5 (**a**) Chiral iridium salen complex. (**b**) Chiral iridium half-sandwiches

Table 23 Substrate scope for enantioselective amination by iridium half-sandwich catalysts (A) and enantio-determining transition state with Ir(**S3**) controlled by H-bonding (B)

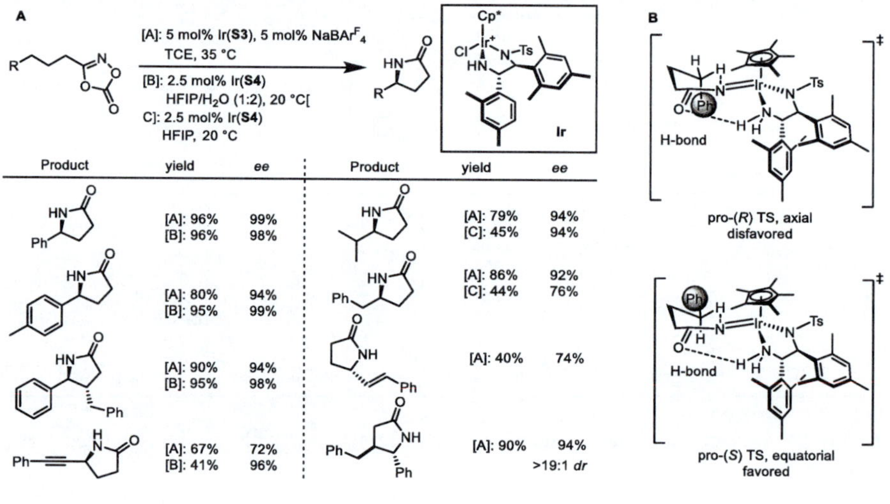

hydrophobic pocket. Substrate NCIs include π–π stacking and C–H/π interactions (Table 23B). As a pair, these two nitrene transfer reactions were able to accommodate diverse substrates, including a variety of benzylic, allylic, propargylic, alkyl, and methylene C–H bonds with moderate-to-high yields and excellent *ee*.

The Yu group reported Ru-catalyzed enantioselective cyclizations of dioxazolones to give γ-lactams in up to 97% yield and 98% *ee* (Table 24). The best ligands proved to be chiral diphenylethylene diamine (dpen) ligands bearing electron-poor arylsulfonyl groups [125].

The Meggers group showed that a chiral-at-metal complex could also be used to give enantioselective C–H amination from dioxazolones (Table 25). Interestingly, a non-C_2 symmetric (**25.1**) and a C_2 symmetric (**25.2**) Ru catalyst showed very different behaviors; the symmetric catalyst promoted Curtius rearrangement, while the non-symmetric catalyst **25.1** furnished the γ-lactam in 92% yield and 90% *ee*

Table 24 Selected scope for enantioselective C–H amination by a ruthenium-dpen catalyst

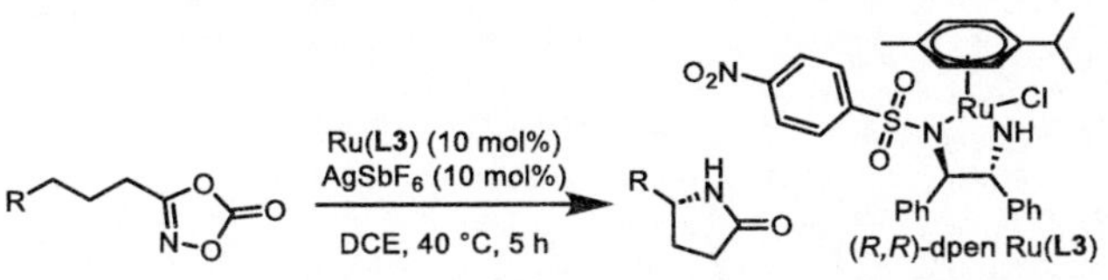

entry	substrate	product	yield [%]	*ee* [%]
1	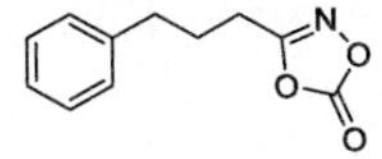	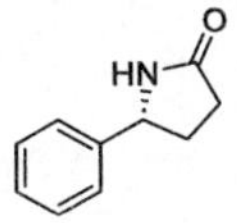	92	94
2	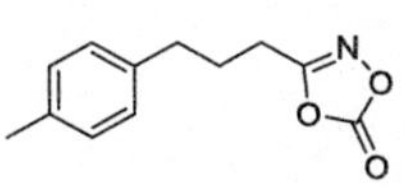	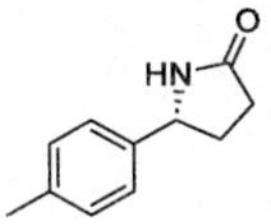	89	95
3	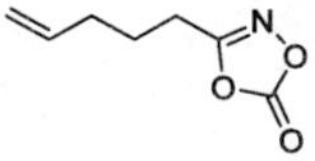	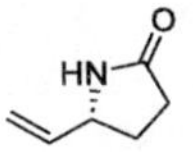	80	77
4	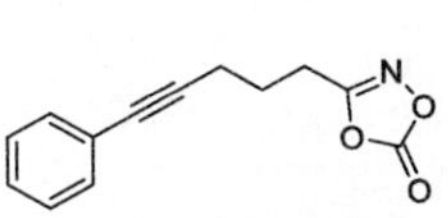	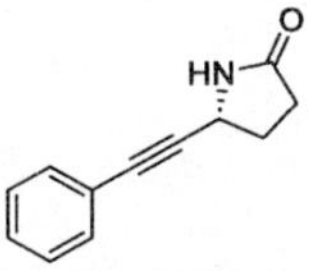	90	41
5	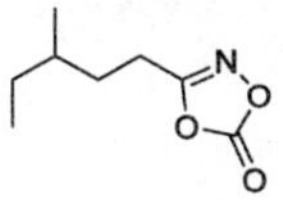	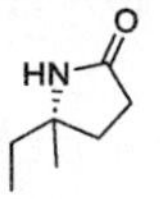	77	20

[94]. Selectivity is hypothesized to result from the strongly electron-donating groups on the NHC ligands, coupled with an NCI between the carbonyl O of the acyl nitrene intermediate and the C–H bond of the pyridyl ligand, both of which modulate nitrene nucleophilicity. The former effect was also observed by Chang and coworkers using their iridium catalysts [123].

Table 25 Selected scope for enantioselective C–H amination for the C_2-asymmetric Ru catalyst

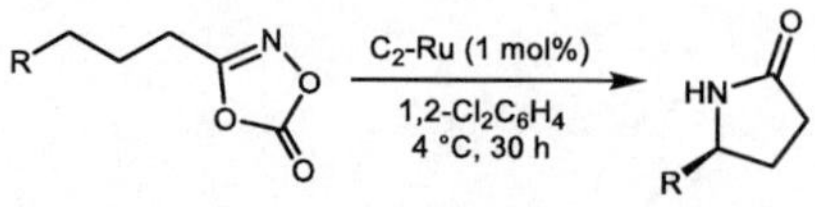

entry	R group	yield (%)	*ee* (%)
1	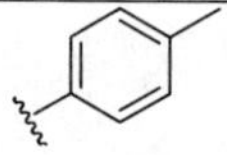	96	90
2	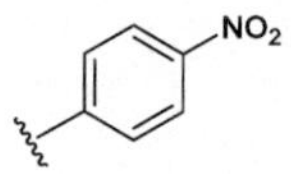	84	90
3		61	86
4	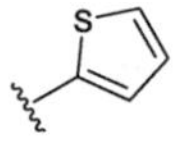	66	88
5		42	60

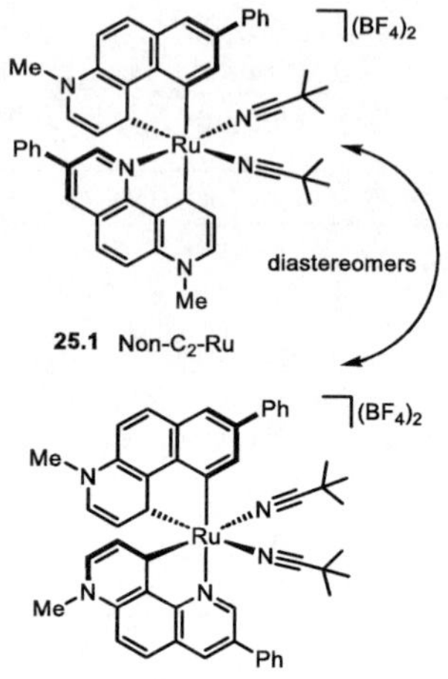

25.1 Non-C_2-Ru

25.2 C_2-Ru

4 Heterocycles from Ring Expansion Reactions Involving Nitrene Transfer as a Key Step

4.1 Background

The previous sections of this chapter discuss the situation where the heterocyclic products are formed directly through alkene/allene aziridination or insertion into a C(sp^2)–H or C(sp^3)–H bond. However, heterocycles can also be formed from tandem reactions that are initiated by nitrene transfer processes to give reactive intermediates that undergo further reactions. The following sections describe the utilization of groups like alkynes and allenes to perform nitrene alkyne metathesis or aziridination followed by ring expansion to generate valuable 5-, 6-, and 7-membered N-heterocycles.

4.2 Reactions of Nitrenes with Alkynes and Subsequent Ring Expansion

Blakey and coworkers described the development of a Rh-catalyzed (tfacam ligand; trifluoroacetamide) formation of metallonitrenes that can engage alkynes in an

Table 26 Intermolecular alkyne oxidative amidation

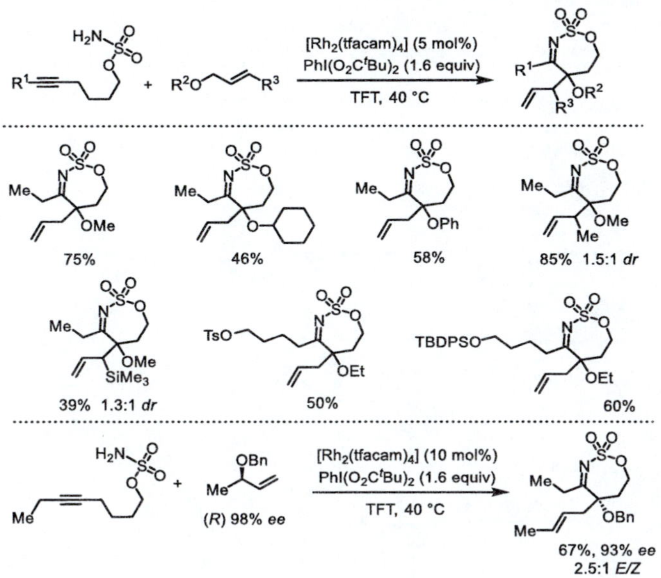

oxidation cascade, followed by intramolecular trapping of the intermediate with allyl ethers (Table 26) [126]. The α-oxyimine products could be obtained in an enantioenriched form if an enantioenriched allyl ether was employed in an intramolecular NT, followed by intermolecular trapping.

The Blakey group proposed that this metathesis reaction proceeds through an initial attack of the alkyne on the electrophilic Rh–nitrene intermediate to generate a high-energy vinyl cation intermediate (Scheme 25a) [127]. This is followed by a [1,3]-rhodium shift to furnish the key metallocarbene intermediate, which is then attacked by the appended ether oxygen to form an oxonium ylide. The ylide undergoes rearrangement to produce the *N*-sulfonyl imine, which can be subjected to a subsequent reduction to deliver the *N*-heterobicyclic products in good yields. The hydride reductant can also be replaced with a Grignard reagent, such as allylMgBr (Scheme 25b). Furthermore, if the tether length of the benzyl alkyne was increased by one carbon, the reactive intermediate can instead undergo an intramolecular Friedel–Crafts reaction to terminate the cascade event (Scheme 25c).

The Blakey group also showed that 1,2-disubstituted olefins with a sufficiently long tether will undergo cyclopropanation as the termination event following initial cyclization to the six-membered ring (Scheme 26) [128]. While *trans*-olefins worked well, a *cis*-olefin resulted in poor regioselectivity in the initial cyclization between the distal carbon versus the proximal carbon of the alkyne, resulting in a low yield of

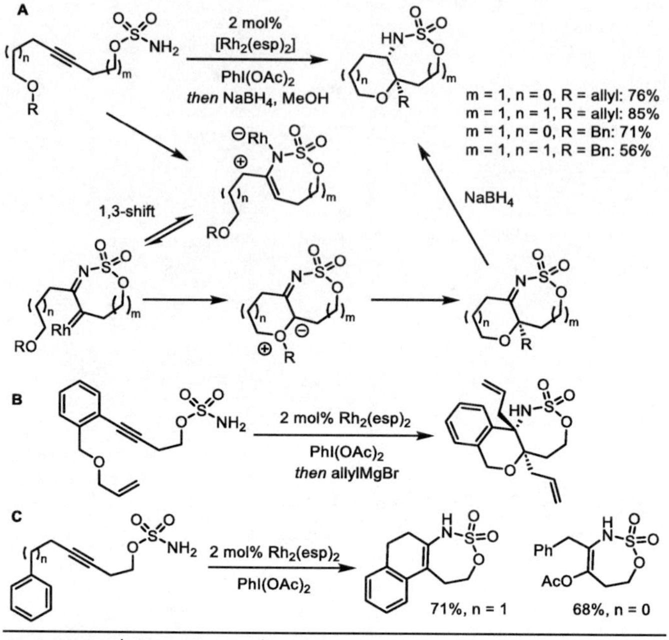

[a] Isolated yield; [b] *syn:anti*; [c] X = Cl; [d] X = OPh.

Scheme 25 Ru-catalyzed nitrene/alkyne metathesis to generate *N*-hetero- bi- and tricycles

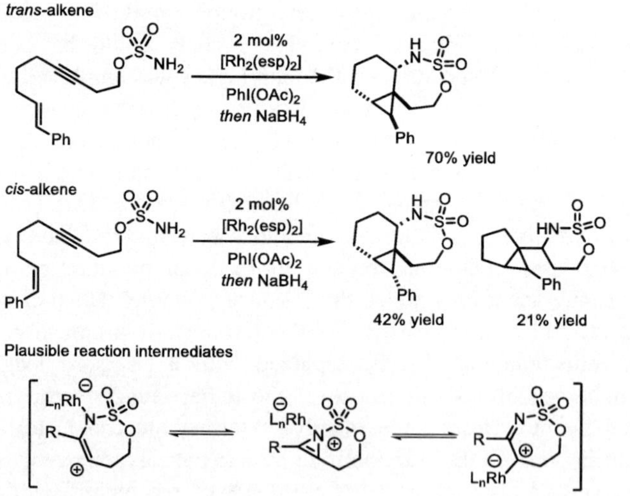

Scheme 26 Rh-catalyzed nitrene/alkyne metathesis and subsequent cyclopropanations of tethered olefins

Table 27 Enantioselective Rh-catalyzed nitrene/alkyne metathesis for *N*-heterocycle synthesis

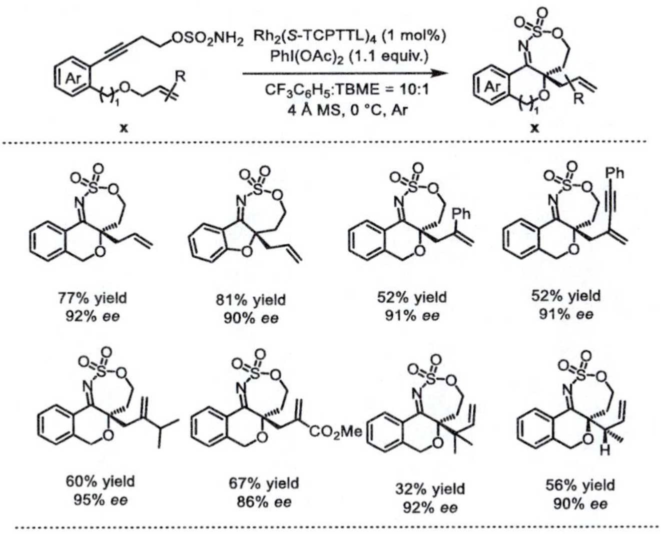

the desired cyclopropane. A variety of reaction intermediates were proposed for this intriguing transformation.

Xu and coworkers have described a rhodium-catalyzed nitrene/alkyne metathesis that delivers enantioenriched *N*-heterocycles [129]. The chiral rhodium catalyst (see Fig. 4 for the catalyst structure) first promotes the nitrene/alkyne metathesis. This is followed by attack of a tethered oxygen on the resultant α-imino Rh carbene to generate an oxonium ylide. A final enantio-determining [2,3]-sigmatropic rearrangement of the oxonium ylide delivers the products (Table 27). The yields are moderate-to-good with high *ee*. While the authors did not show a final reduction of the sulfamoyl imine moiety, addition of either a hydride or organometallic reagent would be expected to furnish the amine products in high diastereoselectivities.

The Xu group also reported a similar tandem reaction involving a Rh-catalyzed nitrene alkyne metathesis and formal C–N bond insertion to furnish 3-iminoindolines [130]. The nitrene alkyne metathesis reaction is terminated by a Stevens [1,2]-acyl shift (Table 28).

Table 28 Rh-catalyzed nitrene alkyne metathesis/formal C–N bond insertion cascade

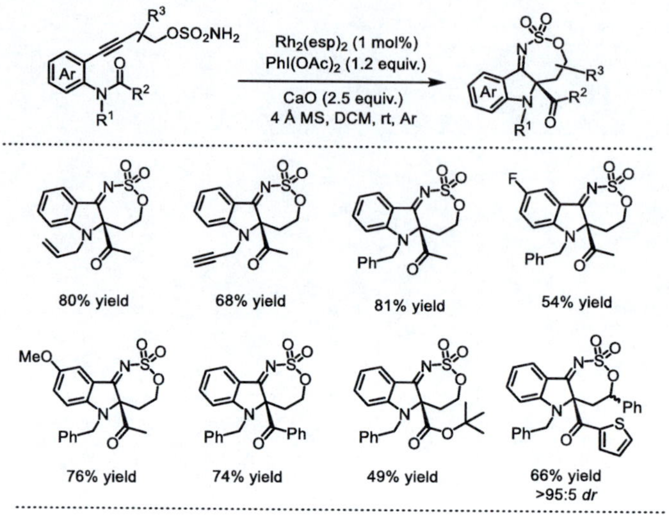

4.3 *Reactions of Nitrenes with Allenes and Subsequent Ring Expansion*

The Blakey group reported the Rh-catalyzed reaction of sulfamates with allenes to produce proposed intermediate 2-amidoallyl cations that first furnished strained cyclopropylimines. These could then undergo subsequent insertion with carboxylates or Grignard reagents to furnish cyclopropylamines in high diastereoselectivities (Table 29) [131, 132]. The Blakey group postulated that the intermediate 2-amidoallyl cation undergoes intramolecular cyclization to form the cyclopropane ring. Efforts to trap the 2-amidoallyl cation intermediate with an aldehyde instead of cyclopropanation were successful. A [3 + 2] cycloaddition of the carbonyl of benzaldehyde with the 2-amidoallyl cation gave rise to a trisubstituted 3-aminotetrahydrofuran, albeit in a low 40% yield. Finally, the authors noted that the use of *N,α*-diphenyl nitrone as a trapping agent generated an unusual bridged heterocycle.

The Robertson group reported the intramolecular Rh-catalyzed nitrene transfer of homoallenic sulfamates to generate acyloxy-enamines, aminocyclopropanes, and methylene aziridines (Table 30) [133]. Insertion of the rhodium nitrenoid into the central carbon of the allene generates a reactive 2-amidoallyl cation intermediate that furnishes diverse products, depending on the substitution pattern on the allene.

The Schomaker group reported a one-pot sequence that transforms homoallenic sulfamates into substituted cycloheptenes that bear five contiguous stereocenters

Table 29 Rhodium-catalyzed intramolecular amination and trapping of allenes

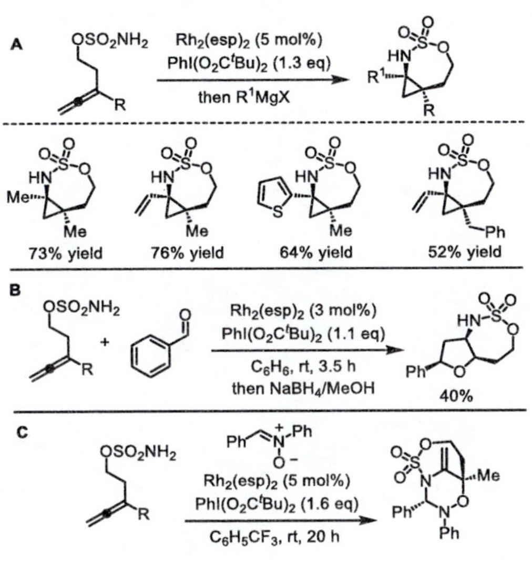

(Table 31) [134]. A tandem allene aziridination/[4 + 3] reaction is followed by a stereoselective reduction to deliver the products. The relative stereochemistry at the five contiguous stereocenters is controlled by the choice of solvent and the conditions for the initial aziridination of the allene, while the choice of reductant can be employed to set the stereochemistry at the amine-bearing carbon. Table 31A highlights results using $NaBH_3CN$ as the reductant, while Table 31B showcases the result of employing AlH_3Me_2NEt as the terminal reductant.

More recently, the Pèrez group reported the first intermolecular aziridination of allenes (Scheme 27A) [135]. A silver salt supported by a scorpionate ligands gave either the methyleneaziridine or further ring expansion to the azetidine. Computational studies led to a detailed understanding of how silver is able to catalyze the formation of the four-membered heterocycle via a key cyclopropanimine intermediate (Scheme 27B).

Metal-catalyzed nitrene transfer reactions can also be coupled to subsequent carbene transfers to generate highly substituted and stereochemically complex *N*-heterocycles. While many of these examples start from isolated aziridines, the use of *N*-tosyloxycarbamate nitrene precursors in the presence of a $Rh_2(OAc)_4$ catalyst can promote both the nitrene and carbene transfer in a single reaction vessel. The Schomaker group showed that bicyclic methyleneaziridines **28.1** formed from allene aziridination can engage with diverse diazoesters **28.2** that form Rh-supported carbenes in a formal [3 + 1] ring expansion to furnish methyleneazetidines **28.3** with good scope, yields, and diastereoselectivities [136, 137]. Selected results are

Table 30 Rhodium-catalyzed intramolecular amination of allenes

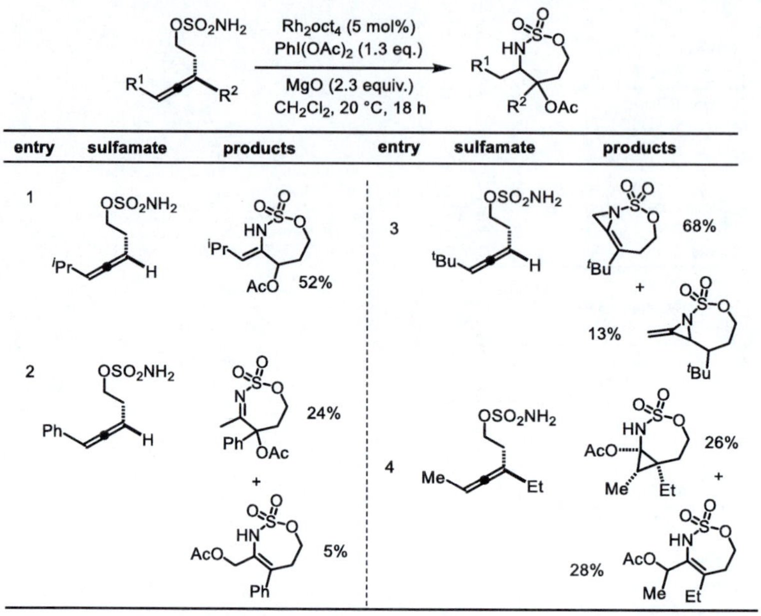

shown in Scheme 28A. The electronics of aryl groups on the diazoester carbene precursors do not significantly affect the reaction outcome, while styryl diazoester precursors can furnish a fully substituted methyleneazetidine in 89% yield and 3:1 *dr*; separation of the diastereomers gave the *syn*-Me/CO_2Me isomer in 54% yield and 15:1 *dr*. A concerted mechanism was suggested by excellent transfer of chirality from enantioenriched methyleneaziridine to the product methyleneazetidine, where formation of the aziridinium ylide intermediate by nucleophilic addition of the ring nitrogen to a rhodium-bound carbene.

4.4 Tandem Catalysis Involving Nitrene Transfer to Furnish Heterocycles

The Schomaker group further explored sequential nitrene/carbene transfer to achieve a formal [3 + 3] ring expansion of bicyclic aziridines **29.1** to highly substituted dehydropiperidines **29.3** in good yields and diastereoselectivities (Scheme 29a) [138]. The aziridinium ylide was proposed to arise from the reaction of a bicyclic aziridine with a vinyl diazoacetate-derived rhodium carbene. Delocalization of the negative charge through the vinyl group precluded competitive cheletropic extrusion

Table 31 Sequential NT/ring-opening/cycloaddition/reduction to transform allenes to aminated cycloheptenes

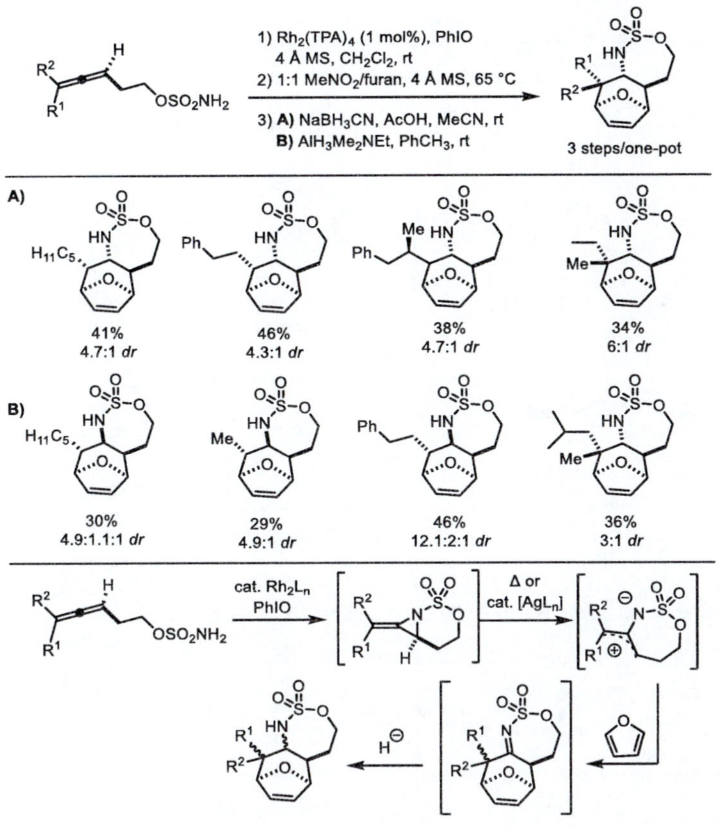

to favor the desired ring expansion to furnish the dehydropiperidine. The chemistry tolerated alkyl groups, halides, and ethers. A series of aryl-substituted diazoesters with varying steric and electronic features were surveyed; diazoesters with electron-donating and neutral substituents gave dehydropiperidines in similar yields, demonstrating that the electronics of the styrene in the carbene precursor do not heavily affect the reaction outcome. Dehydropiperidines could also be obtained in good yield and excellent *dr* from alkyl-substituted diazoacetates, which highlighted the extension of the chemistry beyond aryl diazoacetates.

A chirality transfer experiment (Scheme 29b) supported that the ring expansion reaction takes place in a concerted fashion, likely via a pseudo-[1,4]-sigmatropic rearrangement. An enantioenriched bicyclic aziridine precursor, prepared from an asymmetric silver-catalyzed nitrene transfer reaction, translated the stereochemical information from aziridine to the dehydropiperidine product with good fidelity.

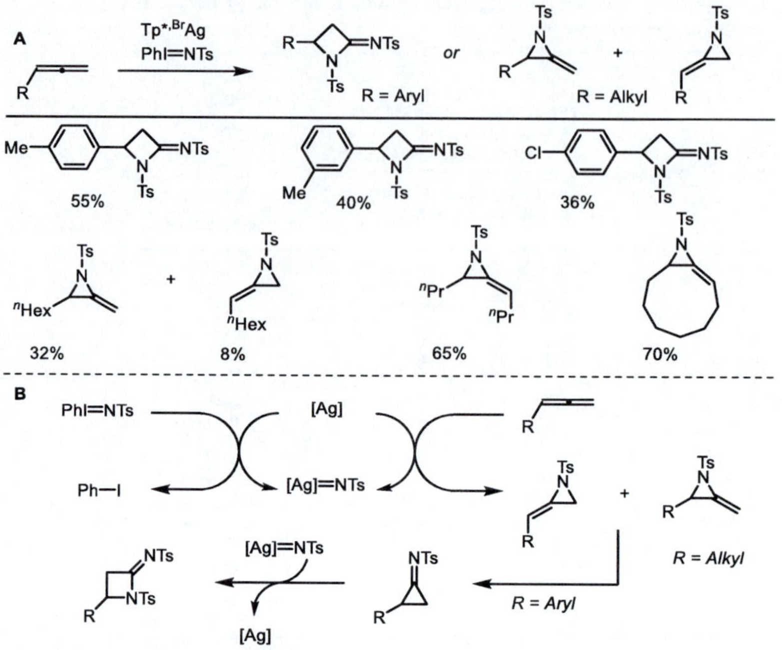

Scheme 27 Ag-scorpionate-catalyzed NT to aryl allenes and proposed mechanism

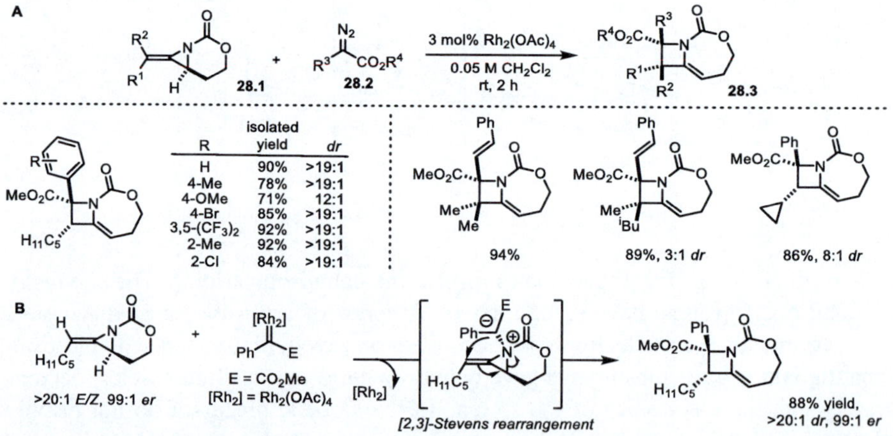

Scheme 28 Intermolecular Rh-catalyzed [3 + 1] ring expansion of bicyclic methyleneazetidines and chirality transfer experiment

The Pérez group reported a transformation of a 2-methylfuran to a 1,2-dihydropyridine via a silver-scorpionate-induced tandem catalysis (Scheme 30) [139]. The reaction is believed to occur via an initial aziridination of the more electron-rich alkene, followed by ring-opening and alkene isomerization to generate

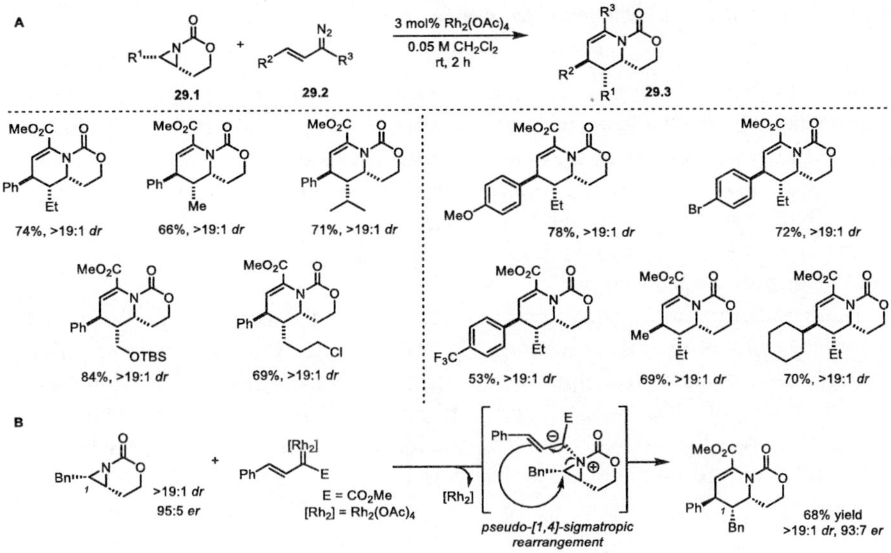

Scheme 29 Intermolecular Rh-catalyzed [3 + 3] ring expansion of bicyclic aziridines to dehydropiperidines and a chirality transfer experiment using enantiopure aziridine

Scheme 30 Selective transformation of furan by silver-induced concurrent tandem catalysis

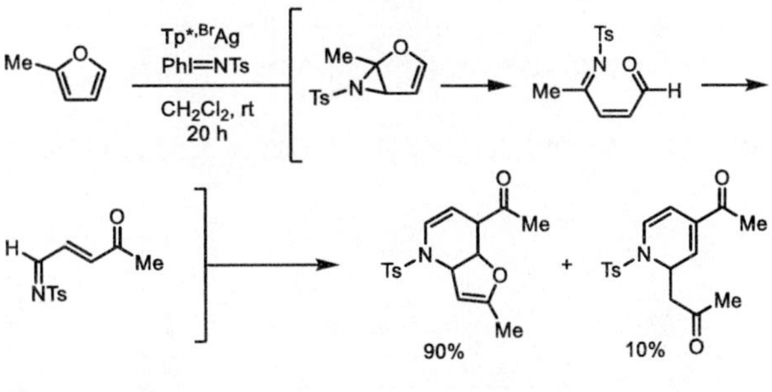

a 1-azadiene. This engages in reaction with another 2-methylfuran molecule to form the dihydropyridine as the major product.

5 Additions of Nitrenes to Heteroatoms to Furnish Heterocycles

In addition to more traditional modes of nitrene reactivity, there are scattered examples of nitrenes adding to various heteroatoms. While the focus of this chapter is on cases that form *N*-heterocycles, selected examples of acyclic compounds are included here to hopefully stimulate the further development of methods to engage nitrenes with diverse heteroatoms in synthetically useful transformations.

5.1 Additions of Nitrenes to Amines

Romo and coworkers reported a structure–activity study based on a Rh-catalyzed C–H amination [140]. The Rh-catalyzed NT reaction gave mainly olefin aziridination, allylic and benzylic C–H amination products; interestingly, NT to tertiary amines were also noted. For example, (*S*,*R*)-noscapine (Scheme 31) reacted with **31.1** to afford hydrazine sulfamate zwitterion **31.2**, along with **31.3** generated by C–H amination and subsequent oxidation. No further optimization for selective N^{+}–N^{-} bond formation was carried out.

The Perez group reported Ag-catalyzed N–N bond formation via nitrene transfer to tertiary amines in 2017 (Scheme 32) [141]. Tertiary amines bearing allylic/benzylic C–H bonds and olefins were selectively transformed into aminimides in the presence of $[Tp^{*,Br}Ag]_2$ and PhI=NTs. This method featured high chemoselectivity toward N–N bond formation over both olefin aziridination and C–H amination. Computational studies were employed to rationalize the formation

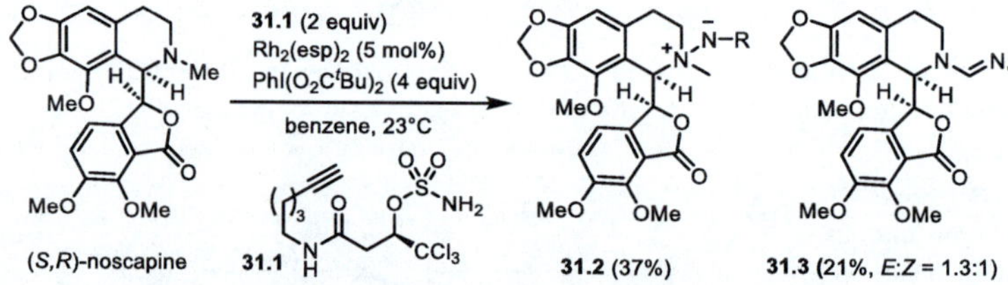

Scheme 31 An unusual Rh-catalyzed reaction of a tertiary amine with a nitrene

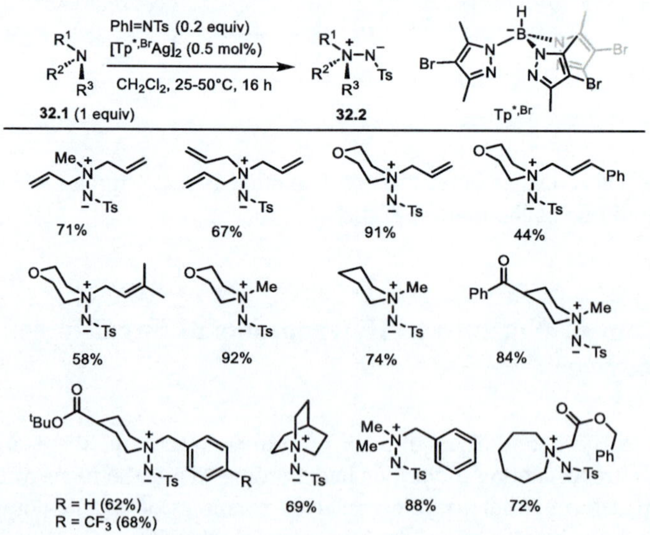

Scheme 32 Rh-catalyzed reaction of tertiary amines with nitrenes

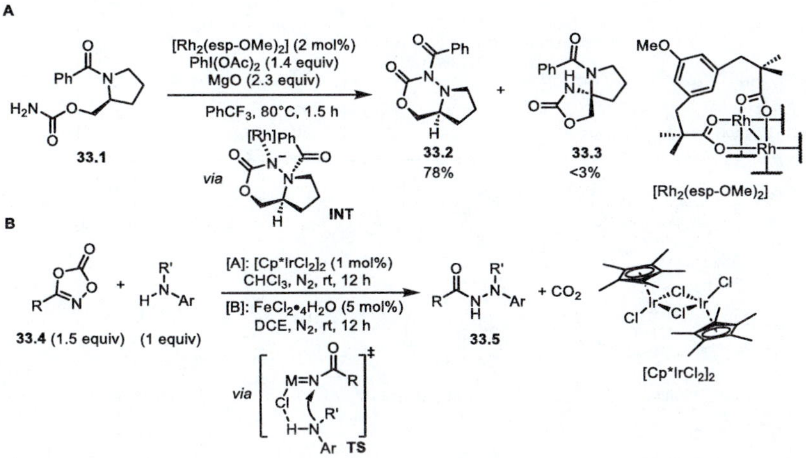

Scheme 33 Examples of amides and diaoxazolones as precursors for N–N bond formation via NT

of the kinetically favored aminimide over the thermodynamically favored aziridine due to the higher nucleophilicity of the tertiary amines and irreversible formation of aminimide from the metal–nitrene complex.

The precedent by Perez showing that the nucleophilicity of the tertiary amine is a key factor in achieving high chemoselectivity stimulated efforts to expand reactivity to include electron-deficient amides. In 2019, the Harada group reported chemoselective intramolecular N–N bond formation over C–H insertion by Rh-supported nitrenes to afford diazacyclic scaffolds (Scheme 33a) [142]. The method features high chemoselectivity toward N–N bond formation over competing C–H amination with electronically deficient carbamates and sulfamates. Mechanistic and computational studies suggested a mechanism involving initial formation of ylide **INT** from **33.1**, followed by acyl migration to rationalize the chemoselectivity for the kinetically favored N^{+}–N^{-} ylide over the thermodynamically favored C–H aminated product.

The Chen group reported Ir- and Fe-catalyzed intermolecular N–N couplings via nitrene transfer from 3-phenylpropyl-dioxazolones **33.4** to aryl amines for the synthesis of hydrazides **33.5** (Scheme 33b) [143]. The method features high chemoselectivity toward N–N bond formation over competing Curtius-type rearrangement and C–H amination. Mechanistic and computational studies suggested the intermediacy of a hydrogen-bonded complex **TS**, where a Cl ligand on the metal complex coordinates to the N–H bond of the aryl amine; this key transition state is believed to be responsible for the selective N–N bond formation over the competing Curtius rearrangement.

Maiti and coworkers reported the photocatalytic generation of nitrenes for rapid formation of diaziridines (Scheme 34) [144]. This method employs aliphatic amines **34.1** and 1,2-diols **34.2** in the presence of an oxidant to achieve an in situ generation of nitrene and imine intermediates, followed by stereoselective cyclization. The

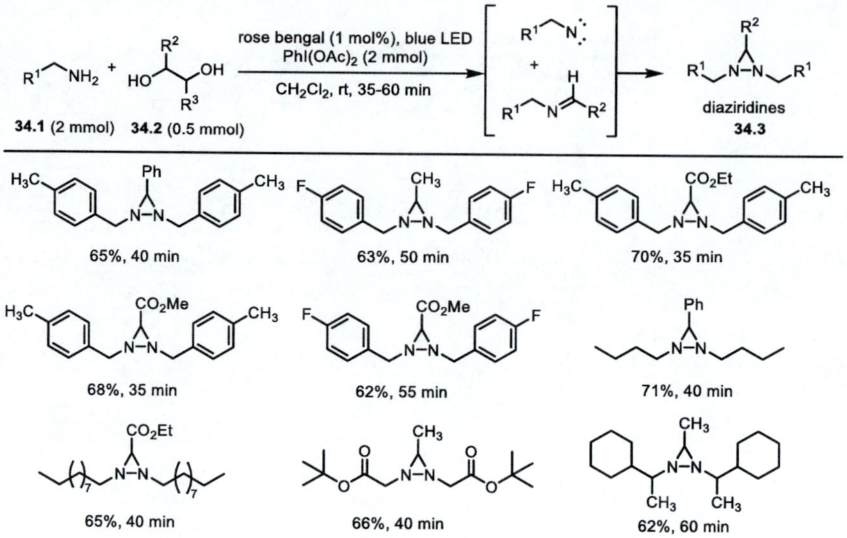

Scheme 34 Formation of diaziridines via photochemical NT

Scheme 35 Quinoline N–N bond cyclization via NT

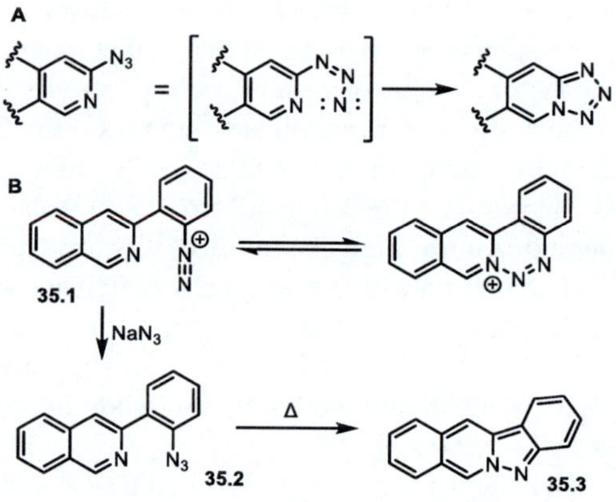

rapid photocatalytic approach delivers a unique scope of diaziridines with 1,2-disubstitution and 1,2,3-trisubstitution patterns.

Another example of the thermal generation and reaction of a nitrene with a nitrogen atom to form a *N*-heterocycle is highlighted in Scheme 35 [145, 146]. The Messmer group reported a ternary equilibrium of tetrazole-azide systems (Scheme 35a) later building on this observation to prepare polyfused heteroaromatics through a Sukuzi coupling between a quinoline triflate and an aryl boronic acid (not shown). Diazotization of an amine on the phenyl group of the Suzuki product gave **35.1**, followed by treatment with sodium azide to give **35.2**. Heating of this intermediate triggers the loss of nitrogen gas and ring-closure to furnish **35.3**, presumably due to the formation of a nitrene intermediate that is attacked by the quinoline nitrogen.

5.2 Additions of Nitrenes to Silicon

Similar to the addition of nitrenes to amines, the addition of nitrenes to silicon atoms lead to acyclic compounds, as opposed to heterocycles; however, there is potential for this chemistry to be used more generally in the future to deliver unusual heterocycles. In 2022, Perez and coworkers reported catalytic amination of silanes via nitrene insertion (Scheme 36) [147]. Unprecedented N–Si bond formation was observed using various Si–H sources that include silanes, dihydrosilanes, disilanes, and disiloxanes in the presence of PhI=NTs and Tp^{Br3}Cu(NCMe) to afford corresponding silylamines. Based on computational studies, the authors suggested that the mechanism involves homolytic cleavage of the Si–H bond with Tp^{Br3}Cu (NTs), followed by radical rebound.

5.3 Additions of Nitrenes to Oxygen and Sulfur Atoms

Nitrenes have been reported to add to the oxygen atoms of carbonyls, nitro group and even ethers (Scheme 37) [148–150]. Hassner and coworkers reported that the treatment of dibromides of α,β-unsaturated ketones (Scheme 37a) with sodium azide gave the isoxazole from the *Z* alkene isomer [148]. Loss of nitrogen gas is proposed to give rise to an intermediate nitrene, which is attacked by the carbonyl oxygen to furnish the heterocyclic product.

Another unusual example (Scheme 37b) involves intramolecular reaction of an intermediate azidated nitroalkene; presumably, loss of nitrogen gas yields a nitrene that is attacked by the oxygen of the nitro group [149]. De and Ghosh showed that heating a vinyl azide, also bearing a vinyl methyl ether (Scheme 37c), gave 1,2-oxazines through formation of a nitrene, addition of oxygen to this electrophilic species, and followed by a methyl group migration to furnish the products [150].

Nitrenes can be attacked by sulfur atoms, where the thermolysis of either a heteroaryl or vinyl azide in the presence of a sulfur atom results in trapping of the intermediate nitrene to form fused 1,2-thiazines (Scheme 38) [151]. Several of these

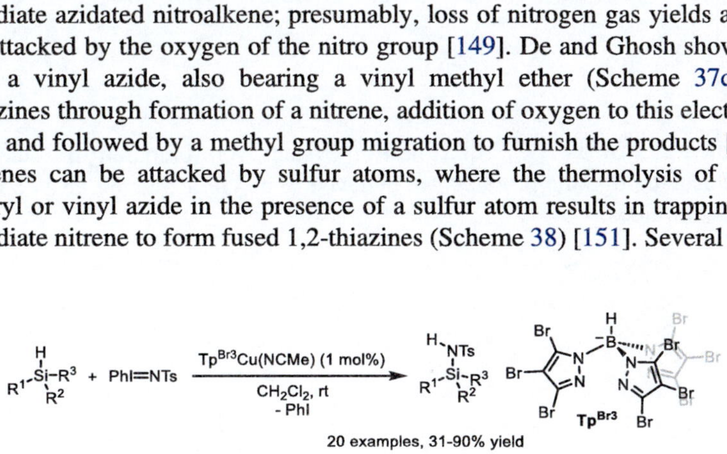

Scheme 36 Addition of nitrenes to silicon

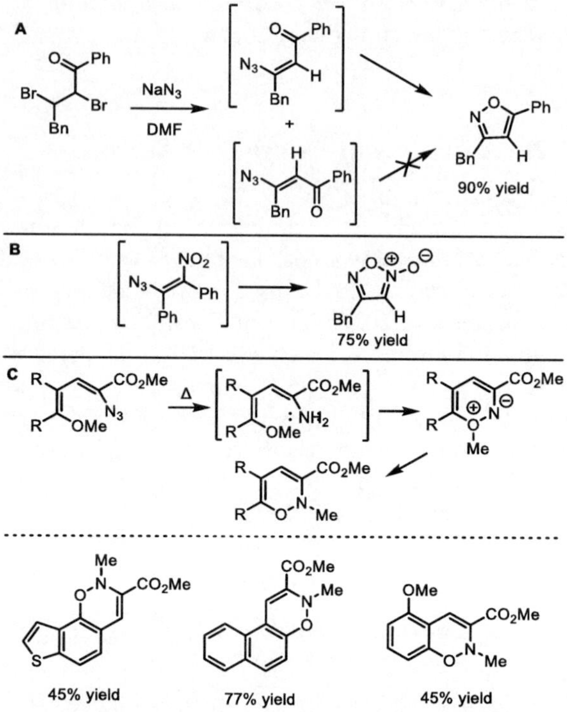

Scheme 37 Addition of nitrenes to oxygen atoms

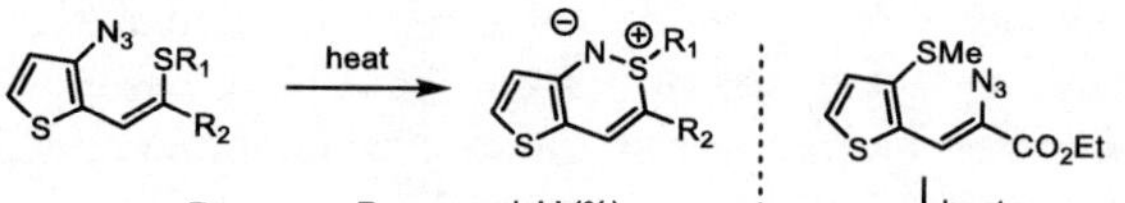

R_1	R_2	yield (%)
Me	CO_2Et	84
Et	CO_2Et	30
Ph	CO_2Et	89
Ph	COMe	90
Ph	CHO	75
Ph	CN	90

Scheme 38 Addition of nitrenes to sulfur

fused derivatives of the 1λ^4,2-thiazine ring system were prepared in moderate-to-good yields, although the products are believed to be ylidic and are best considered as cyclic sulfimides.

Table 32 Formal [4 + 1] cycloaddition of 1,3-dienes and nitrenes to 3-pyrrolines

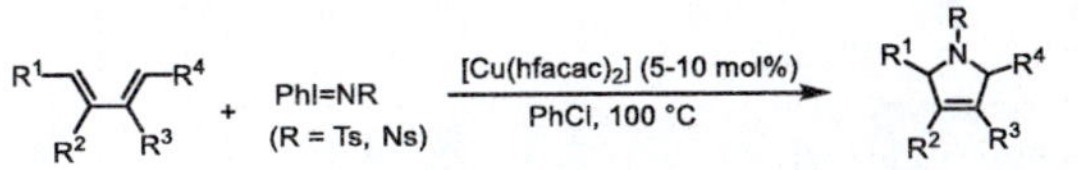

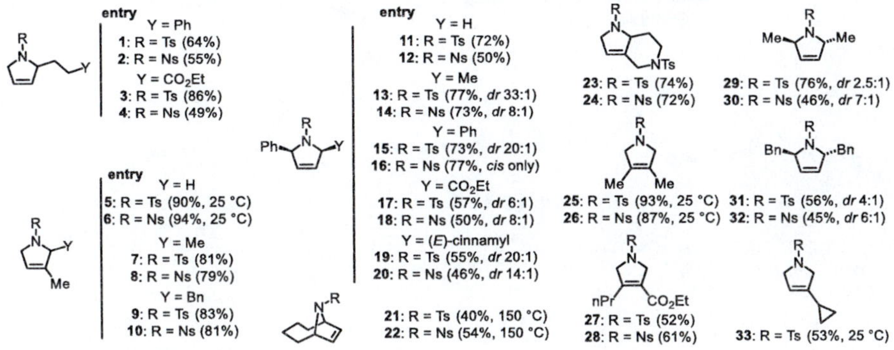

6 Cycloadditions of Nitrene Not Involving Aziridination Processes

In 2011, Zhou and coworkers reported Cu-catalyzed [4 + 1] cycloaddition of 1,3-dienes and a nitrene precursor (PhI=NTs) (Table 32) [152] to achieve direct formation of 3-pyrrolines with broad scope and in good-to-excellent yields. Interestingly, cycloaddition with 1,4-disubstituted 1,3-dienes showed good-to-excellent *cis/trans*-selectivity, where the *cis/trans* preference was dictated by the identity of the substituents, as opposed to the geometry of the diene. Based on their mechanistic studies, the authors suggested that the mechanism involves diene aziridination, followed by subsequent ring expansion.

In 2014, the Yoon group reported visible light-mediated formation of new C–N bonds from vinyl azides by transition metal photocatalysis to generate highly substituted pyrroles (Table 33) [153]. The reaction occurs under mild and selective photocatalytic conditions to favor the formation of vinyl azides over competitive photodecomposition. The scope is broad, with essentially quantitative yields. The authors propose that the mechanism involves an initial formation of the azirine **INT** from ring-closure of vinyl nitrene, followed by rearrangement to the pyrrole.

In 2019, Hilinski and coworkers reported a Rh-catalyzed nitrene transfer [5 + 1]-cycloaddition of aryl-substituted vinylcyclopropanes **34.1** (Table 34) to form functionalized tetrahydropyridines **34.2** with high regioselectivity [154]. Mechanistic studies suggested that the reaction involves an initial nitrene transfer to the olefin of the vinylcyclopropane **34.1** to afford a benzylic carbocation, followed by ring-opening of the cyclopropane and subsequent ring-closure to afford the tetrahydropyridine products.

Table 33 Formal [4 + 1] cycloaddition of 1,3-dienes and nitrenes to 3-pyrrolines

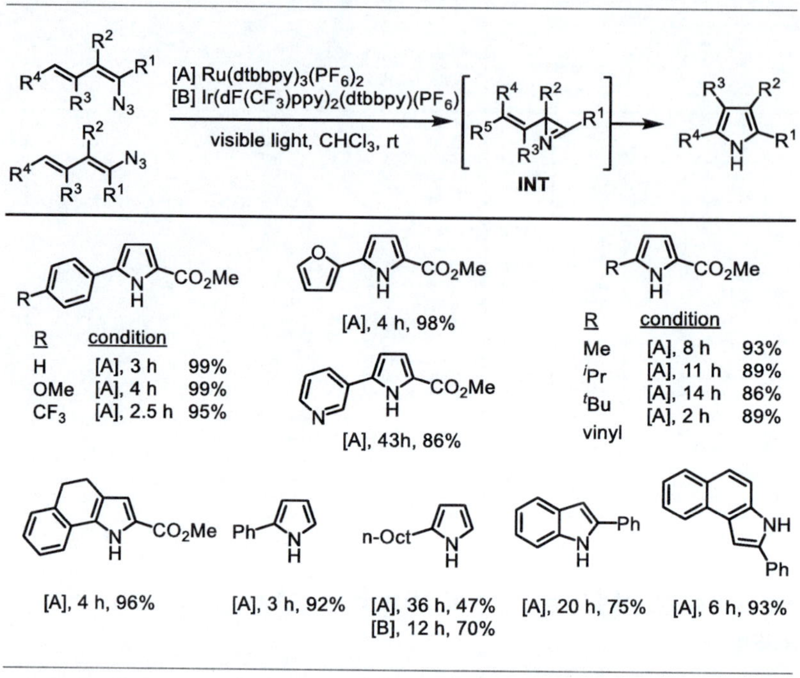

In 2021, Jain and coworkers reported a metal-free synthesis of anthranils through a PhIO-mediated heterocyclization of *ortho*-carbonyl anilines (Table 35) [155]. It featured on the synthesis of anthranils in high efficiency without external additives and transition metal catalyst, as well as a broad reaction scope. Based on their mechanistic studies, the author suggested initial nitrene formation from iminoiodane and subsequent nucleophilic addition of neighboring carbonyl oxygen.

7 Conclusion and Outlook

The field of nitrene transfer has seen a resurgence of interest over the past few decades, particularly in the development of new asymmetric transition metal catalysts able to generate nitrenes that engage in the formation of *N*-heterocycles with high levels of chemo-, site-, and stereoselectivity. Diverse heterocycles can be prepared through NT, including pyrroles, pyridines, indoles, pyrrolidines, piperidines, and rings containing multiple heteroatoms. In this chapter, five major strategies to secure heterocycles via NT were discussed, including both thermal and transition metal-catalyzed aziridinations and subsequent ring expansions, insertion of nitrenes into X–H or C–H bonds, addition of various heteroatoms (N, O, S, Si) to

Table 34 Synthesis of unsaturated piperidines from reactions of nitrenes with vinylcyclopropanes

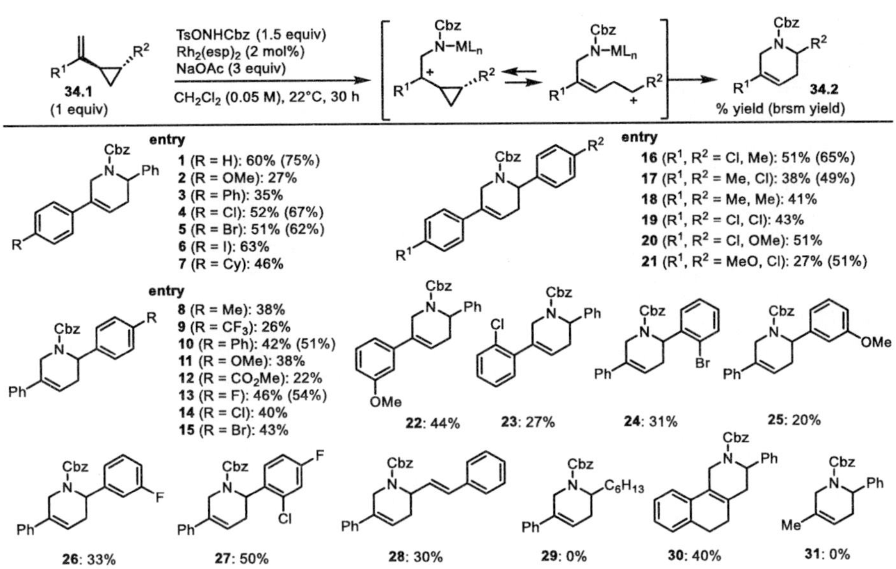

nitrenes, and cases where acyl nitrenes and related species serve as dipoles in cycloaddition reactions. Future advances are likely to occur through the development of more general catalysts for asymmetric aziridination and by coupling aziridination to new modes of ring expansions to furnish modular approaches that generate stereochemically complex and densely functionalized novel amine chemical space. Challenges also remain in the rational design of enantioselective C–H and X–H bond functionalizations that target typically unreactive sites or generate *N*-heterocycles beyond the standard five- and six-membered rings. Environmentally sustainable methods to generate nitrenes from simple precursors continue to be of interest and advances that employ electrochemical, photochemical, or electrophotochemical methods are highly likely to unlock new modes of reactivity in the area of nitrene chemistry.

References

1. Zhang J, Pérez-Temprano MH (2020) Intramolecular C(sp^3)–H bond amination strategies for the synthesis of saturated *N*-containing heterocycles. Chimia 74:895–903
2. Chu JCK, Rovis T (2018) Complementary strategies for directed C(sp^3)–H functionalization: a comparison of transition-metal-catalyzed activation, hydrogen atom transfer, and carbene/nitrene transfer. Angew Chem Int Ed 57:62–101

Table 35 Synthesis of unsaturated piperidines from reactions of nitrenes with vinylcyclopropanes

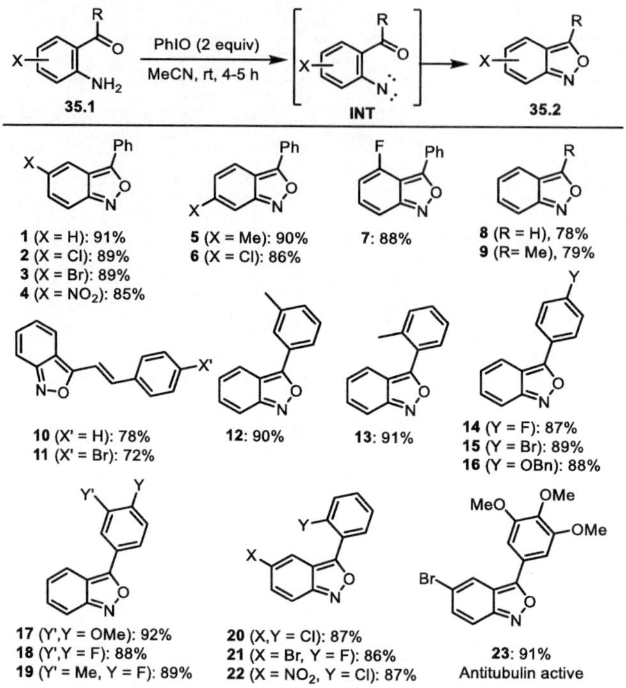

3. Ju M, Schomaker JM (2021) Nitrene transfer catalysts for enantioselective C–N bond formation. Nat Rev Chem 5:580–594
4. Wang Y-C, Lai X-J, Huang K, Yadav S, Qiu G, Zhang L, Zhou H (2021) Unravelling nitrene chemistry from acyclic precursors: recent advances and challenges. Org Chem Front 8:1677–1693
5. Hayashi H, Uchida T (2020) Nitrene transfer reactions for asymmetric C–H amination: recent development. Eur J Org Chem 2020:909–916
6. Díaz-Requejo MM, Pérez PJ (2008) Coinage metal catalyzed C–H bond functionalization of hydrocarbons. Chem Rev 108:3379–3394
7. Collet F, Dodd RH, Dauban P (2009) Catalytic C–H amination: recent progress and future directions. Chem Commun:5061–5074
8. Fantauzzi S, Caselli A, Gallo E (2009) Nitrene transfer reactions mediated by metalloporphyrin complexes. Dalton Trans:5434–5443
9. Zalatan DN, Du Bois J (2009) Metal-catalyzed oxidations of C–H to C–N bonds. Top Curr Chem 292:347–378
10. Davies HM, Du Bois J, Yu JQ (2011) C–H functionalization in organic synthesis. Chem Soc Rev 40:1855–1856
11. Dequirez G, Pons V, Dauban P (2012) Nitrene chemistry in organic synthesis: still in its infancy? Angew Chem Int Ed 51:7384–7395

12. Che CM, Lo KY, Zhou CY (2014) 7.02 oxidation by metals (nitrene). Comprehensive organic synthesis II. 2nd edn., vol 7, pp 26–85
13. Park Y, Kim Y, Chang S (2017) Transition metal-catalyzed C–H amination: scope, mechanism, and applications. Chem Rev 117:9247–9301
14. Hazelard D, Nocquet PA, Compain P (2017) Catalytic C–H amination at its limits: challenges and solutions. Org Chem Front 4:2500–2521
15. Alderson JM, Corbin JR, Schomaker JM (2017) Tunable, chemo- and site-selective nitrene transfer reactions through the rational design of silver(I) catalysts. Acc Chem Res 50:2147–2158
16. Darses B, Rodrigues R, Neuville L, Mazurais M, Dauban P (2017) Transition metal-catalyzed iodine(III)-mediated nitrene transfer reactions: efficient tools for challenging syntheses. Chem Commun 53:493–508
17. Soderberg BC (2000) Synthesis of heterocycles *via* intramolecular annulation of nitrene intermediates. Curr Org Chem 4:727
18. Hajos G (2009) Ring closures to heterocycles of nitrenes. Curr Org Chem 13:791–809
19. Prier CK, Zhang RK, Buller AR, Brinkmann-Chen S, Arnold FH (2017) Enantioselective, intermolecular benzylic C–H amination catalysed by an engineered iron-haem enzyme. Nat Chem 9:629–634
20. Goldberg NW, Knight AM, Zhang RK, Arnold FH (2019) Nitrene transfer catalyzed by a non-heme iron enzyme and enhanced by non-native small-molecule ligands. J Am Chem Soc 141:19585–19588
21. Kuchner O, Arnold FH (1997) Directed evolution of enzyme catalysts. Trends Biotechnol 15: 523–530
22. Arnold FH (2018) Directed evolution: bringing new chemistry to life. Angew Chem Int Ed 57: 4143–4148
23. Yang Y, Cho I, Qi X, Liu P, Arnold FH (2019) An enzymatic platform for the asymmetric amination of primary, secondary and tertiary C(sp 3)–H bonds. Nat Chem 11:987–993
24. Chen K, Arnold FH (2020) Engineering new catalytic activities in enzymes. Nat Catal 3:203–213
25. Ye LW, Sahani RL, Xu Y, Qian PC, Liu RS (2021) Nitrene transfer and carbene transfer in gold catalysis. Chem Rev 121:9039–9112
26. Kwart H, Khan AA (1967) Copper-catalyzed decomposition of benzenesulfonyl azide in cyclohexane solution. J Am Chem Soc 89:1951–1953
27. Degennaro L, Trinchera P, Luisi R (2014) Recent advances in the stereoselective synthesis of aziridines. Chem Rev 114:7881–7929
28. Halfen JA (2005) Recent advances in metal-mediated carbon-nitrogen bond formation reactions: aziridination and amidation. Curr Org Chem 9:657
29. Roizen JL, Harvey ME, Du Bois J (2012) Metal-catalyzed nitrogen-atom transfer methods for the oxidation of aliphatic C–H bonds. Acc Chem Res 45:911–922
30. Jones JE, Ruppel JV, Gao GY, Moore TM, Zhang XP (2008) Cobalt-catalyzed asymmetric olefin aziridination with diphenylphosphoryl azide. J Org Chem 73:7260–7265
31. Tao J, Jin LM, Zhang XP (2014) Synthesis of chiral *N*-phosphoryl aziridines through enantioselective aziridination of alkenes with phosphoryl azide via Co(II)-based metalloradical catalysis. Beilstein J Org Chem 10:1282–1289
32. Jin LM, Xu X, Lu H, Cui X, Wojtas L, Zhang XP (2013) Effective synthesis of chiral *N*-fluoroaryl aziridines through enantioselective aziridination of alkenes with fluoroaryl azides. Angew Chem Int Ed 52:5309–5313
33. Hudlicky T, Frazier JO, Seonane G, Tiedje M, Seoane A, Kwart L, Beal C (1986) Topological selectivity in the intramolecular [4+1] pyrroline annulation. Formal total stereospecific synthesis of (±)-supinidine, (±)-isoretronecanol, and (±)-trachelanthamidine. J Am Chem Soc 108:3755–3762
34. Hennessy ET, Liu RY, Iovan DA, Duncan RA, Betley TA (2014) Iron-mediated intermolecular N-group transfer chemistry with olefinic substrates. Chem Sci 5:1526–1532

35. Liu Y, You T, Wang HX, Tang Z, Zhou CY, Che CM (2020) Iron- and cobalt-catalyzed C (sp^3)-H bond functionalization reactions and their application in organic synthesis. Chem Soc Rev 49:5310–5358
36. Damiano C, Intrieri D, Gallo E (2018) Aziridination of alkenes promoted by iron or ruthenium complexes. Inorg Chem Acta 470:51–67
37. Liu Y, Che CM (2010) [Fe^{III}(F_{20}-tpp)Cl] is an effective catalyst for nitrene transfer reactions and amination of saturated hydrocarbons with sulfonyl and aryl azides as nitrogen source under thermal and microwave-assisted conditions. Chem A Eur J 16:10494–10501
38. Isbill SB, Chandrachud PP, Kern JL, Jenkins DM, Roy S (2019) Elucidation of the reaction mechanism of C2 + N1 aziridination from tetracarbene iron catalysts. ACS Catal 9:6223–6233
39. Chandrachud PP, Bass HM, Jenkins DM (2016) Synthesis of fully aliphatic aziridines with a macrocyclic tetracarbene iron catalyst. Organometallics 35:1652
40. Hopmann KH, Ghosh A (2011) Mechanism of cobalt-porphyrin-catalyzed aziridination. ACS Catal 1:597–600
41. Olivos Suarez AI, Jiang H, Zhang XP, de Bruin B (2011) The radical mechanism of cobalt(II) porphyrin-catalyzed olefin aziridination and the importance of cooperative H-bonding. Dalton Trans 40:5697–5705
42. Kawabata H, Omura K, Uchida T, Katsuki T (2007) Construction of robust ruthenium(salen)(CO) complexes and asymmetric aziridination with nitrene precursors in the form of azide compounds that bear easily removable *N*-sulfonyl groups. Chem Asian J 2:248–256
43. Padwa A, Stengel T (2002) Stereochemical aspects of the iodine (III)-mediated aziridination reaction of some cyclic allylic carbamates. Org Lett 4:2137–2179
44. Espino CG, Fiori KW, Kim M, Du Bois J (2004) Expanding the scope of C−H amination through catalyst design. J Am Chem Soc 126:15378–15379
45. Liang JL, Yuan SX, Chan PW, Che CM (2003) Chiral rhodium(II,II) dimers catalyzed enantioselective intramolecular aziridination of sulfonamides and carbamates. Tetrahedron Lett 44:5917–5920
46. Hayes CJ, Beavis PW, Humphries LA (2006) Rh(II)-catalysed room temperature aziridination of homoallyl-carbamates. Chem Commun 43:4501–4502
47. Rigoli JW, Weatherly CD, Alderson JM, Vo BT, Schomaker JM (2013) Tunable, chemoselective amination *via* silver catalysis. J Am Chem Soc 135:17238–17241
48. Ju M, Weatherly CD, Guzei IA, Schomaker JM (2017) Chemo- and enantioselective intramolecular silver-catalyzed aziridinations. Angew Chem Int Ed 56:9944–9948
49. Dauban P, Dodd R, Estéoule A, Durán F, Retailleau P (2007) Enantioselective intramolecular copper-catalyzed aziridination of sulfamates. Synthesis:1251–1260
50. Boquet V, Nasrallah A, Dana AL, Brunard E, Di Chenna PH, Duran FJ, Retailleau P, Darses B, Sircoglou M, Dauban P (2022) Rhodium(II)-catalyzed enantioselective intermolecular aziridination of alkenes. J Am Chem Soc 144:17156–17164
51. Jat JL, Paudyal MP, Gao H, Xu QL, Yousufuddin M, Devarajan D, Ess DH, Kürti L, Falck JR (2014) Direct stereospecific synthesis of unprotected N–H and N–Me aziridines from olefins. Science 343:61–65
52. Cheng QQ, Zhou Z, Jiang H, Siitonen JH, Ess DH, Zhang X, Kürti L (2020) Organocatalytic nitrogen transfer to unactivated olefins via transient oxaziridines. Nat Catal 3:386–392
53. Lebel H, Trudel C, Spitz C (2012) Stereoselective intermolecular C-H amination reactions. Chem Commun 48:7799–7801
54. Lebel H, Spitz C, Leogane O, Trudel C, Parmentier M (2011) Stereoselective rhodium-catalyzed amination of alkenes. Org Lett 13:5460–5463
55. Lebel H, Parmentier M (2010) Copper-catalyzed enantioselective aziridination of styrenes. Pure Appl Chem 82:1827–1833
56. Lebel H, Huard K, Lectard S (2005) *N*-Tosyloxycarbamates as a source of metal nitrenes: rhodium-catalyzed C−H insertion and aziridination reactions. J Am Chem Soc 127:14198–14199

57. Deng T, Mazumdar W, Yoshinaga Y, Patel PB, Malo D, Malo T, Wink DJ, Driver TG (2021) Rh_2(II)-catalyzed intermolecular *N*-aryl aziridination of olefins using nonactivated N atom precursors. J Am Chem Soc 143:19149–19159
58. Smith PA, Brown BB (1951) The synthesis of heterocyclic compounds from aryl azides. I. Bromo and nitrocarbazoles. J Am Chem Soc 73:2435–2437
59. Sundberg RJ, Russell HF, Ligon WV, Lin LS (1972) *O*-Styrylnitrene route to 2-substituted indoles. Pyrolysis of *o*-azidostyrenes. J Org Chem 37:719–724
60. Sundberg RJ, Lin LS, Blackburn DE (1969) Formation of 2-substituted indoles by pyrolysis of *o*-azidostyrenes. J Heterocylic Chem 6:441
61. Garanti L, Zecchi G (1980) Thermochemical behavior of *o*-azidocinnamonitriles. J Org Chem 45:4767–4769
62. Rozhkov VV, Kuvshinov AM, Gulevskaya VI, Chervin II, Shevelev SA (1999) Synthesis of 2-aryl-and 2-hetaryl-4, 6-dinitroindoles from 2,4,6-trinitrotoluene. Synthesis 12:2065–2070
63. Moody CJ (1991) Oxidation by nitrene insertion. In: Trost BM, Fleming I (eds) Comprehensive organic synthesis, vol 7, pp 21–38
64. Knittel D (1985) Verbesserte synthese von α-azidozimtsäure-estern und 2*H*-azirinen. Synthesis 2:186–188
65. Smolinsky G, Pryde CA (1968) Vinyl azide chemistry. Thermally induced reactions. J Org Chem 6:2411–2416
66. Isomura I, Okada M, Taniguchi H (1969) Azirine formation by decomposition of terminal vinyl azides. Tetrahedron Lett 10:4073–4076
67. Pudlo M, Csanyi D, Moreau F, Hajos G, Riedl Z, Sapi J (2007) First Suzuki-Miyaura type cross-coupling of *ortho*-azidobromobenzene with arylboronic acids and its application to the synthesis of fused aromatic indole-heterocycles. Tetrahedron 63:10320–10329
68. Shen M, Leslie BE, Driver TG (2008) Dirhodium(II)-catalyzed intramolecular C-H amination of aryl azides. Angew Chem Int Ed 47:5056–5059
69. Stokes BJ, Jovanovic B, Dong H, Richert KJ, Riell RD, Driver TG (2009) Rh_2(II)-catalyzed synthesis of carbazoles from biaryl azides. J Org Chem 74:3225–3228
70. Welch M, Phillips RS (1999) Improved syntheses of [3, 2-*b*]-and [2, 3-*b*]-fused selenolo-and thienopyrroles, and of furo [3, 2-*b*] pyrrole. Heterocyclic Commun 5:305–310
71. Pelkey ET, Chang L, Gribble GW (1996) An abnormal Barton-Zard reaction leading to the pyrrolo[2,3-*b*]indole ring system. Chem Commun:1909–1910
72. Shafiee A, Mazloumi A, Cohen VI (1979) Selenium heterocycles. XXVIII. Synthesis of pyrrolo [3,2-*d*] selenazole and pyrrolo [3,2-*d*] thiazole. Two novel heterocycles. J Heterocyclic Chem 16:1563–1566
73. Krutosikova A, Ramsden CA, Dandarova M, Lycka A (1997) Synthesis and reactions of Furo [2,3-*b*]pyrroles. Molecules 2:69–79
74. Hemetsberger H, Knittel D (1972) Synthesis and thermolysis of α-azidoacrylates (ene-azides, IV) Enazide, 4 Mitt. Monat Chem 103:194–204
75. Aratani T, Yoshihara H, Suzukamo G (1989) A synthesis of 4, 7-dihydro-1*H*-dipyrrolo [3,2-b:2′, 3′-*d*] pyrrole and 4,7-dihydro-4*H*-thieno [3,2-*b*] pyrrolo [2,3-*d*] pyrrole systems. Tetrahedron Lett 30:1655–1656
76. Eras J, Galvez C, Garcia F (1984) Reactivity of thienopyrroles. Synthesis of isomeric nitro and bromothienopyrroles. J Heterocyclic Chem 21:215–217
77. Krutositova A, Dandarova M, Alföldi J (1994) Substituted vinyl azides in the synthesis of condensed nitrogen heterocycles. Chem Papers 48:268–273
78. Fresneda PM, Jones RA, Voro TN (1990) Pyrrole studies 42. The synthesis of pyrrolo [3,2-*b*] indoles and of 5-*H*-pyrido[4,3-*b*] indoles 1. Synthetic Commun 20:2011–2017
79. Junk L, Kazmaier U (2016) Synthesis of indoles and tryptophan derivatives via photoinduced nitrene C–H insertion. Org Biomol Chem 14:2916–2923
80. Fauché K, Nauton L, Jouffret L, Cisnetti F, Gautier A (2017) A catalytic intramolecular nitrene insertion into a copper(I)–N-heterocyclic carbene bond yielding fused nitrogen heterocycles. Chem Commun 53:2402–2405

81. Lu H, Lang K, Jiang H, Wojtas L, Zhang XP (2016) Intramolecular 1,5-C (sp^3)–H radical amination *via* Co(II)-based metalloradical catalysis for five-membered cyclic sulfamides. Chem Sci 7:6934–6939
82. Hennessy E, Betley TA (2013) Complex *N*-heterocycle synthesis *via* iron-catalyzed, direct C-H bond amination. Science 340:591–595
83. Iovan DA, Wilding MJ, Baek Y, Hennessey ET, Betley TA (2017) Diastereoselective C-H bond amination for disubstituted pyrrolidines. Angew Chem Int Ed 56:15599–15602
84. Han X, Shan LX, Zhu JX, Zhang CS, Zhang XM, Zhang FM, Wang H, Tu YQ, Yang M, Zhang WS (2021) Copper-nitrene-catalyzed desymmetric oxaziridination/1,2-alkyl rearrangement of 1,3-diketones toward bicyclic lactams. Angew Chem Int Ed 60:22688–22692
85. Hu Y, Lang K, Tao J, Marshall MK, Cheng Q, Cui X, Wojtas L, Zhang XP (2019) Next-generation D_2-symmetric chiral porphyrins for cobalt(II)-based metalloradical catalysis: catalyst engineering by distal bridging. Angew Chem Int Ed 58:2670–2674
86. Cui X, Xu X, Jin LM, Wojtas L, Zhang XP (2015) Stereoselective radical C-H alkylation with acceptor/acceptor-substituted diazo reagents *via* Co(II)-based metalloradical catalysis. Chem Sci 6:1219–1224
87. Jiang H, Lang K, Lu H, Wojtas L, Zhang XP (2017) Asymmetric radical bicyclization of allyl azidoformates *via* cobalt(II)-based metalloradical catalysis. J Am Chem Soc 139:9164–9167
88. Lu H, Zhang XP (2011) Catalytic C–H functionalization by metalloporphyrins: recent developments and future directions. Chem Soc Rev 40:1899–1909
89. Hu Y, Lang K, Li C, Gill JB, Kim I, Lu H, Fields KB, Marshall M, Cheng Q, Cui X, Wojtas L, Zhang XP (2019) Enantioselective radical construction of 5-membered cyclic sulfonamides by metalloradical C-H amination. J Am Chem Soc 141:18160–18169
90. Yang CJ, Zhang C, Gu QS, Fang JH, Su XL, Ye L, Sun Y, Tian Y, Li ZL, Liu XY (2020) Cu-catalysed intramolecular radical enantioconvergent tertiary β-C(sp 3)–H amination of racemic ketones. Nat Catal 3:539–546
91. Lang K, Li C, Kim I, Zhang XP (2020) Enantioconvergent amination of racemic tertiary C–H bonds. J Am Chem Soc 142:20902–20911
92. Li L, Han F, Nie X, Hong Y, Ivlev S, Meggers E (2020) Complementing pyridine-2,6-bis (oxazoline) with cyclometalated *N*-heterocyclic carbene for asymmetric ruthenium catalysis. Angew Chem Int Ed 59:12392–12395
93. Nie X, Yan Z, Ivlev S, Meggers E (2021) Ruthenium pybox-catalyzed enantioselective intramolecular C–H amination of sulfamoyl azides en route to chiral vicinal diamines. J Org Chem 86:750–761
94. Zhou Z, Chen S, Hong Y, Winterling E, Tan Y, Hemming M, Harms K, Houk KN, Meggers E (2019) Non-C2-symmetric chiral-at-ruthenium catalyst for highly efficient enantioselective intramolecular C(sp3)–H amidation. J Am Chem Soc 141:19048–19057
95. Zhou Z, Tan Y, Yamahira T, Ivlev S, Xie X, Riedel R, Hemming M, Kimura M, Meggers E (2020) Enantioselective ring-closing C-H amination of urea derivatives. Chem 6:2024–2034
96. Qin J, Zhou Z, Cui T, Hemming M, Meggers E (2019) Enantioselective intramolecular C-H amination of aliphatic azides by dual ruthenium and phosphine catalysis. Chem Sci 10:3202–3207
97. Wang G, Zhou Z, Shen X, Ivlev S, Meggers E (2020) Asymmetric catalysis with a chiral-at-osmium complex. Chem Commun 56:7714–7717
98. Omura K, Murakami M, Uchida T, Irie R, Katsuki T (2003) Enantioselective aziridination and amination using *p*-toluenesulfonyl azide in the presence of Ru(salen)(CO) complex. Chem Lett 32:354–355
99. Fukunaga Y, Uchida T, Ito Y, Matsumoto K, Katsuki T (2012) Ru(CO)-salen-catalyzed synthesis of enantiopure aziridinyl ketones and formal asymmetric synthesis of (+)-PD 128907. Org Lett 14:4658–4661

100. Ichinose M, Suematsu H, Yasutomi Y, Nishioka Y, Uchida T, Katsuki T (2011) Enantioselective intramolecular benzylic C-H bond amination: efficient synthesis of optically active benzosultams. Angew Chem Int Ed 50:9884–9887
101. Ju M, Zerull EE, Roberts JM, Huang M, Guzei IA, Schomaker JM (2020) Silver-catalyzed enantioselective propargylic C–H bond amination through rational ligand design. J Am Chem Soc 142:12930–12936
102. Ju M, Huang M, Vine LE, Dehghany M, Roberts JM, Schomaker JM (2019) Tunable, catalyst-controlled syntheses of β- and γ-amino alcohol motifs enabled by silver-catalysed nitrene transfer. Nat Cat 2:899–908
103. Davies HM, Long MS (2005) Recent advances in catalytic intramolecular C-H aminations. Angew Chem Int Ed 44:3518–3520
104. Zalatan DN, Du Bois J (2008) A chiral rhodium carboxamidate catalyst for enantioselective C-H amination. J Am Chem Soc 130:9220–9221
105. Reddy RP, Davies HML (2006) Dirhodium tetracarboxylates derived from adamantylglycine as chiral catalysts for enantioselective C-H aminations. Org Lett 8:5013–5016
106. Liang JL, Yuan SX, Huang JS, Yu WY, Che CM (2002) Highly diastereo- and enantioselective intramolecular amidation of saturated C-H bonds catalyzed by ruthenium porphyrins. Angew Chem Int Ed 41:3465–3468
107. Harvey ME, Musaev DG, Du Bois J (2011) A diruthenium catalyst for selective, intramolecular allylic C-H amination: reaction development and mechanistic insight gained through experiment and theory. J Am Chem Soc 133:17207–17216
108. Miyazawa T, Suzuki T, Kumagai Y, Takizawa K, Kikuchi T, Kato S, Onoda A, Hayashi T, Kamei Y, Kamiyama F, Anada M (2020) Chiral paddle-wheel diruthenium complexes for asymmetric catalysis. Nat Catal 3:851–858
109. Alderson JM, Phelps AM, Scamp RJ, Dolan NS, Schomaker JM (2014) Ligand-controlled, tunable silver-catalyzed C-H amination. J Am Chem Soc 136:16720–16723
110. Burley SK, Petsko GA (1985) Aromatic-aromatic interaction: a mechanism of protein structure stabilization. Science 229:23–28
111. Dougherty DA (1996) Cation-π interactions in chemistry and biology: a new view of benzene, Phe, Tyr, and Trp. Science 271:163–168
112. Huang M, Yang T, Paretsky JD, Berry JF, Schomaker JM (2017) Inverting steric effects: using "attractive" noncovalent interactions to direct silver-catalyzed nitrene transfer. J Am Chem Soc 139:17376–17386
113. Milczek E, Boudet N, Blakey S (2008) Enantioselective C–H amination using cationic ruthenium(II)–pybox catalysts. Angew Chem Int Ed 47:6825–6828
114. Paradine SM, White MC (2012) Iron-catalyzed intramolecular allylic C-H amination. J Am Chem Soc 134:2036–2039
115. Yang Y, Arnold FH (2021) Navigating the unnatural reaction space: directed evolution of heme proteins for selective carbene and nitrene transfer. Acc Chem Res 54:1209–1225
116. Noda H, Asada Y, Shibasaki M (2020) *O*-Benzoylhydroxylamines as alkyl nitrene precursors: synthesis of saturated *N*-heterocycles from primary amines. Org Lett 22:8769–8773
117. Nie X, Ye CX, Ivlev SI, Meggers E (2022) Nitrene-mediated C−H oxygenation: catalytic enantioselective formation of five-membered cyclic organic carbonates. Angew Chem Int Ed 61:e202211971
118. Scamp RJ, Scheffer B, Schomaker JM (2019) Regioselective differentiation of vicinal methylene C-H bonds enabled by silver-catalysed nitrene transfer. Chem Commun 55:7362–7365
119. Kim M, Mulcahy JV, Espino CG, Du Bois J (2006) Expanding the substrate scope for C−H amination reactions: oxidative cyclization of urea and guanidine derivatives. Org Lett 8:1073–1076
120. Jarrige L, Zhou Z, Hemming M, Meggers E (2021) Efficient amination of activated and non-activated C (sp^3)−H bonds with a simple iron–phenanthroline catalyst. Angew Chem Int Ed 60:6314–6319

121. Grelier G, Rey-Rodriguez R, Darses B, Retailleau P, Dauban P (2017) Catalytic intramolecular C (sp^3)–H amination of carbamimidates. Eur J Org Chem 14:1880–1883
122. Hong SY, Park Y, Hwang Y, Kim YB, Baik MH, Chang S (2018) Selective formation of γ-lactams *via* C–H amidation enabled by tailored iridium catalysts. Science 359:1016–1021
123. Park Y, Chang S (2019) Asymmetric formation of γ-lactams *via* C–H amidation enabled by chiral hydrogen bond-donor catalysts. Nat Catal 2:219–227
124. Wang H, Park Y, Bai Z, Chang S, He G, Chen G (2019) Iridium-catalyzed enantioselective C (sp^3)-H amidation controlled by attractive noncovalent interactions. J Am Chem Soc 141: 7194–7201
125. Xing Q, Chan CM, Yeung YW, Yu WY (2019) Ruthenium(II)-catalyzed enantioselective γ-lactams formation by intramolecular C-H amidation of 1,4,2-dioxazol-5-ones. J Am Chem Soc 141:3849–3853
126. Mace N, Thornton AR, Blakey SB (2013) Unveiling latent α-iminocarbene reactivity for intermolecular cascade reactions through alkyne oxidative amination. Angew Chem Int Ed 52: 5836–5839
127. Thornton AR, Blakey SB (2008) Catalytic metallonitrene/alkyne metathesis: a powerful cascade process for the synthesis of nitrogen-containing molecules. J Am Chem Soc 130: 5020–5021
128. Thornton AR, Martin VI, Blakey SB (2009) π-Nucleophile traps for metallonitrene/alkyne cascade reactions: a versatile process for the synthesis of α-aminocyclopropanes and β-aminostyrenes. J Am Chem Soc 131:2434–2435
129. Hong K, Su H, Pei C, Lv X, Hu W, Qiu L, Xu X (2019) Rhodium-catalyzed nitrene/alkyne metathesis: an enantioselective process for the synthesis of *N*-heterocycles. Org Lett 21:3328–3331
130. Hong K, Zhou S, Hu W, Xu X (2020) Rh-Catalyzed nitrene alkyne metathesis/formal C–N bond insertion cascade: synthesis of 3-iminoindolines. Org Chem Front 7:1327–1333
131. Stoll AH, Blakey SB (2010) Rhodium catalyzed allene amination: diastereoselective synthesis of aminocyclopropanes via a 2-amidoallylcation intermediate. J Am Chem Soc 132:2108–2109
132. Stoll AH, Blakey SB (2011) Rhodium-catalyzed allene amidation: a facile entry into 2-amidoallylcations for unusual [3+ 3] annulation reactions. Chem Sci 2:112–116
133. Feast GC, Page LW, Robertson J (2010) The intramolecular amination of allenes. Chem Commun 46:2835–2837
134. Gerstner NC, Adams CS, Tretbar M, Schomaker JM (2016) Stereocontrolled syntheses of seven-membered carbocycles by tandem allene aziridination/[4+3] reaction. Angew Chem Int Ed 55:13240–13243
135. Rodriguez MR, Besora M, Molina F, Maseras F, Díaz-Requejo MM, Perez PJ (2020) Intermolecular allene functionalization by silver-nitrene catalysis. J Am Chem Soc 142: 13062–13071
136. Schmid SC, Guzei IA, Schomaker JM (2017) A stereoselective [3+1] ring expansion for the synthesis of highly substituted methylene azetidines. Angew Chem Int Ed 56:12229–12233
137. Schmid SC, Guzei IA, Fernández I, Schomaker JM (2018) Ring expansion of bicyclic methyleneaziridines *via* concerted, near-barrierless [2,3]-Stevens rearrangements of aziridinium ylides. ACS Catal 8:7907–7914
138. Eshon J, Nicastri KA, Schmid SC, Raskopf WT, Guzei IA, Fernández I, Schomaker JM (2020) Intermolecular 3+ 3 ring expansion of aziridines to dehydropiperidines through the intermediacy of aziridinium ylides. Nat Commun 11:1273
139. Fructos MR, Álvarez E, Díaz-Requejo MM, Pérez PJ (2010) Selective synthesis of *N*-substituted 1,2-dihydropyridines from furans by copper-induced concurrent tandem catalysis. J Am Chem Soc 132:4600–4607
140. Li J, Cisar JS, Zhou CY, Vera B, Williams H, Rodríguez AD, Cravatt BF, Romo D (2013) Simultaneous structure–activity studies and arming of natural products by C–H amination reveal cellular targets of eupalmerin acetate. Nat Chem 5:510–517

141. Maestre L, Dorel R, Pablo Ó, Escofet I, Sameera WM, Álvarez E, Maseras F, Díaz-Requejo MM, Echavarren AM, Perez PJ (2017) Functional-group-tolerant, silver-catalyzed N–N bond formation by nitrene transfer to amines. J Am Chem Soc 139:2216–2223
142. Kono M, Harada S, Nemoto T (2019) Chemoselective intramolecular formal insertion reaction of Rh–nitrenes into an amide bond over C–H insertion. Chem A Eur J 25:3119–3124
143. Wang H, Jung H, Song F, Zhu S, Bai Z, Chen D, He G, Chang S, Chen G (2021) Nitrene-mediated intermolecular N–N coupling for efficient synthesis of hydrazides. Nat Chem 13: 378–385
144. Mondal RR, Khamarui S, Maiti DK (2017) Photocatalytic generation of nitrenes for rapid diaziridination. Org Lett 19:5964–5967
145. Messmer A, Hajos G, Tamas J, Neszmelyi A (1979) Structure elucidation of tetrazolo [5,1-c] benzo-as-triazine. An interesting ternary equilibrium of tetrazole-azide systems. J Org Chem 44:1823–1825
146. Timári G, Hajós G, Riedl Z, Béres M, Soós T, Messmer A, Gronowitz S (1996) Synthesis of polyfused heteroaromatics by palladium-catalyzed cross-coupling reaction. Molecules 1:236–241
147. Rodríguez AM, Pérez-Ruíz J, Molina F, Poveda A, Perez-Soto R, Maseras F, Díaz-Requejo MM, Pérez PJ (2022) Introducing the catalytic amination of silanes *via* nitrene insertion. J Am Chem Soc 144:10608–10614
148. Hassner A, L'abbe G, Miller MJ (1971) Stereochemistry. LVII. Synthesis of α-azidovinyl ketones. Mechanism and stereochemistry of vinyl bromide substitution. J Am Chem Soc 93: 981–985
149. Stevens TE, Emmons WD (1958) The reaction of dinitrogen tetroxide and iodine with olefins and acetylenes. J Am Chem Soc 80:338–341
150. Ghosh SC, De A (2000) A convenient synthesis of benz-1,2-oxazine derivatives. Chem Commun 11:979–980
151. Gairns RS, Grant RD, Moody CJ; Rees CW, Tsoi SC (1986) Synthesis of fused 1λ 4,2-thiazines (2-azathiabenzenes). J Chem Soc Perk Trans 1:483–489
152. Wu Q, Hu J, Ren X, Zhou J (2011) An efficient, overall [4+1] cycloaddition of 1,3-dienes and nitrene precursors. Chem A Eur J 17:11553–11558
153. Farney EP, Yoon TP (2014) Visible-light sensitization of vinyl azides by transition-metal photocatalysis. Angew Chem Int Ed 53:793–797
154. Combee LA, Johnson SL, Laudenschlager JE, Hilinski MK (2019) Rh (II)-catalyzed nitrene-transfer [5+1] cycloadditions of aryl-substituted vinylcyclopropanes. Org Lett 21:2307–2311
155. Garia A, Grover J, Jain N (2021) Metal-free synthesis of anthranils by PhIO mediated heterocyclization of *ortho*-carbonyl anilines. Eur J Org Chem 2021:4125–4131